Introduction to Engineering

Teacher's Wraparound Edition

Pearson

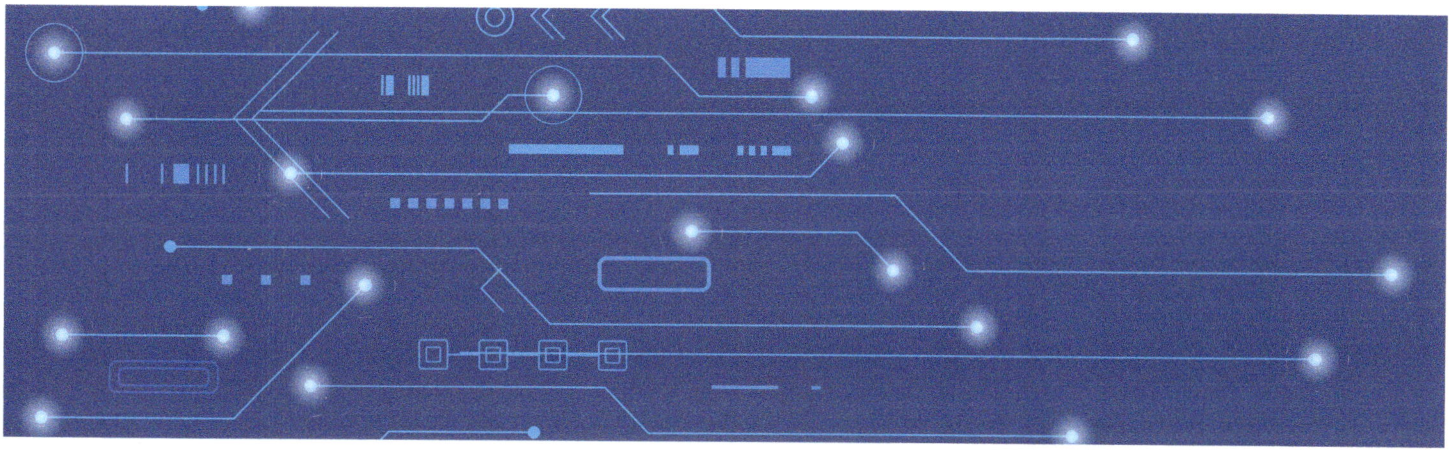

Cover art: Hero Images Imc./Alamy Stock Photo, Hero Images Imc./Alamy Stock Photo, lucadp/123rf.com, stockbroker/123rf.com, stockbroker/123rf.com, Gorodenkoff/Shutterstock, Triff/Shutterstock

Library of Congress Cataloging-in-Publication Data

Cataloging-in-Publication Data is available on file at the Library of Congress

1 2023

Pearson

ISBN-10: 0-13-804293-4
ISBN-13: 978-0-13-804293-6

Authors

Kenneth Reid is Associate Dean and Director of Engineering at the University of Indianapolis. He earned his Ph.D. in Engineering Education in 2009; the seventh in the nation. He is active in engineering within K-12, serving on the Technology Student Association Board of Directors. He and his coauthors were awarded the William Elgin Wickenden award for 2014, recognizing the best paper in the Journal of Engineering Education and the best paper award for the Educational Research and Methods (ERM) Division of ASEE in 2014. He was awarded an IEEE-USA Professional Achievement Award for designing the nation's first B.S. degree in Engineering Education. He is active in e4usa (a high school engineering program), serves on the IEEE Education Society Board of Governors, and has served as a NETI Faculty Fellow. His research interests include success in first-year engineering, engineering in K-12, introducing entrepreneurship into engineering, and international service and engineering. He has co-authored texts in Digital Electronics, Engineering Mathematics, and Engineering Design, has taken multiple teams of students to the Dominican Republic through Solid Rock International, and has a number of teaching awards.

David Reeping is an Assistant Professor of Engineering Education at the University of Cincinnati. Before joining the University of Cincinnati, he was a Postdoctoral Fellow in Electrical and Computer Engineering with the Engineering Education Research Program at the University of Michigan. He earned his Ph.D. in Engineering Education from Virginia Tech and was a National Science Foundation Graduate Research Fellow. He earned his B.S. in Engineering Education with a Mathematics minor from Ohio Northern University. His dissertation received his department's Outstanding Dissertation Award, and the article derived from his dissertation in the *Journal of Engineering Education* was recognized with a William Elgin Wickenden award honorable mention; three other conference papers have received best paper awards. He has extensive experience in curriculum development for K-12 and develops material for the Technology Student Association's annual TEAMS competition. He has co-authored two shorter reference texts on introductory engineering mathematics. His current research focuses on threshold concepts in engineering, streamlining transfer student pathways from community colleges, and measuring the complexities of engineering programs using network analysis.

Scott Orlosky, contributor. Throughout his 30-year career, Scott Orlosky has designed, engineered, developed, marketed, and sold sensors and actuators for industrial and commercial industries. His articles have appeared in *Machine Design*, *Design World*, *Hazardex*, *Control Design*, *Plant Engineering*, and in various online newsletters. He is a co-author of *Encoders for Dummies*, and produced an industrial newsletter for nearly 15 years. He is co-inventor on four patents for the design and manufacturing of inertial sensors. Orlosky holds a master's degree in Manufacturing and Control Theory from the University of California, Berkeley.

Acknowledgements

"I have to begin by thanking my beautiful wife Jenny who is the most supportive person ever. Her love and support mean the world. Special thanks as well to my family: Bryan, Chrissie & Andy, and especially Teddy. Finally, thanks to my colleagues who offer support to tackle the things that are interesting, valuable or fun."

–Kenneth Reid

"I would like to thank everyone who has offered their words of encouragement while developing this text. My parents deserve an additional and special thanks for their undying support of my work, this text, and beyond - I appreciate it more than they could possibly imagine. My dad, Jeff, unexpectedly passed away in June during the development of this text, and I know he would have been proud of what's been accomplished leading up to its publication.

And from both of us, thanks to the team at Pearson and Emergent Learning, and those who helped with so many efforts behind the scenes."

–David Reeping

Overview

Introduction to Engineering

Around 150 B.C., the Romans built above and below-ground **aqueducts**, or channels, to transport water from its sources to the city (Figure 1-4). The Romans developed vast systems of aqueducts over hundreds of years. By the fourth century, the Romans had flowing water, fountains, public and private baths, and used water to wash away waste. Unfortunately, the fall of the Roman Empire meant an end to the development of plumbing, and the practice of dumping waste in the streets returned.

During the Middle Ages in Europe, the knowledge and infrastructure developed during the Roman Empire largely disappeared. There was no effort to separate human waste from drinking water wells. Streets were often dirt roads, and waste mixed with rainwater to create muddy, disease-ridden cities. The unsanitary conditions provided a prime ecosystem for rats and insects, making situations worse. Why did people allow

Introduction to Engineering invites students to explore the essential aspects of the field of engineering, including the history of engineering, the problem-solving process, engineering tools and techniques, and the technological advances that shape the profession today.

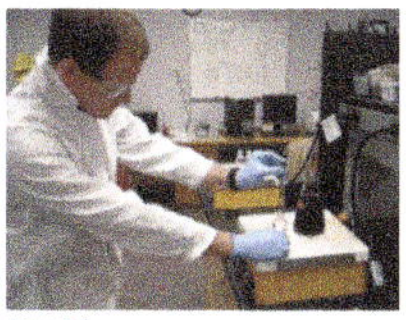
Figure 2-5

"My research is part of a Water Research Foundation project, which is investigating the formation of emerging disinfection byproducts (DBPs) in drinking water treatment.

DBPs are undesirable, toxic compounds that are formed when water is chlorinated. I am investigating the effects of pH, bromide, and iodide concentrations, and preoxidants on the formation of a specific family of DBPs."
D. Jones, BE

Students discover the different types of engineering and engineering-related disciplines, career paths, and the skills and experience necessary for success in engineering professions.

Subsequent chapters help students explore and interact with engineering and STEM disciplines, professional concepts, and various fields of engineering, including how different types of technologies are used.

Engage students with active, hands-on learning

15. Work with a team and answer the following in a shared document. You work closely with someone who owns a pizza shop. Suppose your team is tasked with designing a device that automatically cuts pizza after cooking, after it's placed in a box. Describe how each of Rams' principles of good design would affect the design of such a device.

Numerous projects and activities hold student interest, promote teamwork, and provide meaningful reinforcement.

Comprehensive teaching and learning resources

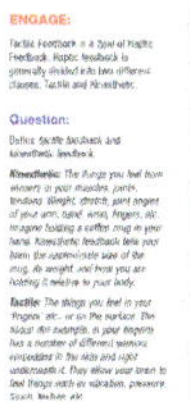

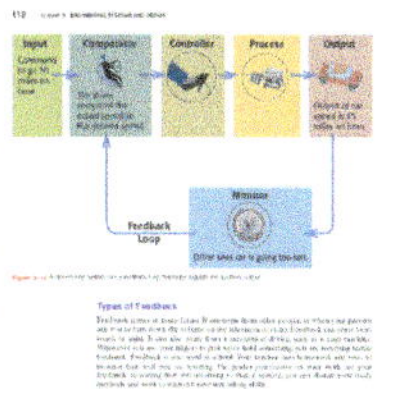

The program also includes a **Student Workbook** with additional activities and learning extensions. Teacher Resources include a **Teacher's Wraparound Edition, Dynamic Class Presentations,** and a **Test Bank** with **TestGen® Assessment software.**

Features and Benefits

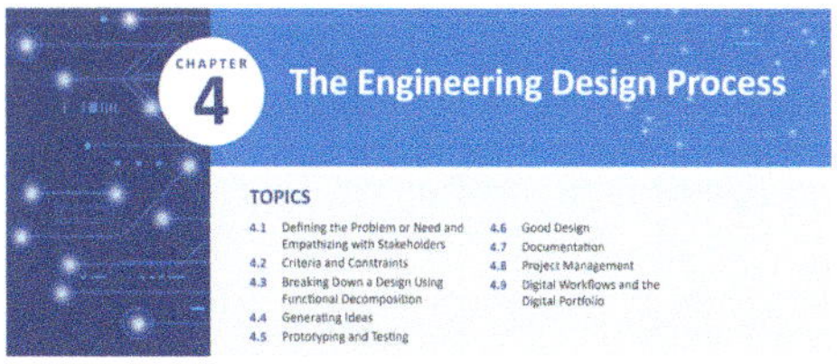

Each chapter begins with a list of the learning objectives to be covered.

Overview

Introduction to Engineering

Features and Benefits

design specifications

To be valid, criteria must be clearly understood and measu we create **design specifications**. For example, when we say

Key terms are highlighted in the margins and within the text for quick reference and review.

TECHNOLOGY IN THE REAL WORLD

Replacement Hearts

Medical inventions have allowed doctors to replace hands, arms, or even entire legs to provide new mobility and independence with prosthetic devices. But replacing organs that lie inside the body is a bit trickier. Every year, people with various organ failures wait for human donors to provide healthy replacements for a kidney, liver, or heart. The list of donors, however, is much shorter than the need for transplants. Even after an organ has been transplanted successfully, patients must take dozens of medications to reduce their body's response to the new organ, and it is still sometimes rejected.

A replacement heart's design mimics the valves and arteries of a real heart.

Side features extend concepts with real world applications.

---

of the design process to help others understand how process steps are connected. Two example design processes for engineering programs are shown in Figure 4-1, but many more images of design processes are easily found using your favorite Internet browser. These two are chosen as examples because of their differences and their similarities. While Figure 4-1a is cyclic and Figure 4-1b is a block diagram sequence, they each communicate the iterative nature of design.

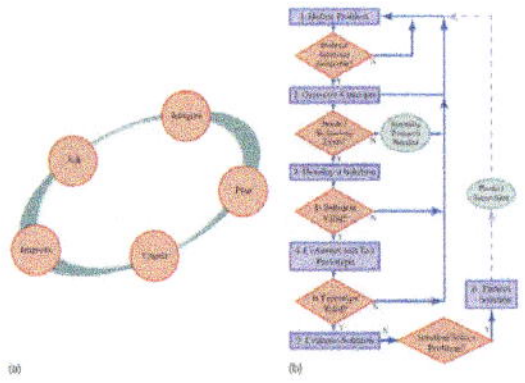

(a) (b)

Diagrams and infographics appeal to visual learners.

Comprehension Check 4-1

We often express criteria in terms that are not clear and measurable. Write a clear criterion to replace each of these vague criteria for the products.

Product	Computer	Automobile	Bookshelf
Inexpensive	Less than $300		
Small			
Easy to assemble			Requires only a screwdriver
Aesthetically pleasing			
Lightweight			
Safe			
Durable			
Environmentally friendly		Has an estimated MPG of at least 50	

The *Comprehension Check* feature provides periodic review before moving on to the next concept.

Example 3-1

If a 2-meter-long horizontal beam deflects 6 centimeters when a load of 2N is hung on the end, how much will a beam of the same material and shape deflect under the same load if it is 4 meters long?

To quickly analyze this, we use the equation for beam deflection and write it twice, once for the initial situation and again for the new condition. Take the ratio of the two equations to find that many parameters will cancel, allowing an easy solution.

Equation of deflection:

$$\delta = \frac{P\ell^3}{3EI}$$

Here δ is the deflection, P is the load, ℓ is the length of the beam, E is Young's modulus, which depends on the material, and I is the moment of inertia. Now simply write this expression for the short beam (subscript s) and again for the longer beam (no subscript) as

$$\delta_s = \frac{(P\ell^3)_s}{(3EI)_s} \quad \text{and} \quad \delta = \frac{(P\ell^3)}{(3EI)}$$

Next, form the ratio of these two equations and note that since the load (P), the shape (I), and the material (E) are all the same, these three variables cancel, and we are left with

$$\frac{\delta}{\delta_s} = \left(\frac{\ell}{\ell_s}\right)^3 = \left(\frac{4\text{ m}}{2\text{ m}}\right)^3 = 8$$

This method is easy to use and shows clearly that there was no need to calculate the moment of inertia or to look up Young's modulus. So, the longer beam will deflect (8)(6 cm) = 48 cm.

Mathematical formulas are broken down and explained to give students additional support on more complex subjects.

Plastic resin (shown in the picture, in pellet form) is used in many types of manufacturing methods including injection molding, extrusion, and blow molding. Injection molding describes the process by which resin is melted and "injected" into a closed mold, then cooled forming the final part. Extrusion is a continuous process by which resin is melted and pushed through an open mold to create shapes like a pipe or rod.

Most cars today use plastic fuel tanks made by blow molding. In blow molding, the resin is melted and pushed through an extrusion head that forms the plastic into a hollow shape. The shape is then pressurized with air and cooled in a mold to form the part. Most fuel tanks are formed with six layers of various plastic to increase toughness and eliminate permeation of the fuel through the tank.

Photo courtesy of E. Stephan

The highly visual design features color photos, line drawings, bulleted lists, and chunked content, creating a clear, easy-to-read presentation.

Overview

Introduction to Engineering

Features and Benefits

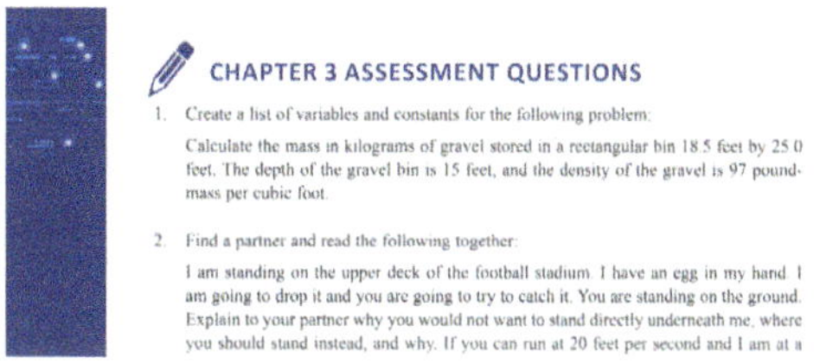
CHAPTER 3 ASSESSMENT QUESTIONS

1. Create a list of variables and constants for the following problem:

   Calculate the mass in kilograms of gravel stored in a rectangular bin 18.5 feet by 25.0 feet. The depth of the gravel bin is 15 feet, and the density of the gravel is 97 pound-mass per cubic foot.

2. Find a partner and read the following together:

   I am standing on the upper deck of the football stadium. I have an egg in my hand. I am going to drop it and you are going to try to catch it. You are standing on the ground. Explain to your partner why you would not want to stand directly underneath me, where you should stand instead, and why. If you can run at 20 feet per second and I am at a

End-of-chapter review and assessment activities such as vocabulary building, content review, knowledge application, and critical thinking questions guide student learning and provide assessment opportunities.

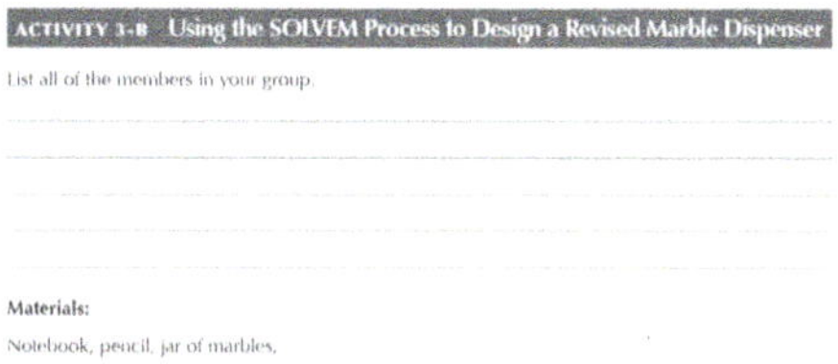
ACTIVITY 3-B Using the SOLVEM Process to Design a Revised Marble Dispenser

List all of the members in your group.

Materials:

Notebook, pencil, jar of marbles.

The Student Workbook provides activities and practice including opportunities to develop sketching skills.

TEACHING TIP

Have students think about a park near them and how this situation could be applicable to that particular park. Doing so will allow the students to make an authentic connection to the scenario.

TEACHING TIP

Have the students work in groups to problem-solve this project with their peers.

0.1 Happy Nature National Park

Streamline class prep and instruction with the annotated Teacher's Wrap Around Edition, Test Bank, and Dynamic Presentations.

Contents

Contents

Contents

CHAPTER 0 A Scenario in the Future

TOPICS

The year is 2032. You earned your engineering degree in Computer Engineering from your favorite university. After that degree, you found a terrific job!

Your relatively new engineering career is going great! You get to use your knowledge of computer engineering in the job you landed with a multinational engineering firm where you have the opportunity to travel, work with diverse teams, and meet new people along the way. You get along well with the team at work; in fact, you even get together for weekend trips from time to time. While you had the chance to work with teams throughout college, it became clear that engineering is often more about teamwork than the mathematical formulas etched into your brain.

Figure 0-1
Congratulations graduate!

One quality of your job that you enjoy is the chance to think outside of the box—to be creative. Your designs have been successful (most of the time), but you have also learned that failure is a natural part of design. In the same way your first draft of a paper is not your final draft, your first iteration of a product will almost never be the design you take to market. Through the constant improvement, your contributions have always met the requirements of the project; you have met the criteria and designed within all constraints.

Now, it's time to take a long weekend!

TEACHING TIP

This purpose of this short chapter is to provide an activity for your students on one of the first days. It should be an active experimentation activity to allow students to interact with one another and begin to conceptualize engineering design. **Once completed, it's vital that you frame this activity as a starting point to begin building the skills necessary to perform the engineering design process systematically.** After introducing the logistics of the course, provide the students with the problem statement and form heterogeneous groups (just ensure that women and underrepresented minority students are not isolated on a team).

TEACHING TIP

Have students think about a park near them and how this situation could be applicable to that particular park. Doing so will allow the students to make an authentic connection to the scenario.

TEACHING TIP

Have the students work in groups to problem-solve this project with their peers.

0.1 Happy Nature National Park

Figure 0-2
A happy family

We all remember the former manufacturing site that was restored a few years ago (keep in mind—it is 2032). When the government came in and cleaned out the site, they discovered beautiful scenery like rock formations, streams, lakes, and small waterfalls. Since it was so beautiful, and they were absolutely sure they cleaned all of the waste, they decided to make it a national park. After a huge contest, the new name was selected: Happy Nature National Park!

Everyone agrees—the park is a magical place where the skies are always sunny, the grass is always green, and the families and friends are always happy. Further, everyone agrees that it is hard to picture this as a formerly contaminated area.

Since you and your team all have vacation time coming up, you decide to explore Happy Nature. "Maybe we should rent a cabin and stay for a few extra days!" Your team agrees and the plans are set.

Your team of four friends, all engineers, packs up and hits the road. You make the drive to Happy Nature. You are all engineers, but decide to leave work behind and just visit nature.

When you arrive at the cabin, a note on the door assures you that this park is SO CLEAN that there is no need for bottled water! The park features streams so clean and fresh that you can drink the water straight from nature. A few empty bottles and a map to the local stream where you can fill them were left on the counter. Your cabin is fortunately also equipped with a beautiful, flat, study table (this will be important later).

Figure 0-3
Fresh, clean water

Figure 0-4
Radiation

The water is clear and seems perfectly safe! One member of your team likes to fish, one likes to hike, one prefers to swim—and there is time for all of it. A beautiful vacation awaits.

One day as your team is hiking, you see a sign warning of radiation in the area. Surely this must be left over from the former manufacturing site, so we won't be concerned with this now. The hike continues.

The days are filled with a mix of relaxation and adventure.

0.2 Trouble in Paradise?

There are a few things that seem a little out of place though…

Figure 0-5
Hiker in a hazmat suit

Later on in the evening, you hear some other hikers walking by, so you give them a friendly wave 'hello!' While they do wave back and respond with a friendly 'hello', you can't ignore that they are wearing hazmat suits with their hiking boots. Their apparel seems strange for hikers. Your team talks with these friendly strangers for a little while, and when asked why they are dressed that way, they brush your concern aside with a gentle chuckle. "You never can be too careful, right?" they say, laughing. While it does seem strange, and you aren't completely satisfied with their explanation, you decide that maybe they just don't understand what they should wear on a hike.

Your teammate heads off for an afternoon of fishing and brings back a bounty of fish. The team decides that a fish dinner over a campfire sounds like a great idea!
When your teammate pulls the fish from the cooler, one looks … strange.

Since this is 2032, we know the government recently found a way to label animals (including wildlife) if they happen to be contaminated. It seems like your friend missed an obviously marked fish among his haul. Yes, one is clearly marked as radioactive. You haven't seen a radioactive fish or animal yet on this trip, so maybe this was left behind by another group. Maybe it swam to the lake from a stream. You decide to investigate and head to the lake, where you see that there is no stream feeding the lake and, now that you pay attention, you notice that there are certainly a lot of these radioactive fish!

Figure 0-6
Radioactive fish

Finally, one more indication that things may not be quite as safe as you thought—while the water tastes fresh and clean, the lake water in the water bottles gives off a slight glow when the lights are off. This, obviously, doesn't seem right.

TEACHING TIP

Make sure that students understand the design requirements. You may ask the students to repeat the requirements in their own words or demonstrate how their robot would perform the task with their arm.

0.3 Problem Solving

To summarize:

a. The water labeled as safe to drink may be radioactive.

b. There is a full bottle sitting exactly 18" from the edge of the table—at this point, you don't want to touch the bottle.

c. You are a member of an engineering team that is capable of solving problems creatively.

Your team decides to meet to determine what to do. In a strange, but fortunate, series of events, one of your teammates happened to pack clothing and three items: a toothbrush, a Geiger counter, and a radiation-proof container that can hold, coincidentally, one water bottle. The Geiger counter confirms that the water in the bottle is radioactive.

The task: The team must design and build a device to lift the radioactive water bottle into a radiation-proof container. You do not want any part of your body to be closer than 18" away from the bottle.

The team decides to build such a device. They scour the cabin and find the following materials (and only these materials):

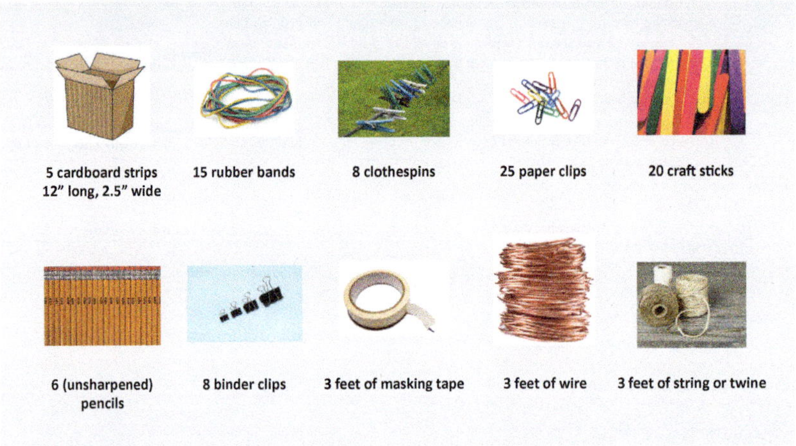

Figure 0-7
Available supplies

(Important—switching back to our current year for a minute: Your teacher may wish to substitute or change quantities of the items in the materials list, and if so, assume you actually find the materials supplied in class. Now, back to the year 2032.)

0.4 Your Task

In teams of 3-4, gather the supplies listed in our story.

Materials list:

(Note: Your teacher may modify the first part of this list if necessary—as long as each team has the same set of supplies, the design goes on.)

5 cardboard strips: 12" long, 2.5" wide

15 rubber bands

8 clothespins

25 paper clips

20 craft sticks

6 unsharpened pencils

8 binder clips

3 feet masking tape

3 feet wire

3 feet string or twine

Also *(not to be incorporated into the radioactive water-lifter)*:

Scissors

Wire clippers

Ruler

Water bottle *(at least ¾ full, not actually filled with radioactive water)*

Bucket *(preferably small, but large enough to hold a water bottle)*

You have 15 minutes to build a radioactive water-lifter able to lift a full water bottle into the air and place it in a bucket. The operator may not get any part of his or her body within 18" of the water bottle. You may test while you build.

After the activity, discuss as a class.

1. **Was this activity "engineering"? Why or why not?**
2. **Did the teams exhibit effective teamwork?**
3. **If you had time to plan, would your design have improved?**

TEACHING TIP

The final chapter in this book will be a repeat of this exercise to show students how far they've come. Take pictures and video to record the students' solutions and testing procedures for comparison later.

TEACHING TIP

These are merely suggested materials. You are free to use different items in your classroom. Note that there should be a combination of sturdy materials and flexible materials. Also, there should be some materials to hold, say a bundle of unsharpened pencils, together.

ENGAGE

When time expires, ask three to five teams (depending on your class size) to demonstrate their robot arm by testing it. This can be a fun way to get students invested in the excitement of testing designs. When designs are successful, you may also choose to alter some conditions, such as changing the water bottle for a thicker one or increasing its weight. At this point, you can discuss how the students would adjust their designs to adapt to changing criteria.

TEACHING TIP

For the debrief discussion, be sure to watch how students partake in the design challenge. Did they dive right into the activity by making a plan? Did they change their design at any point, or were they trying to make their initial idea work at any cost? How are they dividing up the labor to complete the robot arm? Emphasize that engineering design is a **systematic process that leverages ideas from mathematics and science.** Make this a central point of your discussion and ask how the students were systematic in their design process. Also, you can discuss how teams divided up their roles and engaged in the activity.

OVERVIEW

Engineering and technology are all around us. They are used to design and build the vehicles, highways, buildings, and bridges that surround us. They allow us to travel around the globe or communicate with people thousands of miles away. They make possible the exploration of planets and the construction of space stations. Engineering and technology are fundamental to our daily lives. They have changed the way we grow food, treat disease, manufacture complex machinery, commute to school and work, and even heat our homes.

CHAPTER 1

Engineering as a Discipline

TOPICS

1.1 Define Engineering

What is a doctor? Most people can accurately describe what a doctor does. However, a podiatrist, a doctor specializing on the foot, is distinctly different than a cardiac surgeon specializing in heart surgery, who is in turn different than an anesthesiologist, who administers anesthesia.

engineering

It's common to have some confusion around the term **"engineer."** It seems that the name "engineer" is used in many different areas. Perhaps you have heard about mechanical engineering, civil engineering, and electrical engineering—some of the classic **engineering** disciplines. However, looking at potential jobs online, you might find more varied titles like sales engineer, packing engineer, application engineer, usability engineer, and controls engineer. It takes all kinds of engineers to get a job done. A construction engineer at a construction site probably has a very different job than an engineer specializing in nanotechnology.

While there are many different types of engineers: environmental, electrical, ocean, materials, petroleum, and mechanical, for example, they all have characteristics in common. Engineers practice the art of designing products, services, and solutions to problems that use engineering principles. Each discipline has its core body of knowledge, core ideas that all engineers in that discipline understand. Engineers within each discipline have deep knowledge about specific areas in which they work. For example, most software engineers will thoroughly understand computer operating systems, whereas chemical engineers will have expertise in thermodynamics, the study of heat flow.

No engineering discipline stands completely alone, and engineers often need to work together to reach unique solutions to complex problems. As an example to illustrate the connections between different engineering types, contemplate how different engineers might consider the design of a flowering plant.

- A chemical engineer would be most interested in how the plant converts sunlight and water into food.
- A civil engineer would find the structure of the plant of interest, including how thin stems could support heavy buds and how the structure can remain intact after a strong storm.
- An electrical engineer would be interested in how energy is generated as the plant separates water into ions, similar to a battery.
- A mechanical engineer would see how the nested flower petals open from a bud to collect sun and attract bees.
- An environmental engineer would look into the interaction between the plant and its surrounding environment.

OBJECTIVES

- Investigate and report on the history of engineering disciplines, including chemical, civil, electrical, and mechanical engineering
- Conduct and present research on emerging and innovative technology
- Describe how technology has affected individuals, societies, cultures, economies, and environments
- Describe how the development and use of technology influenced past events
- Describe how and why technology progresses
- Predict possible changes caused by the advances of technology
- Assess the risks and benefits of a design solution
- Identify emerging trends in electrical and mechanical systems
- Differentiate among various engineering disciplines
- Identify several different careers that support large scale civil or environmental projects
- Identify several different careers that support the computer industry
- Identify several different careers that support bioengineering or biomanufacturing

TEACHING TIP

At this point, you can emphasize how ABET places a "stamp of approval" on engineering programs. For certain disciplines, such as Civil Engineering, graduating from an ABET accredited program is a requirement for becoming a professional engineer. Graduating from a non-ABET accredited program severely limits job prospects because employers prefer that programs hold the status of ABET-accredited.

We can see why different disciplines are necessary when working toward a complete solution when we consider how they work in teams to design the solution to a problem. Consider the design of a smartwatch that interacts with the wearer to sense vital signs such as their heartbeat.

Engineering Disciplines	Roles
Mechanical engineer	Any moving parts within the watch; how components of the watch fit together
Electrical engineer	Power to the watch (battery life, charging); sensors and interfacing to other components
Software engineer	Software to operate the watch, efficiency to run in a small amount of memory
Computer engineer	Interface of the watch to the outside world (most likely a computer)
Industrial engineer	Study of how the watch can be manufactured most efficiently
Biomedical engineer	Interaction of the sensors to the human wearing the watch
Materials engineer	Materials to be used in the watch
Environmental engineer	How will the watch be disposed of at its end-of-life?

After looking through the engineers' different roles in the example of designing a smartwatch, perhaps you can see where you might be most interested in helping. Maybe you would like to work across two or more areas. If you decide to pursue engineering for your undergraduate degree, you'll have an opportunity to explore these different areas and find what aligns best with your career goals.

accreditation

ABET

Academic programs in engineering typically need to be accredited or certified. The **accreditation** process involves experts in the particular area of engineering visiting an institution that offers the discipline as a major. The experts closely examine the classes, labs, and faculty. The Accreditation Board for Engineering and Technology (**ABET**) is responsible for certifying engineering programs. They define what it means to be an engineer as follows.

> *The profession in which a knowledge of the mathematical and natural sciences gained by study, experience, and practice is applied with judgment to develop ways to utilize, economically, the materials and forces of nature for the benefit of mankind.*

Neil Armstrong, the first astronaut to walk on the moon, describes engineers in this way:

> *Engineers are dedicated to solving problems and creating new, useful, and efficient things.*

design process

ABET further defines the engineering design process, and this **design process** is how engineering ideas become a reality.

Engineering design is the process of devising a system, component, or process to meet desired needs and specifications within certain constraints. It is an iterative and creative decision-making process in which basic and math sciences and engineering sciences are applied to transform resources into solutions. It also includes identifying opportunities, developing requirements, conducting analysis and synthesis, creating multiple solutions, evaluating solutions against requirements, considering risks, and making trade-offs with the aim of obtaining a high-quality solution under certain conditions. Possible constraints may include some or more of the following: accessibility, aesthetics, codes, constructability, cost, ergonomics, scalability, functionality, legal considerations, maintainability, manufacturability, or marketability, policy, regulations, standards, sustainability, or usability.

Let's break their definition down into its core elements:

Devising a system, component, or process. Engineers, regardless of their discipline, use the principles of mathematics and science to address problems. The result can be a new system, part, or process.

An iterative and creative decision-making process applied to transform resources into solutions. The tools for the decision-making process are drawn from the physical sciences, including fields like mathematics, chemistry, and physics. It is vital to note the creative and iterative character of the process. Your first idea is rarely the optimal solution and probably not original. Therefore, the process should involve an initial design plus iterations to refine the design. Designing thinking tools during the phase when you are generating ideas can help push your creative limits.

risk/benefit trade-offs

Identifying opportunities and evaluating solutions against requirements to produce a high-quality solution. This section describes the engineering process—meeting requirements, performing analysis, assessing risk, and making **risk/benefit trade-offs** in pursuit of a high-quality solution.

constraints

Typical constraints. Aside from the functional **constraints** dictated by the design requirements, such as limitations on the product's size, there can be many other considerations. You need to be clear on these at the outset of any project, which involves meeting with the customer to understand their needs. Other constraints can include such requirements as marketability, a preference for certain materials, the use of sustainable materials, and necessary price targets to achieve a center margin of profitability. Risks and benefits of solutions must also be considered.

A typical undergraduate student will study calculus and physics. The mathematical concepts and methods we learn help us solve physical problems. Being an engineer goes beyond textbook exercises; instead, it requires us to solve problems on a grander scale. As stated by Galileo Galilei, "The laws of Nature are written in the language of mathematics." Our ability to harness the natural forces around us is dependent on being able to use mathematics in a practical sense. We are trained to speak the language fluently.

Through an engineer's work, they are meeting a particular need. Whether the need is clean drinking water in a disadvantaged region or packing more capability into a smartphone, the process is fundamentally the same. The things that make a modern lifestyle even possible can be attributed to the work of engineers through the centuries.

Throughout this text, we will delve into the field of engineering and how an engineer goes about meeting the technological challenges of this and future generations.

TEACHING TIP

Consider asking students to construct their own definitions of engineering, then collect the appropriate keywords from their definitions before showing this definition on the board. Possible keywords include *process, system, component, constraints, applying math and science, solutions,* and *benefiting society.*

ENGAGE

Introduce the "Three-Age" system that divides human history into three consecutive time periods, named for their particular tool-making technologies. Discuss the basic tools of each time period and how they were created.

Stone Age: *Hammerstone, handaxe, awls, and bone needles.*

Bronze Age: *Daggers, swords, shields, personal ornaments, and pottery.*

Iron Age: *Ploughs, coinage, weapons, and pottery.*

Engineering Societies

There are several dedicated professional organizations, usually known as societies, for engineering and each discipline. The National Society of Professional Engineers (NSPE) is an organization for all engineers. They are perhaps best known for the NSPE Code of Ethics, a list of rules and guidelines for engineers to follow when faced with challenging ethical problems. We will be exploring engineering ethics later in this chapter.

The IEEE, formerly the Institute of Electrical and Electronics Engineers, is the professional society for Electrical, Computer, and Software Engineers. The IEEE also has chapters for students studying these disciplines in college, training programs, and journals where engineers publish research results to share with other interested engineers. Mechanical engineers have the ASME, or the American Society of Mechanical Engineers, which promotes the art, science, and practice of multidisciplinary engineering and allied sciences around the globe. Civil engineering has the ASCE, or the American Society of Civil Engineers. Other disciplines have similar societies, most of which publish works of interest to those in the discipline, host meetings, and write standards that help engineers solve problems.

1.2 The Historical Development of Engineering

Smartphones are prominent today. These devices provide a wide variety of functions we access during our day. Smartphones serve as tools to send text messages through our cellular carriers, watch movies and TV shows through streaming services, transact with merchants using a digital wallet, take high-resolution pictures, and play games online. The smartphone, in many ways, can be seen as an extension of ourselves. It carries so much information about us, and these devices can snuggly fit in our hands. Often, we might forget that the smartphone can be used to make and receive calls, given the several features packaged within the product. Each of the smartphone's features used to require separate devices. If we look back 20 to 40 years, phones were landlines tethered to a phone jack. Many homes had wireless phones, but they were large devices that had to be close to the base unit. If you were out and someone called, you missed the call. Such a scenario may seem like ancient history, but telecommunications have only been around since the 1870s, when Alexander Graham Bell demonstrated his telephone at the Centennial Exhibition in Philadelphia. The telephone is a significant engineering achievement, enabling communications worldwide. It may seem impossible today that there were times before smartphones, or even before telephones.

If we consider engineering to be a profession where science, math, nature, and materials are used to solve problems, we can see engineering evolve over history. Much of engineering history has been defined by three main characteristics:

- The materials available for constructing tools.
- The prevailing socio-political structure of civilization.
- The maturity of the engineering disciplines available at the time.

The materials that people used in a particular era have given us a way to classify broad historical periods, including the Stone Age. Three critical periods of early history are the Stone Age, the Bronze Age, and the Iron Age.

The socio-political structure refers to a set of conditions where the engineer's work is needed to solve society's problems. Examples are the Roman Empire, Middle Ages, Renaissance, and the Industrial Revolution.

Finally, the major engineering disciplines, in order of development, include mechanical, civil, electrical, and chemical engineering.

Stone Age

Engineering dates to prehistoric times, also known as the **Stone Age**. People made rudimentary tools like stone, wood, and animal hide to hunt, build shelter, and create and contain a fire. No structure we can see throughout the rest of history, like villages or towns, dictated how people would formally organize themselves. People lived alone or in groups but didn't organize beyond this.

Bronze Age

alloy

The **Bronze Age** is thought to be around 3200–600 B.C. Humans had discovered and used copper, but the discovery of bronze, an **alloy** of copper and tin, resulted in the hardest common metal known to humankind. Bronze lent its name to this historical period, which occupies the gap between the stone and iron ages. The manufacture of bronze is one of the first recognized industrial processes, as it required the mining and smelting of tin, which was added to molten copper. As copper and tin ores were typically not found together, early trade in raw materials developed. Bronze meant that humankind had more advanced weapons and cookware, and vessels to hold water (Figure 1-1).

Figure 1-1 Because bronze is lighter than stone, it was used to make urns, pots, and tools.

Iron Age

smelting

The **Iron Age** began as the Bronze Age faded and ended as written history began. The start of the Iron Age varied depending on the country or continent; in Europe, it began around 400 B.C., while in the Middle East it began around 1200 B.C. It lasted to roughly the year 100 A.D. Iron had already been used as the Iron Age appeared, but it was considered inferior to bronze. Iron was made from iron ore through a process called **smelting**, where iron ore and charcoal were placed in a clay furnace. Air was pumped into the furnace with a bellows, which heated the charcoal enough to soften the ore. The red-hot ore could then be pounded into shape (Figure 1-2). People began to create steel for tools and weapons, and more advanced farming tools were developed, such as plows and chains, hammers, and saws.

Figure 1-2 Iron-smelting furnaces were used to soften iron ore. The blacksmith could then hammer the iron into different shapes.

After the Iron Age, materials were developed that we use today. Glass was more widely used from about 1300 A.D., while steel and aluminum were used from the 1800s through today. One interesting note about glass is that it can be recycled endlessly without any loss in quality; however, only about 30% of glass is recycled in the United States.

Plastic was developed in the early 1900s and was widely used in packaging throughout the century. Plastic beverage bottles are commonplace today. Although most plastic is recyclable, very little plastic produced today is recycled or reused. Plastic breaks down very slowly if it's not recycled, whether buried in landfills or dumped in waterways. Therefore, efforts to reduce the use of single-use plastics, such as bags at grocery stores, have been prominent in the current century.

TEACHING TIP

Outline and discuss a list of world changing innovations and technologies that have occurred throughout time. Begin with simple items like fire, language, and farming. Continue with the wheel, iron, steel, and the alphabet. The discussion should then lead students to more modern innovations like the Internet, microprocessors, and smartphones.

TEACHING TIP

Compare *potable water* today to what passed for drinking water throughout history.

Discuss irrigation, sewage, cesspools, and lead poisoning. Watch the History Channel video entitled "Ancient Plumbing."

URL

http://www.history.com/topics/ancient-history/ancient-egypt/videos/lost-worlds-toilets

ELL

As an expansion to the preceding Teaching Tip, consider asking students to take notes on the video. Review them for accuracy to ensure that ELLs are demonstrating listening comprehension of spoken English. Perhaps pass out a worksheet with questions pertaining to the video beforehand, and have students record the answers as the video plays.

1.2.1 Key Engineering Advancements

Before and early in recorded history, people in different parts of the world used stone, bronze, and iron to address many of their needs. People developed tools, shelter, and weapons for hunting.

Agriculture

Prehistoric people were nomadic; they lived on the move rather than living in a town. Their shelter may have consisted of a cave or an overhang, and food was gathered or hunted.

About 10,000 years ago, our way of life changed. People learned how to grow their own food. They cultivated seeds and began to process grains. Plants were used to make clothing. With the invention of the wheel, transporting people and goods became possible. Stones or packed-down dirt were used to make roads. Shelters were built of stone and wood. People lived off of the land and sea, but technological advancement meant increasing instances of community.

The plow, invented around 4000 B.C., represented a huge advance, making farming much easier (Figure 1-3). This was a time when mechanical engineering, the use of simple machines, was the dominant form of engineering. The six fundamental machines that were understood at that time were:

- Wedge
- Inclined plane
- Lever
- Wheel and axle
- Pulley
- Screw

Figure 1-3 The plow allowed most of the heavy work of turning the soil to be done by strong animals such as oxen or horses.

The simple machines allow the user to perform tasks that require considerable force, such as moving a heavy object, by changing the direction or magnitude of an applied force. Considering their utility, these basic machines had been around for thousands of years. With the growth of agriculture, many applications and discoveries were focused on growing and harvesting crops.

Water and Sewer

As people developed tools and learned to farm and produce food rather than gather and hunt, towns began to form. Rather than living by a stream as a source of water and leaving waste where it was created, people needed to be able to move water to where it was needed and, hopefully, move waste away. The development

infrastructure

of roads, waterways, and means to rid the town of sewage all fall under **infrastructure**, which civil engineers primarily design.

Early towns dealt with waste in ways that, today, would seem unimaginable. For the most part, waste was either left where it was created, or swept or dumped into the street. Disease was the leading cause of death, and streets with flowing waste contributed to this.

The Minoan Culture on the island of Crete had a surprisingly modern sewage solution, which would be considered civil engineering today. Royalty in this society had benches with holes (very similar to an outhouse) where waste would go. Underneath these seats were canals lined with clay, designed to carry waste away. The floor had a device very similar to today's floor drain where water could be dumped to get the waste to flow. This technology didn't spread widely to other cultures until it was "invented" in 1596.

aqueducts Around 150 B.C., the Romans built above and below-ground **aqueducts**, or channels, to transport water from its sources to the city (Figure 1-4). The Romans developed vast systems of aqueducts over hundreds of years. By the fourth century, the Romans had flowing water, fountains, public and private baths, and used water to wash away waste. Unfortunately, the fall of the Roman Empire meant an end to the development of plumbing, and the practice of dumping waste in the streets returned.

Figure 1-4 The Pont du Gard is the largest and best preserved of the Roman aqueducts.

During the Middle Ages in Europe, the knowledge and infrastructure developed during the Roman Empire largely disappeared. There was no effort to separate human waste from drinking water wells. Streets were often dirt roads, and waste mixed with rainwater to create muddy, disease-ridden cities. The unsanitary conditions provided a prime ecosystem for rats and insects, making situations worse. Why did people allow this to continue? The connection between waste and disease was poorly understood; people didn't realize that the conditions allowed plagues to spread and thrive.

Many of the waste management practices were brought to the New World (America) from Europe. In 1860, Jean-Louis Mouras invented the septic tank, where waste could be stored and separated before dumping.

Civil engineers have designed water and sewage systems throughout the United States. Public water supplies are filtered and treated with chemicals to make them safe to drink. This system is not perfect, however. Many cities do not have adequate sewage infrastructure to handle large storms with significant rain, leading to sewage being dumped into waterways without treatment. Some cities have specific water treatment methods based on the type of pipes and the water source being delivered to homes. A recent example of failure can be seen in Flint, Michigan in 2014-2016. The source of the drinking water serving Flint changed from the Detroit Water Department to the Flint River without adjusting the chemicals used to treat the water. As a result, Flint residents found that water from their faucets had dangerous levels of bacteria and lead, leading to significant health and safety issues. The issues in Flint provide a clear example of how engineering solutions (or failures) affect society and the environment.

Writing and Printing

Today, the ability to write on paper is taken for granted. However, before the invention of paper and writing instruments, writing was often on clay or stone. As early as the fourth century B.C., papyrus-based sheets which could be written upon were invented in Egypt, and by the year 105, paper was created in China from plant fibers. Years later, paper made from wood pulp was invented.

The printing press transformed the way information was collected and spread. Books used to be copied painstakingly by hand, which could take months. In the mid-1400s, Johann Gutenberg invented the printing press, a device that could use small, metal molds to build words and stories. The metal molds were formed, then pressed onto paper, allowing books and papers to be made in minutes instead of months.

Today, documents are written on a computer or other electronic device using a word processor and are often stored "in the cloud." Many engineering disciplines are involved with this process, from software and computer engineers who design the software and operating systems to systems with vast storage in the cloud. Computer scientists also play a role in developing the programs used to do word processing. Electrical engineers design the devices to power and transmit the data. Materials engineers design the medium in the storage mechanism to store as much data as reliably as possible.

TEACHING TIP

The Machine Age has an approximate dating from about 1880 to 1945. Considered to be at a peak in the time between the First and Second World Wars, it forms a late part of the Industrial Age. Compare the positive and negative impacts on society with the innovation of assembly lines, skyscrapers, radio, and submarines.

TEACHING TIP

Introduce students to the innovations and technology of the Industrial Revolution by listing the machines, processes, and products that were created to serve the people of the time period. *(steam engine, telegraph, cotton gin, etc.)*

Introduce students to the factory system, the new way of organizing labor made necessary by the development of machines which were too large to house in a worker's cottage.

DISCUSSION QUESTION

What were the negative effects of the factory system? *(long hours; low wages; unsatisfactory, even dangerous working conditions; child labor; and women's labor)*

ELL

To provide further support to ELLs via the Discussion Question, consider passing out a list of vocabulary words that will be helpful to the discussion, and allow students to define them as the conversation moves along. Ensure that students are actively engaging in the conversation and are able to accurately use the vocabulary verbally.

The Industrial Revolution

Industrial Revolution

factory system

The **Industrial Revolution** (1600s) marked a shift from human and animal power to the widespread use of machines to perform tasks more efficiently. With the invention of these new machines came factories replacing individual laborers. In the **factory system**, people made goods using machines instead of by hand. Before this time, skilled people had used their own tools and workshops to make shoes, cloth, and tools.

Richard Arkwright, a British inventor and cotton manufacturer, developed an early manufacturing system. Using a sequence of steps, each machine carried out one job and prepared the material to be processed by the next machine. The assembly line became vital for manufacturing in England and the United States. Assembly lines were based on a system of interchangeable parts. In a car, for example, identical copies of each part were assembled in the same way to build many identical cars, a process made famous by Henry Ford (Figure 1-5).

Figure 1-5 Assembly lines modernized the assembly of automobiles.

In the early 1700s, Thomas Newcomen designed the first modern steam engine. The original design was improved by James Watt in the 1760s, followed by Matthew Boulton, who added a rotary motion. As engineering was used to improve the steam engine, it became more prominent in other manufacturing applications, including paper, flour, and cotton mills (Figure 1-6).

Steam engines were powered by burning coal, meaning the demand for coal grew. Steam engines became the driving force in transportation, powering ships and railroads. Coal was burned, meaning tons of smoke poured into the atmosphere.

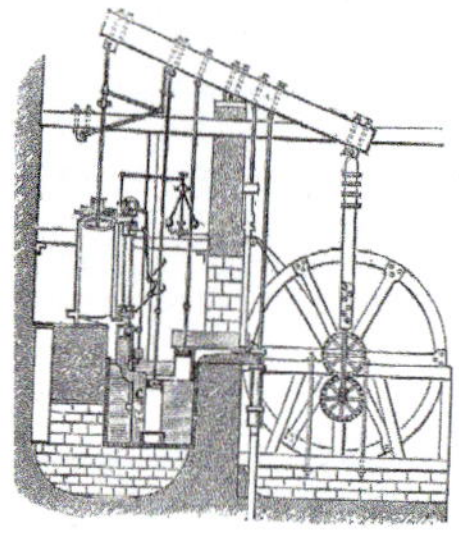

Figure 1-6 During the Industrial Revolution, many machines, such as this early engine, were powered by steam.

Changes to Society Brought on by the Industrial Revolution

The factory system improved the production of goods. Since people who used to manufacture things by hand were losing their livelihood, many moved to the city and started to work in factories. The rise in the number of factory workers led to changes in society. Men, women, and children could work 14-16 hours per day in crowded, hot, and unsafe factories. Workers were simply replaced when they quit, were injured, or killed. The inhumane treatment led to the formation of labor unions, advocating for better and safer working conditions. Politically, laws were passed to ensure that children were not used as labor in factories. As society realized that a smarter workforce would be a better workforce, public school systems were formed. Schooling became mandatory for children in the early 1900s.

Information Age

The advances in technology happened slowly throughout history. Sewage and safe water moved forward and backward over hundreds of years. Agriculture indeed saw advances, but many were relatively small and happened over a long period. On the other hand, we have seen explosive technological growth over the last few decades, from the printing press to mass-produced newspapers and magazines; from landline phones to smartphones; short message service to social media. **Semiconductors** have led the way in computers, processors, and storage—making all of these developments of the **Information Age** possible. At the core of these developments is the transistor in 1947, which is one of the basic building blocks of integrated circuits that make electronics work (Figure 1-7). These are common in everyday devices you interact with, including TVs, computers, and smartphones.

semiconductors

Information Age

Figure 1-7 Information Age Inventions: (a) The transistors used in this early radio were an important invention of the Information Age. (b) Another important invention of the Informtion Age has been the Internet.

1.2.2 The Rate of Technological Change

The rate of technological change has increased over the course of history. During prehistoric times, it took thousands of years to progress from stone tools to metal tools, and for infrastructure to develop. People gradually moved to towns, where advancements in farming and agriculture meant people were fed more efficiently. Print materials allows people to communicate more effectively.

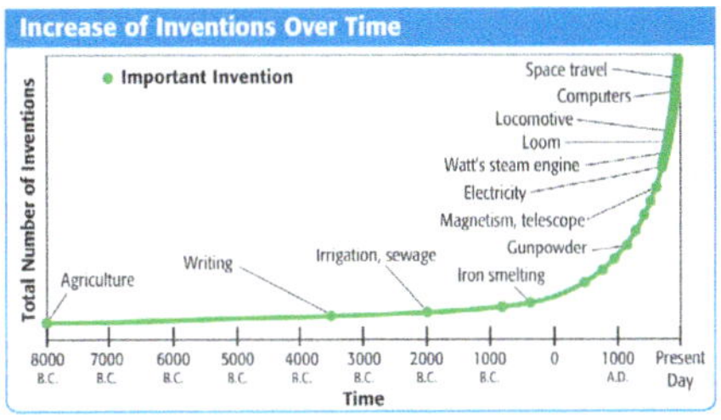

Figure 1-8 The pace of invention has continually increased over time.

If we compare the rate of change in agriculture to that of computers, we see that advances in farming happened over thousands of years. There were long stretches of time where little advancement seemed to happen (Figure 1-8). In computers, we can find that the first PC (personal computer) was sold in the early 1980s. In less than 50 years, we have seen large computers that only displayed text morph into graphical systems using Windows or Mac operating systems, storage go from 128 kilobytes (131,072 bytes) to terabytes (1,099,511,627,776 bytes, or over 8 million times the smaller amount), and sizes go from large multi-component systems to small laptops that fit easily in a book bag. A computer from 30 years ago would be laughably slow. In fact, just like the smartphone, we're quickly moving beyond computers being a separate product. Appliances, TVs, and doorbells are all becoming "smart" devices that are connected to the Internet. The incorporation of processing abilities, software, or sensors into physical, everyday objects to form a network of connected devices is known as the **Internet of Things** (IoT).

Internet of Things

When we plot the number of transistors on an integrated circuit over time, we can see a clear trend. The number of tranistors appears to double every year, which is often described as **exponential growth**. This observation has come to be known as **Moore's Law**, which originated as "the number of transistors in a processor will double every 2 years." Engineers have been able to shrink electrical components further and further, allowing for additional computational capacity. In fact, the trend is accelerating. Instead of two years, the timeframe for the number to double 18 months. It is thought that other aspects of computing will follow the trend of Moore's Law as well: storage, memory, and computing speed. Simultaneously, cost is expected to decrease by one half in the same time periods.

exponential growth

Moore's Law

TEACHING TIP

1984 saw Motorola release the first cell phone—the DynaTAC—to the public. The phone included a basic numeric keypad, a single-line display, and a battery that only allowed one hour of talk time. It was revolutionary at the time, and almost unaffordable at $50.00 per month, not including $0.40 per minute for calls. Compare today's smartphones and create a list of upgrades the mobile phone has seen since 1984. *(battery life, touch screens, hands-free features, Internet-ready, etc.)*

ELL

Consider turning the preceding Teaching Tip into a more targeted activity for ELLs. Take the discussion from a broad overview of the differences to a more specific and detailed exploration of differences as the students become more familiar with the vocabulary pertinent to the discussion.

INTEGRATE CONTENT AREAS

Algebra: Exponential Growth and Decay. Many phenomena can be shown by functions that describe how things grow or decay as time goes by.

Examples include the studies of populations, bacteria, the AIDS virus, radioactive substances, electricity, temperatures, and credit payments.

DISCUSSION

Introduce Moore's Law (the observation that the number of transistors in a dense integrated circuit doubles approximately every two years). Moore's Law is named after Gordon E. Moore, the co-founder of Intel and Fairchild Semiconductor. Describe how this affects the rate of technological change. Discuss the theory that Moore's Law may be slowing down. There are many video resources to choose from:

Intel: 50 Years of Moore's Law

URL

https://www.intel.com/content/www/us/en/silicon-innovations/moores-law-technology.html

Michio Kaku: Tweaking Moore's Law and the Computers of the Post-Silicon Era

URL

http://bigthink.com/videos/tweakingmoores-law-computers-of-the-post-silicon-era-2-2

ELL

Consider actually graphing these examples or passing out activities to encourage students to actively engage with the vocabulary and language.

Connecting To Stem
STEM = Science, Technology, Engineering, and Mathematics

Exponential Rate of Change

What does "exponential rate of change" mean? Let's look at an example. Jenny's mother agreed to let her download 10 songs a month. Her brother, Scott, wanted to download some, too. He got his mother to agree to this plan: He would download two songs the first month and then double the number he downloaded each month thereafter. Who do you think would have more songs after six months?

Jenny: Linear Rate of Change

At the end of the first month, Jenny had 10 songs. Each month, she downloaded another 10 songs. After six months, she had 60 songs, as shown in the table. Because her supply increased by the same rate each month, the number of songs showed a **linear rate of change**. On a graph, this rate looks like a straight line.

Monthly Song Downloads		
Month	Jenny's purchases (number of songs)	Scott's purchases (number of songs)
1	10	2
2	10	4
3	10	8
4	10	16
5	10	32
6	10	64

Scott: Exponential Rate of Change

Scott started out by buying two songs. He doubled his purchases each month thereafter. In the second month, he bought four, and in the third month, he bought eight. Scott's song supply increased faster each month. As a result, his supply increased exponentially. In a graph, an **exponential rate of change** shows a sharp curve. Their mother soon realized she would be spending all her money on songs for Scott, so she put an end to the arrangement.

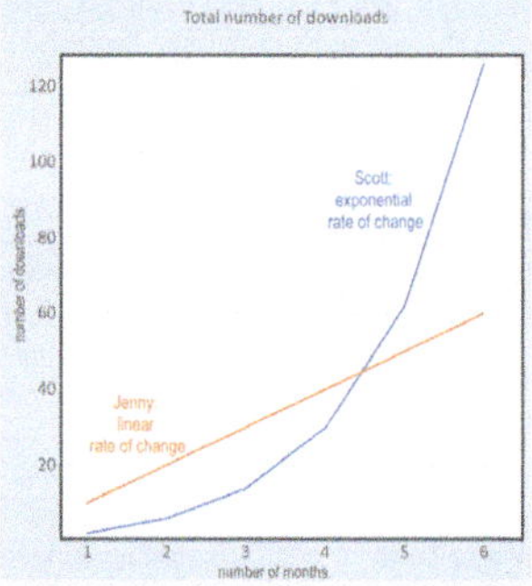

1.3 Engineering in Today's Society

Science, engineering, technology, and mathematics are forces that continue to change society. We continue to see better smartphones, automobiles, medical technology, manufacturing processes, and agricultural processes. As we get a next generation of smart devices like phones, watches, and computers, the models from a few years ago fall out of date and often end up in landfills.

Consider how we watch movies. By 1987, most households owned a videocassette recorder, or VCR. Roughly 37 million VCRs were in homes that year. By 2016, the last manufacturer of VCRs announced that it was no longer producing the machines; DVDs and Blu-ray systems had taken over. The 15 million units sold that year have, for the most part, found their way to landfills or have been disposed of in the ocean. If the average unit weighed 6 pounds, that is 45,000 tons of VCRs in landfills from just one year of sales. The **product life cycle** dictates that, no matter how new and exciting the product is, it eventually becomes outdated and discarded.

product life cycle

The Environmental Protection Agency (EPA) has developed a non-hazardous materials and waste management hierarchy. This shows that no single waste management approach is entirely suitable for waste. Ranking solutions from least to most environmentally friendly, they describe:

Treatment and disposal: This is the least preferred, but perhaps most common, means to dispose of waste (such as VCRs).

Energy recovery: This is the conversion of non-recyclable waste to energy, often by burning the waste to create electricity.

Recycling and composting: This is the collection and processing of waste into usable material. This can also include composting food and other compostables into usable material for planning.

Source reduction and reuse: The most effective way to deal with waste is to not produce waste in the first place. Reducing the production of waste, reusing items, reducing packaging, and reducing toxicity of products all help reduce waste.

Some advancements mean that a person with different skills must be hired to build, make, or create a device, meaning people who worked on the outdated technology may find themselves unemployed. One example is seen in the newspaper industry. The public used to get their information from the daily newspaper. Today, the Internet and television give 24/7 access to information, and the newspaper industry is only a fraction of what it was years ago. Those employed in the newspaper industry need to retrain and adapt to a new paradigm.

Similar changes have resulted from the Information Age. In the 1960s, transistors replaced vacuum tubes. Semiconductor manufacturers prospered, but vacuum tube manufacturers saw their market shrink dramatically.

More recently, the Internet created a new industry almost overnight. The dot.com industry, made up of companies doing business only through websites. This created many new jobs, as people rushed to build online businesses, creating an economic "bubble" as profits failed to materialize and investor cash was eaten up by the business's operating costs. This bubble burst, causing many businesses to fail. This pattern of technology advancement, resulting in both winners and losers, is observed time and again (Figure 1-9).

Workforce Sector	Pre-Industrial Revolution	Industrial Revolution (mid-18th century, early 19th)	Information Age (mid 20th century)	Current Day (early 21st century)
Agriculture (e.g., farming)	High employment	Moderate employment	Low employment	Low employment
Industry/Factory	Low employment	High employment	Moderate employment	Low employment
Information Technology	Low employment	Low employment	High employment	High employment

Figure 1-9 Changes in technology have led to the development of different types of jobs in different eras.

DISCUSSION QUESTION

Ask students to create a list of jobs that are likely to be eliminated by 2030.

DISCUSSION QUESTION

Will cab drivers and delivery trucks be replaced by driverless cars and drones? Will voice recognition advances replace the need for translators? Will shipping, receiving, and warehouse workers be eliminated when 3-D printers are installed in your home?

Consider sharing the following video from Vox about the future of work. This tip will also be repeated in Chapter 10 in case you find it more appropriate to show there.

The big debate about the future of work, explained: https://www.youtube.com/watch?v=TUmyygCMMGA

ACTIVITY

Engineers around the globe are creating new technology that will enable the disabled. Research and discuss innovations like the ReWalk Personal exoskeleton, the UPnRIDE upright wheelchair system, and the Sesame touch-free smartphone that were all designed to benefit the disabled.

Technology Increases Comfort, Safety, and Accessibility

Using technology, products have been redesigned to make our lives more comfortable (Figure 1-10). Chairs are ergonomically shaped for back support. Hiking boots are lighter yet stronger and more waterproof. Kitchen appliances are connected to the Internet and run more efficiently. Buses have bike racks on the front and can kneel at curbs to allow people with wheelchairs to board. As cities become more densely packed, planning engineers design more accessible pathways, green space, and traffic control.

Figure 1-10 Created using technology, specially designed equipment allows people with all types of disabilities to enjoy an active lifestyle.

TECHNOLOGY IN THE REAL WORLD

Replacement Hearts

Medical inventions have allowed doctors to replace hands, arms, or even entire legs to provide new mobility and independence with prosthetic devices. But replacing organs that lie inside the body is a bit trickier. Every year, people with various organ failures wait for human donors to provide healthy replacements for a kidney, liver, or heart. The list of donors, however, is much shorter than the need for transplants. Even after an organ has been transplanted successfully, patients must take dozens of medications to reduce their body's response to the new organ, and it is still sometimes rejected.

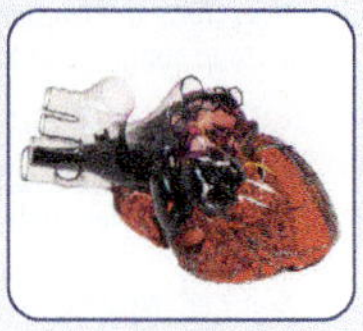

A replacement heart's design mimics the valves and arteries of a real heart.

Instead of Muscle, Titanium

A new direction in medical technology is to replace the diseased organ with a completely synthetic organ. Over the years, many companies have been working toward a solution with the heart, with various degrees of success.

A lot of information has been gathered through these attempts. Materials must be compatible with tissue, so most solutions choose titanium and plastics. One company has successfully 3-D printed tissue that behaves like heart muscle. The size of the replacement must be about the same as the original human heart, and it must have an external power supply. The external power drastically reduces the chance of infection. Lastly, it must have the capability to regulate flow for various activities (resting vs. exercise). We are learning about the constraints on the problem with every attempt.

Making a Real Difference

Every year, more than 300,000 people in the United States die from congestive heart failure. Although synthetic hearts are a new, experimental technology, they hold the promise of changing the lives of thousands of people who suffer from heart disease. As medical technology grows, the list of human-made organs may grow as well.

1.3.1 Technology and the Environment

Advances in technology often have a negative effect on the environment. This conflict has existed for thousands of years. Throughout history, forests were cut down for fuel or to clear land for farming. Garbage was burned outside in the open. Human waste was poured into streets and rivers.

Technology Has Changed Our Landscape

Today, technological advancement has allowed us to use less plastic in packaging, biodegradable packaging has been developed, and wide-scale recycling offers a chance to reduce waste. However, some manufacturers choose not to rework their manufacturing line to reduce packaging; biodegradable packaging is rarely used as it's more difficult and more expensive to work with, and recycling efforts often fail since people simply don't participate, or mix garbage in with recycling.

Tons of plastics are dumped into the oceans each year, and it takes years to break down. This has resulted in, for example, the Great Pacific Garbage Patch, where currents from the Pacific Ocean carry wastes to a spot where an island of trash has formed. Further, as plastic begins to break down, microplastics are now prevalent in fish and water. This is a situation which is ready for an engineering solution.

Using Technology to Solve Environmental Problems

Advances in technology have helped address pollution problems. Pollution control devices on cars and scrubbers on factory smokestacks have reduced pollution. Electric automobiles which do not emit pollutants are now readily available. **Hybrid electric vehicles** (HEVs) combine a conventional engine with a battery and electric motor (Figure 1-11).

hybrid electric vehicles

Replacing gasoline cars with electric or hybrid cars is no simple matter, however. Most families cannot afford to replace a car without planning to do so. Once gasoline cars are phased out, they must be disposed of. Many components can be recycled, but disassembling a car to recycle components is slower and more difficult than smashing it and putting it in a landfill. Further, electric cars need to be charged, and the grid (the infrastructure that delivers power from the power plant to homes and businesses) faces issues. The power grid is behind schedule for maintenance, and vast improvement will be needed if it is asked to power automobiles in addition to homes and businesses.

Figure 1-11 This hybrid vehicle uses both electricity and gas. It captures energy from braking and uses it to recharge its batteries.

Our environment must be treated with care. We must use resources wisely and look for alternatives whenever possible, especially for scarce resources. We must develop solutions that reduce environmental harm. For example, solar energy, wind power, nuclear power, hydrothermal energy, or biofuels may provide an answer in our search for more power delivered more efficiently.

ENGAGE

Introduce students to the Intergovernmental Panel on Climate Change (IPCC), the Potsdam Institute for Climate Impact Research, the United Nations Environment Programme (UNEP), and the World Wide Fund for Nature (WWF).

URL
http://www.ipcc-nggip.iges.or.jp

URL
https://www.pik-potsdam.de/en

URL
http://www.unep.org/climatechange/Introduction.aspx

URL
http://www.worldwildlife.org/initiatives/climate

ELL

Consider collecting some of the more common symbols and signs representing recycling together in a PowerPoint presentation or worksheet, and present them to the class. These symbols might include items such as the "recycle" symbol, as well as bins labeled differently according to what kind of material they recycle. This helps ELLs derive meaning from environmental print.

DISCUSSION QUESTION

Ask the students to provide examples of each organization's response to climate change.

IPCC: *Provides an understanding of human induced climate change, potential impacts of climate change, and options for mitigation and adaptation.*

PIK: *Integrates climate impacts across sectors and scales in novel ways, including climate extremes, socioeconomic damages, and costs. Assesses sector-specific adaptation options and explores synergies between adaptation, mitigation, and human development.*

UNEP: *Creates a new and vitally important interface between the scientific community and decision makers involved in vulnerability, impacts, and adaptation issues, and improves the availability and accessibility of such knowledge for the people that need it most.*

WWF: *Engages with companies to reduce carbon pollution and adopt more renewable energy, both in their own operations and along their supply chain. Helps America's largest companies use their collective purchasing power to scale up renewable energy while at the same time extend affordable access to renewable energy by offering bulk purchase discounts on home solar to employees nationwide.*

ENGAGE

Visit the websites of the World Health Organization (WHO) and the United Nations Children's Fund (UNICEF) with your students. Navigate to the pages on drinking water quality.

URL

http://www.who.int/water_sanitation_health/en/

URL

https://www.unicefusa.org/mission/survival/water

DISCUSSION QUESTION

What steps are the WHO and UNICEF taking to improve access to clean water and sanitation facilities?

- Restoring access to clean water after a disaster
- Promoting safe hygiene practices in schools and communities
- Mobile app challenges users to provide water to a child in need
- Drinking water quality management
- Water supply and sanitation monitoring
- Cholera surveillance and prevention

DISCUSSION QUESTION

Your school recycles waste in several ways. What steps has your school taken to recycle waste? What are some other ideas for recycling that your school could complete?

TEACHING TIP

Point out that the large flat roof of a school building is a great place for solar panels. Using the space to generate clean solar electricity will cut energy costs. Although they are still rather expensive, solar panels would eventually pay for themselves in energy savings.

Living Green

If we assume the typical family car gets 22 miles per gallon and drives 11,500 miles per year, each car in each family emits about 4.6 metric tons of carbon dioxide per year. Every gallon of gasoline burned creates about 8,887 grams of CO_2.

Make a Change

Can you think of possible engineering solutions to this problem?

1.3.2 Technology and Ethics

We have seen some of the effects of engineering innovation, both positive and negative. Ethics must be considered routinely in engineering solutions. Some major issues we have seen so far include the increased use of plastics and the additional pollution it can create, and how introducing a new technology results in economic consequences like massive unemployment in relevant sectors.

Note that ethics are different from morals, even though some use the terms interchangeably. Morals are personal and define what actions you deem appropriate, whereas ethics are defined by a community or organization like the National Society of Professional Engineers. For example, say you were a software engineer at a large privately-owned social media company that put you on a team to develop a content moderation feature that automatically deletes posts containing profane material. You may believe the practice is inappropriate because it can be considered censorship, but the company's code of conduct for users (their ethics) takes a stance on not allowing certain content on the platform. Your morals may conflict with certain codes of ethics, and it's not necessarily rare.

ethical dilemmas

Ethical dilemmas are decisions engineerings face when different values come into conflict with one another during the design process. Ethics is often cast as a way to distinguish "right" from "wrong" in these situations, especially in textbook examples. In practice, these decisions are often more complicated than right or wrong; there can be multiple paths forward for what is appropriate to do, and some will be more appropriate than others. For example, consider biological or biomedical engineers who work on genetic modification research like CRISPR, a highly precise method of gene editing. What are the implications of gene editing? A noble goal of gene editing is eliminating diseases like cancer linked to DNA segments—legitimate medical purposes. What about removing mild lactose intolerance, which is connected to the LCT gene? Where is the line where medical use becomes cosmetic or non-essential? One phenomenon is the rise of "designer babies," where an embryo can be edited to have specific features, like eye or hair color. Note that this is all before we consider the possible negative consequences of editing genes, such as the overall health of the person who had treatment and the inequality of who has access to these services and who does not.

There are a few ethical decisions that have become well-known among engineers as classical dilemmas. For example, a car in the 1970s called the Ford Pinto had a very inexpensive design of its fuel tank, and if the car were hit from behind, it could explode. This seems like an obvious situation where the proper, ethical decision was not made. However, the engineers who designed the systems had to weigh the danger with a strict production schedule in addition to cost and weight requirements because of pressure from Ford's management. The Pinto was less safe than other cars, but how "less safe" is acceptable? Were other cars in that time designed with the same shortcuts but not discovered? Often, a decision may be 'ethical' or not, but ethics are considered in each decision.

In 2015, engineers discovered that other engineers and designers from Volkswagen had intentionally designed software to know whether the car was driving or under emissions testing. Cars were passing emissions tests, but seemed to be polluting more than other cars. An investigation showed that the software

DISCUSSION QUESTION

Ethical dilemmas in technology span from genetically-modified organisms to drone operation. Although our laws and ethical practices have evolved over centuries, today, technology is on an exponential curve and is touching practically everyone everywhere. How can our policy makers keep up with technology and create laws that protect everyone everywhere?

TEACHING TIP

Introduce the students to the U.S. Environmental Protection Agency and its website.

URL

http://www3.epa.gov

What environmental issues does the agency concentrate on?

- Air quality
- Water quality
- Land and cleanup
- Climate change
- Greener living

was written to reduce emissions while the car was in emissions testing, but return to normal while driving. Eventually, the company was fined $4.3 billion and those responsible lost their jobs.

As you explore engineering and engineering design, always keep any ethical issues in mind. We will explore the importance of ethics through other examples.

1.4 Grand Challenges for Engineering

Grand Challenges

History is full of examples of innovations that forever changed the course of human society: the mastery of fire, the development of agriculture, the wheel, metallurgy, mathematics, the printing press, electricity, powered flight, nuclear power, and many others. The NAE (National Academy of Engineering) has established a list of 14 challenges for the 21st century, each of which has the potential to transform the way we live, work, and play. Your interest in one or more **Grand Challenges** for Engineering may help you select your major in college. Details are available at *http://www.engineeringchallenges.org*.

A burgeoning planetary population and the technological advances of the last century are exacerbating many current problems as well as engendering a variety of new ones, for example:

- Relatively inexpensive and rapid global travel make it possible for diseases to quickly span the globe. A century ago, they could spread, but much more slowly.
- The reliance of the developed world on computers and the Internet makes commerce and government vulnerable to cyberterrorism.
- Increased demand for limited resources not only drives up prices for those commodities, but also creates strain among the nations competing for them.

These same factors can also be a force for positive change in the world:

- Inexpensive global travel allows more people to experience different cultures. This helps promote a more tolerant attitude toward those who live by different social norms.
- Modern communications systems—smartphones, the Internet, etc.—make it difficult to isolate the members of a population from the political realities in other parts of the world.
- Increased demand and rising prices of limited resources is driving innovation in alternative energy solutions.

As shown from these few examples, technology not only solves problems, but also creates them. A significant portion of the difficulty in the NAE Challenges is to solve critical problems in the world that do not create other problems. Some of these challenges are well known in society, such as "make solar energy economical," "provide energy from fusion," "secure cyberspace," and "enhance virtual reality."

In addition to the NAE Challenges, the X Prize Foundation, over many years, has provided financially incentivized challenges. An objective of the financial incentives is to create a competitive environment to accelerate the development of these challenges. Often, teams are put together and financed for the purpose of pursuing these challenges. In past years this has included such varied targets as "ocean health," "water abundance," "adult literacy," and "gigaton scale carbon removal." Go visit their website, *https://www.xprize.org/* for more information.

1.4.1 The Nitrogen Cycle

One of the lesser known NAE Challenges concerns the nitrogen cycle. Nitrogen is an element required for all known forms of life. It is part of every one of the amino acids. Amino acids are combined in various ways to form proteins, all five bases used to construct RNA and DNA, and numerous other common

ENGAGE

Introduce the students to the National Academy of Engineering website. Navigate to the Grand Challenges page.

URL

http://www.engineeringchallenges.org

Have the students create a list of the current challenge categories. *(advance personalized learning, enhance virtual reality, engineer better medicines, restore and improve urban infrastructure, provide access to clean water, prevent nuclear terror, develop carbon sequestration methods, make solar energy economical, reverse-engineer the brain, advance health informatics, secure cyberspace, provide energy from fusion, manage the nitrogen cycle, and engineer the tools of scientific discovery)*

ENGAGE

The nitrogen cycle is the process by which nitrogen is converted between its various chemical forms. Define the important processes in the nitrogen cycle, including fixation, ammonification, nitrification, and denitrification.

Fixation: *Nitrogen fixation, natural and synthetic, is essential for all forms of life because nitrogen is required to biosynthesize basic building blocks of plants, animals, and other life forms, e.g., nucleotides for DNA and RNA and amino acids for proteins. It is essential for agriculture and the manufacturing of fertilizer.*

Ammonification: *When a plant or animal dies or an animal expels waste, the initial form of nitrogen is organic. Bacteria or fungi convert the organic nitrogen within the remains back into ammonium (NH4+), a process called ammonification.*

Nitrification: *The conversion of ammonia to nitrate is performed primarily by soil-living bacteria and other nitrifying bacteria. Due to their very high solubility and because soils are largely unable to retain anions, nitrates can enter groundwater. Elevated nitrate in groundwater is a concern for drinking water use because nitrate can interfere with blood-oxygen levels in infants and cause methemoglobinemia or bluebaby syndrome.*

Denitrification: *The process of reducing nitrates and nitrites to nitrogen, especially nitrous oxide. This is beneficial because it removes nitrite from wastewater, but also removes nitrogen from fertilized soil.*

biological molecules such as chlorophyll and hemoglobin. The supply of nitrogen is—for all practical purposes—inexhaustible, constituting over 75% of the Earth's atmosphere. However, nitrogen is mostly in the molecular form N_2. The two nitrogen atoms are held together by a very strong chemical bond. This renders them chemically unavailable to biological systems.

For atmospheric nitrogen to be biologically available, it must be converted, or "fixed," by the addition of hydrogen, into ammonia, NH_3. In that form it may be used directly or converted into other nitrogen compounds for use by microorganisms and plants. The term nitrogen fixation includes conversion of N_2 into both ammonia and these other reactive compounds. Eventually the cycle is completed when these more readily available forms of nitrogen are converted back to N_2 in a process called denitrification (Figure 1-12). Prior to human development, essentially all nitrogen fixation was performed by bacteria that had an enzyme capable of splitting N_2 and adding hydrogen to form ammonia. And small amounts of fixed nitrogen are produced by lightning and other high-energy processes. In the early 20th century, a process called the Haber-Bosch process was developed that would allow conversion of atmospheric nitrogen into ammonia and related compounds on an industrial scale. Today, slightly more than a century later, approximately one-third of all fixed nitrogen is produced using this process.

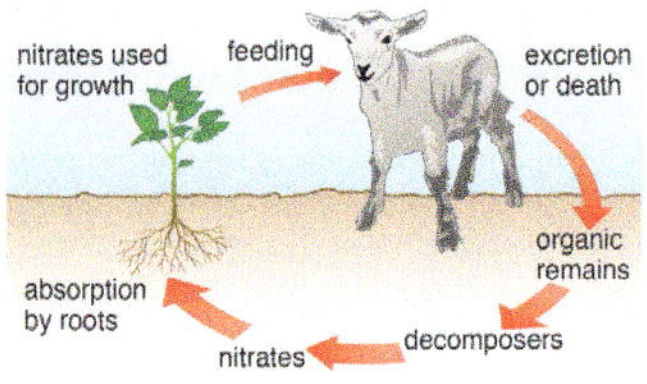

Figure 1-12 Illustration of the Nitrogen cycle.

The ready availability of relatively inexpensive nitrogen fertilizers has revolutionized agriculture, allowing people to increase yields dramatically and to grow crops on previously unproductive lands. By contrast, the widespread use of nitrogen compounds have caused numerous environmental problems. It has caused water pollution, air pollution, numerous human health problems, and disruption of various ecosystems to the extent that entire populations of organisms have died off.

Nitrogen fixation is only one part of the nitrogen cycle problem, however. Many human activities, especially those involving the combustion of fossil fuels, pump huge quantities of various nitrogen compounds into the atmosphere. Nitrous oxide (N_2O), also of some notoriety as the anesthetic commonly known as "laughing gas," is particularly problematic. It is about 200 times more effective than carbon dioxide as a greenhouse gas, and persists in the atmosphere for over a century.

Altogether, human-caused conversion of nitrogen now accounts for about half of all nitrogen fixation. In other words, there is twice as much nitrogen fixed today than there was about a century ago. We have done little to augment the natural denitrification process, so the damaging effects of excessive fixed nitrogen continue to increase. We have overwhelmed the natural nitrogen cycle. If we continue along this path, we must learn to manage the use of these products more efficiently and plan strategies for denitrification to bring the cycle back into balance.

1.4.2 Reverse-Engineering the Brain

The development of artificial intelligence (AI) holds possibly the most overall potential for positive change in the human race as well as the most horrendous possible negative effects. This is reflected in science fiction, where the concept of thinking machines is a common plot device. If history is any guide, the potential for disastrous consequences seldom deters technological advances, so consider what is involved in the development of AI.

Although great strides have been made in creating machines that seem to possess "intelligence," most such systems either rely on brute-force calculations, such as the chess-playing computer, Deep Blue, that defeated the world champion in 1997. An alternative approach is reliance on incredibly fast access to massive databases, such as the Jeopardy-playing computer, Watson, that defeated the highest performing humans to ever win that game. These are simplifications, as there are many more aspects to both of these systems. However, one would be hard-pressed to argue that these computers are truly intelligent—that they are self-aware and contain the spark of creativity, which is the hallmark of humans and other highly intelligent creatures on Earth.

There is a tendency to conflate humanoid robots with artificial intelligence. Robots are machines generally designed to deliver a specific service or perform a specific task—welding, painting, vacuuming and so forth. All of these systems operate within certain restrictions and are limited when confronted with unexpected situations.

Researchers working on AI solutions see the challenge as a progression of four capabilities in software—Reactive Machines, Limited Memory, Theory of Mind, and Self-Awareness. Reactive Machines produce an output based on the most recent input. Deep Blue is a prime example. The second level of AI systems have Limited Memory. This is to say that, to some degree, they have the ability to guess a likely result based on recent events. These are the sort of capabilities built into self-driving machines. Events in the recent past can inform decisions in the very near future. Next on the AI ladder is Theory of Mind. This is the ability to perceive or infer social and emotional states of mind and to respond accordingly. This is in the realm of science fiction for now. Even though in-home digital assistants, chatbots, and some robots appear to respond emotionally, they are following an algorithm rather than a self-generated emotional reaction. Finally, Self-Awareness is what we mean when we think of intelligence. Awareness of one's own existence, self-determination, thinking, and emotional response are all hallmarks of true AI.

To date, we have almost exclusively attempted to construct systems that mimic behavior and thought, not design systems that actually store and process information in a manner like a brain. The human brain utilizes a network of connections that makes even the most advanced computers look very rudimentary. Despite some advances, the means of encoding information and its transfer and storage in the brain is almost completely a mystery.

Gaining even a basic understanding of brain function might allow us to develop prosthetic limbs that actually function as well as the originals, restore sight to the blind, repair brain damage, or even enhance human intelligence.

1.5 Engineering Disciplines

We have seen some specific engineering disciplines thus far. There are many disciplines in engineering, and most of these work in conjunction with others. For example, pulleys and levers (mechanical) can be employed to build complex aqueduct systems (civil). Water wheels (mechanical) eventually were replaced by electric motors (electrical) and so on. Smartphones have replaced walkie-talkies and landlines (electrical, materials, computer, software). Manufacturing processes make things we need astoundingly fast (industrial and systems engineering). The following paragraphs briefly introduce several different types of engineering majors. By no means is this list completely inclusive.

Bioengineering or Biomedical Engineering

Bioengineering (BioE) and biomedical engineering (BME) apply engineering principles to the understanding and solution of medical problems. Bioengineers are involved in research and development in all areas of medicine, from investigating the physiological behavior of single cells to designing implants for the replacement of diseased or traumatized body tissues.

DISCUSSION QUESTION

The National Academy of Engineering promotes reverse engineering of the human brain as one of its Grand Challenges.

URL

http://www.engineeringchallenges.org/challenges/9109.aspx

What are some of the impacts it may one day have on human health and artificial intelligence?

How the living brain works may offer the best guide to engineering the artificial variety. Discovering how by reverse-engineering the brain promises enormous opportunities for reproducing intelligence. Neurological disorders may someday be circumvented by technological innovations that allow wiring of new materials into our bodies to do the jobs of lost or damaged nerve cells. Implanted electronic devices could help victims of dementia to remember, blind people to see, and crippled people to walk.

TEACHING TIP

Define the term biosystem for the students. *(a group of molecules that interact in a biological system)*

ENGAGE

Visit the National Center for Biotechnology Information (NCBI) website with your students. Discuss the impact the organization has had on biomedical and genomic information.

URL
https://www.ncbi.nlm.nih.gov/

NCBI's mission is to develop new information technologies to aid in the understanding of fundamental molecular and genetic processes that control health and disease. More specifically, the NCBI has been charged with creating automated systems for storing and analyzing knowledge about molecular biology, biochemistry, and genetics; facilitating the use of such databases and software by the research and medical community; coordinating efforts to gather biotechnology information both nationally and internationally; and, performing research into advanced methods of computer-based information processing for analyzing the structure and function of biologically important molecules.

Bioengineers design new instruments, devices, and software, assemble knowledge from many scientific sources to develop new procedures, and conduct research to solve medical problems. Typical bioengineers work in the medical and research industries, in such areas as artificial organs, automated patient monitoring, blood chemistry sensors, advanced therapeutic and surgical devices, clinical laboratory design, medical imaging systems, biomaterials, bioinstrumentation, biomechanics, rehabilitation engineering, systems physiology, and sports medicine.

Bioengineers are employed in universities, industry, hospitals, research facilities, and government. In industry, they may be part of a team serving as a liaison between engineers and clinicians. In hospitals, they select appropriate equipment and supervise equipment performance, testing, and maintenance. In government agencies, they are involved in safety standards and testing.

Biosystems Engineering

Biosystems engineering (BE) is the field of engineering most closely allied with advances in biology. BE emphasizes two main areas: (1) bioprocess engineering, with its basis in microbiology, and (2) ecological engineering, with its basis in ecology. The field focuses on the sustainable production of biorefinery compounds (biofuels, bioactive molecules, and biomaterials) using metabolic pathways found in nature and green processing technologies. Further, BE encompasses the design of sustainable communities utilizing low-impact development strategies. This includes bioretention basins, rainwater capture for storm-water retention and treatment—and ecologically sound food and energy-crop production. Scientific emphasis is shifting toward the biosciences. Biosystems engineers apply engineering design and analysis to biological systems and incorporate fundamental biological principles to engineering designs to achieve ecological balance.

Here are some activities of biosystems engineers:

- Design bioprocesses and systems for biofuels (biodiesel, hydrogen, ethanol), biopharmaceutical, bioplastics, and food processing industries
- Develop ecological designs (permeable pavement, bioswales, green infrastructure) to integrate water management into the landscape
- Integrate biological sustainability concepts into energy, water, and food systems
- Provide engineering expertise for agriculture, food processing, and manufacturing
- Pursue medical or veterinary school or graduate school in the fields of BE, BME, or ecological engineering

Chemical Engineering

Chemical engineering (ChE) emphasizes three sciences: chemistry, physics, and mathematics. Chemical engineers are involved in the research and development, manufacture, sales, and use of chemicals, pharmaceuticals, electronic components, food and consumer goods, petroleum products, synthetic fibers and films, pulp and paper, and many other products. They work on environmental remediation and pollution prevention, as well as in medical and health-related fields.

The term chemical engineer was not common until around the mid-1800s. Still, the mainstream view was that it was a form of mechanical processing. Chemical engineering began with methods called "unit operations." This term meant any process which physically changes a substance, without a substantial chemical change. Examples include processes that crush, blend, boil, dehydrate, and pressurize. By 1880, chemical engineering classes appeared at the university level, and chemical engineering grew rapidly as more graduates entered the field.

TEACHING TIP

Define the term environmental remediation for your students. *(the removal of pollution or contaminants from water and soil)*

DISCUSSION QUESTION

How can chemical engineers solve these types of environmental problems?
Environmental remediation uses a wide variety of technologies and methods for cleaning up contaminated areas. The methods used at a particular site depend on the type and extent of the pollution, as well as the characteristics of the site itself. Areas where pollution has seeped deep into the ground and groundwater require different methods than areas where a small amount of chemical has been spilled on the topsoil. All projects begin with a site assessment. Some problems can be eliminated by excavation or dredging on site. Some require pumping of fresh water (or pumping away contaminated water) and treating the water to keep it safe. Solidification and stabilization methods can be used to chemically alter the water or soil. Oxidation, soil vapor extraction, and bioremediation are also methods used by chemical engineers to solve contamination problems.

Chemical engineers:

- Conduct research and develop new chemical products
- Develop and design new manufacturing processes
- Earn additional degrees to practice medicine or patent, environmental, or corporate law
- Sell and provide technical support for sophisticated chemical products to customers
- Solve environmental problems; work in biotechnology
- Troubleshoot and solve problems in chemical manufacturing facilities

Civil Engineering

Civil engineering (CE) involves the planning, design, construction, maintenance, and operation of facilities and systems to control and improve the environment for modern civilizations. This includes projects of major importance such as bridges, transportation systems, buildings, ports, water distribution systems, and disaster planning.

The evolution of civil projects led to new subsets of civil engineering. Examples include structural, construction, geotechnical, environmental, surveying, transportation, and water resources engineering. Civil engineering is considered the second oldest form of engineering. It dates to 2000-4000 B.C. As humans moved from a nomadic to an agrarian society, they needed shelter, transportation, and water control. These were the first civil engineering projects. Before the 18th century, mathematicians, stonemasons, and carpenters worked together to build structures. They were called artisans. Planning and drawings for structures were the jobs of architects. The role of architect and civil engineer were the same in some areas. In France in the late 1700s, civil engineers asked to be recognized for their work. This idea spread to London in the early 1800s. The first civil engineering degree in the United States was awarded in 1835. Since then, there has been significant growth and evolution in this area.

Here are just a few of many opportunities available for civil engineers (Figure 1-13):

- Design and analyze structures ranging from small buildings to skyscrapers to offshore oil platforms
- Design dams and building foundations
- Develop new materials for pavements, buildings, and bridges
- Design improved transportation systems
- Design water distribution and removal systems
- Develop new methods to improve safety, reduce cost, speed construction, and reduce environmental impact
- Provide construction and project management services for large engineered projects throughout the world

ENGAGE

Ask the students to visit the American Society of Civil Engineers (ASCE) website, search for bridge repair, and navigate to the report card for America's infrastructure.

URL

http://www.infrastructurereportcard.org/bridges/

Inform them that one in nine of the nation's bridges are rated as structurally deficient, while the average age of the nation's 607,380 bridges is currently 42 years. The Federal Highway Administration (FHWA) estimates that to eliminate the nation's bridge deficient backlog by 2028, we would need to invest $20.5 billion annually, while only $12.8 billion is being spent currently. The challenge for federal, state, and local governments is to increase bridge infrastructure investments by $8 billion annually to address the identified $76 billion in needs for deficient bridges across the United States.

ELL

Consider having students create their own descriptions of engineering mimicking those on page 29. This aids ELLs in comprehending and reproducing common language structures they'll find in written classroom materials.

DISCUSSION QUESTION

What role can civil engineers play in repairing our bridges and reducing the cost of doing the job?

- *Creating and using Ultra-High Strength Concrete (UHSC), a strong material used to repair bridges much more quickly at a much lower cost*
- *Creating and using new construction techniques to increase safety and reduce construction time and traffic congestion*
- *Creating and using new nondestructive bridge evaluation technologies and techniques*
- *Building bridges that will withstand natural disasters, including alternative processes for the seismic retrofit of bridges*

26 Chapter 1 Engineering as a Discipline

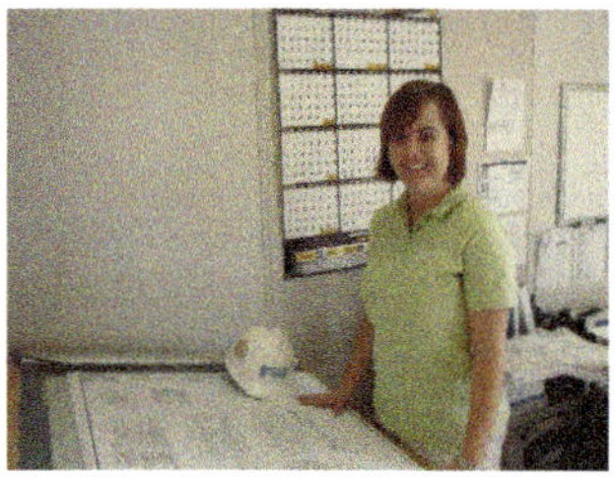

Figure 1-13

"I am responsible for assisting in the management of commercial and healthcare projects for a large general contractor. Working closely with the owner and architect, I maintain open lines of communication and aim to provide exceptional service to the entire project team from the preconstruction phase of the project through construction. I assist in establishing and monitoring procedures for controlling the cost, schedule, and quality of the work in accordance with the construction contract."
L. Edwards, CE

Computer Engineering

Computer engineering (CpE) spans the fields of computer science, engineering, and mathematics, giving a balanced view of hardware, software, hardware-software trade-offs, and basic modeling techniques that represent the computing process involving the following technologies (Figure 1-14):

- Communication system design
- Computer hardware design and interface design
- Computer networking, network management, and network design
- Digital signal processing applications
- Digital system design
- Embedded computer design
- Process instrumentation and control
- Software design and programming
- Systems support

Figure 1-14

"As a Radiation Effects Engineer I test the performance of electronic components in a specific application exposed to different types of radiation. Responsibilities include interfacing with design and system engineers, creating test plans, performing testing and data analysis, and authoring test reports."
A. Passman, EE

Electrical Engineering

Electrical engineering (EE) ranges from the generation and delivery of electrical power to the use of electricity in integrated circuits. It involves different circuit types, devices, and equipment that use electricity. There is also a lot of interaction with computer programming. Electrical engineering broke out in the mid-19th century. Its emergence was attributable to the use of the telegraph, telephone, and electric power. Electrical engineering divided into sub-categories as the field expanded. Some of these include communications, control, data handling, sensors, and computers. From the 1600s to the mid-1800s, the field grew quickly. There were discoveries and breakthroughs in static electricity, magnetism, and the ideas of voltage and current. These led to the invention of batteries, electric switches, and electric motors. People began to understand electrical properties. James Maxwell famously developed a set of equations unifying electricity and magnetism. One of the defining inventions of this time was the telegraph in the early 1800s. It enabled communication around the world. It also led to the formation of electrical engineering societies in both the United States and the UK. Universities started to offer electrical engineering degrees. The work of inventors such as Edison, Hertz, Faraday, and Tesla still guides the work of today's engineers. Their inventions include motors, generators, lights, power, transistors, radios, and telephones.

The rapid development of technology based on integrated circuit devices has enabled the pervasive use of computers in command, control, communication, and computer-aided design. Some systems electrical engineers work with includes the following (Figure 1-15):

- Communication system design
- Control systems—from aircraft to automotive
- Electrical power generation and distribution
- Electromagnetic waves
- Integrated circuit design
- Process instrumentation and control
- Robotic systems design
- Telecommunications

Figure 1-15

"My group supports [a major automotive manufacturer's] decisions pertaining to where to put new plants around the world, what products to build in them, and at what volumes. In particular, my work involves understanding what the other auto manufacturers are planning for the future (footprint, capacity, technology, processes, etc.), so that information can be used to affect decisions about how to compete around the globe."
M. Peterson, EE

Environmental Engineering

Environmental engineering (EnvE) is an interdisciplinary field of engineering that is focused on cleaning up environmental contamination, as well as designing sustainable approaches to prevent future contamination. Environmental engineers apply concepts from basic sciences (including chemistry, biology, mathematics, and physics) to develop engineered solutions to complex environmental problems.

TEACHING TIP

When introducing electrical engineering, it may be helpful to ask the class what they believe the difference is between electrical engineering and computer engineering—perhaps even the difference between computer engineering and computer science.

Electrical engineering, simply put, is about designing any equipment or device that uses electricity, whereas computer engineering is a subset of electrical engineering that involves designing software and hardware for those devices. Computer science emphasizes the theory of computing and algorithms, which includes newer fields like cybersecurity.

TEACHING TIP

- Define electromagnetic waves for your students. *(Electromagnetic waves can travel not only through air and solid materials, but also through the vacuum of space.)*
- Define the term integrated circuit design for your students. *(Integrated circuit design is the creation of electronic components, such as transistors, resistors, capacitors, and the metallic interconnect of these components onto a piece of semiconductor, typically silicon.)*
- Define the term robotic systems design for your students. *(Robotic systems design is an integrated system of devices that automate production and manufacturing of goods and services.)*

DISCUSSION QUESTION

Environmental engineering focuses on issues with environmental contamination. What are the five basic types of contamination? *(air, water, noise, soil, and light pollution)*

TEACHING TIP

Visit the Underwriters Laboratories (UL) website with your students.

URL
http://ul.com/aboutul/

UL is a global independent safety science company with more than a century of expertise innovating safety solutions from the public adoption of electricity to new breakthroughs in sustainability, renewable energy, and nanotechnology. Dedicated to promoting safe living and working environments, UL helps safeguard people, products, and places in important ways, facilitating trade and providing peace of mind.

Environmental engineers design, operate, and manage both engineered and natural systems to protect the public from exposure to environmental contamination and to develop a more sustainable use of our natural resources. These activities include the following:

- Production of safe, potable drinking water
- Treatment of wastewater so that it is safe to discharge to surface water or reuse in applications such as landscape irrigation
- Treatment of air pollutants from mobile (e.g., automobiles) and stationary (e.g., power plants) sources
- Characterization and remediation of sites contaminated with hazardous wastes (e.g., polychlorinated biphenyls, or PCBs)
- Disposal of municipal solid wastes
- Management of **radioactive wastes**, including characterization of how radioactive materials move through the environment and the risks they pose to human health

radioactive wastes

- Evaluation of methods to minimize or prevent waste production and inefficient use of energy by manufacturing facilities
- Reduce human health risks by tracking contaminants as they move through the environment
- Design a more sustainable future by understanding our use of resources

With a BS degree in EnvE, students will find employment with consulting engineering firms, government agencies involved in environmental protection, and manufacturing industries.

I work with scientists and engineers to protect their innovations by writing patent applications describing their inventions and presenting the applications before the United States Patent & Trademark Office. I also assist clients in determining whether another party is infringing their patents and help my clients to avoid infringing other's patents.
M. Lauer, EnvE

Industrial Engineering (Also called Manufacturing Engineering)

Industrial engineering (IE) deals with the design and improvement of systems, rather than with the objects and artifacts that other engineers design. A second aspect of IE is the involvement of people in these systems—from the people involved in the design and production to the people who are ultimate end users. A common theme is the testing and evaluation of alternatives that may depend on random events. Industrial engineers use mathematical, physical, social sciences, and engineering combined with the analytical and design methods to design, install, and improve complex systems that provide goods and services to our society. Industrial engineers are called upon to (Figure 1-16):

- Analyze and model complex work processes to evaluate potential system improvements
- Analyze how combinations of people and machines work together
- Analyze how the surroundings affect the worker, and design to reduce the negative effects of this environment
- Develop mathematical and computer models of how systems operate and interact
- Improve production and service processes from the perspectives of quality, productivity, and cost
- Work on teams with other professionals in manufacturing, service industries, and government agencies
- Use process control to improve manufacturing, economic, or ergonomic processes

Figure 1-16

"My primary job responsibilities include maintaining, upgrading, and designing all the computer systems and IT infrastructure for the Vermont Railroad. I handle all the servers and take care of network equipment. When needed, I also program customized applications and websites for customers or our own internal use. I also serve as a spare conductor and locomotive engineer when business needs demand."
S. Houghton, CpE

As the Business Leader for Central Florida at a major power company, I develop and manage a $42 million budget. I ensure that our engineering project schedule and budget match, and report on variances monthly. I also conduct internal audits and coach employees on Sarbanes–Oxley (financial) compliance requirements.
R. Holcomb, IE

Currently I am working on a project to determine patient priorities for evacuations from healthcare facilities during emergencies. The assumption of an evacuation is that there will be enough time to transfer all of the patients, but in the event of limited resources, there may not be enough time to move all of the patients to safety. Further—and depending on the emergency type—it may be an increased risk to transport some patient types. Based on certain objectives, we are developing guidelines to most ethically determine a schedule for choosing patients for emergency evacuations.
A. Childers, IE

Materials Engineering and Materials Science

Materials engineering, metallurgical engineering, and materials science focus on the properties and production of materials. Nature supplied only 92 naturally occurring elements to serve as building blocks to construct all modern conveniences. A materials engineer works to unlock the relationship between atomic, molecular, and larger-scale structures and the resultant properties. This category includes majors such as ceramic engineering, metallurgical engineering, and polymer science and engineering.

Here is a partial list of products designed and manufactured by material engineers:

- Brick, tile, and large appliance research and manufacturing for the home and workplace
- Ceramic spark plugs, oxygen sensors, and catalytic converters that optimize engine performance
- Metal and ceramic materials that enable biomedical implants and prosthetics
- Microwave responsive ceramics that stabilize and filter cellular phone reception
- Nanotechnology, including silver nanoparticles used as antibacterial agents in socks and t-shirts and carbon nanotubes used to reinforce the fork of racing bicycles
- Plastics found in bulletproof vests, replacement heart valves, and high-tension wires on bridges
- Superconducting metals that are used in medical imaging devices like magnetic resonance imaging (MRI)
- Ultrapure glass optical fibers that carry telephone conversations and Internet communications

DISCUSSION

Visit the International Materials Information Society (ASM) website with your students. Navigate to the strategic plan page for the society.

URL
https://www.asminternational.org/about/strategicplan

Consider showing the following video from Business Insider about Lab-Grown Steak. How does materials engineering factor into developing new ways we consume foods? *(Examples could include mimicking the consistency/texture of steak, the color, and cooking properties.)* What other disciplines of engineering would be involved here? *(Biomedical is one potential crossover; the video mentions Tissue Engineering, which is a Biomedical Engineering sub-discipline.)*

We suggest stopping the video at 5:16. Otherwise, you may also discuss the ethics of plant-based/lab-grown meat if it fits into your lesson plan, or you may come back to it later during the ethics lesson (we will repeat this tip at an appropriate time). Currently the lab-grown steak uses stem cells from slaughtered pregnant cows, but they do not plan to use these stem cells at scale.

Can Lab-Grown Steak be the Future of Meat? | Big Business | Business Insider

https://www.youtube.com/watch?v=UQejwvnog0M

DISCUSSION QUESTION

Visit the American Society of Mechanical Engineers (ASME) website with your students. Navigate to the education link.

URL
https://www.asme.org/communities/university-students

What does the ASME do for student members of their society?

- Engineering internships and jobs
- Scholarships, loans, and grants
- Competitions and conferences
- Networking and mentoring
- Student recognition
- Leadership opportunities

As a Metallurgical Engineer, my duties include:

- *Consulting firm management/administration*
- *Failure analysis*
- *Subcontracted metals testing services*
- *Metallurgical quality systems design/auditing*
- *Metallurgical expert in litigation cases*
- *Materials selection and design consultant, in-process and final inspection and testing services*

E. Basta, Materials Engineer

Mechanical Engineering

Mechanical engineering (ME) involves areas related to machine design, manufacturing, energy production and control, materials, and transportation.

The oldest engineering solutions have dominated engineered problems from prehistory through the early Middle Ages. During this period, levers were used to lift water from the river to the field of crops. The famous hanging gardens of Babylon used water wheels, buckets, and pulleys. In Egypt, pulley cranes enable workers to move large blocks of stone. Artillery, including moveable siege engines, developed. Later in this period, clocks, earthquake detectors, gears, and long-range signaling emerge. This period slowed down as the Roman Empire started to collapse around 200-300 A.D. Of course, invention still progresses, such as in the Middle East and India. In the Middle Ages, Muslim inventors produced wind-powered machines and steam engines. India created the cotton gin and spinning wheel. In 1206, Al-Jazari invented programmable robots and built a small automaton orchestra. The same inventor created an astronomical clock—the first known programmable computer. Great strides in metallurgy happened during this time. The Iron Age was dominant and smelters invented ways to generate the high heat needed for forging high-quality steels. Around 1400, the center of commerce moved to Italy during the Renaissance and Scientific Revolution. Leonardo da Vinci is the most notable inventor of this time. He worked on transportation and military devices. The pendulum clock came into use. Newton published his Laws of Motion and invented calculus.

Finally, at the close of the Renaissance came the Industrial Revolution. With all the inventions that preceded it, the Industrial Revolution was ready to happen. It only required a reason to put all this knowledge to use. This meant building power sources, developing and building the machines for manufacture, and improving logistics. Expanding populations, the rise of a consumer class in Europe and the Americas, and the need for weapons of war were all drivers of the Industrial Revolution. The Industrial Revolution continues today. The emerging term for this continuation is Industry 4.0, to indicate that we are in the fourth wave of invention. This wave uses the digital tools of computers and wireless communication to replace analog solutions.

Areas and emerging trends supported by mechanical engineers include:

- Construction
- Energy production and control
- Environmental systems
- Food production
- Management
- Materials processing
- Medicine
- Military service
- Propulsion and transportation systems
- Technical sales

In my job as a management consultant, I address CEO-level management decisions as part of a project team by helping clients identify, analyze, and solve business-related problems. My responsibilities include generating hypotheses, gathering and analyzing data, conducting benchmarking and best practices assessments, recommending actions, and working with clients to develop implementation plans.
M. Ciuca, ME

As Plant Engineering Manager I report directly to the Plant Manager. My primary responsibilities are managing all the capital investments, providing technical support and expertise to the plant leadership team, and mentoring and developing the plant's engineering staff and technical resources.
J. Huggins, ME

I am a salesman, so at the end of the day I'm looking to grow my market share while trying to protect the market share I already have. I help companies maintain a safe, reliable, and efficient steam and condensate system by utilizing the many products and services that we have to offer. This is mostly done by designing and installing upgrades and improving my customer's existing steam systems.
T. Burns, ME

It is important to understand the influence of specific disciplines of engineering in many areas of history, but almost impossible to limit a contribution to only one discipline. For example, delivering fresh, drinkable water to a community requires input from chemical and materials engineers, mechanical engineers, electrical and software engineers for the controls, environmental engineers, etc.

Write a report on the history of some specific engineering disciplines (chemical, civil, mechanical and electrical engineering) and their contribution to an important historical development. You might write this report as part of a team in a shared, online document.

DISCUSSION QUESTION

What will a successful engineer need in the future?
(According to the National Academy of Engineering, beyond education, he or she will require strong analytical skills, practical ingenuity, creativity, excellent communication skills, knowledge of both business and management, leadership skills, unwavering ethical standards, resilience, flexibility, and the pursuit of lifelong learning.)

ENGAGE

Research the phrase "zero delivery time" and discuss ideas that will help products get to consumers faster in the future. *(drive through grocery stores, drone delivery, 3-D manufacturing on demand, mobile 3-D printers)*

Wise Words: *What Did You Do Yesterday Morning at Work?*

I worked on completing a failure analysis report for an industrial client.
E. Basta, Materials Engineer

I reviewed the results of the overnight simulation runs. There were several failures, so I analyzed the failures and devised fixes for the problems.
E. D'Avignon, CpE

On any given day, my morning might be spent this way: in meetings, at my computer (e-mail, drafting documents/reports), making phone calls, talking to other project members, running a test on the manufacturing lines. Not glamorous, but necessary to solve problems and keep the project moving forward.
S. Forkner, ChE

I continued to design a warehouse/office building on a nuclear expansion project.
T. Hill, CE

Yesterday I designed a spreadsheet to assist in more precisely forecasting monthly expenditures.
R. Holcomb, IE

I attended the Plant Morning Meeting and the Boardmill Leadership Team Meeting, followed by the Plant Budget Meeting. In between meetings I returned e-mails and project-related phone calls. Typically I spend about 50% of my time in meetings. I use the information I receive at these meetings to direct and focus the efforts of the engineering staff.
J. Huggins, ME

Testing some failed prototype biostimulators returned by a trial user, to determine why and how they failed and how to prevent it from happening in coming production versions.
J. Kronberg, EE

Yesterday, I worked on a patent infringement opinion involving agricultural seeding implements, a Chinese patent office response for a component placement and inspection machine used in circuit board manufacturing, and a U.S. patent office response for database navigation software.
M. Lauer, EnvE

In my current position, I spend much of my time reading technical manuals and interface control documents. I attended a meeting detailing lightning protection for the Ares rocket.
E. Styles, EE

1.6 The Future of Engineering

Engineering has solved many problems and helped society advance in many areas. As new issues emerge or old problems hang on, engineering will continue to play a role in solving those problems. Emerging and innovative technology offer promise in solving current problems, and addressing ideas which, for now, lie outside of our imagination. Engineers often need to research emerging technology, conduct and present research on emerging and innovative technology, and incorporate innovative technology as they develop innovative solutions to problems that they strive to solve (Figure 1-17).

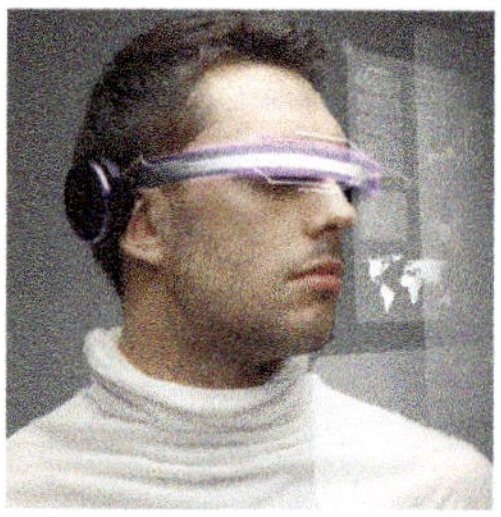

Figure 1-17 Virtual reality can change the way we perceive the world around us.

Engineering in the Future

What effects will technology have on society and our lives? What amazing new inventions will we see in the next few years?

- Computer and phone screens may be lightweight, flexible, and rolled or folded for storage.
- Eco-friendly products may be implemented to reduce pollution.
- Individual manufacturing may become widespread (individual 3-D printers, for example).
- Personalized learning may be implemented, where students learn at their own pace instead of being in a class.
- Flying cars?

As technology changes and becomes more complex, these possibilities may become a reality. Part of preparing for the future will be deciding how we want technology to change.

Computers

Moore's Law has held true for decades, and if it continues, computers will continue to evolve to be faster and have extreme amounts of memory. The electronic components that make up computers will continue to shrink. Camera and screen resolution will continue to improve. With a fast enough Internet connection, we can stream movies and videos now, but in the near future, what will we do? Download and interact with holographic images, where we seem to be immersed in a 3-D world surrounding us? Will computers be able to transmit taste and smell? Consider the explosion in the technical capabilities of smartphones which have evolved to be small computers. Will we have full-power computers that fit in our pocket, perhaps with expandable screens? Electrical engineers, computer and software engineers, and materials engineers will be involved in these advancements.

Artificial Intelligence

artificial intelligence (AI)

Artificial intelligence (AI) is any use of algorithmic machines to imitate human intelligence. AI can be used to recognize and re-create speech, to make humanlike robots, and to create complete virtual reality environments. We see examples of video games and apps which listen to the user and seemingly intelligently respond. As these systems evolve, will they become self-aware and be able to program themselves?

Speech Synthesis

speech synthesis

speech recognition

Speech synthesis applications convert text into speech; **speech recognition** applications convert speech into text. Anyone calling a customer service number has probably had a conversation with synthesized speech; this often sounds real before you realize that there is not another person on the call. Synthesized speech is widely used in apps and video games.

Robotics

robotics

Robotics have been around for decades, and we now have robotic dogs that can stand, jump, and play. We have very human-like robots that speak and look like a human, which are relatively new. Of course, robotics that automate factories, etc. are still in place and plentiful. Robots will continue to evolve and continue to work in areas where their speed, and love of repetitive tasks, will make them ideal for factory work,

DISCUSSION QUESTION

Imagine a world in which typical interfaces were replaced by voice activation. A computer, an automobile, or a wearable device that can listen to our conversations and understand what we need will eventually be a reality. *(Siri, Cortana, Amazon, and Google now won't be the only intelligent voice assistants. As computing devices of all shapes and sizes increasingly surround us, we'll come to rely more on natural interfaces such as voice, touch, and gesture.)*

Why will government agencies, banks, and telecommunications companies require new types of security in the future if computers continue to get more powerful? *(Quantum computers promise to be so fast and powerful that our modern supercomputers will be little more than toys in years to come. The problem is that their sheer capacity for processing data means that they will likely be able to readily crack even the most advanced encryption methods through brute force in minutes.)*

DISCUSSION QUESTION

What will the future of voice recognition and artificial intelligence bring consumers? *(As more voice usage data becomes available, speech recognition accuracy will get better and better. This is what is known as the "virtuous cycle of AI." The more people use voice interfaces, the more data is gathered. As more data is gathered, the better the algorithms work, thus delivering dramatic improvements in accuracy. Computers will start listening to us nonstop. Machines already see better than humans, recognize objects better, and can listen and hear better. Eventually they will also understand meaning better. A conference room, automobile, or wearable device that can listen to our conversations and understand what we need will eventually become the norm. This new world will emerge since we will all expect to have information at our fingertips at any time no matter where we are.)*

TEACHING TIP

You may take the time to discuss the idea of the metaverse with students, which has been widely critiqued. The following article from Wired can be a resource for a discussion.

What Is the Metaverse, Exactly?
https://www.wired.com/story/what-is-the-metaverse/

What technological challenges can your students identify? (*Examples include the socialization of new technology (is it cool/fashionable to wear), how people actually interact with the metaverse, motion sickness while using the technology, privacy concerns.)*

DISCUSSION QUESTION

"Smart" cars are dependent upon network control, making them vulnerable to many computer-related problems. What are some of the potential problems that engineers will have to prepare the "smart" cars for?

(Hackers were able to hack a vehicle's cellular connection to find the car's IP address and gain access from anywhere in the country. The hackers are able to gain access to the chip controlling entertainment and rewrite firmware so that they could implant code to take over things like the engine and brakes. In addition, if your connected car is in a serious accident, who owns the data stored in the crash? You? The auto manufacturer? Your dealer? Or your insurance company?)

nanotechnology

work in dangerous areas like space, and with the incorporation of **nanotechnology** (Figure 1-18), may mean robotic implants that work inside human bodies. Where else will robots be found in the future?

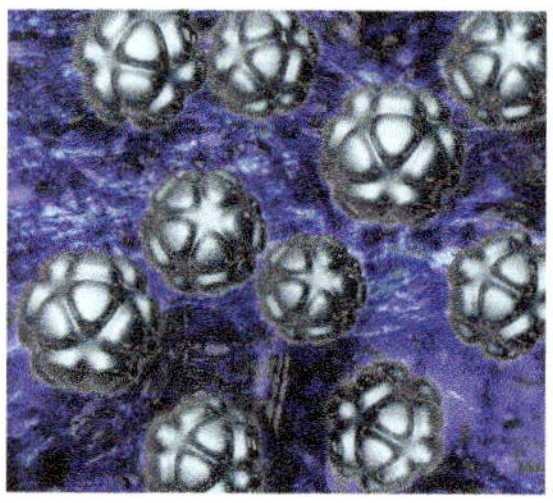

Figure 1-18 These interlocking gears were assembled using many individual carbon atoms.

Machine Learning

In the search for better ways to educate AI systems, engineers have tried to mimic the learning approach of humans. Even though no two dogs are exactly alike, humans are pretty good at identifying a dog when they see one. That is because we have encountered a variety of dogs and know what characteristics make a dog. The AI equivalent is something called Machine Learning or ML. The thinking is, if you show a computer vision system enough different pictures and videos of a variety of dogs it will develop a "concept" of a dog. Then you can show it a dog that it has never seen before and it will still recognize it as a dog. The system isn't perfect, but it has produced some remarkable results. Naturally this is of limited use if the robot has to be programmed using this technique for every type of thing that it may encounter. However, ML is moving from the Reactive Machine stage to the Limited Memory stage, where it can learn through repeated testing and correction. This is similar to how humans learn.

Virtual Reality and Augmented Reality

virtual reality

augmented reality

Virtual reality (VR) and **augmented reality (AR)** are both intended to create an immersive environment. This is done in VR by fully enclosing your field of view and projected an alternative "real" world for you to interact with (Figure 1-19). AR by contrast retains the real world but augments it with an overlay of additional, valuable information. The real world portion in AR can simply be provided using a smartphone as a window to the world. If you were using your phone for navigation while walking, then the AR enhancement might simply be an arrow that "floats" in your field of view and directs you to where you are going.

Figure 1-19 Virtual reality games are one application of artificial intelligence.

It takes a lot of computational power and AI principles to create a three-dimensional computer-generated environment into which a person is projected. Interacting in the VR world also requires wearing special eyewear and specialized wireless joysticks or gloves to create actions—at least for now. The promise of VR is that you can create any place or scenario in the world and feel as if you were transported there. The reality is that since your vision is limited and you lose your references to the outside world, VR must be confined to a tightly controlled environment for now. One of the unusual uses for VR is to take the world of the miniature—bacteria, cells, and molecules—and blow them up so that scientists can see how they behave, in full 3-D without squinting through a microscope. This can lead to a better understanding of how these small and complex parts of our world work.

"Smart" Houses and Cars

We are rapidly moving into a world where Wi-Fi and Bluetooth access is available virtually 24/7 in urban areas. At the same time, appliances and household systems are being developed with added capabilities. This allows for "smart" houses which are locally interconnected with what has been called the Internet of Things (IOT). A simple thermostat can control itself based on "learned" settings from the homeowner. It may even have a small camera and a motion detector embedded as a simple security system that you can access from your phone. Your refrigerator can keep track of its contents and alert you if a food is getting near the end of its shelf life and maybe generate a recipe for you to use that food before it expires. The possibilities are endless.

Nanobots

Microscopic robots, so-called nanobots, may travel through the bloodstream to provide a picture of the body's circulatory system and connected organs. As these evolve, patients will swallow a capsule with sensors that may examine the digestive track. It will transmit signals to small antennas embedded in a wearable shirt and record data directly in a small data "safe" worn on the waistband. A visual representation of the pill's journey will help doctors check overall health of the patient and pinpoint any potential trouble spots.

Personal Manufacturing

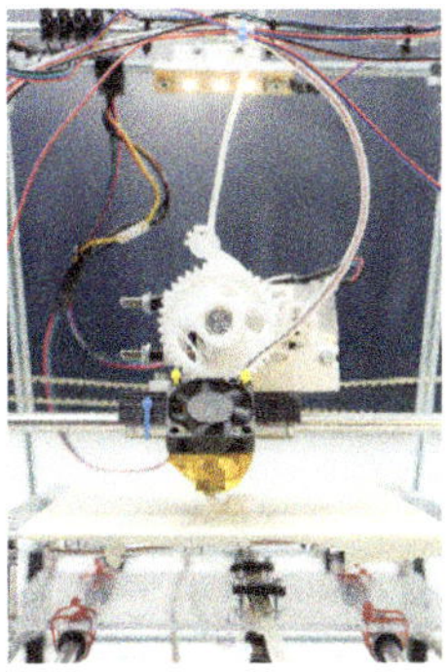

Figure 1-20 A 3-D printer can be used to manufacture objects that are personalized by and for you.

If you want a new smartphone cover, you can go to a store or buy one online. If you owned a 3-D printer (Figure 1-20) and had the proper files, you could print one. What possibilities can you imagine as more people use individual 3-D printers? You could print a new smartphone cover, but what else? Can these printers be used to print nutrition or medicine into food or a manufactured pill? Could you print a microchip for a new puppy without needing to visit a vet?

Could we use large-scale printers to print affordable housing?

Imagine the Future with Engineering

One of the challenges in space exploration is the time it takes to travel from body to body. We would need equipment that could repair itself and its environment automatically, independent of human intervention. Thus, self-replicating robots could create spare parts and can effect repairs on long space flights.

Can the world of art and engineering intersect? As engineers work with advanced materials, will those materials have some aesthetic beauty that inspires creation of artistic pieces?

Can we develop printing suitable for use in the medical field? If a person needs a new hip, could we print a new hip? Can we discover methods to make low-cost and functional prosthetic limbs?

When natural disaster strikes, could we deploy large 3-D printers to replicate the houses that were destroyed? Can we print new housing to help with homelessness? What other large-scale printing would benefit society?

TEACHING TIP

Tell your students that 3-D printers can have ethical implications. Discuss the issues of harmful emissions, increased dependence on plastic, gun control loopholes, and the possibility of 3-D printed drugs.

DISCUSSION QUESTION

In the future, 3-D printers could benefit from a continuous liquid interface, which would speed up the printing process up to 100 times. UV light triggers photopolymerization and oxygen inhibits it. By carefully balancing the interaction of light and oxygen, the continuous liquid printers continuously grow objects from a pool of resin. How would this affect the manufacturing industry?

(Continuous liquid printed objects have smooth sides, unlike typical 3-D printers, whose sides are typically rough to the touch. Some resins produce objects that are rubbery and flexible, something that could not be produced with earlier methods.)

ENGAGE

Visit the International Association of Nanotechnology website.

URL

http://www.ianano.org/Site/Training/ index.html

Navigate to the training page and research the courses offered for professionals in Nanotech and Clean Tech.

1. **Certified Nanotech & Clean Tech Professional (CNCP)** - *programs for engineers, scientists, technical and business managers to prepare for senior level of responsibilities in technology and business development in the emerging field of nanotechnology and applications in clean energy.*
2. **Clean Tech Engineer (CCTE)** - *designed to train a new generation of engineers highly skilled in the design and applications of renewable energy, process engineering, and manufacturing of state of the art clean tech processes, including: thin film photovoltaics, wind turbines, and nanomaterial batteries used in electric vehicles.*

DISCUSSION QUESTION

Compare the costs of hydrogen-fueled vehicles to the costs of electric vehicles. Include research costs for refueling stations, the efficiency of combustion engines, and the maximum range for each type.

HOW TECHNOLOGY WORKS

What will you be driving? Alternative Engines Online

Technology is everywhere in today's cars. Global positioning systems, Internet connections, and smartphones are common features, but what about technology under the hood?

Most major car manufacturers are concentrating on the Battery Electric Vehicle (BEV) market and it will dominate in the near-term. However, for light-duty vehicles, one potential advance is fuel cell technology. In fuel cells, free hydrogen is combined, through a catalytic process, to atmospheric oxygen to produce electricity. The only by-product of the process is clean water, H_2O. This technology is just starting to become available, but it is an intriguing challenger to BEV technologies.

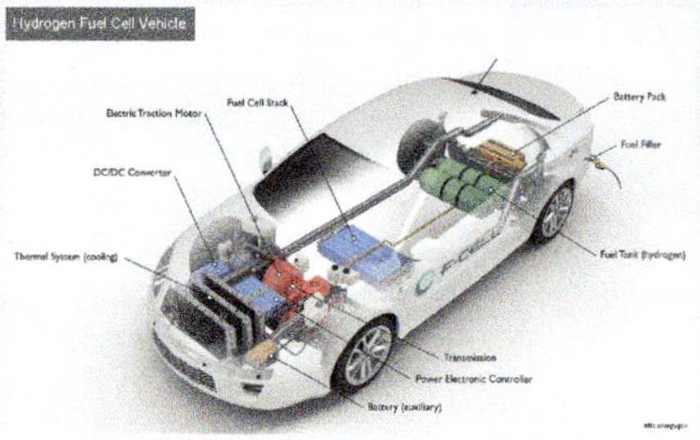

CHAPTER 1 ASSESSMENT QUESTIONS

1. Write your own definition of engineering. What key concepts does your definition contain?

2. What are some of the earliest examples of engineering? List three and explain why they should be considered engineering.

3. Explain how the invention of printing led to three of the greatest technological advances we have today.

4. Explain why the rate of technological advances has increased over time.

5. We know that the burning of fossil fuels causes pollution. Answer the following questions about the burning of fossil fuels in one paragraph each:
   (a) Why do we continue to burn fossil fuels?
   (b) What alternatives should be investigated or implemented?
   (c) Why are these alternatives not already in place?

6. Partner or team assignment: Get with another student or team and use a shared document editor such as Google Docs to do the following.

   Identify one innovation we use today that resulted from engineering. Investigate the history behind that innovation and write a 2-page report on the innovation. Be sure to include relevant illustrations, showing the history of this device. Finally, predict the future of this device: What will it look like in 10 years?

7. Describe a technological device or process that poses an ethical dilemma.

8. Smartphones have had major impacts on how we communicate with each other. However, there are negative consequences from the technology, such as texting during class or while driving. Analyze the positive and negative impacts on society and include anticipated and unanticipated effects.

9. List 10 types of common transportation in chronological order. For each, identify why this was the preferred technology at that time, and identify one reason it was replaced. What will the most common type of transportation be in the year 2050? What are some risks and benefits of this type of transportation? How has this affected the economy?

10. How has the advancement of technology affected you individually? Your community? Your state? The world? Your culture? The environment?

11. Write a paragraph to explain how and why technology advances and progresses.

12. In 2008, experts convened by the National Academy of Engineering (NAE) met and proposed a set of Grand Challenges for Engineering, a list of the 14 key goals for engineers to work toward during the 21st century. For any aspiring engineer, reading this list should feel like reading a description of the challenges you will face throughout your career. To read a description of each Challenge, as well as a description of some connected areas within each, visit the NAE's Grand Challenges for Engineering website at *http://www.engineeringchallenges.org*.

    After reading through the list, write a job description of your dream job. The job description should include the standard components of a job description: title, responsibilities (overall and specific), a description of the work hierarchy (whom you report to, whom you work with), as well as any necessary qualifications. Do not include any salary requirements, as this is a completely fictional position.

    After writing the description of your dream job, identify which of the 14 Grand Challenges for Engineering this position is instrumental in working toward solving. Cite specific examples of the types of projects your fictional job would require you to do and discuss the impact of those projects on the Grand Challenges you have identified.

13. Consider the Engineering Grand Challenges.
    (a) Which do you think is the most difficult to solve in the next 10 years, and why?
    (b) Which do you think is the easiest to solve in the next 10 years, and why?

14. Identify one additional problem for humanity that you feel should be added to the Engineering Grand Challenges. Why do you feel it should be included?

15. A company is working on a design to prevent flood damage during or after a hurricane. Identify how engineers from each of these disciplines could be involved and what they might propose. Don't be afraid to think outside of the box; the solution could be wild or impractical—but it needs to be possible. (For example, your answer should not be, "Invent a time machine to go back in time before the hurricane.")
    (a) Electrical engineer
    (b) Mechanical engineer
    (c) Chemical engineer
    (d) Materials engineer

16. Describe 5 technological innovations we will see in the "smart house" 10 years from now. For each, describe why it doesn't yet exist and how it will change our everyday life.

17. Video game designs have gone from corded controllers to cordless, from cartridges with primitive graphics to disks and almost realistic graphics, from games you play at home to those that you can challenge people worldwide in game play. What will in-home video game systems feature in 5 years? Describe the system.

18. What is the difference between speech recognition and speech synthesis?

19. All engineering disciplines work on engineering designs, or designs to solve specific problems. Investigate the history of each of the following disciplines, including an early example of someone considered to be working in that discipline. What makes this discipline different from the others?
    (a) Chemical engineering
    (b) Civil engineering
    (c) Electrical engineering
    (d) Mechanical engineering
    Note: Your teacher may assign groups for this problem.

20. STEM stands for Science, Technology, Engineering, and Mathematics. How are these similar, and how are they different? Compare preparation for the career path of engineering, technology, and science. Include apprenticeships, community/technical colleges, and university education.

21. Compare the career paths of engineering, technology, and science in the military, in entry-level employment, and long-term employment.

22. The invention and design of public sewage systems is widely regarded as an engineering feat that has been one of the most beneficial to humankind. Write a paragraph on how this invention has affected each of the following:
    (a) An individual—yourself, a family member, or a friend
    (b) Society within your hometown
    (c) Cultures within the United States vs. a developing country
    (d) The economy of the United States vs. a developing country

23. Write a paragraph on how a lack of public sewage systems would affect the environment.

24. Investigation: Define each of these engineering areas. Then, identify disciplines that are related to the field. Finally, identify career(s) that are related to that field.
    (a) Robotics
    (b) Process control
    (c) Process automation
    (d) Mechatronics
    (e) Automation systems
    (f) Electrical systems
    (g) Mechanical systems

25. Investigation: Use the Internet to investigate fuel cells as a fuel source.
    (a) What is the fuel source for a fuel cell?
    (b) What is the waste product?
    (c) Why aren't we driving fuel-cell cars today?
    (d) Can this be considered an emerging trend in electrical engineering? Why or why not?
26. Investigation: Use the Internet to distinguish between medical imaging, rehabilitation engineering, and systems physiology within bioengineering.
27. Write one paragraph describing a career related to Civil Engineering. Choose one among structural, geotechnical, and water resources engineering.
28. Investigate computer engineering using the Internet. What differentiates a network designer, a network manager, and an engineer working in systems support?
29. Design and give a brief presentation on an existing or an emerging Civil Engineering technology.
30. Design and give a brief presentation on an existing or an emerging Computer Engineering technology.
31. Design and give a brief presentation on an existing or an emerging Mechanical systems technology.
32. Describe how each of these careers supports large scale civil or environmental projects.
    (a) Transportation engineering
    (b) Structural engineering
    (c) Construction engineering
    (d) Environmental engineering
    (e) Geotechnical engineering
    (f) Water resources engineering
33. Describe how each of these careers supports the computer industry.
    (a) Computer hardware design
    (b) Network design
    (c) Network management
    (d) Programming
    (e) Systems support
34. Describe how each of these careers supports bioengineering or biomanufacturing.
    (a) Bioinstrumentation
    (b) Biomechanics
    (c) Biomaterials
    (d) Medical imaging
    (e) Rehabilitation engineering
    (f) Systems physiology

OVERVIEW

Engineers are men and women who create new products. It is estimated that there are over 2 million practicing engineers in the United States. They work in fields such as biomedicine, energy, automotive, aerospace, computers, and many others that require people to create products that didn't exist before.

CHAPTER 2

The Profession of Engineering

TOPICS

What is meant by the *profession* of engineering? You may know someone who works as an engineer, but whether they approach their work as a job or professionally as a career makes a big difference. A job is a series of tasks that one would regularly perform in exchange for a certain amount of money for each hour worked. Someone who views being an engineer as their profession or career path has a deeper investment than just time in exchange for money. They often invest time outside of regular work hours to increase their value. They can obtain additional schooling, volunteer to help school kids with science projects, be a tutor for students that would like to enter engineering as a profession, or obtain professional certifications. In short, they view their engineering work in a broader context. It's an opportunity to continue growing, learning, solving difficult problems and learning to grow a business.

A common association with those that enjoy engineering is people "who are good at math and science" or people who "love math." Certainly, math and science are tools engineers use, but for engineers, math and science are the tools to solve problems. Engineering is a profession that involves helping people and helping society; a profession in which designs are created within a set of constraints—realistic boundaries and/or requirements.

The expansion of technology is visible every day and all around us. With the better of modern engineering achievements, we can swim underwater with scuba tanks, float through the air, and communicate worldwide. We can fly to the moon, and back to Earth again. Our phones have evolved beyond their initial purpose to incorporate features that enable us to take pictures, play our entire music library, send messages, and even make video calls. Engineering is the process of designing, creating, or advancing technology.

The National Academy of Engineering describes engineering as the "application of science," because engineers build tangible products from abstract ideas. Engineers are people who create new products and "design under constraint," which means that a design must conform to the requirements and should do exactly what it is expected to do without any unexpected consequences. Albert Einstein once said, "Scientists investigate that which already is. Engineers create that which has never been." The laws of nature constrain engineering; for instance, energy cannot be created or destroyed. Engineering is typically conducted by teams in which each team member contributes his or her specialized knowledge.

OBJECTIVES

- Investigate different types of occupational training and educational opportunities
- Demonstrate positive work ethics and leadership skills
- Recall education requirements for professional success as a designer/engineer

OBJECTIVES

- Demonstrate knowledge of how to dress, speak, and conduct oneself in a manner appropriate for the profession
- Cooperate, contribute, and collaborate as a member of a group in an effort to achieve a positive collective outcome
- Present written and oral communication in a clear, concise, and effective manner
- Demonstrate punctuality, dependability, reliability, and responsibility in performing assigned tasks
- Compare engineering, science, and technology career paths, including entry-level employment, military service, apprenticeships, community and technical colleges, and universities
- Master relevant safety tests
- Follow lab safety guidelines as prescribed by instructor in compliance with local, state, and federal regulations
- Identify industry safety terminology related to the personal work environment such as Occupational Safety and Health Administration (OSHA), American Society of Mechanical Engineers (ASME), and personal protective equipment (PPE)
- Recognize the classification of hazardous materials and wastes
- Describe appropriate ways to dispose of hazardous materials and wastes
- Maintain, safely handle, and properly store laboratory equipment
- Describe the implications of negligent or improper maintenance
- Describe career opportunities in electrical and mechanical systems
- Identify fields and career opportunities related to robotics, process control, and automation systems
- Identify potential safety hazards and follow general laboratory safety practices
- Identify occupations related to engineering

QUESTION

Define the terms *career, profession, salaried employee,* and *hourly employee* with your students. A career is a job that someone works to attain; it usually requires specialized knowledge and training. A profession is more than a job. People who belong to a profession are sufficiently competent in their abilities, have the necessary training to be successful, and achieve with the intent to help the general public. Hourly employees get paid for each hour that they work. Typically, if you don't work one day, you don't get paid. Salaried workers earn a steady salary, usually a given amount per year.

ENGAGE

Visit the DiscoverE website with your students. The DiscoverE (formerly National Engineers Week Foundation) mission is to sustain and grow a dynamic engineering profession through outreach, education, celebration, and volunteerism.

URL
http://www.discovere.org/about-us/mission-and-history

DISCUSSION

According to DiscoverE, what are the top 10 reasons to become an engineer? *(Love your work. Be creative. Work with great people. Solve problems. Design things that matter. Never be bored. Earn a big salary. Enjoy job flexibility. Travel. Make a difference. Change the world.)*

A career in engineering is more than a job. People who enter this profession are sufficiently competent in their abilities, have the necessary training to be successful, and achieve with the intent to help the general public. Members of a professional group are typically paid with a salary (dollars per year) instead of hourly (dollars per hour). Usually, they have a higher level of commitment to the responsibilities of their chosen profession and to their professional code of ethics.

What's the difference between a salaried and an hourly employee? An hourly employee gets paid for each hour that he or she works. Typically, if you don't work one day, you don't get paid. Salaried workers earn a steady salary, usually a given amount per year. To calculate an hourly rate for a salaried employee, dividing by 2080 (52 weeks per year × 40 hours per week = 2080 hours per year) gives a good estimate. For example, an engineer earning $80,000/year is earning roughly equivalent to $38.46 per hour.

2.1 What Is a Career?

A career may start out as a job that grows into something more over time. Creating a career requires specialized knowledge and training; a person desiring a career must invest in themselves to move along a career path. The individual must commit to learning and growth. This may involve college, other training, licensure, etc.

Today, the careers available to you upon your graduation are numerous and diverse. It is often difficult as a young adult to determine exactly what occupation you want to work in for the rest of your life because you have so many options. You cannot make a good decision without accurate information. No one can (or should). You must make this decision for yourself—not your relatives, teachers, advisors, or friends. You may not know all the answers to your questions right away. That means you will have to get them by gathering more information from outside resources and through your personal experience. Be aware that during your working years, opportunities will change, and your interests may also shift. Prepare to be flexible and look out for new opportunities that align with your current interests or look like new challenges that would be interesting. One can never predict precisely what the trajectory of a career will be so you need to make the best decision you can make at the time. There will always be new challenges if you stay open-minded.

Choosing your major and ultimately your career is a process. You constantly evaluate and reevaluate what you learn and experience. A key component is whether you feel challenged or overwhelmed. True success in a professional career is not measured in monetary terms; it is measured in job satisfaction (Figure 2-1). As you gain experience, you can adjust your career path so that you find new opportunities.

"My team is responsible for implementing the engineered design in the field. We install, tune, test, and accept into operations all of the electronics that allow customers to use our state-of-the-art fiber optic network to run voice, video, and data for their residential needs."

L. Gascoigne, CE

Figure 2-1

2.1.1 Engineering Careers

Engineering careers are traditionally some of the highest paying, especially for new college graduates. Numerous sites on the Internet show potential starting salaries for different engineering disciplines, especially compared to other starting salaries for recent college grads. Two of the more popular fields are electrical and mechanical engineering.

Figure 2-2 Wind turbines generating electricity.

What does an electrical engineer actually do? Suppose you read that electrical engineers might choose to go into solid state design, analog or digital communications, wireless technology, or power distribution. In that case, you would likely understand very little of the differences between these areas. Similarly, suppose you see a list of possible mechanical engineering careers, including transportation systems, power production, or the production of energy or development of machines. In that case, it could look like there is a lot of overlap with electrical engineering. In many ways, this is the case. There is more to each engineering discipline than you might typically think. Electrical engineers do more than just design electronic systems; mechanical engineers do more than just design robots. In fact, lots of engineering opportunities require cooperation from several different engineering disciplines. Consider these emerging trends in engineering:

power grid

- **Power generation and distribution:** You have probably seen wind turbines growing in popularity, and you probably know that the mechanical action of the spinning blades generates electricity (Figure 2-2). That power needs to be transmitted throughout the community to reach each house and business. This transmission happens through the **power grid**, usually just referred to as "the grid." Both mechanical and electrical engineers play an active role in power and energy distribution.
- **Wireless communication:** Look at smartphone technology of 10 years ago and notice the difference between that and today's technology (Figure 2-3). Your smartphone is now a music library, browsing device, game system, computer and, also, a phone. It contains remarkable power within a very small unit. Electrical engineers play a substantial role in developing this technology. In fact, the organization that specifies exactly how these devices communicate is the IEEE, the Electrical Engineering professional society.
- **Computers:** If you look inside a computer, you see many electronic components, as well as cooling units, which are crucial to the successful operation of your device. Electrical and mechanical engineers play a crucial role in the design and development of computers.

Figure 2-3 Smartphones allow us to communicate using wireless technology.

- **Internet of Things (IoT):** More household appliances can connect to a network. Making these devices communicate effectively requires engineers from different disciplines.

Emerging trends and solid engineering fields requiring multiple disciplines beyond electrical and mechanical engineering include:

- **Robotics:** Robots have been around for decades, in fields like manufacturing and surgery, even disinfecting areas like the interior of airplanes during a pandemic. Electrical and computer engineers are needed to program the robotics; mechanical engineers design the systems giving the robot motion; materials engineers may work on materials to make the systems lighter and stronger, etc.
- **Process control:** Almost all systems require process control to work with manufacturing processes, processes within a system, or even at a very low level, working with electrical or mechanical processes

ENGAGE

Visit the Electrical Engineering professional society's (IEEE) website with your students. Navigate to the pre-university education page.

URL
https://www.ieee.org/education/preuniversity/index.html

IEEE and IBM have created a portal that combines information on engineering careers with interactive activities.

URL
http://tryengineering.org

Intended for pre-university parents, teachers, school counselors, students, and the general public, TryEngineering.org lets visitors explore how to prepare for an engineering career, find accredited engineering degree programs, download lesson plans, read profiles of engineers and engineering students, search for opportunities, play interactive games, and more.

ENGAGE

Visit the American Society of Safety Engineers (ASSE) website with your students. Navigate to the code of professional conduct page.

URL

http://www.asse.org/about/code-ofprofessional-conduct/

Ask the students for examples of ways that engineers can work with professional conduct. *(Serve the public, employees, employers, clients, and the profession with fidelity, honesty, and impartiality. In all professional relationships, treat others with respect, civility, and without discrimination. Abstain from behavior that will unjustly cause harm to the reputation of the Society, its members, and the profession.)*

QUESTION

Define the term *professional conduct.*

(The professional is highly trained and often has a specialty. A professional must be a good communicator. A professional typically does whatever it takes to get the job done. A professional works toward advancing his or her profession. A professional maintains a professional attitude.)

within a system like a robot, a car, or other larger devices. Process control ensures that, for example, manufacturing lines operate at peak efficiency. This field also works to specify where personnel belong in a process, how to slow down or speed up some subprocesses if temperature or humidity get out of control, or if some components in one part of the assembly are in short supply.

- **Automation systems:** Automation is a rapidly growing field in engineering, most often studied in industrial engineering programs. Consider companies that sell and ship millions of items daily. Automation is required to identify an item, retrieve it, transport it to a shipping site, package it, and ship the item as efficiently as possible. Automation is also used in manufacturing to ensure that devices are manufactured properly, step-by-step.
- **Other systems:** Space travel, transportation, amusement park rides, etc., all rely on engineering from different disciplines, including electrical and mechanical engineering.

2.1.2 Professional Conduct

A professional is more than a person who works for a salary or dresses more formally for work. There are many aspects of a professional that are generally assumed to be true. The professional is highly trained and often has a specialty. Teachers are excellent examples of trained professionals. They attend college for a degree in education and usually must continue their education throughout their career. Teachers have areas of specialization—for example, your history teacher probably isn't qualified to be your math teacher.

A professional must be a good communicator (Figure 2-4). A professional needs to effectively and politely communicate with their customers or clients, with few exceptions. Professional managers—whether they are sports managers, personnel managers, or company-level managers—need to communicate with their employees and their managers or bosses. It is worth taking special training in communications and listening skills if you have your sights on a manager role. A professional typically does whatever it takes to get the job done. Doctors go through rigorous training, sometimes working 24-hour shifts. In times of tragedy, doctors are called upon to go to hospitals and clinics to help. The Hippocratic Oath, an oath taken by doctors as they enter their profession, requires them to treat the sick to the best of their ability, preserve patient privacy, and teach the secrets of medicine to the next generation, among other things. Doctors are expected to uphold this oath.

A professional works toward advancing his or her profession. One of the goals of a professional is typically to enhance that profession. Veterinarians, for example, have a responsibility toward their patients, but as they discover more effective techniques or new or unusual issues, they are generally expected to spread their expertise to others in their profession.

A professional is able to demonstrate these positive skills as part of a strong work ethic and a professional attitude:

- Responsibility
- Reliability
- Dependability
- Effective communication
- An ability to effectively delegate
- Cooperation
- Teamwork
- Integrity
- Punctuality

Figure 2-4 Professionalism at work.

Admittedly, the concept of a "professional attitude" is somewhat hard to define. If you work as a professional and you encounter someone who does not maintain a professional attitude, you will probably recognize it. This may mean anything from how they dress to writing clear, concise emails, and responding quickly and appropriately to urgent matters. For example, you would rarely, if ever, see a law enforcement official wearing anything other than a uniform or suit. You would rarely see a politician without a suit, or a superintendent without a tie. Strive for your best work—appear and act in a professional manner when in public or with clients, enhance others who work with your employer or in your profession, and uphold any oaths or standards of conduct or codes of ethics.

2.1.3 Preparing Job Documents For an Application

When you apply for a professional position, your potential employer needs a clear picture of you as a potential employee. Most employers will require a cover letter, resume, and references (among other paperwork). They will use these documents to understand your experience, skills, and training, as well as aspects such as your work ethic and motivators. The employer usually considers hiring the potential employee as part of a long-term relationship and expects the employee to make a solid contribution over several years. As part of a career search, you will need to prepare and organize the materials you will need.

For every job search, you should be prepared to provide the following:

- A **resume**, which is a written summary of your work-related skills, experience, and education. It introduces you to the prospective employer by presenting a snapshot of your qualifications. It should be brief and to the point, printed on white paper, true and accurate, and have no typographical, grammatical, or spelling errors. A good resume attracts the reader's interest, so they want to learn more about you.
- A **cover letter**, or letter of interest, which is a letter of introduction that you send with a resume.
- A **list of references** includes the names and contact information of people who know you and your qualifications and are willing to speak about you to potential employers. Potential employers will likely call or email these references to obtain a richer understanding of you before deciding whether to offer you an interview, or the position.
- A **thank-you letter** that you send to the interviewer(s) after an interview to show your interest in the position and to thank them for their time. While this isn't a required document, it is almost always a good idea.

We will cover each of these job documents in more detail, below.

Resume

A resume, which summarizes your training and experience, is absolutely necessary when searching for a job. It is likely the most critical document in the hiring process because it is your ticket to an interview. Employers will usually want to review all resumes before inviting any applicants for an interview. Employers often must sort through large numbers of resumes, so make yours informative, easy to read, and concise. In fact, a 2019 eye-tracking study by Ladders, Inc., a career development site, found that hiring managers spend 7.4 seconds on average looking at a single resume. A resume should list your education, experience, and skills but should never (or very rarely) exceed two pages. To make your resume stand out from the rest in a professional way, consider these tips:

- Make it easy to read. Leave space between lines with enough white space to avoid the page looking cluttered. Optimize the text for an F-style reading pattern, where the reader scans the page in a horizontal movement, followed by a vertical scan down the left side until they find something interesting that stands out, leading them to read horizontally again.
- Use one, easy-to-read font, and apply different font styles and sizes for emphasis.
- Use bullets and bolded job titles to make lines of text stand out.
- Include keywords related to the position, but don't overdo it.
- Use proper spelling, punctuation, and grammar. Be careful to use or define any appropriate technical terms. Avoid jargon.

QUESTION

Define the terms *resume, cover letter,* and *job interview* with your students.

*(The **resume** is a document used to present a person's backgrounds and skills. Resumes can be used for a variety of reasons, but most often they are used to secure new employment. Job seekers frequently send a **cover letter** along with their resume and application for employment as a way of introducing themselves to potential employers and explaining their suitability for the desired positions. A **job interview** is a type of employment test that involves a conversation between a job applicant and a representative of the employing organization.)*

ENGAGE

Visit Monster.com with your students. Monster.com is one of the most visited employment websites in the United States and one of the largest in the world. Monster is primarily used to help those seeking work to find job openings that match their skills and location. Navigate to the page on engineering interview questions.

URL

http://career-advice.monster.com/jobinterview/interview-questions/engineering-interview-questions/article.aspx

QUESTION

According to Monster.com, what are the key questions an engineer is likely to hear during the interview?

(Which of your past projects was your favorite one? How do you ensure that you complete a project on time? How do you deal with clients who have unrealistic expectations? How do you keep pace with new developments in engineering? What kind of problems have you encountered in your work? Which engineering trends do you find most interesting?)

QUESTION

If cover letters are not being read by hiring managers, what other ways can job seekers stand out from the crowd? *Often it is a well-crafted professional portfolio. Monster.com provides advice for including your best work based on three questions: "(1) Are you proud of the work?, (2) Does the work in the portfolio align with the work needed by the employer, and (3) Do you want to keep doing this work?" These portfolios should include 10-20 work samples and showcase a diverse skillset.*

URL

https://www.monster.com/career-advice/article/building-great-portfolio-interview-game-changer

- Keep it to one page, if possible; two pages at the most. If you use two pages, be sure to put your name in the header or footer on page 2, in case it becomes separated from page 1.
- For each job listed in your experience section, you should include a few concrete, measurable accomplishments. For example, "increased sales by $100,000 in the first quarter." Avoid listing responsibilities; highlight what impact you had in your role.
- Similarly, avoid buzzwords that only "tell" but don't "show." Examples of these words include, "team player," "self-motivated," "goal-oriented," "including, but not limited to," "responsible for," "entrepreneurial," "excellent communication skills," "detail-oriented," "creative," "results-driven," and "top-notch."
- Do not include vague measurements of your expertise. For example, saying you have "intermediate" knowledge in Python may not align with the hiring manager's definition of "intermediate." Even more unclear are scales or ratings, such as saying you are a 6/10 in Python; the same principle applies.
- Get feedback from friends, teachers and faculty, relatives, and your institution's career office. Everyone will have a different perspective and can provide diverse feedback to improve your resume. Make sure to tell them what kind of feedback you'd like so they can focus on the elements you're most concerned about.
- Especially for programming jobs where you are expected to have a portfolio of projects, link to where those projects are stored. A common ask is for your GitHub, a version control website where programmers manage and share their various software projects.

When hiring managers have a lot of resumes to sort through, usually they will ask their HR team to use an Applicant Tracking System (ATS). The ATS will automatically scan and log in your resume. If the potential employer is using an ATS, it is especially important to notice any keywords in the job description. As much as possible, use those keywords in your resume to improve your chances of passing this initial screening.

Cover Letter

A cover letter is a strong and succinct way to introduce yourself to a potential client or employer. It also helps convince employers to review your resume by focusing attention on specific resume information; for example, credentials, skills, or specific projects in your portfolio of design work. It is also a chance for you to make the case that you understand the company objectives, the specific job requirements, and would be a good match to their existing team. The letter should be concise and well written, with no grammar or spelling mistakes. The cover letter is your first point of contact, so craft one that makes an excellent impression.

Note that some companies are no longer requiring cover letters or explicitly do not allow you to upload one. For example, when last checked on August 2022, Amazon had this to say on their Online Application FAQ, "Being a peculiar company, we don't accept cover letters. Just ensure your resume is up-to-date and you're all set." This might not be a peculiarity for long. Survey research by the Addison Group (a professional services firm) of 505 hiring managers revealed that only 18% of hiring managers view the cover letter as a vital component of the application package. Considering the fall of the cover letter appears to be a trend toward extinction, it is especially crucial to ensure your resume is well-aligned to the position for which you're applying.

In addition to a cover letter and resume, some potential employers may ask you to fill out an application form, usually found online, but sometimes on paper. These are often a "copy and paste" task that duplicates your resume information—but take note of the prompts to ensure you are addressing them fully. Because resumes can take a variety of forms, employers want your information in a standardized format such that their ATS can better manage all the applicant information. Application forms require you to enter specific information about your education and past employment, including dates and locations. As with your resume, it is important to be truthful and accurate. In some cases, a company may have an automated site that fills out an application from the information on a resume. In these cases, be sure to check the accuracy of the application.

List of References

If a list of references is called for, be sure to list their names, positions, company they work for, and your professional relationship with that person. References can be anyone with whom you have a professional relationship, who is familiar with your work, and who will give you a strong positive recommendation. Make sure to include both an email contact and a phone contact. If you already have a colleague that works at the company that you are applying for, make sure to include them on the list. An "inside contact" makes the hiring manager's job easier to check your reference and can give you a slight advantage over other applicants. It is recommended that you limit your references to three people and that they are fairly recent contacts. Make sure that you have informed your references that you are job hunting so that they won't be surprised by a phone call from the company.

Thank-You Letter

After the interview, you should write a thank-you note within 24 to 48 hours. A thank-you note reminds the interviewer that you are serious about wanting the job. Use the thank-you note to restate your interest in the job, highlight your key qualifications, and to thank the interviewer for spending time with you. Refer to something specific that you discussed during the interview. Address the note to the person or people who interviewed you.

2.1.4 Phone, Virtual, and In-Person Interviews

If an employer thinks you have the qualifications for the job, you will be invited for a job interview. A preliminary interview may be by phone or video conference, but almost all employers will expect a face-to-face meeting at some point. The interview is an opportunity for you and the interviewer to ask questions and decide if the position is right for you. The following tips can help you through the phone interview.

- You should prepare for the interview by researching the company and the position and by practicing the answers to questions you think the employer might ask. If you know the employer is looking for a specific skillset, be sure to highlight specific examples of how you meet those requirements in your answers.
- Even though the person on the other side of the phone can't see you, try to smile during your conversation. Smiling helps you project a positive tone in your voice.
- Be an active listener. Sometimes it can hard to know when a person has finished their thought over the phone. Therefore, you should be attentive and ready to ask follow-up questions based on what the interviewer mentioned during the interview. Try practicing with friends and/or family to get comfortable.
- At this stage, an employer might ask you what your salary range is. Be sure to understand what people at the level you're interviewing for in your industry are being paid on average (called the market rate). Provide a range you would be comfortable with that is $5,000 wide.
- You might be asked about details on your resume, so be sure to have your job documents ready to address those inquiries.
- There are often common interview questions that you can find online; you may find it helpful to prepare answers to these. If you do, be sure to speak as though you are not reading from your notes. The pace and inflection used when reading is noticeable over the phone.
- You should also prepare a few questions that you can ask the employer. The absolute worst response to "Do you have any questions for us?" is "No." You can ask about next steps at this point.

ENGAGE

Having students conduct mock-interviews with one another can be a way to engage students in class. You may also include prerecorded videos of good and bad examples of any type of interview. Have students identify the positives and negatives of each interaction based on the recommendations in the text.

You can also ask about students' interview experience, such as if anyone has ever had a summer job. Inquire about what strategies they used when putting their resume together and interacting during the interview. You can use these experiences to compare to the multiple-round interviews for engineering positions.

DISCUSSION

Have students discuss the purpose of behavioral, situational, and competency based interview questions. Why would a potential employer use all three? *Behavioral questions attempt to understand the interviewee's character, including personality traits that would impact job performance. Situational questions attempt to understand how you might react to hypothetical work situations (note that behavioral questions concern events that actually occurred at some point in the past). Competency-based questions attempt to understand your technical expertise, including problem-solving abilities.*

ENGAGE

Indeed has a list of seven behavioral and situational interview questions with sample answers; consider using these as examples in class.

URL
https://www.indeed.com/career-advice/interviewing/behavioral-interview-questions

TEACHING TIP

You should emphasize that whiteboard problems are standard in computing roles. So, if the role requires programming, interviewees should prepare for a whiteboard interview.

Google has an example whiteboard problem posted on their company YouTube page. You may consider showing a portion of this video as a concrete example of what these interviews involve.

How to: Work at Google — Example Coding/Engineering Interview
https://www.youtube.com/watch?v=XKu_SEDAykw&t=900s

virtual interviews

In the wake of the pandemic that began in 2020, more and more companies have shifted to **virtual interviews** for some or all of the hiring process. These require a bit more care and preparation to have a successful experience if there is an expectation for your video to be on during the interview. Aside from the usual preparation you would put into a typical phone or in-person interview, the following pointers apply to virtual interviews:

- Ensure that you have access to a good Internet connection and a high-quality webcam.
- Download the appropriate video conferencing platform at least a day before and perform a test run with your equipment the day of the interview to ensure you're ready.
- Find a neutral, quiet area where your background will be free from clutter. Alternatively, you can also use a neutral filter offered through video conferencing platforms that blur your background.
- Make sure your lighting is appropriate; the light source should be over your head and not behind it.
- Set up your camera at eye level and avoid looking at the screen to mimic the eye contact you would normally make in an in-person interview.
- If you run into technical issues, report them to the organizer as soon as possible.
- In terms of sound equipment, if you are not using headphones or earbuds, ensure you mute yourself when you are not speaking to avoid echoes.
- Dress professionally and log on a few minutes early.

The pointers from the phone interview still apply when you move into the in-person interview or virtual interview stage. On the day of the interview, make sure your appearance is clean and neat. Normally, it is suggested you be prepared to shake hands and make eye contact with everyone you meet. However, the pandemic changed the given social norm of shaking hands to make it fuzzier. Be aware that some people may not be comfortable with shaking hands.

Some companies may ask you to create a video as part of the application or interview process. You should treat these videos with the same decorum you would with a virtual interview.

behavioral interview question

situational interview question

competency-based interview question

During a phone interview, virtual interview, or in-person interview, you may be asked a combination of questions that are classified into broad categories like behavioral, situational, and competency-based. A **behavioral interview question** attempts to understand how you addressed different situations in the past, including difficulties at work, such as "tell me about a time you went above and beyond for your job responsibilities" and "tell me about a time you had to manage several competing deadlines." A **situational interview question** asks you to demonstrate how you would handle a specific hypothetical scenario in the future, such as "what would you do if you made a mistake no one else noticed?" Finally, a **competency-based interview question** attempts to understand if you have the appropriate background knowledge to perform the job successfully, for example, "tell me about a time you helped a colleague solve a problem" or "how do you typically approach programming in a team."

You can use the STAR method to structure your responses to any type of interview question, which stands for Situation, Task, Action, and Result. The (S)ituation should give the interviewer the context of your story, such as "In my previous role as a materials engineer, we were tasked with sourcing a specific polymer for a home appliance. However, the cost was much higher than anticipated." Next, the (T)ask will describe what you needed to do to respond to the challenge, "As the lead for the project, I needed to determine how to meet the customer's criteria while managing the increased cost of the material we proposed." Next, the (A)ction details what you did to address the challenge, "I researched different composites that would behave similarly to the one we proposed. Using our existing inventory, I ran tensile tests on three candidate substitutes." Finally, tell them the (R)esult, "After comparing the results, I found a suitable alternative with

TEACHING TIP

Common pitfalls that interviewees encounter can be shared with students or incorporated into the discussion. For example: *Interviewees may go overboard with providing or asking for too much or too little context, which limits the time available to actually present a solution to the problem. Alternatively, jumping right into the solution is not ideal either because interviewees can miss key details important in the problem. Interviewees may also forget to vocalize their thoughts; interviewers want to hear the problem-solving process just as much as they'd like to visualize it on the board. Interviewees can also get lost in the details and forget their place in the problem-solving process; it's best to understand the big picture of the problem (or algorithm), then sort out the details.*

similar tensile strength within a 2% margin that was 12% less expensive. This substitute allowed my team to meet the client's criteria and budget within 5% of our estimate."

2.1.5 Technical Interviews and "Whiteboard Problems"

When you claim a skill on your resume, expect to be asked about it to prove you are competent in that area. Further, you will likely be tested on it in certain industries and companies. You might be tasked with a small technical project before or between in-person interviews, especially in programming-centric jobs. These small projects intend to test basic programming skills and are usually scoped to be completed in a few hours; occasionally these are compensated but that is not guaranteed. Moreover, you are likely to be tested with a so-called "whiteboard interview" during the in-person interview activities. These assessments are done in real time with at least one of the interviewers or hiring manager. You will be given a question relevant to a core skillset for the position and you will be expected to solve the problem by drawing out your solution on the whiteboard, which includes writing code. The purpose of these interviews is not necessarily *only* to see how fast you can solve the problem. The interviewer wants to see *how* you solve problems and how clearly you can communicate your solution, so you are encouraged to verbalize your thoughts. These types of interviews are common in the tech industry. Several educational websites have emerged to help hopeful applicants to practice the kinds of questions that might be asked in the interview, such as LeetCode and AlgoExpert. There are also several resources online where content creators have done mock interviews, walk step-by-step through problems, and provide general advice. As mentioned previously, remember to send a thank-you note within 24 to 48 hours after the interview.

2.1.6 Clear and Concise Communication

What if you have a list of accomplishments that would push your resume to a third page, or you have an amazingly outgoing personality and desire for a position pushing your cover letter into two or three pages? Wouldn't your potential employer want to know each detail about your qualifications? Should you just ignore the fact that you assembled your own bicycle at the age of 8 years old, or that you are able to recite the entire dialogue of your favorite movie? Probably. Your writing in these instances and other, especially technical, writing should be concise. **Concise writing** is more than making sure the information is short, it should also be brief and comprehensive. Your cover letter, for example, should be one page, which may mean you need to leave out items you want to include, or may need to edit sections to use fewer, but possibly more descriptive words.

concise writing

For example, a cover letter may describe an applicant's prior experience in extracurricular activities as shown here:

> From the age of 16 through graduation, I was very involved in many extracurricular activities. One of my favorite extracurricular activities was when I was part of a team that built a competitive robot for our school's FIRST Robotics team. I also was involved with three competitions in the Technology Student Association, including Video Game Design, and the CO_2 racer design where I helped the team place third. I was also involved in a new competition for our school, the TEAMS competition, a competition involving extensive mathematics. Our team placed fourth in the state for TEAMS. I was also a member of the Spanish Club, where we had to only speak Spanish during each meeting. I was also a member of the Student Engineering Club.

QUESTION

Define the term *technical writing.*

(Technical writing is performed by a technical writer and is the process of writing and sharing information in a professional setting. A technical writer's main task is to convey information to another person or party in the most clear and effective manner possible. The information that technical writers convey is often complex, and it is one of their main tasks to analyze the information and present it in a format that is easy to read and understand.)

ENGAGE

Visit the ASME website with your students. Navigate to the page on how engineers can improve technical writing.

https://www.asme.org/topics-resources/content/how-engineers-can-improve-technical-writing

QUESTION

According to the article, what are the two key competencies that are involved in technical writing?

1. The ability to understand technical language.

2. The ability to express that knowledge in a clear, concise, and coherent manner.

QUESTION

Define the term *profession. (The dictionary defines profession as a type of job that requires special education, training, or skill.)*

Ask the students to consider engineering as a profession.

Does it require a special education? *(yes)*

Does it require special training? *(yes)*

Does it require special skills? *(yes)*

TEACHING TIP

To highlight the differences between college and technical (or trade) school, the following article and video from Community College Review and Study.com would be helpful.

URL

https://www.communitycollegereview.com/blog/community-college-vs-trade-school

URL

https://study.com/blog/community-college-vs-technical-college-know-the-differences.html

Be sure to emphasize that a trade school can still lead to well-paying jobs, including some that pay more than those requiring a college education (before accounting for forgone wages and student debt).

To further contrast options, the following video from Apprentice and Skills TV's YouTube channel shows the advantages of apprenticeships and entry-level employment.

URL

https://www.youtube.com/watch?v=W-Crgw0HM2z4

As an activity, you may want to have the students create a table of advantages and disadvantages in small groups. Then you can debrief with the entire class.

TEACHING TIP

To show students engineering opportunities in the military, you may show this short clip from GoArmy's YouTube channel.

URL

https://www.youtube.com/watch?v=UQxggnFTutw

What disciplines of engineering do you see represented in the various jobs discussed in the video?

Nothing in this 125-word paragraph is negative, so there is nothing wrong with including any of that information. The mention of Spanish Club doesn't seem to add to the description. It's not bad, but it could be deleted unless it was part of the job description. A condensed rewrite might look something like this:

> During High School, I was involved in a number of activities related to engineering, including the Engineering Club and competitions for the Technology Student Association (TSA) and FIRST Robotics, including a third-place finish in TSA's CO_2 racer competition. I was a member of the first team to compete in the TEAMS competition, involving extensive mathematics.

This paragraph comes in at a concise 55 words and offers the same information, which can help keep that cover letter to one page. It is shorter, but also clear and effective. Each positive item from the original paragraph is emphasized. Technical documentation is different than writing a book or story where rich, deep, and descriptive prose is appropriate.

2.2 Engineering as a Profession

Other professions are related to engineering; perhaps the closest is the field of engineering technology. We described engineering as applying math and science toward problem solving intended to benefit the public. University programs in Engineering and Engineering Technology are assessed for their effectiveness by an organization called ABET, one organization that offers descriptions of both fields. Engineering programs typically focus on theory and conceptual design and require higher-level mathematics, including multiple semesters of calculus and beyond. Engineering technology programs focus more on application and implementation of designs and systems. Engineering is more typically associated with design, while engineering technology is more typically associated with implementation and manufacture (Figure 2-5).

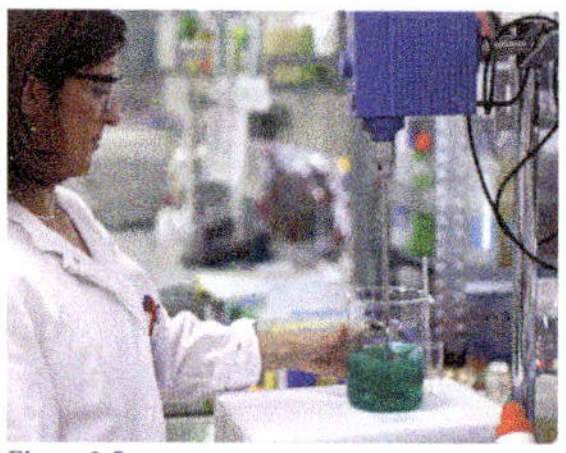

"My research is part of a Water Research Foundation project, which is investigating the formation of emerging disinfection byproducts (DBPs) in drinking water treatment.

DBPs are undesirable, toxic compounds that are formed when water is chlorinated. I am investigating the effects of pH, bromide, and iodide concentrations, and preoxidants on the formation of a specific family of DBPs."
D. Jones, BE

Figure 2-5

Whether working in entry-level employment or deep into a career, engineers and those working in the science are typically required to have at least a Bachelor's degree in engineering. Engineering Technology professionals are usually required to have an Associate's or Bachelor's and are much more likely to have had an apprenticeship. Military service usually helps someone gain experience in engineering technology or engineering.

While earning an Associate's degree in Engineering Technology is possible from a community or technical college, it makes more sense to compare the BS degrees in Engineering and Engineering Technology to observe the formal differences. Student outcomes of both programs are shown in Table 2-1 (from *https://www.abet.org*). Note that the criteria do not appear in the same order in the two sets of criteria. In some cases, the wording may be different between the two criteria, but the intent is similar. There is no reason to suggest that these are different outcomes.

ENGAGE

Visit the IEEE Professional Communication Society website with your students. Navigate to the page on writing clearly and concisely.

URL

https://procomm.ieee.org/communication-resources-for-engineers/style/write-clearly-and-concisely/

Ask the students to prepare a list of techniques an engineer can utilize to write concisely. *(Choose your words deliberately. Use simple words. Eliminate unnecessary words. Pay attention to sentence length.)*

ENGAGE

Visit the Bureau of Labor Statistics website with your students. Navigate to the page on architecture and engineering occupations.

URL

http://www.bls.gov/ooh/architecture-and-engineering/home.htm

Define the term *median wage*, and ask the students to identify the median wage for people in these occupations. Compare that wage to the median for all occupations.

Table 2-1 ABET Criteria for Engineering and Engineering Technology Programs

Engineering	Engineering Technology
1. an ability to identify, formulate, and solve complex engineering problems by applying principles of engineering, science, and mathematics.	1. an ability to apply knowledge, techniques, skills and modern tools of mathematics, science, engineering, and technology to solve broadly-defined engineering problems appropriate to the discipline;
2. an ability to apply engineering design to produce solutions that meet specified needs with consideration of public health, safety, and welfare, as well as global, cultural, social, environmental, and economic factors.	2. an ability to design systems, components, or processes meeting specified needs for broadly-defined engineering problems appropriate to the discipline;
3. an ability to communicate effectively with a range of audiences.	3. an ability to apply written, oral, and graphical communication in broadly-defined technical and non-technical environments; and an ability to identify and use appropriate technical literature;
4. an ability to recognize ethical and professional responsibilities in engineering situations and make informed judgments, which must consider the impact of engineering solutions in global, economic, environmental, and societal contexts.	
5. an ability to function effectively on a team whose members together provide leadership, create a collaborative and inclusive environment, establish goals, plan tasks, and meet objectives.	5. an ability to function effectively as a member as well as a leader on technical teams;
6. an ability to develop and conduct appropriate experimentation, analyze and interpret data, and use engineering judgment to draw conclusions.	4. an ability to conduct standard tests, measurements, and experiments and to analyze and interpret the results to improve processes.
7. an ability to acquire and apply new knowledge as needed, using appropriate learning strategies.	

Generally, Engineering program curricula are more academic, focusing on theory and concepts, whereas Engineering Technology program curricula are more practical, focusing on applications and skills.

Engineering graduates have wide-ranging jobs ranging from design to analysis, office work to field work, from companies that make things to companies that design things that are made by others. Graduates of four-year engineering technology programs are more often found in jobs where things are made or sold, and are more often engaged in field work. Four-year technology graduates are called technologists—the term "technician" is appropriate for two-year engineering technology graduates. Thus, engineering careers are more flexible whereas engineering technology careers tend to result in more tangible accomplishments (rather than accomplishments on paper).

QUESTION

Define the term *engineering disciplines. (In the contemporary era, it is generally considered to consist of the major basic branches of chemical engineering, civil engineering, electrical engineering, and mechanical engineering. There are numerous other engineering subdisciplines and interdisciplinary subjects that are derived from concentrations, combinations, or extensions of the major engineering disciplines.)*

ENGAGE

Consider two different college degrees: a bachelor's degree in engineering and a bachelor's degree in engineering technology. Ask the students to research colleges in the local area to see which of them offers either one of the degrees. Compare the differences between the schools that offer the degrees. Are they large schools, private schools, junior colleges, or four-year institutions? Develop a two-column list. Are there any schools near you that offer both degrees? *(Answers may vary.)*

QUESTION

Define the term *safety factor.*
(A term describing the structural capacity of a system beyond the expected loads or actual loads. Basically, the factor of safety is how much stronger the system is than it usually needs to be for an intended purpose. Safety factors are often calculated using detailed analysis because comprehensive testing is impractical on many projects, such as bridges and buildings. However, the structure's ability to carry load must be determined to a reasonable accuracy.)

DISCUSSION QUESTION

What factors do you believe would impact the safety factor used? *Cost, how critical the part is to the design, impact of product failure on the user, possible variations in manufacturing, environmental considerations, and how users might interact with the product.*

2.3 Safety

A product, system, or process must be designed to ensure that it is safe to use. For example, cooking utensils made of heat-conducting materials should have handles that do not conduct heat. Similarly, electrical devices must be designed to shield the user from the internal electrical parts. Safety codes and regulations must be met. These are especially important for devices that can pose a risk to humans—components of devices that sustain life, such as medical devices, and large equipment, such as automobiles or amusement rides.

safety factor

Everything has a life span; living organisms die and physical devices fail. A critical responsibility of a designer is to ensure that a product's failure does not endanger people. Engineers are responsible to ensure their designs are safe. As engineers design, they may consider a safety factor in their design. A **safety factor** is a multiple of some critical criteria used in a design. For example, if a bridge is designed to hold 8,000 tons, an engineering firm may design to a safety factor of 1.5 and design the bridge to accommodate 12,000 tons.

Figure 2-6 Safety features are an important part of good design.

Safety factors are developed based on the risks (both known and unknown), which are based on experience or past history of similar designs (Figure 2-6). Some things that make up a safety factor include the material used for building, whether it will be maintained and inspected periodically, and if it is subjected to repeated stresses. The thickness of hot water piping in your house probably has a moderate safety factor. The brakes on your car may have a higher safety factor, as they can get worn quickly, overheat, and then fail to stop the vehicle, which would be catastrophic. Manned spacecraft would likely have higher safety factors, since failure can easily result in death and there is no easy way to maintain or inspect a lot of the systems on the craft. Specialized engineering disciplines will often publish a code of safety factors explaining when and how to use them. For example, the ASME (American Society of Mechanical Engineers) publishes codes for safety factors to use in boilers.

Figure 2-7 Printed circuit board.

Considering safety is also essential as we work with different tasks in engineering. For example, manufacturing almost any electronic device involves designing and building circuit boards (Figure 2-7). If you have ever seen the inside of a computer, there is usually a large, green board inside with electronic components—the motherboard. Producing the printed circuit board involves chemically removing copper and epoxy (glue), producing waste. When the components are placed on the board, they are held in place by solder. Until a decade or two ago, almost all solder was a mixture of tin and lead. This means, as older computers are discarded, they contain copper, tin, lead, and a number of other chemicals that may or may not be hazardous.

On a smaller scale, we produce waste every day. Many of us don't think twice about grabbing a bottle of water, but less than 20% of these bottles end up being recycled. Over 80 billion bottles end up in landfills each year. Plastic bottles take up room in a landfill, but they are not considered to be hazardous waste.

hazardous waste

hazardous materials

What is the difference? **Hazardous waste** is classified as waste that is potentially harmful to health or the environment (Figure 2-8). **Hazardous materials** are similar to hazardous waste—these are materials that can be hazardous to health or

Figure 2-8 Hazardous materials, hazardous waste, and even non-hazardous wastes should be disposed of properly.

the environment but are still in their original, usable form. Some household waste is considered hazardous (used cleaning products, broken compact fluorescent, or CFL, light bulbs), while a lot more is generated by industry. Almost every household has some hazardous material (cleaning products, gasoline, and paint thinners) as well. Your classroom probably has some material classified as hazardous.

The Environmental Protection Agency (EPA) is in charge of classifying which wastes are considered hazardous, and how they are to be disposed of. Fortunately, you will not likely come into contact with most of these, but some are common.

Classroom and Laboratory Safety

hazardous
corrosive
explosive
toxic

As a student and perhaps someone who will design, build, manufacture, or recycle material in the future, you need to be familiar with the terminology and regulations at your place of work and your personal work environment. The federal and state Environmental Protection Agencies have specific rules and methods to follow to store and use hazardous materials and dispose of waste products. A material is **hazardous** when it poses a danger to your health and safety. A material is **corrosive** when it is capable of creating a chemical reaction with another material and causes damage to that material (example: salt water). A material is **explosive** when it has the potential of creating a chemical reaction that leads to an excessive release of gas or energy (example: pure hydrogen peroxide). A material is **toxic** when it poses a serious health issue to the point of possible death, if not helped. Common rules include:

- Store flammable hazardous materials in a well-marked, explosion-proof cabinet (if there is a significant amount of the material).
- Clearly post safety rules and guidelines in areas where these materials are stored and used.
- Require safety goggles when working with these materials.

Material Safety Data Sheet

Many hazardous materials come with paperwork explaining how they are to be used and stored. This paperwork is called a **Material Safety Data Sheet**, or MSDS (Figure 2-9). These sheets should be kept in a visible, obvious location in case of an accident, and they need to be reviewed quickly.

Figure 2-9 Material Safety Data Sheets.

Hazardous waste has rules and procedures for handling and disposing similar to those of hazardous materials. For example:

- Do not pour waste products down the drain, including products that may have resulted from a chemistry experiment.
- Do not dispose of known hazardous material in the normal trash can.

There are four key classifications of hazardous materials and wastes:

- A **WHITE** diamond label signifies specific hazards, such as oxidizers or alkali materials. These can cause a chemical reaction that is harmful. One example of this classification is the radiation hazard sign.
- A **YELLOW** diamond label signifies chemical instability, and it has a scale of severity. For example, a yellow diamond sign with the label Oxidizer and number 4 would indicate that the item might detonate.
- A **RED** diamond label signifies a fire hazard/flammability, and it has a numeric scale that signifies the temperature levels. For example, a red diamond with a flame symbol and a 2 indicates a flammable material that combusts below 200 degrees Fahrenheit.

ENGAGE

Visit the Environmental Protection Agency (EPA) website with your students. Navigate to the page on universal waste.

URL
https://www.epa.gov/hw/universal-waste

QUESTION

The EPA lists standards for universal wastes in four main categories. What are they? *(Batteries, pesticides, mercury-containing equipment, and light bulbs)*

TEACHING TIP

At this point, you should emphasize the general rules in the text **but also include whichever additional procedures are standard in your classroom.**

If your class will use the lab frequently, you might dedicate a class period to having students act out the most critical rules you expect them to follow. For ELL students, these acting sessions will allow you to model the correct behavior for engaging in lab activities beyond having these rules posted or reiterated verbally.

- A **BLUE** diamond label signifies a health hazard. It has a numeric scale that communicates the level of severity, from 4 being deadly to 0 being normal materials. An example of a blue diamond label is a "flammable when wet" sign.

With regard to hazard classifications, there are nine classes:

- Class 1: Explosives
- Class 2: Gasses
- Class 3: Flammable liquids
- Class 4: Flammable solids
- Class 5: Oxidizing substances
- Class 6: Poisons
- Class 7: Radioactive materials
- Class 8: Corrosive materials
- Class 9: Miscellaneous hazardous materials

For example, arsenic would be classified in Class 6 as a poison. It would also be labeled with a blue diamond label with its name as identifier and the number 4.

Occupational Safety and Health Administration

Many safety regulations are set by the federal government, by an organization called the **Occupational Safety and Health Administration** (OSHA). To have a federal standard for safety in the workplace and laboratory helps everyone to have the same familiarity and direction with specific ways of doing things properly and in a way that is safest. For example: there is a proper way to lift something heavy. If you lift something using your legs and keeping your back straight, rather than bending down by arching your back, you protect your lower back from being injured.

Best Practice guidelines

OSHA has a very comprehensive list of safety topics, including **Best Practice guidelines**. Best Practice guidelines are a standard way that has been established within the industry to be the best way of doing something. These guidelines have been developed over time based on trial and error, finding the best way to do something. For example, it makes sense to label chemicals or materials you work with so that you know what's in each bottle or container, and you don't mix chemicals that can cause a reaction. OSHA has a standard way of labeling hazardous wastes and materials so that you can label things in a way that another student in another state can recognize, as well.

Your classroom and/or laboratory should have a set of lab safety guidelines. You are responsible for being familiar with these. It's not just a matter of your safety, but the safety of those around you and your local environment.

It is important to handle, maintain, and store laboratory equipment properly and safely. Beakers and test tubes made of glass can easily shatter while they are in use or if they are improperly stored. Sharp instruments which are not maintained or stored properly can lead to cuts. Failure to clean containers could lead to contamination. Hot plates are obviously hot while in use, but storing them before they cool can cause damage. Using a precise scale without calibrating it can result in improper readings, which can give incorrect results. Always follow the procedures with laboratory equipment.

When it comes to personal safety, remember to dress appropriately and to use personal protective equipment (PPE) to keep your body safe from any possible chemical or mechanical harm. Don't wear any loose clothing or jewelry that could snag or get in the way. If you have long hair, it helps to tie it back so that you can see, and your hair does not get it the way. Protect your face (especially your eyes), your hands, your skin, and your clothing with PPE.

It is important to behave calmly and respectfully in the classroom. First, this shows that you can follow directions and are committed to a safe workspace environment. Second, this ensures you are alert and are paying attention. Horseplay puts you and others into danger when handling chemicals or lab equipment.

You may be asked to pass a safety test. Safety tests are ways of finding out whether you are performing your classroom tasks in a safe manner, and in the way you are supposed to do them. Safety tests identify whether you have the proper knowledge to do your tasks safely, and whether you understand the safety instructions and the reasons for their existence. If you are not familiar with parts of the safety test content, you might perform the lab work in a manner that could cause harm to you, those around you, or your environment.

It is important for you to master safety tests for three reasons:

- Your overall safety
- The safety of those around you in the lab
- The safety of the end user

It is also very important to understand how to use precision measuring instruments properly. Many scales, especially scales that are precise, require calibration and zeroing (or taring). Failing to do so may result in a weight that is inaccurate. Beakers, pipettes, and test tubes often allow us to very precisely measure the volume of liquids. Failing to accurately measure liquids could result in erroneous results, which could be dangerous.

Example 2-1

Work in small groups to answer the following questions:

1. Identify three important safety procedures you will need to consider in this class. You are allowed to ask for advice from your teacher on this question.
2. Find a Material Safety Data Sheet, either in your classroom or on the Internet. What is the chemical? How is it to be handled and/or disposed of?
3. Identify a piece of laboratory equipment you may use this year in class. How is it maintained? How is it to be properly handled? How is it properly stored? What could go wrong with it, and what is a good procedure to follow if something does go wrong with it?
4. What are some implications of improper or negligent maintenance of classroom equipment?

ELL

Example 2-1 Expansion: Partner struggling ELL students with students stronger in language skills, and encourage them to seek clarification as needed in order to enhance and confirm their understanding of complex spoken and written language.

QUESTION

Define the term *ethics. (A complex of moral precepts held or rules of conduct followed by an individual)*

Define the term *code of ethics. (The services provided by engineers require honesty, impartiality, fairness, and equity, and must be dedicated to the protection of the public health, safety, and welfare. Engineers must perform under a standard of professional behavior that requires adherence to the highest principles of ethical conduct.)*

2.4 Ethics

We make numerous ethical decisions daily, although most are so minor that we do not even view them as such.

- You know that another student has plagiarized an assignment; do you report that to the teacher?
- When going through a public door, do you habitually look back to see if releasing the door will cause it to slam in someone's face?
- A cashier gives you too much change for a purchase. Do you correct the cashier?

On a grand scale, few of these decisions are particularly important, although some might lead to undesirable consequences. However, as an aspiring engineer, you may face decisions in your career that could affect the lives and well-being of thousands of people. Just like almost everything else, practice makes perfect, or at least better. The more you practice analyzing day-to-day decisions from an ethical standpoint, the easier it will be for you to make good decisions when the results of a poor choice may be catastrophic.

Some ethical decisions are clear-cut. For example, almost everyone would agree that it is unethical to kill someone because you do not like his or her hat. Unfortunately, many real-world decisions we must make are far from clear-cut issues, and instead they have many subtle aspects that must be considered to arrive at what one believes is the "best" decision.

No proven algorithm or set of rules can be followed to guarantee that the most ethical decision possible is being made in any particular situation. However, numerous people have developed procedures that can guide us in considering questions with ethical ramifications. A four-step procedure is discussed here, although there are various other approaches.

Step 1: Determine *What* the issues are and *Who* might be affected by the various alternative courses of action that could be implemented.

stakeholders

We will refer to the *Who* as **stakeholders**. Note that at this point, we are not trying to determine how the stakeholders will be affected by any particular plan of action.

- The issues (What) can refer to a wide variety of things, including, for example, personal freedom, national security, quality of life, economic issues, fairness, and equality.
- The term stakeholders (Who) does not necessarily refer to people, but might be an individual, a group of people, an institution, or a natural system, among other things.

Example 2-2

Consider the question of whether to allow further drilling for oil in the Alaska National Wildlife Refuge (ANWR). List several issues and stakeholders.

Issues:

- Oil independence
- The price of gasoline
- Possible impacts on the ecosystem

Stakeholders:

- Oil companies
- The general population of the United States
- Other countries from whom we purchase oil
- The flora and fauna in ANWR
- The native people in Alaska

Step 2: Consider the effects of alternative courses of action from different perspectives.

Here, we look at three perspectives: consequences, intent, and character.

Perspective 1: Consequences
When considering this perspective, ask how the various stakeholders will be affected by each alternative plan being contemplated. In addition, attempt to assign a relative level of importance (weight) to each effect on each stakeholder. For instance, an action that might affect millions of people adversely is probably more important than an action that would cause equal harm to a dozen people.

Perspective 2: Intent
The intentions of the person doing the acting or deciding are considered in this perspective, sometimes called the "rights" perspective. Since actions based on good intentions can sometimes yield bad results, and vice versa, the intent perspective avoids this possible pitfall by not considering the outcome at all, only the intentions.

Ask yourself the following questions:

(a) Is the action I am taking something that I believe everyone should do?

(b) Do I believe that this sort of behavior should be codified in law?

(c) Would I like to be on the receiving end (the victim) of this action?

Example 2-3

Should you download music illegally over the Internet?

Rephrasing this question using the suggestions above yields:

(a) Should everyone illegally download the music they want if it is there for the taking?

(b) Should the laws be changed so that anyone who obtains a song by any means can post it on the Web for everyone to get for free?

(c) If you were a struggling musician trying to pay the bills, would you like your revenue stream to dry up because everyone who wanted your music got it for free?

Perspective 3: Character
Character is the inherent complex of attributes that determines a person's moral and ethical actions and reactions. This perspective considers the character of a person who takes the action under consideration. There are different ways of thinking about this. One is to simply ask: "Would a person of good character do this?" Another is to ask: "If I do this, does it enhance or degrade my character?" Yet another way is to ask yourself if a person you revere as a person of unimpeachable character (whoever that might be) would take this action.

Step 3: Correlate perspectives.

Now look back at the results of considering the issues from the three perspectives. In many cases, all three perspectives will lead to the same or a similar conclusion. When this occurs, you have a high level of confidence that the action is the best choice from an ethical standpoint.

If the three perspectives do not agree, you may wish to reconsider the question. It may be helpful to discuss the issue with people whom you have not previously consulted in this matter. Did you omit any factors? For complicated issues, it is difficult to make sure you have included all possible stakeholders and consequences. Did you properly assign weights to the various aspects? All three perspectives may converge. If not, go with two out of three.

TEACHING TIP

Is it critical to avoid describing this process as giving the "correct" answer to an ethical dilemma. There is no "right answer" in ethics because different ethical frameworks can often pose drastically different solutions. Ethics should be described as *shades of gray*; each option carries its own consequences and the purpose of the frameworks is to help an engineering student weigh those consequences against one another with a code of ethics as the backdrop for correlating the options with expected behavior of professionals in the discipline.

Step 4: Act.

This is often the hardest step of all to take, since ethical action often requires courage. The whistle-blower who risks losing his or her job, Harriet Tubman repeatedly risking her life to lead slaves to freedom via the Underground Railroad, the elected official standing up for what she knows to be right even though it will probably cost her the next election, or even something as mundane as risking the ridicule of your friends because you refuse to go along with whatever questionable activities they are engaging in for "fun." Ask yourself the question: "Do *I* have the courage to do what I know is right?"

Example 2-4

NOTE: In the interest of brevity, this is not an exhaustive analysis but shows the general procedure.

At the time of this writing in 2022, the electronic cigarette (or vaporizer) called JUUL was temporarily suspended for sale in the United States by the FDA. Your company has been granted a contract to develop the next generation of electronic cigarettes to attempt to recapture some market share by showing how e-cigarettes can benefit public health. You have been assigned to the design team. Can you in good conscience contribute your expertise to this project?

Step 1: Identify the issues (What) and the stakeholders (Who).

Issues:

- *Nicotine is poisonous and addictive.*
- *These devices eliminate many of the harmful components of tobacco smoke.*
- *Laws concerning these devices range from completely legal, to classification as a medical device, to completely banned, depending on the country.*
- *There are claims that such devices can help wean tobacco addicts off nicotine.*
- *The World Health Organization does not consider this to be an effective means to stop smoking.*
- *Whether an individual chooses to use nicotine should be a personal decision, since its use does not generally degrade a person's function in society.*
- *The carrier of the nicotine (80–90% of the total inhaled product) is propylene glycol, which is relatively safe, but can cause skin and eye irritation, as well as other adverse effects in doses much larger than would be obtained from this device.*
- *A profit can be made from nicotine products or anti-smoking devices.*

Stakeholders:

- *You (your job and promotions)*
- *Your company and stockholders (profit)*
- *Cigarette manufacturers and their employees and stockholders (lost revenue)*
- *Tobacco farmers (less demand)*
- *The public (less second-hand smoke)*
- *The user (various health effects, possibly positive or negative)*

Example 2-4 *(continued)*

Step 2: Analyze alternative courses of action from different perspectives.

Consequences

- *You may lose your job or promotion if you refuse.*
- *If you convince management to abandon the project, the company may lose money.*
- *If you succeed brilliantly, your company may make money hand over fist, and you receive a promotion.*
- *If the project goes ahead, the possibility of future lawsuits exists.*
- *Users' health may be damaged.*
- *Users' dependence on nicotine may either increase or decrease.*

Intent

- *Should everyone use electronic cigarettes, or at least condone their use?*
- *Should use of electronic cigarettes be unrestricted by law?*
- *Would I like to risk nicotine addiction because of using these devices?*
- *Would I be able to kick my tobacco habit by using these devices?*

Character

- *Would a person of good character develop this device, use it, or condone its use?*
- *Would work on this project (thus implicitly condoning its use) or use of the device itself enhance or degrade my character?*
- *Would my personal spiritual leader, or other person I respect, condone development or use of this product?*

Step 3: Correlate perspectives.

Here we enter the realm of subjective judgment. The individual author responsible for this example has a definite personal answer, but it is in the nature of ethical decision making that different people will often arrive at different results in good conscience. You would have to weigh the various factors (including any that have been overlooked or knowingly omitted) to arrive at your own conclusion. We refuse to dictate a decision to you.

Step 4: Act on your decision.

If your decision was that working on this project poses no threat to your character, probably little courage is required to follow through, since your career may blossom, or at least not be curtailed.

On the other hand, if you believe that the project is unethical, you need to have the courage to either attempt to change the minds of management or refuse to work on the project, both of which may put your career at risk.

DISCUSSION

Have students define the difference between ethics and morals. Encourage them to come up with personal morals that you can use to find others in the room who might disagree (respectfully). If you do so, emphasize that the class must engage at the conceptual level and not attack the individual. If you believe this might cause conflict based on your class dynamics, you may elect to collect these anonymously, using sheets of paper that you can screen for discussion, or using an automated tool like Google Forms. *Morals can be thought of as a* ***personal code of ethics****, which is shaped by the person's environment, upbringing, and collective experiences. Ethics are* ***defined by groups*** *that apply to a greater population than oneself. These ethical behaviors are defined using a code of ethics.*

TEACHING TIP

It is worth mentioning that morals and ethics are shaped by cultural values, so classes that are not particularly diverse should have this point emphasized.

2.4.1 Ethics vs. Morals

Because of the common occurrence of unethical behavior and its negative impact on others, almost all societies have developed rules to specify what is acceptable behavior and the punishments for violations.

Through the centuries, such codes and laws have been expanded, modified, and refined so that most forms of severe antisocial behavior are addressed and consequences for violations specified. These codes exist from a local to a global level. Several examples are given below.

- Most countries abide by the Geneva Conventions, which govern certain types of conduct on an international scale.
- Most countries have national laws concerning murder, rape, theft, etc.
- In the United States, it is illegal to purchase alcohol unless you are 21 years of age. In England, the legal age is 18.
- In North Dakota and South Dakota, you can obtain a driver's license at age 14 ½. In most other states, the legal age is 16.
- Nearly all states in America (48 of them) have ordinances completely prohibiting text messaging while driving.

morals
ethics

These laws and guidelines apply to everyone within a population. However, even without a set of laws, people have a sense of right and wrong that guides their actions. One person may feel very strongly that lying is absolutely wrong, while others may lie routinely. If a person finds a bag full of cash, he or she may take it to a local police station while someone else may simply take it—after all, 'finders keepers, losers weepers.' Many of these examples represent **morals** rather than **ethics**. Moral decisions are those that apply to an individual. Ethical decisions are those based on a right vs. wrong as defined for a population. For example, we would generally agree that stealing is wrong. However, some people would argue that taking a few dollars from a parent isn't terribly wrong. Some would say that, although we know stealing is wrong, we would do it in this case. Decisions of this type are based on a person's character. An engineer may receive an offer like the following: In exchange for using a lesser grade of steel in a bridge project, a company may be happy to send that engineer on an all-expenses paid trip to Hawaii. There is certainly a moral component: One engineer may consider this more dishonest than another. However, this specific issue is addressed in the engineering Codes of Ethics. Engineers are not to take bribes and are to put the safety of the public beyond other considerations. These rules are to be followed by a population of people—in this case, a population of engineers.

2.4.2 Codes of Ethics

Code of Ethics

In general, ethical decisions in engineering have a narrow focus specific to the problems that arise when designing and producing products or services of a technical nature. By the nature of their profession, engineers and scientists have a body of specialized knowledge that is understood only vaguely, if at all, by most of the population. This knowledge can be used for tremendous good in society but can also cause untold mischief when used by unscrupulous practitioners. Various engineering organizations have developed codes of conduct specific to the profession. Perhaps the most well-known is the Code of Ethics for engineers developed by the National Society of Professional Engineers (NSPE). The entire NSPE **Code of Ethics** is rather long, so we list only the Engineer's Creed and the Fundamental Canons of the Code here.

Engineer's Creed

As a Professional Engineer, I dedicate my professional knowledge to the advancement and betterment of public health, safety, and welfare.

I pledge:

- To give the utmost of performance;
- To participate in none but honest enterprise;
- To live and work according to the highest standards of professional conduct;
- To place service before profit, the honor and standing of my profession before personal advantage, and the public welfare above all other considerations.

In humility, I make this pledge.

Fundamental Canons

Engineers, in the fulfillment of their professional duties, shall:

1. Hold paramount the safety, health, and welfare of the public.
2. Perform services only in areas of their competence.
3. Issue public statements only in an objective and truthful manner.
4. Act for each employer or client as faithful agents or trustees.
5. Avoid deceptive acts.
6. Conduct themselves honorably, responsibly, ethically, and lawfully so as to enhance the honor, reputation, and usefulness of the profession.

The complete code can easily be found online at a variety of sites. Use an online search tool to find the URL for the Code of Ethics from the NSPE and for your desired discipline.

2.5 Teaming

A team is a group of two or more people who work together to achieve a common goal. When you are part of a team, you have access to all your teammates' knowledge, experience, and abilities. Together you can have more ideas, achieve more goals, and solve more problems.

Most engineering projects require teamwork. They are usually multidisciplinary, which means they involve people from many different areas within a company and with outside resources. For example, a design engineer may work on a team with other engineers who specialize in production, testing, and sales and service. The team is likely to include financial managers, legal advisors, marketing representatives, and, of course, a project manager. There may even be client representatives on the team.

But, no matter who is on the team and what the team is doing, the interactions between team members will be most productive if the members of the team:

- Listen to each other
- Respect each other's opinions
- Recognize each other's skills and abilities
- Share the workload
- Share responsibility

DISCUSSION

Have students explore the National Society of Professional Engineer's code of ethics and other disciplinary codes of ethics. One assignment can be comparing and contrasting the codes of ethics for the different disciplines. The results can be discussed in class by grouping fundamental canons into a visualization to highlight the similarities and differences.

National Society of Professional Engineers
https://www.nspe.org/resources/ethics/code-ethics

Institute of Electrical and Electronics Engineers
https://www.ieee.org/about/corporate/governance/p7-8.html

Association for Computing Machinery
https://www.acm.org/code-of-ethics

American Society of Civil Engineers
https://www.asce.org/career-growth/ethics/code-of-ethics

American Institute of Chemical Engineers
https://www.aiche.org/about/governance/policies/code-ethics

Biomedical Engineering Society
https://assets.noviams.com/novi-file-uploads/bmes/PDFs_and_Documents/Governance/BMES_Code_of_Ethics_2021_-_Revised_October_2021.pdf

American Institute of Aeronautics and Astronautics
https://www.aiaa.org/about/Governance/Code-of-Ethics

American Society of Mechanical Engineers
https://www.asme.org/wwwasmeorg/media/resourcefiles/aboutasme/get%20involved/advocacy/policy-publications/p-15-7-ethics.pdf

QUESTION

Ask your students to create a two-column list of the advantages and the disadvantages of working on a team.

Advantages

Team members have the opportunity to learn from each other.

Potential exists for greater work force flexibility with crosstraining.

Opportunity provided for synergistic combinations of ideas and abilities.

New approaches to tasks may be discovered.

Team membership can provide social facilitation and support for difficult tasks and situations.

Communication and information exchange may be facilitated and increased.

Teams can foster greater cooperation among team members.

Interdependent workflow can be enhanced.

Potential exists for greater acceptance and understanding of team decisions.

Greater autonomy, variety, identity, significance, and feedback for workers can occur.

Team commitment may stimulate performance and attendance.

Disadvantages

Some individuals are not compatible with teamwork.

Workers must be selected to fit the team as well as requisite job skills.

Some members may experience less motivating jobs as part of a team.

Organization may resist change.

Conflict may develop between team members or other teams.

Teamwork may be time-consuming due to need for coordination and consensus.

Teams can stymie creativity and inhibit good decision-making if group thinking becomes prevalent.

Evaluation and rewards may be perceived as less powerful.

"Free-riding" within the team may occur.

Less flexibility may be experienced in personnel replacement or transfer.

2.5.1 Engineering Teams

To prepare for a career as a professional engineer, students can work in teams to complete comprehensive projects. This provides the opportunity to learn, practice, and demonstrate the ability to function as a team member. Specifically, team members can:

- *Apply the design process as a team participant.* Team members count on each other to understand how to apply the design process to the task at hand.
- *Assume different roles as a team member within the project.* Each person on the team must be prepared to step in as necessary to help others and move the project toward successful completion.
- *Maintain an engineering notebook for the project.* An accurate and complete notebook shows the progress of each team member and provides documentation that work is completed.
- *Develop and test a* ***scale model*** *for the project.* Each team member should be prepared to participate in these steps which are vital to a project's success.
- *Prepare and present the project using effective communication skills.* Team members should all be able to use written and verbal communication skills to explain and demonstrate the project, including using tools such as word processing, graphing, and presentation software programs.

scale model

2.5.2 Leadership

Even when all members of a team have an equal role in decision making and problem solving, it is important to have a leader. A leader is a type of person who knows how to use available resources to help others achieve their goals.

Leaders exhibit positive qualities that other people respect, such as self-confidence, honesty, effective listening and speaking skills, decisiveness, respect for others, and open-mindedness. They are organized and supportive of the other team members.

Although you might have heard that someone is a "born leader," that's not usually the case. Becoming a leader takes time and patience. Leaders must prove that they can make decisions, set goals, and solve problems. You can develop leadership characteristics by recognizing and modeling positive leadership qualities. One way to do this is by participating in student leadership and professional development opportunities. For example, you can join clubs and activities such as your school's robotics team, and volunteer for leadership roles. You might also join a career technical student organization (CTSO) such as Technology Student Association (TSA) or SkillsUSA. CTSOs offer members a range of individual and group programs, activities, and competitions designed to build professional skills as well as leadership qualities.

2.5.3 Qualities of an Effective Team Member

Although a strong leader is important to the success of a team, team members must also be committed to the group's success. An effective team member helps teammates if they need help, does not blame teammates for problems or mistakes, and offers ideas and suggestions instead of criticism. Effective team members use critical-thinking skills and interpersonal skills to identify and solve problems and achieve their goals.

You are a good team member if you are:

- Open minded
- Willing to compromise
- Cooperative
- Friendly
- Trustworthy
- Work in the best interests of the team

ENGAGE

Navigate to the article *The End of Fierce Individualism: Why Teams Matter More Than Ever*

URL

https://www.prnewswire.com/news-releases/the-end-of-fierce-individualism-why-teams-matter-more-than-ever-and-what-you-need-to-know-about-them-203199821.html

According to the author, "team" is more than just a business buzzword. It's an important one for engineers to become more comfortable and conversant with. If done well, building and captaining a team will determine whether you merely survive or actually thrive.

QUESTION

What are the seven insights the author tells you are helpful to focus on for team success?

1. Successful teams are led by captains.
2. Fierce individualism has no place in teams.
3. Teams keep expectations high for everyone.
4. Teams must be willing to fail.
5. Teams must be persistent.
6. Successful teams share values, integrity, and a commitment to one another.
7. Teams must feel "at home" with uncertainty and complexity.

2.6 Communication

Engineers spend a great deal of time communicating. They must speak, read, write, and develop documents for communication such as reports, presentations, and, of course, their Engineering Notebook.

Effective communication is a process that involves an exchange of information between a sender and a recipient. The sender transmits the message with a specific intent. The recipient interprets the message and responds. Effective communication occurs when the receiver interprets the message the way the sender intended. Ineffective communication occurs when the receiver misinterprets the message. Sometimes the sender thinks the message is clear, but in fact it might confuse or mislead the receiver. Both the sender and the receiver are responsible for error-free communication.

In order to ensure effective communication:

- The message must be clear.
- The sender must be able to deliver the message in a concise and accurate manner.
- The receiver must be able to hear and receive the message.
- The receiver must be able to interpret the message correctly.

2.6.1 Using the Communication Process

Whether communicating verbally, visually, or in written form, you can communicate effectively by using a six-step process:

1. *Be clear.* The receiver is more likely to get your message if you deliver it in a way he or she can understand. Speak slowly and look at the person you are talking to.
2. *Be personal.* Use the other person's name. Use an "I" statement—a statement that starts with the word "I"—to frame the statement in terms of you and your goals. An "I" statement indicates that you are taking responsibility for your thoughts and feelings. It helps the receiver understand your point of view and respond to you, instead of focusing on herself and how your statement affects her.
3. *Be positive and polite.* Phrase your message in positive terms. Say what you want, not what you don't want.
4. *Get to the point and be concise.* Follow the "I" statement with an explanation of the message you are sending. Explain how or why you feel a certain way, or how or why you think a certain thing.
5. *Actively listen to the response.* Pay attention and make sure you hear the response that makes sense based on your "I" statement.
6. *Think before you respond.* Make sure you understand what you've heard. Repeat it, if necessary, and ask questions for clarification. Use critical thinking to make sure you are not letting emotions and preconceived ideas get in the way.

Once you master the communication process, you will be able to use it effectively for:

- Verbal communication, which includes speaking or writing
- Nonverbal communication, which includes visual messages that help put words into context, such as a smile or the tone of your voice, as well as visual aids such as slides, charts, images, and graphs
- Active listening, which is when you give the message sender your full attention and consideration.

QUESTION

Define the terms *verbal communication, non-verbal communication,* and *active listening.*

Verbal communication is the sharing of information between individuals by using speech. Individuals working on teams need to effectively use verbal communication that employs readily understood spoken words.

Nonverbal communication involves the processes of encoding and decoding. Encoding is the act of generating the information such as facial expressions, gestures, and postures. Decoding is the interpretation of information from received sensations from previous experiences.

Active listening occurs when you make a conscious effort to hear not only the words that another person is saying but, more importantly, try to understand the complete message being sent. In order to do this, you must pay attention to the other person very carefully.

ENGAGE

Visit the Forbes Entrepreneur website and read the article on *5 Habits That Effective Team Members Should Never Outgrow*

URL

https://www.forbes.com/sites/brentgleeson/2014/04/27/5-habits-that-effective-team-members-should-never-outgrow/?sh=6e2d230b6fdf

QUESTION

According to the article, what are the five habits effective team members should never outgrow?

1. Put others before yourself.

2. Reflect on your actions.

3. Be obsessively organized.

4. Assume you don't know enough.

5. Never get comfortable.

QUESTION

Define the term *effective communication. (The process that involves a successful exchange of information between a sender and a recipient)*

2.6.2 Verbal Communication

For most of us, verbal communication is the most common way we stay in touch with other people in our lives; for example, we talk face to face or on the phone. In contrast, we use a wide variety of nonverbal communication when we send text messages, emails, and instant messages. Moreover, we write blogs, pass notes, and send cards.

While written communication can be a useful mode of communication, talking can often be more effective. When speaking, you have the opportunity to send your message clearly, concisely, and effectively. Focus on being polite and professional. When you speak clearly and use language the receiver understands, he or she will interpret the message the way you intend it. This is due to the additional information verbal communication embeds in the message through inflection—the sound of your voice. In addition, verbal communication is usually paired with nonverbal cues such as our posture, hand gestures, eye contact, and so on.

Our natural nonverbal communication helps put words into context. This form of communication includes visual messages that the receiver can see, like a smile when you are talking. It also includes physical messages, like a pat on the back. During a conversation, the tone of your voice and the language you use combine to provide context for the words. You lose some of the context when you write, which can make the communication less effective. Even though context can be added by using emoticons or capitalizing words for emphasis, there is still room for misinterpretation.

Exchanging written messages does not take place face to face; it might even be across great distances. When the receiver cannot hear your voice or see your nonverbal cues, he or she might misinterpret the message because communication involves not only our voice, but our body language. With this in mind, craft any correspondence in the clearest manner possible. Being concise will help in conveying your messages without the aid of nonverbal cues.

2.6.3 Active Listening

Active listening is an important part of effective communication. When you are an active listener, you pay attention to the speaker, and make sure you hear and understand the message.

Figure 2-10 Active listening.

Use these skills to be an active listener:

- Show interest using eye contact and positive nonverbal messages.
- Let the other person finish speaking before you respond.
- Ignore distractions such as smartphones and other people.
- Set your predetermined opinions and emotions aside.
- Repeat the message that you hear out loud, to make sure you received it correctly.

Active listening is a sign of respect. It shows you are willing to communicate and that you care about the speaker and the message. When you listen actively, the other person is more likely to listen when you speak, too (Figure 2-10).

Connecting To Stem
STEM = Science, Technology, Engineering, and Mathematics

Digital Privacy and Security

As our lives become increasingly digital, it is becoming clearer that there are more ethical considerations to determine when creating systems with digital components. When engineers create a product that involves submitting information to an application, for example, data about the user is stored—whether it was entered purposefully by the user or collected automatically. Therefore, when an engineer is designing such a system, it is crucial to understand how user information will be secured. One consideration in particular is how user passwords are stored. An example of an application with particularly vulnerable information is patient management software, where individuals can sign up for appointments, check test results, and message their health care provider. Such an application has mountains of sensitive data, including health records, Social Security numbers, and addresses. Engineers in the digital space must be aware of modern practices for securing user data.

Why Be Concerned about Data?

Data has monetary value, so advertisers will want to buy access to user information to better market their products. Recent state legislation in states like California and New York has established how user data is supposed to be handled by companies that collect personal information from their customers, including how these data are sold to third parties. There have been several high-profile cases of mishandling customer data; for example, the company Meta (formerly known as Facebook) settled a lawsuit in August 2022 about third parties, like the political consulting firm Cambridge Analytica, being granted access to users' personal data without their consent. Engineers must be aware of political and legal constraints on their designs, including best practices for collecting sensitive data.

Data Breaches

Being a good steward of user data is paramount, and poor management can impact hundreds of thousands of people—especially the company itself with eventual legal consequences involving some kind of payout to those affected. For example, in 2017, the credit reporting agency Equifax experienced a data breach that allegedly impacted 143 million people. An Argentinian employee portal into Equifax's system, called Veraz, which was the first line of defense against unauthorized users gaining access to customer data, was only protected by the "admin" username and password. Because there were no additional safeguards, bad actors could easily get access to customer data with one piece of login information.

Search for "data breaches" on the Internet and compile a list of at least five distinct incidents. What are the commonalities among them? What could engineers have done to prevent the breach?

CHAPTER 2 ASSESSMENT QUESTIONS

1. Please address the following questions in approximately one page. You may write, type, draw, sketch your answers...whatever form you would like to use.

   (a) Where are you from?

   (b) What type of engineering are you interested in? Why?

   (c) Are there any engineers in the circle of your friends and family?

   (d) How would you classify yourself as a student (freshman, sophomore, or junior)?

   (e) What are your activities and hobbies?

   (f) What are you most proud of?

   (g) What are you fantastic at?

   (h) What are you passionate about?

   (i) What has been the most difficult aspect of high school so far?

   (j) What do you expect to be your biggest challenge this year?

   (k) Do you have any concerns about this class?

   (l) Anything else you would like to add?

2. In one paragraph, describe what it would mean to "act professionally" in a job interview.

3. Use one sentence per item to describe what this trait would mean for a working engineer:

   (a) Responsibility

   (b) Reliability

   (c) Dependability

   (d) Effective communication

   (e) Delegation

   (f) Cooperation

   (g) Teamwork

   (h) Integrity

   (i) Punctuality

4. Engineers and scientists often collect data to be analyzed in order to address research questions. In this case, researchers realized that we constantly use our phones, iPods, social media, etc. These researchers are working on a study which investigates whether constant screen time results in a higher level of stress and creativity. This excerpt is from an article in development.

   Can we Reduce Stress by Exposure to Nature?

   Most of us use technology throughout the day, from surfing the web through an iPod, to using our phones to surf, text, and message. We may use our devices to seek out beach scenes, videos taken on a hike, or videos of animals.

"Technology, in moderation, can enrich our lives and bring knowledge and happiness. Like anything else, if it is abused, it can cause harm. While it can be used to research nature, it is no substitute," said Andy Reid, technology expert.

Humans have been able to learn about nature through television for decades. This has given them the ability to see far-off, distant lands with nature they would otherwise never be able to experience. With so many opportunities for screen time, children are experiencing more and more virtual, screen-based nature than physical nature. However, humans need to experience nature for their physical and psychological well-being.

Environmental generational amnesia is partially responsible for an increased lack of interest and decreased perceived value in experiencing nature. Children view the conditions in which they grow up as largely normal. Children who grow up in polluted environments, for example, don't believe their environment is particularly polluted. They perceive their current conditions as typical, including experiencing nature via screens.

"Children may not realize that they are not getting the rewards and positive experiences from physically experiencing nature; in other words, they may believe they receive the same benefits whether experiencing nature or viewing nature on a screen," said Sandy Heath, behavioral expert.

from: *"Study of Stress Reduction and Creativity Considering Exposure to Nature, Both Virtual and Physical,"* Publication in development.

Go someplace quiet and spend at least 10 minutes clearing your head. Only after these 10 minutes, get out a piece of paper and sketch something you see; are thinking of; want to create or invent or imagine, etc. On the same sheet of paper, write a poem about your sketch; something you are thinking of; or quiet, wonder of the universe, lack of technology, etc.

Rules:

- Must be done all by hand.
- Must be original work (no copied poems or artwork).
- Ability does not count—draw like you are 5 years old!
- Draw in any medium you want: use pencil, pen, colored pencils, markers, crayons, watercolors.
- Poetic form does not matter: use rhyme, no rhyme, haiku, whatever you want it to be.
- It does not need to be elaborate; simple is fine.

5. Find a current job advertisement for an engineer on the Internet.
   (a) Identify the company for whom the engineer would work.
   (b) Give the Web address (the URL) for the advertisement.
   (c) Write 1-2 paragraphs describing the main responsibilities of this position. You may have to make some assumptions based on the description in the ad and the company placing the ad, but keep the assumptions realistic. Be concise!
   (d) Describe what the employer might look for on an applicant's resume. Be as specific as possible. The job ad should give valuable hints. Be clear and concise.

(e) Describe what someone applying for this job should wear to the interview. Include a picture from the Internet.

(f) Describe what this person would typically wear to work, and why that would be appropriate. Include a picture from the Internet.

6. Briefly describe the role an electrical engineer and a mechanical engineer would play in the design of the following items and systems. Be sure your answer is clear and effectively distinguishes electrical from mechanical:

   (a) Giant roller coaster

   (b) Generator for drinkable water on the International Space Station

   (c) A driverless automobile

   (d) Small drones or quadcopters

7. Briefly describe the role an engineer, a scientist, and a technologist would play in the design, manufacture, maintenance, and operation of the following items and systems. Be sure your answer is clear and effectively distinguishes engineering from technology and science:

   (a) Giant roller coaster

   (b) Small drones or quadcopters

8. For each of the following situations, indicate whether you think the action is ethical, unethical, or you are unsure. Do not read ahead; do not go back and change your answers.

Situation	Ethical	Unethical	Unsure
Speeding 5 miles per hour over the limit			
Killing a cockroach			
Speeding 15 miles per hour over the limit			
Killing a mouse			
Driving 90 miles per hour			
Driving 90 miles per hour taking an injured child to the hospital			
Killing a person			
Accepting a pen from a company trying to sell a new computer system to your company			
Using time at work to plan your friend's surprise party			
Accepting a wedge of cheese from a company trying to sell a new computer system to your company			
Taking a company pen home from work			

Situation	Ethical	Unethical	Unsure
Borrowing company tools			
Accepting a Hawaiian vacation from a company trying to sell a new computer system to your company			
Letting a friend who has been sick copy your homework			
Cheating on a test			
Shortcutting by walking across the grass behind a house			
Helping people with their homework			
Letting somebody copy your homework			
Shortcutting by walking through a house			

9. For each of the following situations, indicate how great you feel the need is in the world to solve the problem listed. Do not read ahead; do not go back and change your answers.

Situation	Urgent	Great	Somewhat	Little	None
Teaching those who cannot read or write					
Helping starving children in poor nations					
Helping people locked in prisons					
Helping to slow population growth					
Helping to reduce dependence on foreign oil					
Helping to reduce greenhouse gas emissions					
Helping people persecuted for sexual orientation					
Helping to reduce gun ownership					
Helping those who are mentally disabled					
Helping to supply laptops to poor children					
Helping prevent prosecution of "victimless" crimes					
Helping to end bigotry					

Situation	Urgent	Great	Somewhat	Little	None
Helping to prevent development of WMD (weapons of mass destruction)					
Helping prosecute hate crimes					
Helping to eliminate violence in movies					
Helping homeless people in your community					
Helping people with AIDS					
Helping people in warring countries					
Helping endangered species					

10. Find the IEEE Code of Ethics and the NSPE Code of Ethics online. Compare the IEEE Code of Ethics to the Fundamental Canons described in the NSPE Code of Ethics. What differences do you see? What similarities?

11. Research and explore the differences in career paths for someone employed in the engineering, technology, and science field.

    (a) What roles do apprenticeships, entry-level employment, and military service play in these careers? Do any of these experiences contribute to these careers more than the others?

    (b) What roles do community colleges, technical colleges, and universities play in these careers? Are any specifically required to pursue careers in each of these areas? .

12. Choose an ordinary device that you use in your home. Make a timeline of innovations and inventions for this product. What future changes, or innovations, might be made?

13. Look online or in your newspaper for an example of a local, national, or global problem that uses technology as part of its solution. Explain how technological literacy can help you discuss the problem intelligently.

14. Explain why it is unethical to infect or damage other people's computers with viruses or to hack into other computer systems to gather or change information.

15. (a) Find the Internet address for your state Environmental agency (for example, in Texas, this is the Texas Commission on Environmental Quality) and the federal Environmental Protection Agency.

    (b) Identify a material you may have at home or in the classroom (for example, unused paint or a specific cleaning fluid). Identify how you should safely dispose of this material according to each agency.

16. Identify a material classified as a toxin, another that is classified as a flammable solid, another classified as an oxidizing substance, and one classified as a corrosive material.

17. Identify one rule found on the OSHA website, and how it might affect the job of an electrical engineer. Use proper terminology as found in the OSHA website.

18. Identify three key procedures you are to follow for safety in your classroom. You may consult your school's website, other students, or your teacher for help with this question.

19. Identify 5 key differences between a group of people and a team.

20. Consider a time you were part of a team that was highly functional.

    (a) How did you specifically contribute as a member of the team to achieve a positive outcome?

    (b) How did you specifically work with others, or collaborate, to achieve a positive outcome?

    (c) What roles did you serve? Which did you serve most effectively?

21. Describe a time when you engaged as an active listener. What was the situation? What specifically made you an active listener?

22. Describe three things you are currently doing or could do this year to strengthen your leadership ability. For each, describe why this activity would help your leadership ability.

23. Select a discipline—either electrical, mechanical, chemical, or civil. Compare and contrast the career of someone working in one of these disciplines as an engineer, technologist, or scientist. What would each professional do that is similar among engineering, technology, and science? What would be some key differences?

24. What type of whiteboard problem might you be asked to solve for each of the following engineering disciplines? You do not necessarily need to solve the problem; just describe the problem itself.

    (a) Bioengineering

    (b) Chemical engineering

    (c) Civil and environmental engineering

    (d) Computer engineering

    (e) Electrical engineering

    (f) Mechanical engineering

    (g) Materials science

25. Identify a piece of equipment in your classroom. What could happen if this equipment is not properly cared for, handled, maintained, or stored?

26. Identify a precise measurement device in your classroom. This may, for example, be a scale, volumetric flask, or buret. Describe how to accurately and precisely use this in one paragraph.

27. Interview with a partner: You are interested in a career opening for an Environmental Engineer.

    (a) Write your own ad for this Environmental Engineering position. You may use information from online job advertisements and may craft some characteristics on your own. The ad should be no more than ½ page.

    Now, it's time to apply for this position. Swap job ads—you will apply for the position designed by your partner.

    (b) Write a one-page resume, targeting this specific ad as much as possible, and give this resume to your partner.

    (c) Prepare for a mock interview. Figure out a few key things you want to convey to the interviewer and a few questions you can ask your interviewer.

Pre-interview feedback:

(d) Critique the resume of this applicant. Do you see anything on your partner's resume that could be improved? Mark it on the resume.

The interview:

Each partner should interview the other.

(e) The interviewer: Be professional. Ask at least the following questions:

(i) Why are you interested in this position?

(ii) What qualifies you specifically for this position?

(iii) Do you have any questions for me?

(f) The interviewee: Be professional. Either dress for a professional interview or describe what you should wear as part of the reflection. Answer each question. Focus on being polite, concise, clear, and effective with your answers. Engage in active listening. After the interview, write or email a thank-you letter to your interviewer.

Reflection:

(g) What did you do well during the process?

(h) What could you have improved upon?

Be sure to describe your professionalism: Include what you wore (or should wear), your initial greeting, and your conduct and communication during the interview.

28. Why should there be a requirement to pass a basic safety test without error (with a score of 100%) before using any tools or shop equipment?

29. What are different occupational training and education requirements for three careers found in this chapter? Include engineer as one of those careers.

OVERVIEW

The systems engineering process is quite unlike a manufacturing process. A manufacturing process concentrates on repetitive activities that result in high quality outputs with minimum costs and time. The systems engineering process must begin by uncovering the real problems that need to be resolved, and by identifying the most likely or highest impact failures that can occur. Systems engineering involves finding complicated solutions to these problems.

CHAPTER 3

Engineering Problem Solving and Systems

TOPICS

All disciplines in engineering and science are chock full of problems to solve. These cover a myriad of topics; some are easy, some difficult, some tedious, and some straightforward. We often face new and previously unsolved problems. Some require highly accurate solutions, while others require just a "ballpark" value. Sometimes we face familiar problems, and sometimes the problem has absolutely no characteristics we have seen before.

What makes a good problem solver? The answer is simple: someone who can solve problems, often with an ease that belies the difficulty of the problem. Experience tells us that some people seem to have a problem-solving knack, while other, equally bright individuals have to work much harder at it. Determining what makes an effective problem solver is beyond the scope of this work. Suffice it to say that practice helps, and practicing certain techniques can help even more. It is the purpose of this chapter to suggest some simple tools that facilitate problem solving. You provide the rest—practice, practice, practice!

3.1 Types of Problems Engineers Face

NOTE:

Also mentioned in this section are some situations and conditions that inhibit your ability to solve problems—avoid these.

While not all problems are easily pigeonholed, there are several types that occur often in engineering situations. This is a good place to start. If the problem is written, read it over several times before deciding on the type. If it is given orally (as often happens in the work world) then just pause and repeat the basics to yourself before proceeding.

Some common problem types follow.

3.1.1 Ratio

We are interested in how a particular quantity varies when the conditions change.

OBJECTIVES

- Identify the inputs, processes, and outputs associated with technological systems
- Describe the procedure of *estimation,* and define the different types of estimation: estimation by *analogy,* by *aggregation,* and by *upper and lower bounds,* using *modeling*
- Explain how *inputs, outputs, processes,* and *subsystems* work together to produce a specific result
- Define the *open-loop system, closed-loop system,* and *feedback loops*
- Describe the three elements of a feedback loop: *sensor, comparator,* and *controller*
- Define the terms *prototype, scale model,* and *mathematical model*

Example 3-1

If a 2-meter-long horizontal beam deflects 6 centimeters when a load of 2*N* is hung on the end, how much will a beam of the same material and shape deflect under the same load if it is 4 meters long?

To quickly analyze this, we use the equation for beam deflection and write it twice, once for the initial situation and again for the new condition. Take the ratio of the two equations to find that many parameters will cancel, allowing an easy solution.

Equation of deflection:

$$\delta = \frac{P\ell^3}{3\,E\,I}$$

Here δ is the deflection, P is the load, ℓ is the length of the beam, E is Young's modulus, which depends on the material, and I is the moment of inertia. Now simply write this expression for the short beam (subscript s) and again for the longer beam (no subscript) as

$$\delta_s = \frac{(P\ell^3)_s}{(3\,E\,I)_s} \quad \text{and} \quad \delta = \frac{(P\ell^3)}{(3\,E\,I)}$$

Next, form the ratio of these two equations and note that since the load (P), the shape (I), and the material (E) are all the same, these three variables cancel, and we are left with

$$\frac{\delta}{\delta_s} = \left(\frac{\ell}{\ell_s}\right)^3 = \left(\frac{4\text{ m}}{2\text{ m}}\right)^3 = 8$$

This method is easy to use and shows clearly that there was no need to calculate the moment of inertia or to look up Young's modulus. So, the longer beam will deflect (8)(6 cm) = 48 cm.

3.1.2 Approximation

It is not always necessary to obtain an answer to four decimal places. In many situations in industry, an approximate answer (often called "quick and dirty" or "back of the envelope") is all that is needed. There is no need to spend the time or energy to provide more precision than necessary. When this is the case, it is possible to approximate in several ways: (1) use approximate input values, (2) use approximate equations, or (3) use an approximate geometry.

Example 3-2

How far (d) will a falling body fall in 4.8 seconds if it is initially falling at 0.03 miles per second?

1. We may not need precise input values—(assume a fall time of 5 seconds).
2. We may be able to use an approximate governing equation. The exact equation follows:

$$d = v_o t + \frac{1}{2} g t^2$$

By approximating its initial fall speed (v_o) as zero, we can use only the second term. Knowing that $g = 9.8\text{ m/s}^2$ (or approximately 10 m/s^2) we quickly estimate that the distance fallen will be:

$$D = (1/2)(10\text{ m/s}^2)(5\text{ s})^2 = 125\text{ m}$$

3.1.3 Parametric

parametric

Often in design, it is required to study a wide range of input parameters (hence a **parametric** analysis), as well as the corresponding range of outputs. This can often conveniently be done using spreadsheets, graphical representations, and programming.

Example 3-3

How does the range of a rocket vary with the initial speed (never to be greater than 600 miles per second) and launch angle?

Develop a spreadsheet (and corresponding graph) which shows the range for launch angles of 0, 5, 10, 15, 20, . . . 90 degrees for launch speeds of 100, 200, 300, . . . 600 m/s.

3.1.4 Ideal

ideal problems

Sometimes, particularly with straightforward problems, we are required to simply calculate a single, exact answer. We call these **ideal problems.**

Example 3-4

How far does a vehicle travel if it moves at 36 miles per hour for 15 minutes?

distance – (speed)(time)
distance = (36 mph)(0.25 h) = 9 miles

3.1.5 Discussion

In conceptual studies, it is often important to simply develop a better understanding of how various parameters interact. In such situations, we may not need to obtain a numerical answer but rather to determine if certain variables are critical or if they may be neglected by discussing the overall problem statement. Such an analysis may then lead to a new design, quantification of material needed, or selection of parameters to be studied further.

Example 3-5

Does the diameter of a bridge pier affect the amount of scour that occurs around its base during flooding conditions?

It is reasonable to assume that if larger piers are used, then fewer are needed to support the bridge? If more but smaller piers are used, how much of the cross section of the river is blocked compared to fewer but larger piers? We must find information relating the total bridge load on the piers and their strength (that would include the diameter, material, etc.). As more of the river cross section is blocked, the speed of the water between the piers increases, which would increase the scour (erosion). Once we have examined these variables, it may be possible to tell the bridge designer that it is better to use fewer large piers for support (or more smaller piers).

The problem solver first decides on problem type, picking from the five generic types:

Ratio	Compare a quantity with a known quantity
Approximation	Estimate values/use approximate equations
Parametric	Input parameters are studied over ranges
Ideal	Exact solution
Discussion	Talk about how parameters affect each other

This has the easily remembered acronym of RAPID and often provides a rapid means to get started solving a problem.

The RAPID process can help classify problems. Even when the type of problem type is known, different people are likely to approach solving problems in different ways. There are, however, some procedures and techniques that can help in developing a generally successful problem-solving approach. Although it is not possible to write down a specific recipe that will always work, some broad approaches will help. One approach to solving problems is presented here.

QUESTION

Define the term *problem solving. Problem solving is used when products or processes fail, so corrective action can be taken to prevent further failures. It can also be applied to a product or process prior to an actual fail event, i.e., when a potential problem can be predicted and analyzed, and mitigation applied so the problem never actually occurs.*

QUESTION

Define the term *sketch. A sketch (from Greek schedios, "done extempore") is a rapidly executed freehand drawing that is not usually intended as a finished work.*

ACTIVITY

Ask the students to sketch a simple floor plan of their bedroom at home on one half sheet of paper. Label it "before." Have the students sketch a simple floor plan of their bedroom at home–with upgrades. Label it "after."

3.2 Structured Problem Solving

SOLVEM

One problem-solving approach has been given the acronym **SOLVEM**:

Sketch
Observations or Objectives
List
Variables and
Equations
Manipulation

Note that this approach is equally useful for problems involving estimation and more precise calculations. Each step is described next.

3.2.1 Sketch

Figure 3-1 illustrates how a drawing can help you visualize a problem. In sketching a problem, you are subconsciously thinking about it. Be sure to draw the diagram large enough so that everything is clear, and label the things that you know about the problem in the diagram. For some problems, a before-and-after set of diagrams may be helpful. In very complex problems, you can use intermediate diagrams or subdiagrams as well.

Comprehension Check 3-1

We use SOLVEM to complete this problem in the Comprehension Checks in this chapter. Create a sketch for the following problem.

Calculate the mass in kilograms of gravel stored in a rectangular bin 18.5 feet by 25.0 feet. The depth of the gravel bin is 15 feet, and the density of the gravel is 97 pound-mass per cubic foot.

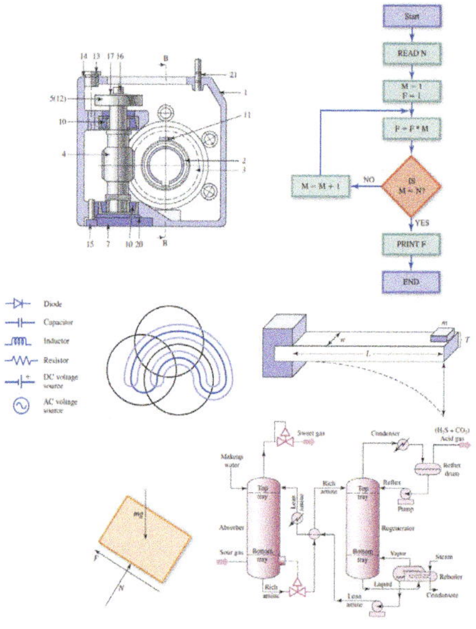

Figure 3-1 Sketches for seven problems.

3.2.2 Objectives, Observations

These can be in the form of simple statements, questions, or anything else that might acquaint you with the problem at hand. It often helps to divide your observations and objectives into several categories. Some of the easiest to remember follow:

NOTE: An essential part of stating observations is to state the objectives. When making observations, do not forget that you have five senses.

- Objective to be achieved
- Observations about the problem geometry (size, shape, etc.)
- Observations about materials and material properties (density, hardness, etc.)
- Observations about parameters not easily sketched (temperature, velocity, etc.)
- Other miscellaneous observations that might be pertinent

You will almost always find after writing down some observations that you know more about the problem than you originally thought

Here are some typical examples:

Objectives

- Find the velocity, force, flow rate, time, pressure, etc., for a given situation.
- Profitably market the device for less than $25.
- Fit the device into a 12-cubic-inch box.

Observations

Problem Geometry

- The liquid has a free surface.
- The submerged plate is rectangular.
- The support is vertical.
- The tank is cylindrical.
- The cross-sectional area is octagonal.
- The orbit is elliptical.

Materials and Material Properties

- The gate is steel.
- The coefficient of static friction is 0.6.
- The specific gravity is 0.65.
- Ice will float in water.
- The alloy superconducts at 97 kelvin.
- The alloy melts at 543 degrees Fahrenheit.

Other Parameters

- If depth increases, pressure increases.
- If temperature increases, resistance increases.
- The flow is steady.
- The fluid is a gas and is compressible.
- The pulley is frictionless.
- The magnetic field is decreasing.
- Temperature may not fall below 34 degrees Fahrenheit.

Miscellaneous

- The force will act to the right.
- Gravity causes the ball to accelerate.
- The sphere is buoyant.
- Drag increases as the speed increases.

Remember to include those quantities whose value is zero! Often such quantities are hidden with terms such as:

- Constant (implies derivative = 0)
- Initially (at time = 0)
- At rest (no motion)
- Dropped (no initial velocity)
- At the origin (at zero position)
- Melts or evaporates (changes phase, temperature is constant)

NOTE:

Quantities whose value is zero contain valuable information about the problem.

QUESTION

Define the term *observation.*

Observation (from Latin onservare, "respectful adherence to rules") is the action or process of observing something or someone carefully in order to gain information.

ACTIVITY

Ask the students to observe the desk or table at which they are seated. Ask them to record the problem geometry (size, shape, etc.), the material (steel, ceramic, etc.), and the material properties (density, hardness, etc.).

QUESTION

Define the terms *variable* and *constant*.

A variable is something that may change in value. Constants are quantities whose values do not change.

QUESTION

Ask the students to make a two-column list. Label one column "variable" and the other "constant." Then ask them to observe and label each of the following items:

- The temperature in the classroom/lab *(variable)*
- The amount of light in the classroom/lab *(variable)*
- The number of humans in the classroom/lab *(variable)*
- The number of floor tiles or ceiling tiles visible in the classroom/lab *(constant)*
- The number of electrical outlets in the classroom/lab *(constant)*
- The gravity in the classroom/lab *(constant)*

Comprehension Check 3-2

State the objective and any relevant observations for the following problem.

Calculate the mass in kilograms of gravel stored in a rectangular bin 18.5 feet by 25.0 feet. The depth of the gravel bin is 15 feet, and the density of the gravel is 97 pound-mass per cubic foot.

Finally, one of the most important reasons to make many observations is that you often will observe the "wrong" thing. For example, write down things as you read this:

A bus contains 13 passengers.
At the first stop, four get off and two get on.
At the next stop, six get off and one gets on.
At the next stop, nobody gets off and five get on.
At the next stop, eight get off and three get on.
At the next stop, one gets off.
At the last stop, four get off and four get on.

After putting your pencil down and without looking again at the list, answer the question given below the text box discussing "The Importance of Observations."

THE IMPORTANCE OF OBSERVATIONS

An excerpt adapted from The Crooked Man *by Sir Arthur Conan Doyle*

Dr. Watson writes: I looked at the clock. It was a quarter to twelve. This could not be a visitor at so late an hour. A patient, evidently, and possibly an all-night sitting. With a wry face I went out into the hall and opened the door. To my astonishment, it was Sherlock Holmes who stood upon my step.

"Ah, Watson, I hoped that I might not be too late to catch you."

"My dear fellow, pray come in."

"You look surprised, and no wonder! Relieved, too, I fancy! Hum! You still smoke the Arcadia mixture of your bachelor days, then! There's no mistaking that fluffy ash upon your coat. It's easy to tell that you've been accustomed to wear a uniform, Watson; you'll never pass as a pure-bred civilian as long as you keep that habit of carrying your handkerchief in your sleeve. Could you put me up for the night?"

"With pleasure."

"You told me that you had bachelor quarters for one, and I see that you have no gentleman visitor at present. Your hat-stand proclaims as much."

"I shall be delighted if you will stay."

"Thank you. I'll find a vacant peg, then. Sorry to see that you've had the British workman in the house. He's a token of evil. Not the drains, I hope?"

"No, the gas."

"Ah! He has left two nail marks from his boot upon your linoleum just where the light strikes it. No, thank you, I had some supper at Waterloo, but I'll smoke a pipe with you with pleasure."

I handed him my pouch, and he seated himself opposite to me, and smoked for some time in silence. I was well aware that nothing but business of importance could have brought him to me at such an hour, so I waited patiently until he should come round to it.

"I see that you are professionally rather busy just now."

"Yes, I've had a busy day. It may seem very foolish in your eyes, but I really don't know how you deduced it."

"I have the advantage of knowing your habits, my dear Watson. When your round is a short one you walk, and when it is a long one you use a hansom (a carriage). As I perceive that your boots, although used, are by no means dirty, I cannot doubt that you are at present busy enough to justify the hansom."

"Excellent!"

"Elementary. It is one of those instances where the reason can produce an effect which seems remarkable to his neighbor, because the latter has missed the one little point which is the basis of the deduction."

While we cannot all be as observant as Sherlock Holmes, we can improve our powers of observation through practice. This will pay dividends as we seek to be engineers with high levels of analytical skills.

Source: AF Fotografie/Alamy

QUESTION: For the bus problem, how many stops did the bus make?
The lesson here is that often we may be observing the wrong thing.

3.2.3 List of Variables and Constants

Go over the observations previously determined and list the variables that are important. It may help to divide the list into several broad categories—those related to the geometry of the problem, those related to the materials, and a properties category—although for some types of problems those categories may not be appropriate. Include in your list the written name of the variable, the symbol used to represent the quantity, and if the value of the variable is known, list the numeric value, including units. If a value is a constant you had to look up, record where you found the information.

Initial and Final Conditions

- Initial temperature (T_0) 60 [°F]
- Initial radius (r_i) 5 [cm]
- Mass of the object (m) 23 [kg]

Constants

- Acceleration of gravity (g) 32.2 [ft/s^2]
- Ideal gas constant (R) 8314 [(Pa L)/(mol K)]

Geometry

- Length of beam (L) [m]
- Cross-sectional area of a pipe (A) [cm^2]
- Volume of a reactor vessel (V) [gal]

Materials

- Steel
- Polyvinyl chloride (PVC)
- Plasma
- Gallium arsenide
- Medium-density balsa wood

Properties

- Dynamic viscosity of honey (μ) 2500 [cP]
- Density of PVC (p) 1380 [kg/m^3]
- Spring constant (k) 0.05 [N/m]
- Specific gravity (SG) 1.34

Comprehension Check 3-3

Create a list of variables and constants for the following problem.

Calculate the mass in kilograms of gravel stored in a rectangular bin 18.5 feet by 25.0 feet. The depth of the gravel bin is 15 feet, and the density of the gravel is 97 pound-mass per cubic foot.

TEACHING TIP

Understanding what is given in a problem is one of the most critical steps in the problem-solving process. Emphasize that understanding your knowns and unknowns will help them form appropriate assumptions to fill in necessary blanks to solve the problem.

ENGAGE

Hooke's law is a principle of physics that states that the force *(F)* needed to extend or compress a spring by some distance *(x)* is proportional to that distance. The law is named after 17th century British physicist Robert Hooke. He first stated the law in 1660. Hooke's law is only a first-order linear approximation to the real response of springs and other elastic bodies to applied forces. It must eventually fail once the forces exceed some limit, since no material can be compressed beyond a certain minimum size, or stretched beyond a maximum size, without some permanent deformation or change of state. Many materials will noticeably deviate from Hooke's law well before those elastic limits are reached.

QUESTION

Ask the students for real-world examples of Hooke's law in action. *(Answers could include: slinky, clickable pen spring, electric garage door opener spring, a simple rubber band, trampoline springs, etc.)*

DISCUSSION

Basic rules for manipulating equations:

1. You can add, subtract, multiply, and divide by anything, as long as you do the same thing on both sides of the equals sign. In an equation, the equals sign acts like the fulcrum of a balance: If you add 5 of something to one side of the balance, you have to add the same amount to the other side to keep the balance steady. The same thing applies to an equation: Doing the same operation on both sides keeps the meaning of the equation from changing.
2. To move or cancel a quantity or variable on one side of the equation, perform the "opposite" operation with it on both sides of the equation.

3.2.4 Equations

Only after completing the preceding steps (S-O-L-V) should you begin to think about the equations that might govern the problem. It is useful to make a list of the pertinent equations in a broad sense before listing specific expressions. For example:

- Conservation of energy
- Conservation of mass
- Conservation of momentum
- Frequency equations
- Ideal gas law
- Newton's laws of motion
- Stress-strain relations
- Surface areas of geometric solids
- Volumes of geometric solids
- Work, energy relations

You may need "subequations" such as:

- Distance = (velocity) (time)
- Energy = (power) (time)
- Force = (pressure) (area)
- Mass = (density) (volume)
- Voltage = (current) (resistance)
- Weight = (mass) (gravity)

For an equation, list the broad category of the equation (Hooke's law) and then the actual expression ($F = kx$) to help with problem recognition.

Do not substitute numerical values of the parameters into the equation right away. Instead, manipulate the equation algebraically to the desired form.

Comprehension Check 3-4

Create a list of equations for the following problem.

Calculate the mass in kilograms of gravel stored in a rectangular bin 18.5 feet by 25.0 feet. The depth of the gravel bin is 15 feet, and the density of the gravel is 97 pound-mass per cubic foot.

3.2.5 Manipulation

Most of the time, you need to manipulate pertinent equations before you can obtain a final solution. ***Do not substitute numerical values of the parameters into the equation right away.*** Instead, manipulate the equation algebraically to the desired form. Often you will discover terms that will cancel, giving you a simpler expression to deal with.

By doing this, you will:

- Obtain general expressions useful for solving other problems of this type
- Be less likely to make math errors
- Be able to judge whether your final equation is dimensionally consistent
- Better understand the final result

The SOLVEM acronym does not contain a word or step for numerical solution. In fact, this process helps you analyze the problem and obtain an expression or procedure so that you can find a numerical answer. The thought here is that engineers need training to be able to analyze and solve problems. If you can do everything except substitute numbers, you are essentially finished. As an engineer, you will "be paid the big bucks" for analysis, not for punching a calculator.

3.3 Estimation Techniques

Enrico Fermi was a Nobel laureate and one of many brilliant scientists and engineers involved in the Manhattan Project, which developed the first nuclear weapons during World War II. He wrote the following after witnessing the first test of an atomic bomb, called the Trinity Test (Figure 3-2). Of particular note is the final paragraph

> *On the morning of the 16th of July, I was stationed at the Base Camp at Trinity in a position about ten miles from the site of the explosion.*
>
> *The explosion took place at about 5:30 A.M. I had my face protected by a large board in which a piece of dark welding glass had been inserted. My first impression of the explosion was the very intense flash of light, and a sensation of heat on the parts of my body that were exposed. Although I did not look directly towards the object, I had the impression that suddenly the countryside became brighter than in full daylight.*
>
> *I subsequently looked in the direction of the explosion through the dark glass and could see something that looked like a conglomeration of flames that promptly started rising. After a few seconds the rising flames lost their brightness and appeared as a huge pillar of smoke with an expanded head like a gigantic mushroom that rose rapidly beyond the clouds probably to a height of the order of 30,000 feet. After reaching its full height, the smoke stayed stationary for a while before the wind started dispersing it.*
>
> *About 40 seconds after the explosion the air blast reached me. I tried to estimate its strength by dropping from about six feet small pieces of paper before, during and after the passage of the blast wave. Since at the time there was no wind I could observe very distinctly and actually measure the displacement of the pieces of paper that were in the process of falling while the blast was passing. The shift was about 2½ meters, which, at the time, I estimated to correspond to the blast that would be produced by ten thousand tons of T.N.T.*

Citation: U.S. National Archives, Record Group 227, OSRD-S1 Committee, Box 82 folder 6, "Trinity."

Before the test, no one knew what would happen. Speculation among the many people involved concerning the results of the test ranged from nothing (no explosion at all) to setting the planetary atmosphere on fire and destroying all life on Earth. When all the data from the blast were analyzed, the true strength of the blast was calculated to be 19 kilotons. By simply observing the behavior of falling bits of paper 10 miles from ground zero, Fermi's estimation of 10 kilotons was in error by less than a factor of 2.

Figure 3-2 The first atomic bomb test, Alamogordo, New Mexico.

After the war, Fermi taught at the University of Chicago, where he was noted for giving his students problems in which so much information was missing that a solution seemed impossible. Such problems have been named **Fermi problems**, and in general, they require the

Fermi problems

ENGAGE

Visit the Science Olympiad website about Fermi Questions with your students. Start with the tutorial and read the directions as a group.

URL

https://www.soinc.org/fermi-questions-c

Instruct the students to then use the link to practice these types of problems.

ENGAGE

Visit the National Aeronautics and Space Administration (NASA) website with your students. Navigate to the page explaining Enrico Fermi's piano problem.

URL

https://www.grc.nasa.gov/www/k-12/Numbers/Math/Mathematical_Thinking/fermis_piano_tuner.htm

Ask your students to create a simple diagram that will define the problem and produce the solution. *(Be sure students include relevant data in image form and the formulas needed to solve the Fermi problem.)*

person considering them to determine an answer with far less information than would really be necessary to calculate an accurate value. Engineers are often faced with solving problems for which they do not have all the information. They must be adept at making initial estimates. This skill helps them identify critical information that is missing, develop their reasoning skills to solve problems, and recognize what a "reasonable" solution will look like.

Most practical engineering problems are better defined than Fermi problems and can be estimated more easily and accurately. The following are just a few examples of real engineering problems. See if you can estimate answers for these problems. These should be done without reference to the Web or any other source of information.

- How many cubic yards of concrete are needed to pave 1 mile of interstate highway (two lanes each direction)?
- How many feet of wire are needed to connect the lighting systems in an automobile?
- What is the average flow rate in gallons per minute of gasoline moving from the fuel tank to the fuel injectors in an automobile cruising at highway speed?

An accomplished engineer knows the answer to most problems before doing any calculations. This does not mean an answer to three significant figures, but a general idea of the range of values that would be reasonable. For example, if you throw a baseball as high as possible, how long will it take for the ball to hit the ground? Obviously, an answer of a few milliseconds is unreasonable, as is several months. Several seconds seems more realistic. When you actually do a calculation to determine this time, you should ask yourself, "Is my answer reasonable?"

Sample Fermi Problems

- Estimate the total number of hairs on your head.
- Estimate the number of drops of water in all of the Great Lakes.
- Estimate the number of piano tuners in New York City.

Plastic resin (shown in the picture, in pellet form) is used in many types of manufacturing methods including injection molding, extrusion, and blow molding. Injection molding describes the process by which resin is melted and "injected" into a closed mold, then cooled forming the final part. Extrusion is a continuous process by which resin is melted and pushed through an open mold to create shapes like a pipe or rod.

Most cars today use plastic fuel tanks made by blow molding. In blow molding, the resin is melted and pushed through an extrusion head that forms the plastic into a hollow shape. The shape is then pressurized with air and cooled in a mold to form the part. Most fuel tanks are formed with six layers of various plastic to increase toughness and eliminate permeation of the fuel through the tank.

Photo courtesy of E. Stephan

Example 3-6

Every year, numerous people run out of fuel while driving their vehicle on the road. Determine how many gallons of gasoline are carried to vehicles with empty fuel tanks each year in the United States so that the vehicles can be driven to the nearest gas station. Do this without reference to any other material, such as the Internet or reference books.

Estimations

In almost all cases, the first step is to estimate unknown pieces of information that are not available. In general, there are numerous paths to a solution, and different people may arrive at different answers. Often, the answer obtained is only accurate to within an order of magnitude (a factor of 10) or less. Nonetheless, such problems can provide valuable insight not only into the problem itself but also into the nature of problem solving in general.

Remember, these are estimates, not accurate values. Someone else making these estimates might make different assumptions.

- *Number of people in the United States: 500,000,000 persons.*
- *Fraction of people in the United States that drive: seven drivers per 10 persons.*
- *Average times a person runs out of gas per year: one "out of gas" per 4 years per driver.*
- *Fraction of "out of gas" incidents in which gas is brought to the car (rather than pushing or towing the car to a station): 24 "bring gas to car" per 25 "out of gas."*
- *Average amount of gas carried to car: 1.5 gallons per "bring gas to car."*

Calculation

This is where you combine your estimates to arrive at a solution. In the process, you may realize that you need further information to complete the computation.

$$\text{Drivers in United States} = (5 \times 10^8 \text{people})\left(\frac{7\,\text{drivers}}{10\,\text{people}}\right) = 3.5 \times 10^8 \text{drivers}$$

$$\text{Number out of gas per year} = (3.5 \times 10^8 \text{drivers})\left(\frac{1\ \text{out of gas}}{(4\ \text{years})(1\ \text{driver})}\right)$$
$$= 8.75 \times 10^7 \frac{\text{out of gas}}{\text{year}}$$

$$\text{Number bring gas to car per year} = \left(\frac{8.75 \times 10^7 \text{out of gas}}{\text{year}}\right)\left(\frac{24\ \text{bring to car}}{25\ \text{out of gas}}\right)$$
$$= 8.4 \times 10^7 \frac{\text{bring gas to car}}{\text{year}}$$

$$\text{Amount of gas to cars} = \left(\frac{8.4 \times 10^7 \text{bring gas to car}}{\text{year}}\right)\left(\frac{1.5\ \text{gallons of gas}}{\text{bring gas to car}}\right)$$
$$= 1.26 \times 10^8 \frac{\text{gallons}}{\text{year}}$$

Thus, we have estimated that about 125 million gallons of gas are taken to "out of fuel" vehicles each year in the United States. It would be perfectly valid to give the answer as "about 100 million" gallons, since we probably have only about one significant digit worth of confidence in our results, if that.

A few things to note about this solution:

- *"Units" were used on all numerical values, although some of these units were somewhat contrived (e.g., "out of gas") to meet the needs of the problem. Keeping track of the units is critical to obtaining correct answers and will be highly emphasized, not only in this text but also throughout your engineering education and career.*
- *The units combine and cancel according to the regular algebraic rules. For example, in the first computation, the unit "persons" appeared in both the numerator and the denominator, and thus canceled, leaving "drivers."*
- *Rather than the computation being combined into one huge string of computations, it was broken into smaller pieces, with the results from one step used to compute the next step. This is certainly not an immutable rule, but for long computations it reduces careless errors and makes it easier to understand the overall flow of the problem.*
- *Fermi problems, like other engineering work, can benefit from the perspective of others. In this case, it is likely that if a team were working on this problem, someone on the team would know that the population of the United States is closer to 300,000,000.*

ENGAGE

Visit the crowd-sourcing website entitled "Tank on Empty" with your students. Click on the links for the most active models.

URL
http://www.tankonempty.com

QUESTION

According to the results, on average, which vehicle will travel the furthest when the warning light comes on indicating low fuel? Which vehicle will go the shortest distance on empty?

QUESTION

Define the term *estimation.* Estimation (Latin *aestimare*, "determine, appraise") is the process of finding an estimate, or approximation, which is a value that is usable for some purpose even if input data may be incomplete, uncertain, or unstable. The value is nonetheless usable because it is derived from the best information available.

You can make simple assumptions if you're dealing with common objects, such as the height of one story when estimating the size of a building.

QUESTION

Define the term *order of magnitude. If numbers differ by one order of magnitude, x is about ten times different in quantity than y. If values differ by two orders of magnitude, they differ by a factor of about 100. Two numbers of the same order of magnitude have roughly the same scale: The larger value is less than ten times the smaller value.*

ENGAGE

Visit the Molecular Expressions website, created by Florida State University and the National High Magnetic Field Laboratory, with your students. Navigate to the page on powers of ten.

URL
http://micro.magnet.fsu.edu/primer/java/scienceopticsu/powersof10/index.html

Notice how each picture is actually an image of something that is 10 times bigger or smaller than the one preceding or following it. The number that appears on the lower right just below each image is the size of the object in the picture.

3.3.1 General Hints For Estimation

As you gain more knowledge and experience, the types of problems you can estimate will become more complicated. Here we give you a few hints about making estimates.

- **Try to determine the accuracy required.**
  - Is order of magnitude enough? Is it ±25%?
  - What level of accuracy is needed to calculate a satellite trajectory?
  - What level of accuracy is needed to determine the amount of paint needed to paint a specified classroom?

orders of magnitude

The term **orders of magnitude** is often used when comparing things of very different scales, such as a small rock and a planet. By far the most common usage refers to factors of 10; for example, three orders of magnitude refer to a difference in scale of $10^3 = 1000$. If we wanted to consider the order of magnitude between 10,000,000 and 1000, we would calculate the logarithm of each value (log(10,000,000) = 7 and log(1000) = 3), thus there are 7 – 3 = 4 orders of magnitude difference between 10,000,000 and 1000.

- Remember that a **"ballpark" value is often good enough** for an input parameter.
  - What is the square footage of a typical house?
  - What is the maximum high temperature to expect in Dallas, Texas, in July?
  - What is the typical speed of a car on the highway?
- Always ask yourself **if it is better to err on the high side or the low side.**
  - **Safety and practical considerations**. Will a higher or lower estimate result in a safer or more reliable result?
    - If estimating the weight a bridge can support, it is better to err on the low side, so that the actual load it can safely carry is greater than the estimate.
    - For the bridge mentioned above, if estimating the load a single beam needs to support, it is better to err on the high side, thus giving a stronger beam than necessary. Be sure you understand the difference between these two points.
- **Estimate improvement.** Can the errors cancel each other?
  - If estimating the product of two numbers, if one of the terms is rounded low, the other should be rounded high to counteract for the lower term.
  - If estimating a quotient, if you round the numerator term on the low side, should the denominator term be rounded low or high?
- Do not get bogged down with **second-order** or **minor effects.**
  - If estimating the mass of air in the classroom, do you need to correct for the presence of furniture?
  - In most instances, can the effect of temperature on the density of water be neglected?

The best way to develop your ability to estimate is through experience. An experienced painter can more easily estimate how much paint is needed to repaint a room because experience will have taught the painter such things as how many coats of one paint color it will take to paint over another, how different paint brands differ in their coverage, and how to estimate surface area quickly. In *Outliers: The Story of Success,* Malcolm Gladwell provides examples from diverse career pathways that demonstrate 10,000 hours of practice are required to develop world-class expertise in any area. Fortunately for aspiring engineers, much of this expertise can be developed starting at a young age and outside of formal schooling. For example, how many hours have you spent observing the effects of gravity? Of course, some important engineering concepts stem from phenomena that are not so easily observed, and some lend themselves to misinterpretation. As a result, it helps to have a systematic approach to developing estimates—particularly where we have less experience to guide us. Estimating an approximate answer of a calculation including known quantities is a valuable skill, such as approximating the square root of 50 as about 7, approximating the value of pi as 3 for quick estimates, etc. These mathematical approximations, however, assume that you have all the numbers to begin with and that you can use shortcuts to estimate the precisely calculated answer to save time or as a check against your more carefully calculated answer. Estimation is discussed here in a broader sense—estimating quantities that cannot be known, are complicated to measure, or are otherwise inconvenient to obtain. It is in these cases that the following strategies are recommended.

3.3.2 Estimation By Analogy

One useful strategy for estimating a quantity is by comparison to something else we have measured previously or otherwise know the dimension of. The best way to prepare for this approach to estimation is to learn a large number of comparison measures for each type of quantity you might wish to estimate. Each of these comparison measures becomes an anchor point on that scale of measurement. This book provides some scale anchors for various measurable quantities—particularly in the case of power and energy, concepts with which many people struggle.

Example 3-7

Estimate the size of a laptop computer using analogy.

Laptop computers come in different sizes, but it is not difficult to estimate the size of a particular laptop. Laptops were first called notebook computers—a good starting estimate would be to compare the particular laptop to notebook paper, which is 8.5 inches by 11 inches in the United States.

Example 3-8

Estimate the size of an acre and a hectare of land using analogy.

American football field—playing area is 300 feet by 160 feet—approximately 50,000 square feet. An acre is 43,560 square feet. Using these data, we have a better sense of how much land an acre is—about 90% of the size of the playing area of an American football field. Soccer fields are larger, but vary in size. The largest soccer field that satisfies international guidelines would be about 2 acres.

A hectare, or 10,000 square meters, is equivalent to 108,000 square feet and is more than two acres—about the maximum size of the pitch in international rugby competition.

3.3.3 Estimation By Aggregation

Another useful strategy is to estimate the quantity of something by adding up an estimate of its parts. This can involve multiplication in the case of a number of similarly sized parts, such as estimating the size of a tile by comparing it to your foot (estimation by analogy), counting the number of floor tiles across a room, and multiplying to estimate the total length of the room. In other cases, aggregation may involve adding together parts that are estimated by separate methods. For example, to estimate the volume of a two-scoop ice cream cone, you might estimate the volume of the cone and then separately estimate the volume of each scoop, assuming they are each spheres.

Example 3-9

Estimate by aggregation the amount of money students at your school spend on pizza each year.

Ask students around you how often they purchase a pizza and how much it costs;

Convert this estimate into a cost per week;

Multiply your estimate by the number of weeks in an academic year;

Multiple that result by the number of students at your school.

QUESTION

Define the phrase *estimation by analogy.*

Analogous estimating is a technique that uses the values of parameters from historical data as the basis for estimating the same or similar parameter for a future activity. Examples of parameters include scope, cost, duration, or measures of scale such as size, weight, or complexity.

QUESTION

Define the phrase *estimation by aggregation.*

Aggregation is a group or mass of distinct or varied things, persons, etc. Estimating by aggregation is a technique of estimating that involves adding up the parts.

QUESTION

Would it be possible to estimate the length and width of your classroom based on the size of a single item?

Ask your students to study the classroom floor, walls, and ceiling. Could students estimate the square footage or volume of the room?

ENGAGE

Explain the phrase *estimation by upper and lower bounds* to your students. *This requires the engineers to prepare for maximums and minimums in their designs. Simple examples of upper and lower bounds include the maximum amount of weight something can hold before breaking and the minimum width something can be to fit a certain space.*

3.3.4 Estimation By Upper and Lower Bounds

An important part of estimating is keeping track of whether your estimate is high or low. In the earlier example of estimating the volume of a two-scoop ice cream cone, we would have over-estimated, because one of the scoops of ice cream is pressed inside the cone. The effect of pressing the scoops together is not that important, because the same amount of ice cream is still there, but if the scoop is pressed into an ellipsoid, it may be difficult to estimate the original radius of the scoop.

Engineers frequently make conservative estimates, which consider the worst-case scenario. Depending on the situation, the worst case may be a lower limit (such as estimating the strength of a structure) or an upper limit (such as estimating how much material is needed for a project).

Example 3-10

If you are to estimate how many gallons of paint are needed to paint the room you are in, what assumptions will you need to make? Where will you need to make assumptions to ensure that you have enough paint without running out?

In estimating the wall area, you should round up the length and height.

Noting that paint (for large jobs) is sold in five-gallon pails, you will want to round your final estimate to the next whole five-gallon pail.

Close estimates allow for subtracting 21 square feet per doorway (if the doors are not being painted the same color). In making a rough estimate, if there are not a lot of doorways, it would be conservative to leave in the door area.

3.3.5 Estimation Using Modeling

mathematical models

In cases that are more complicated or where a more precise estimate is required, **mathematical models** and statistics might be used. Sometimes dimensionless quantities are useful for characterizing systems, sometimes modeling the relationship of a small number of variables is needed. At other times, extrapolating even a single variable from available data is all that is needed to make an estimate.

Example 3-11

You would like to enjoy a bowl of peas, but they are too hot to eat. Spreading them out on a plate allows them to cool faster. Describe why this happens and devise a model of how much faster the peas on the plate will cool. Mice have a harder time keeping warm than elephants. Explain how this is related to the bowl of peas. How does this relate to the fact that smaller animals have higher heart rates? Canaries and hummingbirds can have heart rates of 1200 beats per minute, whereas human heart rates should not exceed 150 beats per minute even during exercise.

The peas cool faster when spread out because of the increase in surface area. The ratio of surface area (proportional to cooling) to volume (proportional to the heat capacity for a particular substance) is therefore important. Similarly, smaller animals have a harder time staying warm because they have a higher ratio of surface area to volume. The higher heart rate is needed to keep their bodies warm. This also relates to why smaller animals consume a much larger amount of food compared to their body mass.

Example 3-12

A large sample of sunflower seeds is collected and their lengths are measured. Using that information, estimate the length of the longest sunflower seed you are likely to find if you measure one billion seeds.

Given a large sample, its average and standard deviation can be calculated. Assuming that the length of sunflower seeds is normally distributed, the one-in-a-billion largest sunflower seed would be expected to be six standard deviations greater than the sample average.

3.4 Representing Final Results

NOTE:

Don't plug values into the equation until the final step.

When you have completed all the steps to SOLVEM, plug in values for the variables and constants and solve for a final answer. Be sure to use reasonableness. The final answer should include both a numeric value and its unit. In addition, it is often useful to write a sentence describing how the answer meets the objectives. Box your final answer for easy identification.

Repeated use of SOLVEM can help you develop a better gut-level understanding about the analysis of problems by forcing you to talk and think about the generalities of the problem before jumping in and searching for an equation into which you can immediately substitute numbers.

Comprehension Check 3-5

Calculate the mass in kilograms of gravel stored in a rectangular bin 18.5 feet by 25.0 feet. The depth of the gravel bin is 15 feet, and the density of the gravel is 97 pound-mass per cubic foot.

3.4.1 Significant Figures

significant figures

decimal places

Significant figures or sig figs are the digits considered reliable as a result of measurement or calculation. This is not to be confused with the number of digits or decimal places. The number of **decimal places** is simply the number of digits to the right of the decimal point. Example 3-13 illustrates these two concepts.

Example 3-13

NOTE:

Decimal places is the number of digits to the right of the decimal point.

Significant figures are the digits considered reliable.

Number	Decimal Places	Significant Figures
376	0	3
376.0	1	4
376.908	0	6
3,760,000	0	3
3,760,000.	0	7
0.376	3	3
0.37600	5	5
0.0037600	7	5
376×10^{-6}	0*	3

There is no universal agreement concerning whether numbers in scientific or engineering notation should be considered to have the number of decimal places indicated in the mantissa (as shown), or the number that would be present if the number were written out in standard decimal notation (6 in the last example).

For those who have never seen the term "mantissa" (or have forgotten it) —the two parts of a number expressed in either scientific or engineering notation are the mantissa and the exponent. The mantissa is the part that gives the numerical values of the significant figures; the exponent specifies the location of the decimal point, thus the magnitude of the overall number. In the last row of the preceding table, the mantissa is 376 and the exponent is −6. The mantissa can also contain a decimal point, for example, 3.76×10^{-4}: in this case the mantissa is 3.76 and the exponent is −4.

ENGAGE

Discuss the difference between two-dimensional and three-dimensional figures. Remind the students that two-dimensional figures have area, while three-dimensional figures have volume. Provide the students with opportunities to model real world situations with geometric solids and then to use the model to find other quantities such as volume or density.

ENGAGE

Define *significant figures* and discuss the concise rules that identify them. *The significant figures of a number are digits that carry meaning, contributing to its measurement resolution. This includes all digits except the following: Numbers are often rounded to avoid reporting insignificant figures. Numbers can also be rounded merely for simplicity rather than to indicate a given precision of measurement, for example to make them faster to read out loud.*

Concise Rules

1. *All non-zero digits are significant.*
2. *Zeros between non-zero digits are significant.*
3. *Leading zeros are never significant.*
4. *In a number with a decimal point, trailing zeros, those to the right of the last non-zero digit, are significant.*
5. *In a number without a decimal point, trailing zeros may or may not be significant.*

QUESTION

When is "0" significant?

If a zero is found between significant digits, it is significant. Zeros can be used as (insignificant) placeholders to the left of significant digits if the number is a decimal. For example, a mass of 42g has two significant digits.

TEACHING TIP

Emphasize that students can always transform a number into scientific notation to expose any insignificant zeros in their numbers. This is especially helpful in multiplication problems.

ENGAGE

Discuss the *calculation rules* engineers utilize in estimating and ultimately creating a product.

Multiplication/Division Rule: The product cannot be greater than the number of significant figures in the factor having the fewest significant figures.

Addition/Subtraction Rule: The sum of the numbers cannot be more precise than the least precise number. You do not really use significant digits.

ENGAGE

Discuss the two types of *reasonableness* outlined in your text.

Physically reasonable. Does the answer make sense when we compare it to the physical situation being explored or the estimates that we can make?

Reasonable precision. Is the number of digits in the answer the same as the level of accuracy and precision available to us in the parameters of the problem?

3.4.2 The Meaning of "Significant"

All digits other than zero are automatically considered significant. Zero is significant when:

- It appears between two nonzero numbers.
  - 306 has three significant figures
  - 5.006 has four significant figures
- It is a terminal zero in a number with a decimal point.
  - 2.790 has four significant figures
  - 2000.0 has five significant figures

Zero is not significant when:

- It is used to fix a decimal place.
  - 0.0456 has three significant figures
- It is used in integers without a decimal point that could be expressed in scientific notation without including that zero.
  - 2000 has one significant figure (2×10^3)
  - 35,100 has three significant figures (3.51×10^4)

Comprehension Check 3-6

Determine the number of significant figures and decimal places for each value.

(a) 0.0050
(b) 3.00
(c) 447×10^9
(d) 75×10^{-3}
(e) 7,790,200
(f) 20.000

3.4.3 Calculation Rules

As an engineer, you will likely find that your job involves the design and creation of a product. It is imperative that your calculations lead to the design being reasonable. It is also important that you remember that others will use much of your work, including people with no technical training.

Engineers must not imply more accuracy in their calculations than is reasonable. To assist in this, there are many rules that pertain to using the proper number of significant figures in computations. These rules, however, are cumbersome and tedious. In your daily life as an engineer, you might use these rules only occasionally. The rules that follow provide a reference if you ever need them; in this text, however, you are simply expected to be *reasonable*. In general, asking yourself if the number of significant figures in your answer is reasonable is usually sufficient. However, it is a good idea to be familiar with the rules, or at least know how to find them and use them if the need ever arises.

Multiplication and Division

A quotient or product should contain the same number of significant figures as the number with the fewest significant figures. Exact conversions do not affect this rule.

Example 3-14

$$(2.43)\,(17.675) = 42.95025 \cong 43.0$$

- *2.43 has three significant figures.*
- *17.675 has five significant figures.*
- *The answer (43.0) has three significant figures.*

Example 3-15

$(2.479 \text{ h})(60 \text{ min/h}) = 148.74 \cong 148.7 \text{ min}$

- *2.479 hours has four significant figures.*
- *60 minutes/hour is an exact conversion.*
- *The answer (148.7) has four significant figures.*

Addition and Subtraction

The answer resulting from an addition or subtraction should show significant figures only as far to the right as the least precise number in the calculation. For addition and subtraction operations, the least precise number should be that containing the lowest number of decimal places.

Example 3-16

$1725.463 + 489.2 + 16.73 = 1931.393$

- *489.2 is the least precise.*
- *The answer should contain one decimal place: 1931.4.*

Example 3-17

$903{,}000 + 59{,}600 + 104{,}470 = 1{,}067{,}070$

- *903,000 is the least precise.*
- *The answer should be carried to the thousands place: 1,067,000*

Rounding

If the most significant figure dropped is 5 or greater, then increase the last digit retained by 1.

Example 3-18

Quantity	Rounded to	Appears as
43.48	3 significant figures	43.5
43.48	2 significant figures	43
0.0143	2 significant figures	0.014
0.0143	1 significant figure	0.01
1.555	3 significant figures	1.56
1.555	2 significant figures	1.6
1.555	1 significant figures	2

TEACHING TIP

When calculations combine operations, it's best to perform the calculations first, then apply the significant figure rules step by step. This avoids the tendency to round at an intermediate step.

You can also emphasize that when addition/subtraction calculations reduce the number of decimal places available in the final answer, the remaining digits are the only ones available to become significant figures.

To check your solutions, the following calculator provides step-by-step solutions to mixed operation problems.

URL
https://www.omnicalculator.com/math/sig-fig

TEACHING TIP

Consider showing this specific example in class to highlight how rounding can go wrong if done too early.

3.4.4 At What Point in a Calculation Should I Round My Values?

Calculators are quite adept at keeping track of lots of digits—let them do what they are good at. In general, it is neither necessary nor desirable to round intermediate values in a calculation, and if you do, maintain at least two more significant figures for all intermediate values than the number you plan to use in the final result. The following example illustrates the risk of excessively rounding intermediate results:

$$\text{Evaluate } C = 10{,}000[0.6 - (5/9 + 0.044)]$$

- Using calculator with no intermediate rounding:

$$C = 4.4$$

- Rounding value in inner parenthesis to two significant figures:

$$C = 10{,}000(0.6 - (0.599555)) = 10{,}000(0.6 - (0.60)) = 0$$

- Rounding 9/5 to two significant figures:

$$C = 10{,}000(0.6 - (0.55555 + 0.044)) = 10{,}000(0.6 - (0.56 + 0.044))$$
$$= 10{,}000(0.6 - 0.604) = -40$$

As you can see, rounding to two significant figures at different points in the calculation gives dramatically different results: 4.4, 0, and −40. *Be very sure you know what effect rounding of intermediate values will have if you choose to do so!*

Some numbers, such as certain unit conversions, are considered exact by definition. Do not consider them in the determination of significant figures. In calculations with a known constant, such as pi (π), which is defined to an infinite number of significant figures, *include at least two more significant figures* in the constant than are contained in the other values in the calculation.

Comprehension Check 3-7

Express the answer to the following, using the correct number of significant digits.

(a) 102.345 + 7.8 − 169.05 =
(b) 20. * 3.567 + 175.6 =
(c) (9.78 − 4.352)/2.20 =
(d) (783 + 8.98)/(2,980 − 1,387.2) =

3.5 Using Structured Problem Solving

Example 3-19

Estimate how many miles of wire stock are needed to make 1 million standard paper clips.

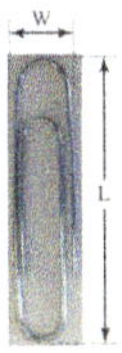

Sketch:

See the adjacent diagram.

Objective:

Determine the amount of wire needed to manufacture a million paper clips.

Observations:

- *Paper clips come in a variety of sizes.*
- *There are four straight segments and three semicircular sections in one clip.*
- *The three semicircular sections have slightly different diameters.*
- *The four straight sections have slightly different lengths.*

List of Variables and Constants:

- L — *Overall length of clip*
- W — *Overall width of clip*
- L_1, L_2, L_3, L_4 — *Lengths of four straight sections*
- D_1, D_2, D_3 — *Diameters of three semicircular sections*
- P_1, P_2, P_3 — *Lengths of three semicircular sections*
- A — *Total amount (length) of wire per clip*

Estimations and Assumptions:

- *Length of clip:* $L = 1.5$ in
- *Width of clip:* $W = 3/8$
- *Diameters from largest to smallest*
  - $D_1 = W = 3/8$ in
  - $D_2 = 5/6$ in
  - $D_3 = 1/4$ in
- *Lengths from left to right in sketch*
  - L_1 = To be calculated
  - $L_2 = 0.8$ in
  - $L_3 \approx L_4 = 1$ in

Equations:

- *Perimeter of semicircle:* $P = \pi D/2$ *(half of circumference of circle)*
- $L_1 = L - D_1/2 - D_2/2$
- *Total length of wire in clip:* $A = L_1 + L_2 + L_3 + L_4 + P_1 + P_2 + P_3$

Manipulation:

In this case, none of the equations need to be manipulated into another form.

Length of longest straight side:	$L_1 = 1.5 - 3/16 - 5/32 \approx 1.2$ in
Lengths of semicircular sections:	$P_1 = \pi 3/16 \approx 0.6$ in $P_2 = \pi 5/32 \approx 0.5$ in $P_3 = \pi 1/8 \approx 0.4$ in
Overall length for one clip:	$A = 1.2 + 0.8 + 1 + 1 + 0.6 + 0.5 + 0.4 = 5.5$ in/clip
Length of wire for 1 million clips:	$(5.5 \times \text{in/clip})(1 \times 10^6 \text{ clips}) = 5.5 \times 10^6$ in
Convert from inches to miles:	$(5.5 \times 10^6 \text{ in})(1 \text{ ft}/12 \text{ in})(1 \text{ mile}/5{,}280 \text{ ft}) \approx 86.8$ miles

One million, 1.5-inch paper clips require about 87 miles of wire stock.

ENGAGE

According to early patent records, the first patent for a bent wire paper clip was awarded in the United States to Samuel B. Fay, in 1867. Fay's invention was intended to replace the pins used to attach pieces of paper to clothing, especially the most delicate fabrics such as "silk, lace, and all the finer class of goods, without injury or leaving the slightest trace upon the goods when removed." Fay received U.S. patent 64,088 on April 23, 1867.

DISCUSSION

Visit the Smithsonian Libraries website with your students. Navigate to the page on U.S. patents granted in 1867. Search in the field provided for the number "64088."

URL

http://library.si.edu/digital-library/book/annualreportofco18673unit

The "Ticket Fastener" is amongst other patents granted that year. Flip through the pages of the volume, observe the sketches, and make a list of the inventions you saw. Would any of these be usable today, or is there now a machine that would do it better?

QUESTION

Visit the Google Patents website with your students. Search for the 1899 patent for the wire paper-clip making machine, designed by William D. Middlebrook.

URL

https://patentimages.storage.googleapis.com/f8/2f/f3/f1428de641b1ff/US636272.pdf

Look closely at the images. (Zoom in on your browser.) The inventor (Middlebrook) and three others have their signatures on the diagram. Who were the other three people and what roles did they play in the patent design? (*They were the law firm of Baldwin, Davidson, and Wright; the other two gentlemen were witnesses.*)

Example 3-20

A spherical balloon has an initial radius of 5 inches. Air is pumped in at a rate of 10 cubic inches per second, and the balloon expands. Assuming that the pressure and temperature of the air in the balloon remain constant, how long will it take for the surface area to reach 1000 square inches?

Sketch:

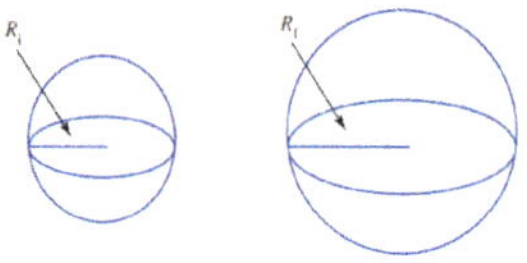

Objective:

Determine how long it will take for the surface area of the balloon to reach 1000 in²

Observations:

- *The balloon is spherical*
- *The balloon, thus its volume and surface area, gets larger as more air is pumped in*
- *The faster air is pumped in, the more rapidly the balloon expands*

List of Variables and Constants:

- *Initial radius:* $R_i = 5$ [in]
- *Final radius:* R_f [in]
- *Initial surface area:* A_i [in²]
- *Final surface area:* A_f [in²]
- *Change in volume:* ΔV [in³]
- *Initial volume:* V_i [in³]
- *Final volume:* V_f [in³]
- *Fill rate:* $Q = 10$ [in³/s]
- *Time since initial size:* t [s]

Equations:

- *Surface area of sphere:* $A = 4\pi R^2$
- *Volume of sphere:* $V = 4/3\ \pi R^3$
- *Change in volume:* $\Delta V = Qt$

Manipulation:

There are a few different ways to proceed. The plan used here is to determine how much the balloon volume changes as air is blown into the balloon and to equate this to an expression for the volume change in terms of the balloon geometry (actually the radius of the balloon).

Radius of balloon in terms of surface area:	$R = \left(\frac{A}{4\pi}\right)^{1/2}$
Final balloon radius in terms of surface area:	$R_f = \left(\frac{A_f}{4\pi}\right)^{1/2}$
Final volume of balloon in terms of surface area:	$V_f = \left(\frac{4\pi}{3}\right)\left(\frac{A_f}{4\pi}\right)^{3/2}$
Volume change in terms of air blown in:	$\Delta V = V_f - V_i = Qt$
Volume change in terms of geometry:	$V_f - V_i = \left(\frac{4\pi}{3}\right)\left(\frac{A_f}{4\pi}\right)^{3/2} - \left(\frac{4\pi}{3}\right)R_i^3$
Solve for time to blow up balloon:	$t = \left(\frac{4\pi}{3Q}\right)\left(\frac{A_f}{4\pi}\right)^{3/2} - \left(\frac{4\pi}{3Q}\right)R_i^3$
And simplifying:	$t = \left(\frac{4\pi}{3Q}\right)\left\{\left(\frac{A_f}{4\pi}\right)^{3/2} - R_i^3\right\}$

It takes just over 4 minutes to increase the volume to 1000 cubic inches.

NOTE:

Whenever you obtain a result in equation form, you should check to see if the dimensions match in each term.

IMPORTANT

Recall that you should manipulate the equations *before* inserting known values. Note that the final expression for elapsed time is given in terms of initial radius (R_i), flow rate (Q), and final surface area (A_f). So, you now have a general equation that can be solved for any values of these three parameters. If you had begun substituting numbers into equations at the beginning and then wanted to obtain the same result for different starting values, you would have to re-solve the entire problem.

Avoiding Common Mistakes

Erroneous or argumentative thinking can lead to problem-solving errors. For example:

- ***I can probably find a good equation in the next few pages.*** Perhaps you read a problem and rifle through the chapter to find the proper equation so that you can start substituting numbers. You find one that looks good. You do not worry about whether the equation is the right one or whether the assumptions you made in committing to the equation apply to the present problem. You whip out your calculator and produce an answer. *Don't do this!*
- ***I hate algebra, or I cannot do algebra, or I have got the numbers, so let us substitute the values right in.*** Many problems become much simpler if you are willing to do a little algebra before trying to find a numerical solution. Also, by doing some manipulation first, you often obtain a general expression that is easy to apply to another problem when a variable is given a new value. By doing a little algebra, you can also often circumvent problems with different sets of units. *Do some algebra!*
- ***It is a simple problem, so why do I need a sketch?*** Even if you have a photographic memory, you will need to communicate with people who do not. It is usually much simpler to sort out the various parts of a problem if a picture is staring you right in the face. *Draw pictures!*
- ***I do not have time to think about the problem, I need to get this stuff finished.*** Well, most often, if you take a deep breath and jot down several important aspects of the problem, you will find the problem much easier to solve and will solve it correctly. *Take your time!*

3.6 Characterizing Engineering Systems

Structured problem solving is a fundamental skill that engineers must have in order to be successful in their career. However, the engineer's day to day isn't necessarily solving simple physics problems using pen and paper. Digging deeper into the ideas of engineering analysis, you need to become fluent in describing, dissecting, and analyzing systems. A **system** is a set of interrelated parts that produce a desired result when they work together.

system

You are surrounded by systems every day, everywhere you go; some examples of systems include vending machines, traffic lights, vehicles of all shapes and sizes, plants, computers like your tablet and smartphone—the list can go on and on. But how? When we break down a system, it will always have three things: inputs, outputs, and processes.

inputs

The **inputs** are the commands, resources, or energy that we provide to the system. Generally, the inputs are whatever the system needs to do what it's designed to do. For example, your smartphone needs electricity in order to function, but it also needs your commands. These commands could be tactile, using your fingers to press on the screen, or audial, such as speaking your command to a digital assistant. The inputs could be materials as well. For example, if we wanted to have a cookout, we would need to ensure we had propane—or charcoal, if you want to "taste the heat and not the meat." Otherwise, we wouldn't be able to start our grill and we would have to eat whatever food we brought raw, which is not ideal for many food products.

QUESTION

Students are likely most familiar with the idea of a mathematical model. Define the *mathematical model*, and discuss how these models can help speed production and research.

Mathematical models offer new possibilities to manage the increasing complexity of technology, which is at the basis of modern industrial production. They can explore new solutions in a very short time period, thus allowing the speed of innovation cycles to be increased.

QUESTION

Define the word *system*.

Systems are a set of interconnected components that serve to transform, store, transport, or control materials, energy, and/or information. These systems exist in the world as the result of human design and function without further human design input.

TEACHING TIP

When we discuss engineering design in Chapter 4, the idea of processes will be crucial. Students need to understand that processes are what the system needs to do to achieve the desired outputs. When generating possible designs, it will be much easier to think about methods to achieve certain functions instead of how the entire product will work. Emphasize the idea of processes during instruction and that they'll be revisiting the idea in the next chapter.

DISCUSSION

Consider showing students the following ad for the failed project, the *Coolest Cooler*. See how many subsystems they can identify.

Music (Bluetooth integration), Charging, LED lighting, movement, blending

Watch the Coolest Cooler's 2014 Record-setting Kickstarter Video

https://www.youtube.com/watch?v=yddkQmyjUVQ

ENGAGE

Ask your students to define *input, output,* and *processes* as part of a system. Ask the students what part people play in the system.

(People provide the input and utilize the output; in some cases, they can actually be involved in the process itself.)

Ask the students what systems they have already used today.

QUESTION

Define *open-loop systems.*

An open-loop system has no way of monitoring or adjusting itself.

Ask the students to name three open-loop systems they use at school or at home.

Light switch, toaster, water faucet, stove, television, clothes dryer, stereo, window shades

output

The **output** (or **outputs**) is what the system produces when it interacts with the inputs. For example, when we turn on a lamp, the output is light so we can see in a dark or dimly lit room. The systems that engineers work on aren't necessarily as simple; most will have multiple outputs. Moreover, although we used the phrase "produce a desired result" in our definition of a system, that doesn't mean a system is incapable of producing *undesired* results. For example, the desired output is energy when we burn fossil fuels, but the process of burning them releases nitrogen oxides into the atmosphere. Sometimes the result may not even be expected. Regardless, whenever we're designing a system, we need to balance the tradeoff between desired and undesired outputs.

processes

The **processes** are what the system does with the inputs in order to create the desired result. We are vague here because the process could be a wide range of things. For example, the relationship between the inputs a plant takes in, carbon dioxide, water, and sunlight, and its outputs, sugars and oxygen, has a specific name—photosynthesis. In practice, the system's processes will ultimately determine what the inputs need to be and what outputs, both expected and unexpected, will result. That is where you come in as the engineer to choose the appropriate functions to weigh the desired and undesired output tradeoff based on the customer's requirements.

subsystems

Often, the system we are working with is a combination of **subsystems**. The definition of subsystem is exactly the same as "system," with a small tweak—the system in question is part of a larger system. Each of the subsystems has its own set of inputs, outputs, and processes. More often than not, the output of one subsystem will become the input for another subsystem. thus, interacting with each other to achieve a common goal: working toward the system. In fact, our example of connecting a set of headphones or wireless earbuds could be seen as a system with two subsystems. The device that is playing the music you chose (one of the device's inputs) transmits the information for the song or video (the device's output) to the earbud (becoming the earbud's input), which can then take those instructions to create the appropriate vibrations for your ear (the earbud's output). Depending on the level of detail you need during your analysis, you may need to decompose the device further and further down to the most fundamental input-process-output relationships.

Often, the system we are working with is a combination of subsystems. The definition of subsystem is exactly the same as "system," with a small tweak—the system in question is part of a larger system. Each of the subsystems has its own set of inputs, outputs, and processes. More often than not, the output of one subsystem will become the input for another subsystem. In fact, our example of connecting a set of headphones or wireless earbuds could be seen as a system with two subsystems. The device that is playing the music you chose (one of the device's inputs) transmits the information for the song or video (the device's output) to the earbud (becoming the earbud's input), which can then take those instructions to create the appropriate vibrations for your ear (the earbud's output). Depending on the level of detail you need during your analysis, you may need to decompose the device further and further down to the most fundamental input-process-output relationships.

3.6.1 Open-Loop and Closed-Loop Systems

We can classify systems into two broad categories, open-loop and closed-loop. The distinguishing feature between them is how the system uses its output. Consider a simple toaster. Once you choose the setting and press down on the handle to engage the heating unit, the toaster will continue to burn your bread to a crisp. It doesn't know at all that you like your toast only lightly browned, nor do you have a way to tell it that. In other words, the toaster can't change its processes, such as its applied heat or timing, based on information about the output, the crispiness or color of the toast. The toaster is an example of what we call an open-loop system. An **open-loop system** is a system that can't adjust its output based on changing conditions.

open-loop system

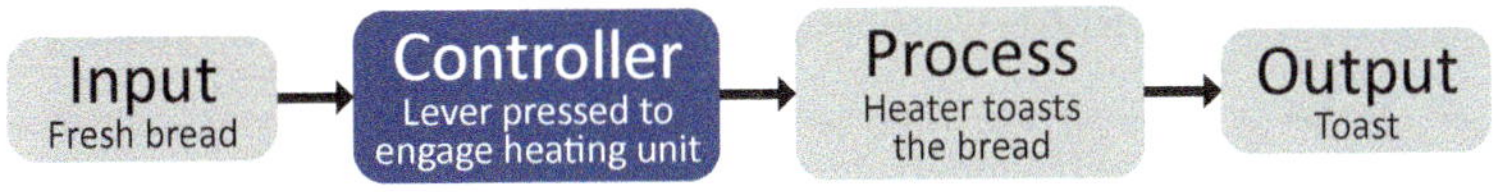

Have you ever plugged in your headphones or connected your earbuds wirelessly and whatever you were watching or listening to plays so loudly that your ears hurt? You quickly had to adjust the volume yourself or pause the audio; the device didn't know that the volume was too loud for you. This situation is an example of open-loop system behavior. An open-loop system can't adjust its output based on new developments in the conditions in which it is operating. In this case, the changing condition is that an external audio device was connected to the device, but the output, the sound, did not change in volume. Some devices are more careful with this changing condition and will instantly pause whatever content you're watching or listening to whenever an external audio device is connected or disconnected.

closed-loop system

feedback

A **closed-loop system** is a system that can respond to changes in the situation and adjust its inputs accordingly to achieve the desired output. Closed-loop systems attempt to automatically output the desired results and maintain it by comparing the desired output with the current condition. A closed-loop system can perform self-monitoring by using **feedback**, which is information about the system's output that is connected back to its input. This connection back to the input of the system is why the system is considered to be a "closed loop."

3.6.2 The Elements of a Feedback Loop

balancing feedback loop

Closed-loop systems can exhibit a variety of behaviors, depending on the system. One type that is useful in engineering contexts is a balancing feedback loop. A **balancing feedback loop** is a closed-loop system that aims to stabilize the output. For example, for coffee lovers, keeping their drink warm throughout the morning is a key priority. Some devices on the market have the capacity to keep a mug warm at a specific desired temperature throughout the day. Using a combination of sensors, it can tell when you fill the mug with a liquid and turn itself on. It also knows the temperature of the liquid, the user's desired temperature, and will use the desired temperature as a reference point as its heating system warms up the liquid until it gets closer and closer to the desired temperature. If we were to abstract what makes the self-warming mug a closed-loop system, we would find three essential elements: a sensor, a comparator, and a controller.

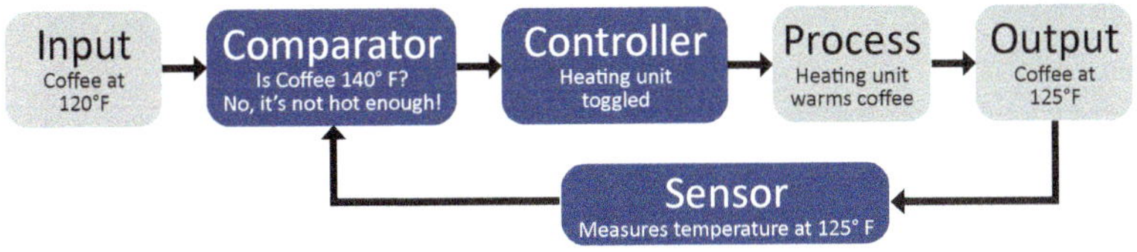

sensor

The sensor in our self-warming mug measures temperature, but a sensor could measure a variety of physical quantities. Generally, a **sensor** is a component of a system that monitors one of the system's outputs. Consider the rise of self-driving cars, which are a goldmine of examples of sensors. They still have the standard sensors included in most modern vehicles, such as pressure sensors in our tires that monitor if they have too much or too little air. For our safety, airbag sensors work off several stimuli to minimize the bodily harm inflicted on everyone involved in an accident, such as an inertia sensor that detects a decrease in acceleration. Cars are also being equipped with several ultrasonic sensors to detect objects around them. These sensors help us ensure we don't scrape the curb or someone else's car while we are trying to parallel park into a tight spot.

comparator

How does the self-heating mug know when we've reached the optimal temperature? That's the job of the comparator. A **comparator** will compare the current output of a system that was measured by the sensor

QUESTION

Define *closed-loop systems.*

A closed-loop system can respond to changes and adjust its inputs or outputs.

Ask the students to name three closed-loop systems they use at school or at home.

Thermostat, toaster setting, refrigerator settings, oven temperature, clothes dryer delicate setting, washing machine water level

QUESTION

Define *feedback.*

The purpose of feedback is to improve the operation of a system. Therefore, some type of feedback can be added to most systems. Ask the students to give examples of appliances that require feedback to be safely operated.

ENGAGE

Have students imagine systems that don't work effectively. Examples might include a traffic light with no adjustments to surrounding traffic, or a coffee maker that continues to brew coffee even if the pot is removed.

Have students brainstorm about how these and other systems might be improved.

ENGAGE

Divide the students into small groups. Give each group a stack of cards, each of which describes a technological system. Have each group sort the cards into open-looped or closed-loop systems. When they present their results to the class, they should be able to explain their reasoning for each type of system.

ENGAGE

Tactile feedback is a type of Haptic feedback. Haptic feedback is generally divided into two different classes: Tactile and Kinesthetic.

QUESTION

Define *tactile feedback* and *kinesthetic feedback.*

Tactile: The things you feel in your 'fingers' etc., or on the surface. The tissue (for example, in your fingers) has a number of different sensors embedded in the skin and right underneath it. They allow your brain to feel things such as vibration, pressure, touch, texture, etc.

Kinesthetic: The things you feel from sensors in your muscles, joints, and tendons. Weight, stretch, joint angles of your arm, hand, wrist, fingers, etc. Imagine holding a coffee mug in your hand.

Kinesthetic feedback tells your brain the approximate size of the mug, its weight, and how you are holding it relative to your body.

ENGAGE

In small groups, have students list examples of systems in which a human is the controller. Then list the systems in which a computer is the controller. See which team can compile longer lists in a set amount of time.

ENGAGE

Introduce your students to how Tesla's autopilot "sees" on the road from Carscoops. This is a contemporary example of a closed-loop system.

This Is What Tesla's Autopilot Sees On The Road

https://www.youtube.com/watch?v=fKXztwtXaGo

You can also expand the discussion to include the levels of driving automation. Below is an article that describes the six levels: no automation, driver assistance, partial automation, conditional automation, high automation, and full automation.

URL
https://www.synopsys.com/automotive/autonomous-driving-levels.html

How do the six levels of automation map to the idea of a closed-loop system? Where does Tesla's autopilot feature fit? *The responsibility of the controller gradually becomes more on the automation system rather than the driver. Tesla's autopilot fits in Level 2 because the driver can take over at any time, and it is not recommended that the driver allow the vehicle to operate completely on its own.*

96 Chapter 3 Engineering Problem Solving and Systems

and compare it to the desired result. This part of the feedback loop could be a physical device or a piece of software. It doesn't even have to be automatic. The sensor is consistently taking in information, and sometimes the desired result may change with time. For example, in some vehicles, if someone suddenly brakes in front of us, then the driver receives an audio and visual alert on the dashboard in addition to the vehicle braking for us. The visual and audio alerts are telling you, the driver, to make an adjustment to the system output. How would the car know that we're about to crash into the person that just slammed on their brakes to avoid a cute animal in the middle of the street? The comparator again. We know from Physics how long it takes us to move from point A to point B and how quickly we could stop if we needed to, which changes with variables like speed, road conditions, and the condition of the tires. Therefore, the comparator is comparing the distance between the cars against the estimated safe following distance.

controller

Finally, what makes the change in the output based on the feedback? The controller. A **controller** is a device that toggles a process on or off depending upon the feedback received during the system's operation. In the case of our self-heating mug, the controller is the trigger for the heating unit that warms the liquid. In the crash detection example, it's the braking system. You may notice that the controller could reasonably work without the need of a sensor or comparator. We could simply turn the mug on and hope for the best. We could use the brakes to adjust our speed at will. You would be right; a controller is present in both open-and closed-loop systems. The difference is how the controller is impacted by the output of the system. If the controller does not get any information in the form of feedback, the system is open-looped. If it does receive feedback, then the system is closed-loop.

CHAPTER 3 ASSESSMENT QUESTIONS

1. Create a list of variables and constants for the following problem:

   Calculate the mass in kilograms of gravel stored in a rectangular bin 18.5 feet by 25.0 feet. The depth of the gravel bin is 15 feet, and the density of the gravel is 97 pound-mass per cubic foot.

2. Find a partner and read the following together:

   I am standing on the upper deck of the football stadium. I have an egg in my hand. I am going to drop it and you are going to try to catch it. You are standing on the ground. Explain to your partner why you would not want to stand directly underneath me, where you should stand instead, and why. If you can run at 20 feet per second and I am at a height of 100 feet, how far away can you stand and still catch the egg if you start running when I let go?

3. Water drips from a faucet at the rate of three drops per second. What distance separates one drop from the following drop 0.65 seconds after the leading drop leaves the faucet? How much time elapses between impacts of the two drops if they fall onto a surface that is 6 feet below the lip of the faucet?
   Your sketch should include the faucet, the two water drops of interest, and the impact surface. Annotate the sketch, labeling each item shown and denote the relevant distances in symbolic form. For example, you might use d_l to represent the distance from the faucet to the first drop.

4. If you are to estimate how many gallons of paint are needed to paint the room you are in, what assumptions will you need to make? Where will you need to make assumptions to ensure that you have enough paint without running out?
   In estimating the wall area, you should round up the length and height.
   Noting that paint (for large jobs) is sold in 5-gallon pails, you will want to round your final estimate to the next whole 5-gallon pail.
   Close estimates allow for subtracting 21 square feet per doorway (if the doors are not being painted the same color). In making a rough estimate, if there are not a lot of doorways, it would be conservative to leave in the door area.

5. Determine the number of significant figures and decimal places for each value.
   (a) 0.0050
   (b) 3.00
   (c) 447×10^9
   (d) 75×10^{-3}
   (e) 7,790,200
   (f) 20.000

6. Express the answer to the following, using the correct number of significant digits.
   (a) $102.345 + 7.8 - 169.05 =$
   (b) $20.0 \times 3.567 + 175.6 =$
   (c) $(9.78 - 4.352)/2.20 =$
   (d) $(783 + 8.98)/(2,980 - 1,387.2) =$

7. (a) Estimate by aggregation how many gallons of gasoline are used by cars each year in the United States.
   (b) Estimate by aggregation the volume of a person. A rough approximation of the volume of a person would be a cylinder approximately 1.75 m tall with a radius of 0.25 m. How different was your estimate from the cylindrical approximation?

8. Use the following materials and procedures for this problem:

   Materials
   Ruler
   Tape measure
   Calipers

   Procedure
   The following measurements and estimations are to be completed individually, not using one set of measurements per team. This activity is designed to help you learn the size of common items to help you with future estimates. Be sure to give both the value and the unit for all measurements.

   Measure the following:
   (a) Your height in meters
   (b) Your arm-span (left fingertip to right fingertip) in meters
   (c) The length of your index finger, in centimeters
   (d) The width of your thumb, in centimeters
   (e) The width of the palm of your hand, in centimeters
   (f) The length of your shoe, in feet
   (g) The length of your "pace," in feet. A pace is considered as the distance between the toe of the rear shoe and the toe of the lead shoe during a normal step. Think about how to make this measurement before doing it.

   Determine the following relationships:
   (h) How does your arm-span measurement compare to your height?
   (i) How does your palm width compare to your overall height?
   (j) How does the length of your index finger compare to the width of your thumb?

   Determine through estimation:
   (k) The height of a door in units of feet
   (l) The length of a car in units of yards
   (m) The area of the floor of the classroom in units of square meters
   (n) The volume of your body in units of gallons

9. In each of the cases below, either display a value to the requested number of significant figures or display a calculated value to the appropriate number of significant figures.
   (a) Round to three significant figures: 0.70973 kilograms
   (b) Area of a box measuring 1.15 centimeters long by 1.62 centimeters wide. Area is determined by length times width.
   (c) You measure something using a ruler that has markings every 1/16 inches. You measure an object to 3 and 11/16 inches, commonly written as 3-11/16". Report its length in decimal format in inches.
   (d) (536,000 meters) × (6,576 meters)

10. The highest mountain in the world is Mount Everest in Nepal. The peak of Mount Everest is 29,029 feet above sea level. Convert the height from feet [ft] to miles [mi].

11. Approximately how far will a falling body fall in 10.2 seconds if it is initially falling at 0.06 miles per second? (Hint: see Example 3-2.)

12. A coyote decides to drop an anvil onto a roadrunner below. Assume the coyote holds the anvil steady until it is dropped, and it takes 2.17 seconds to hit the ground.

    (a) Use approximation to find the final velocity when it misses the road runner but crashes into the ground below.

    (b) Consider this an ideal problem and solve for the exact value.

13. Many toilets in commercial establishments have a value printed on them stating the amount of water consumed per flush. For example, a label of 3 Lpf indicates the consumption of 3 liters per flush. If a toilet is labeled as 2.1 liters per flush, approximately how many flushes would be required to consume 2,200 gallons of water?

14. Given the following problem, state the objective and any observations: Calculate the square footage of fencing required to fence a field that has 3 sides of length that are 120 ft, 90 ft, and 100 ft, and one where a building leaves only 20 ft of required fencing. The fencing should be 8 ft high.

15. Describe a procedure to estimate by aggregation the number of pennies which can be found in all of the cars in the school parking lot.

16. Describe a method you might use to estimate the number of ping pong balls required to fill your classroom.

17. A good friend receives a nice tape measure for their birthday. They come to school and say, "I am exactly 60.000" tall, and if I grow at about 2 ½ inches per year, next year I'll be 62.500" tall." How would you correct this statement?

18. Assume you do your homework on notebook paper every day, and your teacher says that you can submit your work electronically from now on. Also assume you have worked ahead and have 80 pages of notebook paper ready to submit. To make this process easier, you buy a scanner that can automatically feed pages in, scan the pages, break them into individual lessons, and email them to you. Consider this scanner as a system. What are the inputs, processes, and outputs of this scanner?

19. List the inputs, outputs, and processes of the following:

    (a) coffee grinder

    (b) electric bicycle

    (c) battling robot

20. In 2001, the first iPod by Apple had a rated battery life of 10 hours to run audio files. The seventh and final model, introduced in 2019, had a rated battery life of 40 hours to run audio files. If the average song is 3.5 minutes, approximate how many more songs you can listen to using the seventh model iPod rather than the original iPod on a single battery charge.

21. List the variables and constants needed to solve the following problem: Estimate the number of marbles required to fill a one-gallon milk jug.

22. Your friend decides to walk to the mall. She says "I can walk down this hallway in one minute, and the mall is about 200 times farther away than this hallway is long, so it should take about 200 minutes to walk there." You aren't sure of this estimate, so you decide to help her solve the problem. Describe how you might solve this using:

    (a) Estimation by analogy

    (b) Estimation by aggregation

    (c) Estimation by upper and lower bounds

    (d) Estimation by modeling

23. You decide to help your friend calculate how far their new remote-control car can travel in a given amount of time. We know that the equation for distance traveled is

    $$d = vt + \tfrac{1}{2}at^2$$

    where v is the initial velocity, t is time in seconds, and a is acceleration (m/s²).

    Calculate the distance (using the correct number of significant figures) if you are given:

    $t = 13.84$ sec

    $v = 0.22$ m/sec

    $a = 2.83774$ m/sec²

24. What is a system input? What is a system output?

25. Think of a system to squeeze orange juice. Identify the input, process, and output.

26. What is feedback? Explain feedback in the example of driving a car.

27. What systems can you think of that are open-loop systems? Would they be improved if feedback was added to the loop? Why or why not?

28. Does monitoring a system always allow control over it? Why or why not?

29. Give an example of a controllable system, and explain what makes it controllable.

30. Choose a system that you are familiar with (e.g., a toaster, a home heating system, or an automobile cruise control) and diagram it using the universal systems model.

31. Find an example of a complex system which is made up of subsystems. Describe the subsystems, and how they interact to achieve a common goal.

OVERVIEW

This chapter discusses the steps in the design process at a high level, which includes defining criteria, constraints, and specifications, defining and breaking down the problem using functional decomposition, generating ideas, prototyping and testing, documentation, and project management.

CHAPTER 4

The Engineering Design Process

TOPICS

Design is a creative process that requires problem definition, idea generation and selection, solution implementation and testing, and evaluation. Design is inherently multifaceted, so any problem addressed will have multiple solutions. While a particular solution might address some objectives well, other objectives might not be met at all. The goal is to identify a design that meets the most important objectives. To evaluate ideas and communicate them to others, engineers commonly sketch possible solutions and even build models of their work, so those are commonly a part of any design process.

There are many different versions of "design processes," so it is better to discuss "design" or "a design process" rather than "THE design process." When designers or textbook authors describe design, the processes they describe have a lot of similarities, but those features can have a variety of names. To introduce design concepts, we use various examples so that you can see how the design process is applied in different contexts.

It is common to draw a diagram of the design process to help others understand how process steps are connected. Two example design processes for engineering programs are shown in Figure 4-1, but many more images of design processes are easily found using your favorite Internet browser. These two are chosen as examples because of their differences and their similarities. While Figure 4-1a is cyclic and Figure 4-1b is a block diagram sequence, they each communicate the iterative nature of design.

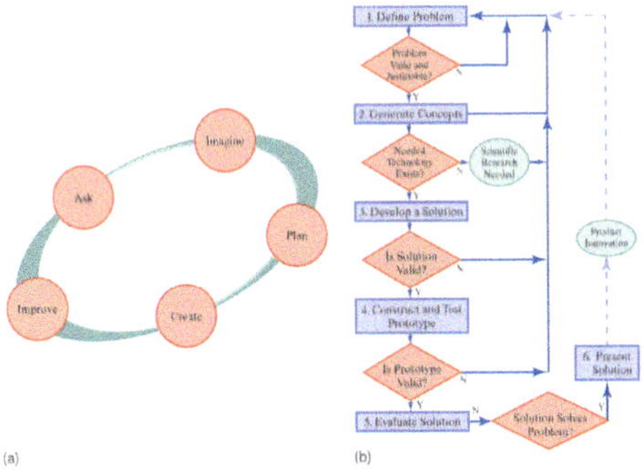

Figure 4-1 Two models of the design process: (a) the processes used in the Boston Museum of Science's Engineering is Elementary curriculum and (b) in Project Lead The Way's Introduction to Engineering Design.

The two processes in Figure 4-1 have a rough correspondence. The "Define Problem" step roughly

OBJECTIVES

- Identify the qualities of good design and their relationship to the design's user
- Examine a design with respect to its quality and usabilty
- Understand that these qualities are the result of choices made and consraints applied during the design process
- Establish and evaluate potential contraints, including health, safety, social, environmental, ethical, political, regulatory, and legal, pertaining to a problem
- Demonstrate time-management skills in prioritzing tasks, following schedules, and performing goal-relevant activities in a way that produces efficient results
- Maintain a design and computation engineering notebook
- Identify and describe an engineering design process needed for a project, including the design process and prototype development and initiating, planning, executing, monitoring, controlling, and closing a project
- Identify and explain the elements of an engineering design process

TEACHING TIP

You may consider showing all of the different design processes used by various companies, design firms, and universities to capture the stages of the design process. Make it clear to students that these processes all share common features, including empathizing with stakeholders, defining the problem, generating ideas, prototyping, testing, and refinement. They are all *iteratve processes.*

TEACHING TIP

Consider spending most of your time on the empathize stage. If a design does not appropriately address the needs of the target audience, *it is a failure.* Using the Rolls Royce example as an anchor for discussion can highlght how communicating with the users is necessary. An engineer can follow a customer's criteria perfectly, yet the design can be suboptimal!

102 Chapter 4 The Engineering Design Process

corresponds to "Ask" and has also been called "Specification." "Generate Concepts" is similar to "Imagine." Other names for that process are "Ideation," "Innovating," and "Identifying Possible Solutions." "Develop a Solution" or "Plan" adds details to some of the concepts generated in the previous step. The process of evaluating the ideas before doing detailed planning is not pictured in either process, but it is an important one—it is not practical to follow all imagined solution pathways, so it is critical to systematically choose those most likely to succeed. While the Project Lead the Way curriculum includes "Identifying Criteria" as part of their "Define Problem" step and "Evaluate Possible Solutions" as part of their "Generate Concepts" step, we will discuss these processes separately. "Construct and Test Prototype" is similar to "Create" and may be called "Implement" in other models of the design process. "Evaluate Solution" coordinates with "Improve," known as "Refine," "Reflect and Redesign," "Product Testing," and "Communicate Results" elsewhere.

What follows describes the tools engineers use in this process, providing detail about each of these processes.

4.1 Defining the Problem or Need and Empathizing with Stakeholders

The process of problem definition is critical because it shapes all the others. Defining the problem, or initiating a project, includes getting input from stakeholders—users, clients, and others who will be affected by your design. As with other parts of the design process, problem definition is iterative because it is important to confirm that the constituencies are pleased with the results. You may also need to adapt to changing needs.

In education, during college and earlier, the problem or need is often identified and described by a teacher or professor. The "problem" might be something to be calculated, a topic for a term paper, a position to debate, or a subject to paint—but it is commonly chosen in advance and provided to the student. Similarly, in the workplace, many "problems" are assigned to employees—to make sales calls to a list of phone numbers, to stock shelves, to gather vital health data from a set of patients, to make a series of deliveries, or to serve vegetarian lasagna to the traveling engineer at table 8 after a long day of field work. In engineering, a problem or need may be provided by a client or by an employer—where certain basic decisions about the design pathway have been made before the engineer makes any decisions. At other times, engineers have much more latitude in defining (or redefining) the problem. If an engineer is asked to "design a process by which the asbestos in a school boiler room can be removed," the problem is already well defined, even though there are still decisions to make. On the other hand, if that same engineer is asked to "design a process by which the school boiler room can be made safe from airborne asbestos," the engineer could choose from multiple ways to define the problem—removing the asbestos, containing it, increasing airflow and filtration to ensure that any loose particulate is swept away and captured, and others.

There are countless examples of how redefining a problem has been commercially profitable. When Dum-Dums® Pops are made, the process is continuous, which causes the manufacture of some lollipops that combine two of the flavors. This might have been considered a problem, and some companies would have designed their manufacturing process so as to stop and clean the machinery between flavor runs. Instead, the Dum-Dums company wraps these "combination" flavor lollipops and labels them "Mystery" flavor. By their defining the problem a different way, a creative solution emerged, saving money and providing a market attraction. In all designs, the way in which the problem is defined will affect the set of solutions explored.

It cannot be understated how essential this step is to the success of an engineering design. Engineers must listen to who they are designing for, otherwise the product they develop is likely to fail. Listening to all stakeholders will reveal issues that may have never crossed the design team's mind during the phase of the design process when ideas are generated. For example, consider the ultra-luxury British car maker, Rolls-Royce. One of the features of a luxury car is not hearing too much of the noise from outside of the car, so a quiet cabin is something customers tend to want in such an expensive vehicle. However, in the second-generation of their Ghost model, the cabin was so quiet that an article in Bloomberg, called "Rolls-Royce's New Car Was So Quiet at First, It Nauseated Drivers," described it as "a tomb" and "so quiet it was disorientating." The engineers needed to do something that seemed counterintuitive, pump sound into the cabin by tuning rear seat frames and trunk components to vibrate at a low frequency and rearrange the layers of sound insulation. Other car designers have implemented similar systems to counteract the uncanny silence that modern sound design has offered in vehicles. Imagine not consulting with the customers and causing them to be nauseated when they drove the car you designed—even after following their specifications exactly! In the next chapter, we will discuss ways to elicit this information to improve the designs created by the engineering team.

Participatory Design

participatory design

Although engineers may initially gather ideas from stakeholders, the team may likely follow by meeting by themselves, generating ideas, and making decisions. In such a model, the stakeholders are only brought back into the process when the engineering team has something to show them or gather feedback. There is a more integrated model, however, called participatory design. The idea of **participatory design** involves inviting all possible stakeholders to work with engineers and other members of the team to collaboratively make design decisions and develop ideas.

Participatory design provides a mechanism for engineers to work more directly with the people who will be impacted by their designs. This is particularly crucial when designing for populations without properly considering their needs or for those who do not often "have a seat at the table," so to speak. For example, activism in the 1980s and 1990s cemented the principle of "nothing about us without us" regarding people with disabilities. In the case of older adults, Riley Gibson describes in the article, "What Design for Aging Gets Horribly Wrong," that a root issue is how the needs of the stakeholders are framed by the design team. In the case of older adults, instead of framing the problem of what the population can't do, reframe the problem to open the range of potential solutions to what is possible in old age. In this case, if the team is designing a product for older adults, participatory design would demand that older adults be involved in the process. Engineers must be aware of their stakeholders and work to involve them wherever possible. Doing so reduces the potential that the chosen designs are not sensitive to the users' needs or desires and might potentially be stigmatizing without the team even knowing it.

4.2 Criteria and Constraints

Starting with a problem definition, we can begin to generate an appropriate set of criteria. At first, those criteria will provide some basic direction as we consider all the possible solutions. Later, those same criteria will be used to compare design options and narrow our choices objectively. We might be particularly concerned about cost, so certain approaches become less attractive. Other proposed solutions might be hazardous, thus further narrowing our options. The criteria for evaluating potential solutions should be discussed before you start thinking of solutions to avoid choosing criteria that favor a popular solution. Criteria must be identified before evaluation begins so that all ideas can be considered fairly—this helps avoid arguments based on hidden criteria—such as criteria that favor giving the contract to someone's sister-in-law.

must criteria
should criteria

Criteria range from **must criteria** that any successful solution has to have to **should criteria**, which are qualities that are generally agreed upon as desirable and that help distinguish one solution as better than another. Some solution parameters are preferences or options where there is no agreement on what is better. Must criteria are commonly called constraints and may be established legally at a regional level, such as California's automotive emission laws and Florida's building code provisions regarding wind-borne debris, and at the national level, such as the Corporate Average Fuel Economy (CAFE) standards, by an independent certifying body. These certifying bodies include the American National Standards Institute (ANSI) or the International Organization for Standardization (ISO), or an international agreement, such as the Basel Convention on the Control of Transboundary Movements of Hazardous Wastes and Their Disposal, a treaty that has 182 United Nations (UN) member states and the European Union as parties (Haiti and the United States have signed, but not ratified the Convention).

design specifications

To be valid, criteria must be clearly understood and measurable. When we assign numbers to our criteria, we create **design specifications**. For example, when we say a car is fuel efficient is a must criterion, we can

DISCUSSION

First, show the "shopping cart" design case study from the design firm IDEO. Then, show the following video from UC Santa Cruz researcher, Tom Dayton, about participatory design at NASA.

IDEO Shopping Cart

https://www.youtube.com/watch?v=M66ZU2PCIcM

CRF- Participatory Design

https://www.youtube.com/watch?v=4npftEf3_n4

How does participatory design differ from tradtional ways of designing products? *Both approaches involve consulting with the end user, but particpatory design includes the end user through ideation.*

TEACHING TIP

When discussing criteria, push students to be thinking in terms of specifications–i.e., measurable and testable goals/target values. Emphasize that all criteria and constraints should be paired with at least one specification.

TEACHING TIP

Students often struggle with the distinction between criteria and constraints, and you will likely find that people have a variety of definitions to delineate criteria from constraints. In fact, some professionals may not even consider the distinction to be practically useful.

The clearest distinction is labeling requirements that are inherent to how the device functions are criteria. If there is an additional requirement that is not related to how the device performs its functions, that is a constraint.

measure fuel efficiency in terms of miles per gallon. This translates our criterion from "the car must be fuel efficient" to "the car must provide a fuel efficiency of at least 25 miles per gallon." For criteria to be considered should criteria, usually called simply criteria as opposed to constraints, it must also be clear what is better. While the temperature of an office work area is important to productivity, there is no general agreement on the best temperature. On the other hand, it is universally agreed that a building floor system should be stiff—that the floor should not sag beneath our feet while we walk on it. While we will face limits based on cost, material selection, and other factors, the stiffness of a floor system is clearly a should criterion. Similarly, the energy consumed by an appliance is a should criterion—we may accept higher energy usage to meet other needs, but we can all agree that appliances that use less energy to accomplish the same task are better.

Other necessary features of a design include preferences or options. These are features that distinguish designs, but that different people, groups, or applications indicate a different choice for what is better. No matter how comfortable a couch is, if it is too large to fit in your living room, it's not the right couch for you. An elderly couple living in an efficiency apartment might want a small refrigerator that uses less space, but if two parents, two children, and two grandparents live in the same house, a larger unit is probably desirable.

constraints

Constraints are limitations on a design. Some resources online muddy the waters between criteria and constraints, such as the difference hinging on how a requirement from the stakeholder is phrased. Constraints often take the form of limitations on size, cost, materials, and time. A simple way to distinguish between criteria and constraints is as follows. If the need is related to the function of the design, then it is a criterion. If the need is not related to the design's function, meaning the design could feasibly work without considering the need, it is a constraint.

For example, consider the design of a vending machine. One function of the vending machine is to accept cash payments; this is a criterion because it is necessary for the vending machine to collect the revenue from the sale of its contents. However, the vending machine cannot feasibly accept $100 bills, because there is unlikely to be enough change available. Therefore, a constraint would be that the machine cannot accept bills larger than $5. This constraint doesn't affect the fundamental functions of a vending machine, but it does limit what it can do. The same argument can be made for size; a vending machine could be any number of sizes and still dispense its contents. However, a stakeholder could require that the vending machine only take up a certain amount of space, which will limit the design but not necessarily its basic functionality. Just like criteria, constraints will be assigned design specifications that should be included in future product decisions.

Example 4-1

In selecting a cordless screwdriver, what criteria are important?

While a particular person might have a variety of criteria, they do not all meet the test of good criteria:

- Battery life is measurable, and longer is clearly better.
- Charging time is measurable, and shorter is clearly better.
- Versatility might be measured by the range of attachments the device can use. This is measurable and more is better, but this fails to distinguish options because the ¼-inch socket is standard and attachments can be purchased separately.
- Torque is measurable, and more seems better, but some users could be injured by excessive torque.
- Weight is measurable, and less is clearly better.
- Comfort is qualitative, but it can be measured. Yet is it is unclear whether users could agree on what is better.

Each possibility listed is only the definition of a problem, leaving much of the design process incomplete.

In general, using fewer criteria keeps things simple. A rule of thumb is that if you can think of 10 criteria that are meaningful, you should consider only the two most important criteria to compare solutions. In this way, you can be sure that the important criteria maintain their importance in your decision making. Seek consensus on what the most important criteria are. Experts, specialists, managers, customers, and research can help you focus on the most important criteria. In certain situations, many criteria are used. For example, when a magazine rates consumer products, extra criteria should be included to be sure that all the criteria important to the magazine's readers have been considered.

Comprehension Check 4-1

We often express criteria in terms that are not clear and measurable. Write a clear criterion to replace each of these vague criteria for the products.

Product	Computer	Automobile	Bookshelf
Inexpensive	Less than $300		
Small			
Easy to assemble			Requires only a screwdriver
Aesthetically pleasing			
Lightweight			
Safe			
Durable			
Environmentally friendly		Has an estimated MPG of at least 50	

Another situation, in which a large number of criteria are included, occurs when very complex decisions are being evaluated. The U.S. Green Building Council has long lists of criteria used for certifying various types of projects according to the Leadership in Energy and Environmental Design (LEED) Green Building Rating System. A subset of items from the LEED for Homes project checklist follows.

LEED is an internationally recognized green building certification system, providing third-party verification that a building or community was designed and built using strategies aimed at improving performance across all the metrics that matter most: energy savings, water efficiency, CO_2 emissions reduction, improved indoor environmental quality, and stewardship of resources and sensitivity to their impacts.

From the U.S. Green Building Council website: http://www.usgbc.org

LEED FOR HOMES

LEED for Homes Simplified Project Checklist

Addendum: Prescriptive Approach for Energy and Atmosphere (EA) Credits

Points cannot be earned in both the Prescriptive (below) and the Performance Approach (pg 2) of the EA section

			Max Points	Project Points Preliminary			Final
Energy and Atmosphere (EA)	(No minimum Points Required)	OR	Max	Y/Pts	Maybe	No	Y/Pts
2. Insulation	2.1 Basic Insulation		Prereq				
	2.2 Enhanced Insulation		2	0	0		0

PREREQUISITES

2.1 Basic Insulation. Meet all the following requirements:

(a) Install insulation that meets or exceeds the R-value requirements listed in Chapter 4 of the 2004 International Energy Conservation Code. Alternative wall and insulation systems, such as structural insulated panels (SIPs) and insulated concrete forms (ICFs), must demonstrate a comparable R-value, but thermal mass or infiltration effects cannot be included in the R-value calculation.

(b) Install insulation to meet the Grade II specifications set by the National Home Energy Rating Standards (Table 16). Installation must be verified by an energy rater or Green Rater conducting a predrywall thermal bypass inspection, as summarized in Figure 3.

Note: For any portion of the home constructed with SIPs or ICFs, the rater must conduct a modified visual inspection using the ENERGY STAR Structural Insulated Panel Visual Inspection Form.

TEACHING TIP

You may consider doing Comprehension Check 4-1 in class as an exercise for students to complete in groups. Once the groups are done, you can have them report out and compare solutions with one another.

TEACHING TIP

It's critical to emphasize that engineers do not consider *how* the design will accomplish the functions; they need to know *what* the design needs to do such that it can produce the desired outputs.

For example, a chair could have the ability to recline. The function, "adjust level of footrest," does not tell us how the footrest will be adjusted (could be using a lever, a switch, or a button with mechanical or electrical power sources).

These checklist items are neither clear nor measurable, however, as shown below. The full LEED for Homes rating system and the reference guide associated with the rating system provide the additional detail needed to make the insulation criteria both clear and measurable. Note that these criteria build from existing standards, meaning that even greater levels of detail can be found elsewhere (e.g., Chapter 4 of the 2004 International Energy Conservation Code and The National Home Energy Rating Standards).

Note that the LEED criteria specify only those characteristics that are important to the overall goal of green building. This leaves the choice of the type of insulation (fiberglass, cellulose) up to the designer and the consumer. Because homes, outdoor facilities, and factories face different challenges, different LEED criteria apply. It is important to recognize which choices are options rather than criteria—even the most energy-efficient factory cannot substitute for a house. Further, the LEED criteria allow different thresholds for different sizes of houses, so as not to penalize multifamily households—even though multifamily households will require more material to build and more energy to operate, the marginal cost is not as high as building another home.

When designing a product or a process, you must first listen to what a client wants and needs. The client might be from a department, such as manufacturing, marketing, or accounting; a customer; your boss; or the government. Subsequently, you will need to translate those wants and needs into engineering specifications. Pay particular attention to separating the wants (criteria) from the needs (requirements). Requirements can be imposed legally (e.g., Corporate Average Fuel Economy regulations, building codes), they can be from within your company (e.g., the finished product must not cost more than $50 to manufacture), or they may come from your client (e.g., the product must be made from at least 90% recycled material).

Comprehension Check 4-2

You have been asked to improve the fuel efficiency of an automobile by 20%. Convert this request into engineering criteria. What changes might be made to the automobile to achieve this objective?

4.3 Breaking Down a Design Using Functional Decomposition

When discussing the fundamentals of systems, we noted that the processes are where engineers will spend a considerable amount of effort and the kinds of processes chosen will drive the inputs and outputs. It is at this point in the design process where engineers begin to understand what the processes need to be in order to achieve the desired outcome. Engineers perform a process called functional decomposition using the criteria and constraints developed as part of the discussions with stakeholders and initial research. A **functional decomposition** involves breaking down a system into its different functionalities. The important point to note here is that we don't care *how* a function is performed when we're doing a functional decomposition. We want to know *what* the function does to contribute to the intended outcome and *how* it relates to other functions in the system. This will be crucial to effectively planning and executing a design project.

functional decomposition

To perform a functional decomposition, first identify the most general task the system needs to complete. What is the desired output of the design? This shouldn't be an overly complicated description and should be summarized in one sentence. For example, consider a pair of scissors. The most general task scissors need to do is "cut material"—two words is all we needed—and the output is "material with cut." However, the function "cut material" can be broken down a bit more.

Next, consider what needs to happen immediately before the general function you identified and keep working backward to understand how previous actions lead into the next necessary action. In the case of scissors, the next immediate step before "cut paper" is "cutting method activates." In order for "cutting method activates" to happen, we need someone or something to prompt that action to begin, such as "engage cutting method." Note, again, that we are being cautious to not accidentally say *how* the device we would know as scissors does these three functions we identified. When you're specifying functions, they should have at least one verb (the action) and a noun (the object the action is directed toward). For example, "cut (verb/action) material (noun/object)." Keep track of the chain of logic taking the system from one function to another; it'll be necessary later.

Once you have all necessary functions, consider the inputs of the system that the functions need to achieve the desired output. In the case of scissors, we need the material we want to cut and some way to engage the cutting method. Think about any energy sources you might need or readings from a sensor for the functions to work as anticipated.

Doing a good job of specifying all of the necessary functionality at this stage will help much later on. Instead of focusing on the product as an entire unit, the engineering team can generate ideas for smaller components of the design. By only considering how the design might perform one function at a time can greatly reduce the cognitive load of accounting for all the different intricacies that might be involved in even moderately sized projects. Skipping this step entirely and going straight into generating design ideas may lead to missing key functions that could not be caught for several iterations until testing begins. At that point, the project may fall behind schedule as the team regroups and begins coming up with different ideas or solutions to address the oversight.

Functional Block Diagrams

functional block diagram

When we perform a functional decomposition, the goal is to come up with a functional block diagram. A **functional block diagram** is a picture that relates the inputs, functions required to perform the system's processes, and outputs. The picture will look like the input-process-output model, but will specify the connections between the inputs, functions, and outputs. First, draw a box to represent the system's process block, which will contain all its functions. On the left-hand side, label all of the inputs you identified, and label all of the outputs on the right-hand side. Using the chain of processes documented during the functional decomposition, connect the inputs to the functions, functions to other functions, and functions to outputs—as appropriate.

Let's walk through an example of a functional block diagram for a coffee maker. What does a coffee maker need to do? We need coffee beans, water, a filter, and a power supply to produce the desired output of coffee. Some coffee machines have a built-in grinder, which will turn the beans into a powder. For the sake of example, we'll assume our coffee maker has this functionality so our inputs are the full beans. While the grinder takes care of the beans, another component of the coffee maker will heat up the water. We place the ground beans into a filter that will sit in a tray above the cup—that will secure the grounds in place and ensure our coffee is free of bean debris. Once the water is hot enough, it will be delivered to the ground coffee beans by dripping it over the tray they occupy. The hot water will flow through the grounds, extracting flavor, and fall into a cup below the tray.

Now that we described the whole process, we have the following functions:

- Grind beans
- Secure bean grounds (in a tray)
- Heat water
- Drip water over grounds

We also have the following inputs:

- Power supply
- Water
- Coffee beans
- Filter

TEACHING TIP

Using the verb and noun "rule" for specifying functions can help avoid issues in students skipping to describing how the design will perform each function.

TEACHING TIP

Functional block diagrams as we discuss here are preliminary descriptions of the product's or system's functionality. As the design evolves, eventually the functions will be replaced by the subsystems responsible for each function.

ENGAGE

Figure 4-2 shows *one* example of how a coffee maker could be conceptualized using a functional block diagram. It should be emphasized that there is no one correct functional block diagram for a given product, especially one that does not exist. These diagrams flow from how the engineers conceptualize the customer requirements and decompose the criteria/constraints into functions.

You may consider having students pick an item in the classroom, perform a functional decomposition, and then create a functional block diagram. Have students complete this in groups, then have each group report out.

DISCUSSION

Possible discussion questions could be: How does your functional block diagram compare to other groups? What similarities and differences do you see? How would you adjust your functional block diagram?

DISCUSSION

When discussing generating ideas, you could pose the following design prompt: *Design a household item for an elderly population to make their lives easier.*

After some discussion, have students read the article, *What design for aging gets horribly wrong.*

URL
https://www.fastcompany.com/90660516/what-design-for-the-elderly-gets-horribly-wrong

How did their ideas align with the principles in the article? The discussion should focus on the *framing* of the problem. How do their ideas focus on what the elderly *can't* do versus what they *can* do?

ENGAGE

Define the term *prototype.*

A prototype is a full-scale, fully operational version of a solution.

The anticipated outcome of the functions when combined with the inputs will yield a cup of coffee plus some used coffee grounds that are a byproduct of the coffeemaking process. Therefore, by ordering the functions as we described, we are left with the functional block diagram in Figure 4-2. Note that there is no "right" functional block diagram; they have varying degrees of correctness. More specifically, the more descriptive your team is when generating the necessary functions for a design and more attentive your team is to the criteria and constraints, the more useful the functional block diagram will be as the design evolves. Eventually, the descriptions of functions will be replaced by the actual subsystems used to achieve the functions. For example, the "heat water" function is handled by a resistive heating element, a coiled wire that gets hotter and hotter by running electricity through it.

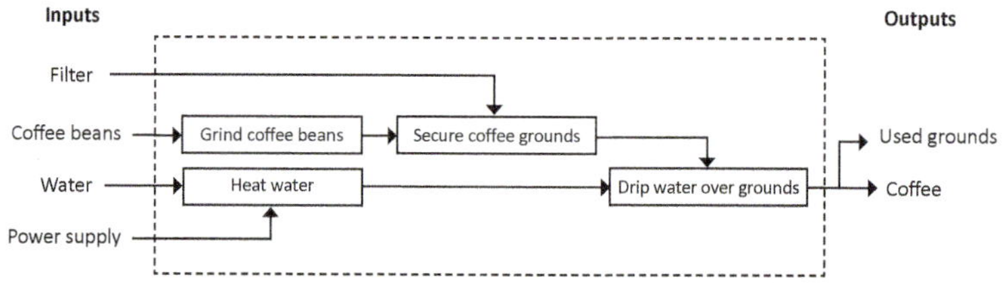

Figure 4-2 Functional block diagram for coffee maker.

4.4 Generating Ideas

At this stage of the design process, the engineers have all of the necessary functions, criteria, constraints, and specifications for their designs. Now, it is time to ideate. There are several tools that engineers can use in order to generate potential concepts. Similarly, engineers can justify moving forward with different ideas using a variety of techniques. We will dive into a selection of these techniques in the next chapter to avoid monopolizing too much space in this chapter. For now, note that ideating should come after the engineering team has a solid understanding of the problem statement, stakeholder needs, and the necessary functions to achieve the criteria, constraints, and specifications.

4.5 Prototyping and Testing

When a solution has been identified as the best fit to the criteria, it is best to stop for a reality check before moving forward. After the evaluation process, some ideas will be left on the cutting room floor. Do any of these merit further consideration? Are there important elements of those ideas that can be incorporated into the chosen design? If the reality check reveals that an idea really should not have been eliminated, then a change in the selection criteria may be appropriate.

prototype

An important reality check is to make sure a particular design will work in practice—to actually try it out—to build a sample, a scale model, or a **prototype** and find out if it performs the way we think it will. Historically, generating prototypes has been resource-intensive—both in terms of materials and construction time—so it has been critical to reduce the number of design options before considering making a prototype. In many applications, a prototype was only made to ensure that a chosen design would function correctly, and no further design options were considered unless the prototype did not function as expected. More recently, advances in rapid prototyping, including 3-D printing and digital fabrication have reduced the cost of materials and labor needed to generate a prototype, so it has become more feasible to use prototypes to explore a larger range of the design space.

Moreover, new platforms have allowed engineers to simulate the viability of certain design solutions. For example, consider a part that will be part of a larger machine that has five different concepts under consideration

QUESTION

Why is a prototype used in the design process?

A prototype allows the designer to work out any flaws. It makes it easier for the designer to make modifications that would lead to refinement of the design.

by the design team. While the machine is operating, the part will be subject to things like vibrations and heat. Computer modeling software will often have the capability to perform finite element analysis. Finite element analysis allows engineers to predict how a product will perform under a variety of conditions, such as sustained exposure to heat and vibrations.

Even when the selection of a particular design is considered final, it will undergo testing. Sometimes, this is intentional testing in laboratories and in user studies. This may include the introduction of the design into a pilot installation or a test market, which yields information about both the performance of the design and its marketability.

After deciding, you will implement your chosen solution. Undoubtedly, both carrying out the design and using the design once it is complete will provide new information about how the design might be improved. In this way, design tends to be iterative—design, build, test, redesign, and so on. Even when a design performs as expected, it may be important to test a model extensively or even build multiple models to be sure that the design is reliable.

The goal of prototyping and testing is to understand how well your design is meeting the design specifications outlined earlier in the design process. When engineers develop their test plans, they will develop an appropriate plan that allows them to measure the performance of their prototypes against the target values assigned for each criterion and constraint. There is no one test plan we can recommend. Instead, you will need to determine what needs to be tested in order to move forward with the ideas developed by the team. Later, we will discuss basic techniques for conducting testing protocols and analyzing the data.

4.6 Good Design

How do we know if our design is a good design? The initial test is ensuring that the design meets all of our established criteria and doesn't violate any of our constraints. Any acceptable design must meet these standards.

Georgia O'Keeffe, an American artist, once described art by saying "I don't know what art is, but I know some things it isn't when I see them." In a similar fashion, we may be able to easily identify bad design when we see it, but it may be difficult to describe good design.

Dieter Rams, chief design officer for Braun, developed 10 principles for good design, which is one of the most widely cited lists in the study of good design. Following these principles should ensure that the design is completed with a relationship to the design's user, will help us examine a design with respect to its quality and usability, and understand that the quality of the design is the result of choices made and constraints applied during the design process.

- **Good design is innovative:** Without innovation, design is not design.
- **Good design makes a product useful:** While this may seem obvious, a useful design is more valuable than one that is not useful.
- **Good design is aesthetic:** While there may be examples where aesthetics is not important (as in an internal component of a larger device, like a battery in a smartphone), a design meant for a user should appeal to the user.
- **Good design makes a product understandable:** A successful design should be well understood by the user.
- **Good design is unobtrusive:** Successful designs should serve their purpose without being obtrusive or distracting.
- **Good design is honest:** A successful design should honestly portray what it will accomplish.
- **Good design is long-lasting:** A good design should be long-lasting and durable.

ENGAGE

Ask the students to research notebook pages and drawings from Da Vinci, Thomas Edison, Alexander Bell, and Samuel Morse. What do all the pages have in common? (*Answer: All of the pages are signed and dated by the author and done in ink. The pages all contain more words than sketches or diagrams, even Da Vinci's.)*

ENGAGE

Ask the students to complete a Google image search for "engineering notebook pages." Study the different types of pages that can be found in an engineering notebook. Point out to the students that pages in today's notebooks have designated sections for page numbers, dates, and names of those contributing to the notebook. Different types and sizes of grids and the pages themselves allow the designer to choose the best notebook for his or her end purpose.

ENGAGE

Visit the Wiki Books website. Complete a search for "General Engineering Introduction Notebooks Drawing."

URL
https://en.wikibooks.org/wiki/General_Engineering_Introduction/Notebooks/Drawing

According to the website, what are the seven categories of content that should be added to every notebook entry? (*Answer: Function, context, setup, follow up, size, scanning, and orthographic drawings.)*

- **Good design is thorough down to the last detail:** A product missing a crucial detail is probably a failed design.
- **Good design is environmentally friendly:** A good design should leave the environment in better condition, or at least, should not damage the environment.
- **Good design is as little design as possible:** Good design should focus on the essential aspects of the solution to the problem.

Focusing on these characteristics should help us identify a good design that solves a problem we're interested in solving.

4.7 Documentation

Designing new creations is one of the highlights of an engineering career. The opportunity to imagine, transform, and create something that addresses a need is one reason people choose to become engineers. There are times when, given a pile of components, engineers (and nearly everyone else) cannot wait to dive right in and start putting things together. As we have seen with the design process, it is to our advantage to understand our constraints, understand the criteria that will define a successful design, plan and assess potential solutions, and—once something is built/designed or prototyped—test the design to ensure it works and meets the criteria.

4.7.1 Engineering Documentation: A Historical Perspective

Leonardo da Vinci (1452–1519) was widely known as an artist, inventor, mathematician, and engineer. Many of da Vinci's designs were eventually realized. For example, he designed a flying machine that was similar to a modern helicopter. While he did not live to see a flying machine, he studied and designed models that would plausibly allow humans to fly. Da Vinci documented these designs famously in a series of journals and loose notes (Figure 4-3). Imagine if he had not documented these ideas—someone widely regarded as one of the most intelligent people to have lived may have been forgotten or turned into some kind of folk legend!

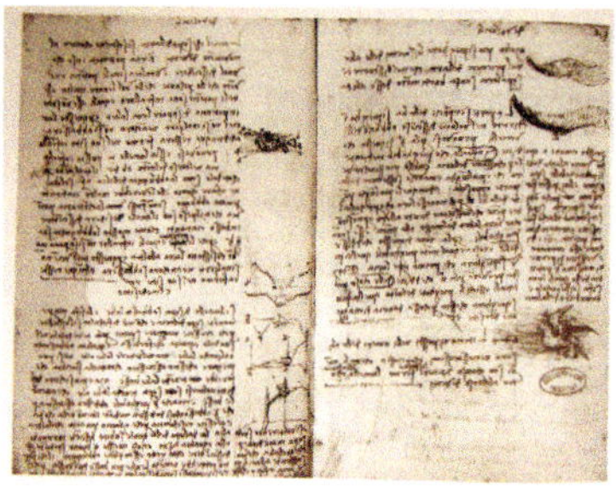

Figure 4-3 Engineering design pages documented by Leonardo da Vinci.

Another one of the most famous inventors and engineers, Thomas Edison (1847–1931), had many inventions that framed life as we know it (Figure 4-4). Imagine life without light bulbs or the recording of sound. Edison held more than 1,000 patents, documenting his inventions and protecting his intellectual property. A patent on the light bulb means that someone cannot steal the idea and claim that they invented it; the honor belongs to Edison alone. Although Edison began to invent and engineer in the late 1860s, he began to systematically keep written documentation in 1871. One of his journals from 1870 contains the following note: "All new inventions I will here after keep a full record." Eventually, Edison and his team produced over 3,000 notebooks with designs and notes to document the design process.

Figure 4-4 Edison's phonograph—one invention that changed society.

Both of these famous engineers/inventors documented their work to prove that they were indeed working toward a solution or an invention. This does not mean that everything worked or that the idea being documented is going to be an important or influential invention by any means. In fact, these notebooks contain all work—including some errors! Fortunately, there are procedures to deal with errors, changes in plans, and even success!

4.7.2 The Engineering Notebook

The engineering notebook provides you the opportunity to document the entire thought process, development, investigation into criteria and constraints faced, mistakes, and successes (Figure 4-5). It is the place to record calculations, ideas, and sketches leading to the final solution. The purpose of the notebook is to outline a detailed, chronological record of the development of the design.

Figure 4-5 Record the entire design process in your engineering notebook.

Why would we care about a notebook of this type? Wouldn't it be easier to use a 3-ring binder or an electronic document? If the design becomes patentable (or it is an idea that can make you rich and famous), this notebook serves as a record that the idea was yours. Since it is chronological, it shows the complete thought process behind the design. Further, if you unexpectedly leave to join a prestigious university or to begin a high-salary career, the notebook will provide a record to pass to the person who will take over the project. Surely the design you are working on is valuable and deserves to be preserved and continued. Without your expertise, this notebook will serve as the tool the new team needs to complete your design.

Documentation—material providing official information or evidence—is incredibly important in the design process. We can think of engineering documentation a few different ways, one of which is as a permanent record as we just discussed. In the same way we can peruse the drawings of da Vinci and Edison and spot mistakes or poor ideas along the beaten path of the design process, your notebook should not only catalogue your success, but your struggles as well. Saving the poor designs could provide inspiration for a redesign, like a phoenix rising from the ashes.

The notebook should be bound. While it would be easier to use a binder, a binder also allows someone to remove and insert pages. As a result, the record isn't chronological anymore. A quad-ruled notebook would be ideal, but a composition book—or even plain paper—will work. The style of the notebook is the preference of the designer.

What are some key features of the engineering notebook? Your teacher may have specific requirements, or the type of project or the team you work with might dictate some specific things in the notebook, but a general set of guidelines is as follows:

- The first page or two should be reserved for a Table of Contents (TOC). When you start a new section, enter it into the TOC.
- Each page should be numbered consecutively, beginning with the first page after the TOC. If you use one side of a piece of paper, don't write on the back of any of the other pages—you should be consistent. If both sides are used, both sides should be numbered.
- Your work should be clear and neat. Remember that one purpose of the notebook is to allow you to share the work or pass it to someone else, who could then continue your work. On the other hand, this isn't meant to be a perfect record suitable for a book—it is meant to capture the ideas and work of the author. Treat the notebook as though your work will lead to a new invention, and the notebook will need to be the documentation when you go for the patent.

- You should only use ink—not pencil. Pencil can be erased, which gives the appearance of tampering with the record.
- Each page should have the date and each page or each major section should have the name of the author.
- Entries should be written in chronological order. Don't try to add sections between existing pages. Use the pages in order. When a new idea starts, enter it in the TOC.
- Start on a new page each day. If there is room at the bottom of a page at the end of the day, cross out the empty section (see below).
- Try not to leave large, blank sections. If the section of work is ended and you are ready to start on a new page, use a single line to cross out the blank area. It's often good to sign the line to show that it was intentionally drawn there.
- In the case of an error, cross out the incorrect work with a single line. Don't try to destroy the work, try to erase it, or use correction fluid. Documenting the mistakes along the way helps to establish that the notebook is a record of your progress.
- The notebook should include all calculations, sketches, and work. For all calculations and measurements, use consistent units.

Other notebook guidelines can be found online; your teacher may also have additional ideas. Example pages are shown in Figure 4-6.

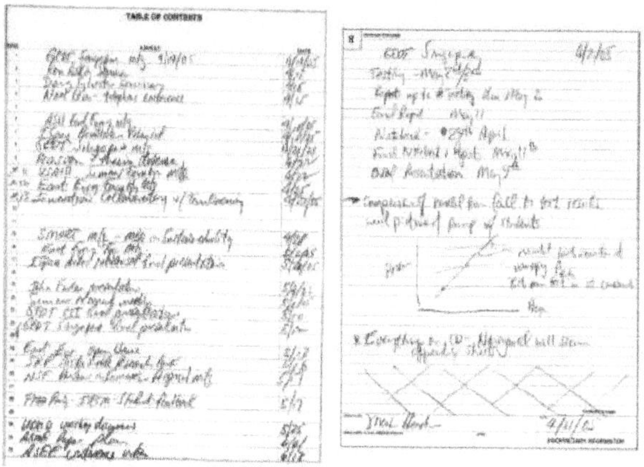

Figure 4-6 Sample page from a Table of Contents and a page within the journal.

4.8 Project Management

To complete a project successfully, on schedule, and satisfying all constraints, careful planning, monitoring and controlling is required. The following steps should help your team plan the completion of project work.

Step 1: Create a Project Timeline

The first consideration is the project's due date. All team members need to note this on a calendar. Examine the due date within the context of other assignments and classes. For example, is there a calculus test in the fourth week? When is the first English paper due?

Your team should consider meeting at least once a week at a consistent time and location for the duration of the semester. More meetings will be necessary, but at least once per week, there should be a time when the entire team can gather and review the project status. A standing meeting time should prevent issues like "I didn't know we were going to meet."

Next, look at the project itself and break it into individual tasks and subtasks. Create a list, making it as specific, detailed, and thorough as possible. Your list should include:

- All tasks needed to complete the project
- Decisions that need to be made at various times
- Any supplies/equipment that will need to be obtained
- Remember to define closing the project: how do you know when it is complete, and what are the last steps?

Carefully consider the order in which the tasks should be completed. Does one task depend on the results of another? Then, working backward from the project due date, assign each task, decision, or purchase its own due date on the calendar.

slack time

For each task, estimate how long it will take to complete. More advanced timelines can include what's known as slack time. **Slack time** is a time period where project activities can be completed before the time the activity needs to be completed. Keep track of which tasks depend upon one another and note them using a number or letter system. For example, consider the tasks for making a sandwich with toasted bread and chopped veggies (Table 4-1). The list of tasks and dependencies might look like so:

Table 4-1 Sample Task Lists and Times

Task Label	Task	Dependencies	Time to Complete
A	Get ingredients	None	5 minutes
B	Toast bread	A	3 minutes
C	Cut up vegetables	A	5 minutes
D	Spread condiments on toast	B	1 minute
E	Assemble sandwich with sliced vegetables	C, D	1 minute

We only need to connect the task with its immediate dependencies, so although E is technically dependent on tasks A-D, adding those dependencies to our list doesn't add any new information. We already know E is dependent upon the previous tasks because its dependencies C and D link back to A and B.

4.8.1 Gantt Charts

Gantt chart

The most critical tool for providing the big picture of the project schedule is called a Gantt chart (Figure 4-7). A **Gantt chart** is a bar chart variant that shows the dependencies between project activities in relation to the status of the project and helps with proficiency of the design process. Depending on the size of the

ENGAGE

Consider having students create a simple Gantt chart of the tasks in Table 4-1.

ENGAGE:

Monday.com outlined 10 common project management challenges. You may consider having students read the following article.

URL
https://monday.com/blog/project-management/project-management-challenges/

What common challenge of project management does the responsibility matrix aim to address? *Absence of accountability.*

What common challenge of project management can the Gantt Chart address? *Impractical deadlines and scope creep.*

TEACHING TIP

You can instruct students to ensure that each item on the Gantt chart be associated with the assignments made in the responsibilty matrix. By doing so, the team can better understand how project responsibilities would overlap over the course of the project.

project, these types of charts can vary in complexity and size. Time runs along the x-axis (i.e., horizontal axis), which is where project milestones are plotted along with project activities. Each activity is a box that begins at the earliest point where it could start, based on its dependencies. The length of the box is the estimated time it would take to complete the activity. The y-axis (i.e., vertical axis) lists all the project activities.

Figure 4-7 Working with a Gantt chart.

To create the Gantt chart, the team can use their responsibility matrix as a basis. Starting from the first task, create a bar corresponding to the length of the time to complete the estimate from the responsibility matrix. Next, incorporate the task or tasks that can be completed after the previous task is complete, beginning the bar where the previous task ended. You can add an arrow connecting the previous task's bar to the next to explicitly show the dependencies. Sometimes tasks may be disconnected and not strictly dependent on one another in a linear fashion. In that case, they can be placed where the team believes that task should begin, within the scope of the project. The same process of adding the tasks for which that task is a dependency would follow. Several software packages have either a Gantt chart template or the capacity to create one for you by entering the information on the responsibility matrix.

Step 2: Create a Responsibility Matrix

List the project's tasks and subtasks one by one down the left side of the paper. Then, create columns beneath each team member's name, written side by side across the top. Put a check mark in the column beneath the name of the member who agrees to perform each task. It then becomes the responsibility of that team member to successfully perform the task by the due date that was agreed upon in Step 1. An alternate grid is shown in Table 4-2.

Table 4-2 Sample Responsibility Matrix

Task	Completed By	Checked By
Purchase supplies	Pat and Chris by 9/15 team mtg	
Write initial proposal	Terry—Email to Robin by 9/22	Robin by 9/25 team mtg
Conduct preliminary calculations on height	Robin—Email to Pat by 9/22	Pat by 9/25 team mtg
Build prototype in lab	All—Lab: 7-9 p.m., 9/28	

In assigning the tasks, consider the complexity and time required for the job. One team member may have five small tasks while another may have one major task, with the goal being an equal distribution of effort. A second team member should be assigned to each task to assist or check the work completed by the first team member. Be sure all team members are comfortable with the assignments.

Step 3: Consider Team Dynamics

Communication: The success of any project depends to a great extent on how well the team members communicate. Do not hesitate to share ideas and suggestions with the group and consider each member's input carefully. Do not be afraid to admit that you are having difficulties with a task or that the task is taking longer than expected. Be ready and willing to help one another.

Trust and respect: Remember that the team is working toward a shared objective. Therefore, you must choose to trust and respect one another. Treat fellow team members with simple courtesy and consideration. Follow through with promises of completed tasks, remembering the team is counting on your individual contributions. Try to deal honestly and openly with disagreements. However, do not hesitate to ask for help from faculty if problems begin to escalate.

Nothing is carved in stone: It is important to plan the project as carefully as possible; however, unforeseen problems can still occur. Treat both the Project Timeline and Responsibility Matrix as working documents. Realize they were created to serve as guides, not as inflexible standards. Watch the project progress relative to the timeline, and do not hesitate to redesign, reallocate, or reschedule should the need arise. Review your matrix each week and adjust as needed.

Finally, do not forget to have fun!

4.8.2 Modern Project Management Frameworks

scrum master

product owner

development team

Modeled after the way that a rugby team works as a team to move the ball down the field toward the goal, scrum is a project management process that was popularized as a method of software development. The scrum method is distinguished from earlier methods in that it is holistic rather than linear, and flexible rather than sequential. This flexibility is particularly with respect to user requirements, which may change during the development process. Another reason for the rugby metaphor is that control of the project is passed back and forth among the team members, which makes the scrum method particularly appropriate in today's engineering workforce, where it is common for engineers to manage certain projects and participate in projects that others manage without any hierarchical management structure. As a result of this more shared leadership structure, the term **scrum master** is used rather than project manager. The scrum master is primarily responsible for ensuring that nothing hinders the team's progress—in a facilitative role. Another role in the scrum process is the **product owner**, who is the primary liaison to customers and clients and represents their voice in the design process. The remaining team members (as many as are needed) comprise the development team. The **development team** focuses on project deliverables—sharing their progress publicly as often as possible in completed, if minor, advancements.

Modern project management has evolved to have several different frameworks for ensuring projects are completed on time, effectively. Like scrum, they have their own intricacies and quirks that are designed to accomplish the same goal. There are far too many to detail in this text; however, you are encouraged to examine other frameworks like Agile, Kanban, Adaptive Project Framework (APF), and Lean.

4.9 Digital Workflows and the Digital Portfolio

digital workflow

automation

business communication platforms

Considering the pandemic's impact on how teams collaborated virtually, it is increasingly vital to be comfortable with digital workflows. A **digital workflow** is the **automation** of a company's (or team's) processes in addition to internal steps to accomplish the company's (or team's) goals. Companies and teams within companies will communicate using a combination of tools—sometimes integrated into one suite to delegate responsibilities, update team members on the progress of different tasks, and share relevant files. According to a 2016 study by Harvard Business Review, workers spend about 80 percent of their time at work in meetings, talking on the phone, and responding to emails. The most common tool within the scope of digital workflows to wrangle with this 80 percent of workers' time and streamline it are labeled **business communication platforms**. The platforms include text, voice, and video communication in addition to file sharing.

Some organizations may subscribe to a particular service that bundles several products together, meaning the products from one company may dominate day to day operations. For example, a company using Microsoft products will likely have Office 365 and Microsoft Teams. On the other hand, another organization could be subscribed to Google products. Or they might use an office messaging platform like Slack. Disciplines of

DISCUSSION

Consider having students read the following article from Asana, which details 12 project management frameworks:

URL
https://asana.com/resources/project-management-methodologies

Have students categorize them based on similarities and differences, the objectives of the frameworks, and which projects would be best suited for each one.

ENGAGE

Visit the Thayer School of Engineering at Dartmouth College website. Navigate to the "careers" link and choose "creating your portfolio" from the selections.

URL
https://career.engineering.dartmouth.edu/students/create-a-design-portfolio

Ask students what the very last suggestion is on the page. (*Back up your files, and then back them up again.*) Ask them why this is so important.

ENGAGE

Ask your students to complete a search for "free Web portfolio sites." The students can create a chart that lists the free sites and the options they make available to the user.

engineering and computing may be more likely to use a particular platform to manage their digital workflows. For example, those in computing with heavy programming responsibilities will use GitHub to host their various projects and version control, which allows multiple individuals to work on the same code without breaking the project with one bad addition.

In many ways, these file repositories function much like an engineering notebook. They contain sketches and can contain chronological notes, ideas, and even mistakes. However, they are different in many ways as well.

A digital portfolio is typically a website or a set of shared documents within a business communication platform detailing a design (Figure 4-8). It may be specific to one design, or it may be a record of multiple designs for a single person or team. In some cases, a person or team may choose to build a PowerPoint file and call this the digital portfolio in its original PowerPoint form. It could also be saved as a read-only document such as a .pdf file.

Because it is digital, the portfolio is expected to be effective as a visual presentation. Therefore, aesthetic ideas like layout, background, and organization can be important as you try to get your message across. In addition, a digital portfolio can be edited and rearranged, so it is not considered a chronological presentation of your design process. Instead, it is more useful as a tool to share your design. The reader doesn't need to see the details to see significant accomplishments or final ideas.

Figure 4-8 The front page of your digital portfolio should give a good overview.

Further, because it is digital, the portfolio can be far more interactive than an engineering notebook. Imagine da Vinci's notebooks if they could have links and animation!

Teams can create a digital notebook using shared document space on the Cloud, such as Google Docs. Some sites allow individuals to build their own websites; one of these could host your digital portfolio. If you're in a computing-based profession, you'll most likely be using GitHub to manage your projects.

If you have browsed through pages on the Internet, you have probably seen some pages that look "right" and others that are simply unappealing or difficult to use. Trust your instincts when designing your Web-based digital portfolio. If you want an audience to discover, view, and share the resource, the layout should not detract from your message. Here are some tips that may be helpful:

- Follow company guidelines for file storage and presentation. If these are flexible, communicate with your team regarding the shared norms for developing and maintaining the digital portfolio.
- Create a visually appealing homepage with clear links to projects.
- Make your best products appear first to draw the attention of the viewer.
- Describe your solution.
- Consider your audience as you describe your solution.
- Be sure to link back to the original problem and/or to reflect on what the solution may mean to you or to society.
- Have a backup copy of files where appropriate. Companies often change their subscriptions to digital workflow solutions, so it is not unusual for you to need to migrate your files to a new service.

CHAPTER 4 ASSESSMENT QUESTIONS

1. If you are doing a design at school and your design will probably be very similar to other designs, why should you document your work in an engineering notebook?
2. Describe the key differences between an engineering notebook and a digital portfolio.
3. Describe how negligent and improper maintenance affects robot performance.
4. Research and describe two applications of robots and automation.
5. Research to identify an emerging trend in robotics and automation. Why is it an emerging trend? What types of engineers would be involved in this area?
6. Research: Find an advertisement for a job involving robotics and automation on the Internet. What type of engineer is invited to apply? What are some of the responsibilities of that position?
7. Research: Identify five inventions credited to Thomas Edison.
   (a) How did one of these inventions change our lives (if it did)?
   (b) Can you find any evidence of Edison's documentation on each invention? If a picture from the design process exists, capture the image.
   (c) If you were working on one of these inventions, how would the entry for this invention begin in your engineering notebook?
   (d) Select one design on which to focus: Describe how this invention affected:
      (i) Individuals
      (ii) Society
      (iii) Economics within our society
      (iv) Culture: your own culture or a different culture
      (v) The environment
   (e) Consider these five inventions. Can you generate an idea of an invention you would like to see, or perhaps would like to invent? Describe the invention as it would exist in 10 years. How will this affect society? On your own or with a partner, design a presentation to present this idea to your class.
8. Perform a functional decomposition for a laser printer. Draw a functional block diagram for a laser printer to demonstratate your result.
9. Draw a functional block diagram for the following:
   (a) Computer mouse
   (b) Pasta maker
   (c) Drone
   (d) Panini maker

118 Chapter 4 The Engineering Design Process

10. An engineer temporarily begins work on a new project, putting the project she worked with for two years to the side. As the newer project wraps up, she is ready to bring the older project back and begin working on it. Which of the following are true statements if she followed all of the guidelines when keeping her notebook?
    (a) Her old engineering notebook was, fortunately, a perfect record of every design decision, every prototype image, and every step taken, so it's a perfect record of her design to this point.
    (b) Every page is numbered and dated.
    (c) The only pages that have been torn out were those with mathematical mistakes.
    (d) The notebook must be stored in a safety deposit box, so she will need the key to access it.
11. Write a paragraph describing the Kanban framework.
12. Are there any engineering disciplines that typically use an Agile framework? Why or why not?
13. Your team has been tasked to design a new smartwatch, which must sell to the consumer for less than $200. What are three reasonable criteria, and what are three reasonable constraints your team may have to work with? Describe the steps in an engineering design process, including the design process and prototype development and initiating, planning, executing, monitoring and controlling, and closing the project.
14. Select a device from your kitchen and describe how well it meets each of Dieter Rams' design criteria. Submit an image of the item you selected as part of your analysis.
15. Work with a team and answer the following in a shared document. You work closely with someone who owns a pizza shop. Suppose your team is tasked with designing a device that automatically cuts pizza after cooking, after it's placed in a box. Describe how each of Rams' principles of good design would affect the design of such a device.
16. Describe why each of these is important to show proficiency in the engineering design process.
    (a) Prioritizing tasks
    (b) Developing and following schedules
    (c) Performing goal-relevant activities
17. Your team has been tasked to design a new video game controller to be worn like a glove. Describe how your team would engage in each of these processes:
    (a) Identify and define the design problem
    (b) Brainstorm solutions
    (c) Create models and build a prototype
    (d) Test the prototype
    (e) Redesign and optimize
18. Describe how this design process would be considered to be iterative and subject to continuous evolutionary improvement.

OVERVIEW

This chapter digs deeper into the specific tools to accomplish the steps of the design process. In particular, we focus on different models of brainstorming, organizing the ideas of brainstorming using a morphological chart, and comparing designs using a decision matrix. A conceptual design project closes the chapter, which takes students step by step through a model design process.

CHAPTER 5

Tools for Design

TOPICS

As we alluded to in the previous chapter, there are several techniques that allow engineers to systematically generate ideas that could be translated into prototypes and rigorously tested. Although there are way too many different techniques to describe here, we will discuss a few general methods for generating concepts and deciding among them.

5.1 The Team as the Foundation for Ideating

Throughout the ideating process, the team is going to serve as the backbone for the design process. This is especially true in terms of coming up with viable solutions. You will all be responsible for using one or more of the tools discussed in this chapter and synthesizing those findings. Therefore, the team needs to function as a cohesive unit. After all, what distinguishes a team from an ordinary group of individuals?

The ability to work in a team is one of the most critical traits an engineer needs. Even if you're the greatest engineer in the world, you will not know all of the answers or have all of the right ideas. We can always learn something from our peers.

A. Thompson, EE

Group: A number of people who come together at the same place, at the same time.
Team: Individuals cooperating to accomplish a common goal .

As a student and in the workplace, you will complete some assignments individually and complete some as part of a team. When you work independently, you are mostly free to choose when and how you will work. When you work as part of a team, make sure the team has ground rules for how it will operate. Diversity within teams is a positive and leads to more effective teams. Interdisciplinary teams ensure that ideas from different perspectives are taken into account. Any time several people are asked to work closely together, there is a potential for much good from a diversity of ideas and skills, but there is also a potential for conflict. Because conflict can be both productive and unproductive, you need to manage it.

5.1.1 Team Behavior

The most critical task for a team, particularly a new team, is to establish its purpose, its process (its way of doing things), and a means of measuring team progress. Here are several topics regarding team behavior that you may wish to consider:

119

OBJECTIVES

- Identify and define an engineering problem
- Understand the importance of inter-disciplinary teams
- Use collaborative tools such as desktop or web-based applications to share and develop information
- Take minutes of a team meeting
- Create and utilitze an engineering notebook per established conventions
- Maintain a design and computation engineering notebook
- Formulate goals, objectives, and requirements to solve an engineering problem
- Predict performance, failure modes, and reliability of a design solution
- Identify the requirements for and role of intellectual property in design
- Demonstrate proficiency of the engineering design process
- Use clear and concise written, verbal, and visual communication techniques
- Maintain a paper or digital portfolio using the engineering documentation process
- Use problem-solving techniques to develop technological solutions such as product, process, or system
- Apply the design process, including decision matrices, as a team participant
- Perform different roles within the project as a team member
- Demonstrate communication skills by preparing and presenting the project, including building consensus setback resolution and decision matrices

ENGAGE:

Teaming is a crucial skill that engineers must have to be effective in the workforce. Indeed.com provides seven examples of more specific teamwork skills.

7 Examples of Important Teamwork Skills
https://www.indeed.com/career-advice/career-development/teamwork-skills

What are the seven important teamwork skills highlighted in the Indeed article? *Communication, responsibility, honesty, active listening, empathy, collaboration, and awareness.*

ENGAGE

If you will be using teams extensively in class, it would be beneficial to have teams complete a *code of cooperation.* In the code of cooperation, the team specifies their goals, norms for how the team is supposed to function (including processes like how shared files are stored, communication styles, and scheduling agreements), individual norms (including expectations on workload and conduct), and consequences for individuals not obeying the established norms. Finally, the code of cooperation specifies a plan to ensure psychological safety, which concerns making everyone comfortable with contributing ideas to the team.

The items in the code of cooperation need to be specific. If the norms are too vague, for example, then it will be difficult to enforce.

TEACHING TIP

When teams are exhibiting dysfunction, you can ask for their code of cooperation to help them navigate disagreements. Use the opportunity to have students craft more specific objectives if you believe the team does not appear to be using them effectively.

We trained hard but it seemed that every time we were beginning to form teams we would be reorganized.

I was to learn later in life that we tend to meet every situation by reorganizing, and a wonderful method it can be for creating the illusion of progress while producing confusion, inefficiency, and demoralization.

Petronius

- **Ground rules:** Each team needs to come to a consensus about acceptable and unacceptable individual behaviors as well as team behavior.
- **Decision making:** Teams by necessity make decisions. Each team needs to decide how these decisions will be made. For example, will they be done through consensus, majority vote (either secret or show of hands), or by other methods?
- **Communication:** This is often one of the hardest parts of working effectively as a team. Team members need to recognize the value of real listening and constructive feedback. During the course of team meetings, *every* team member needs to participate and listen.
- **Documentation:** Whenever the team meets, meeting minutes should be taken and saved. Shared online documents are often a good choice.
- **Roles:** You may adopt various roles on your team. In a long-term project, roles should rotate so that everyone has a chance to learn each role. Some typical roles are a meeting organizer who plans the meeting, including reviewing what tasks team members should complete before the meeting; a recorder who keeps track of decisions made and assignments agreed upon; a timekeeper who alerts the team if a topic is taking too long; and an encourager-gatekeeper-facilitator who helps ensure that everyone contributes and is heard during team meetings.
- **Participation:** Decide as a team how work will be distributed. Your team should also consider how to handle shifts in a workload when a team member is sick or otherwise unavailable.
- **Values:** The team as a whole needs to acknowledge and accept the unique insights that each team member can contribute.
- **Outcomes:** Discuss and agree on what types of measures will be used to determine that the team has reached its final goal.

I work with a chemical engineer, an instrumentation engineer, a mechanical engineer, and a piping engineer.

D. Jones, BE

Every project involves a team. My typical team includes surveyors, structural engineers, environmental scientists (wetlands, endangered species), permit specialists, electrical engineers, land appraisers, archaeologists, and architects.

J. Meena, CE

I work alone, but seek advice from my management team.

C. Pringle, IE

I work on a team including mechanical, electrical, aeronautical and systems hardware designers, software designers, integrators, power engineers, and other liaisons like myself.

E. Styles, EE

The Hubble Space Telescope servicing mission project is made up of a team of mechanical design engineers, human factors engineers, environmental test engineers, thermal engineers, underwater test engineers, safety engineers, mechanical technicians, documentation specialists, space scientists, and systems engineers.

R. Werneth, ME

5.1.2 Teammate Evaluation: Practicing Accountability

Engineering is a self-governing profession. ABET (pronounced with a long A) is charged with accrediting engineering programs in the United States through membership from the profession. Among a set of required outcomes of engineering graduates, ABET requires that engineering students graduate with an ability to function on multidisciplinary teams. To ensure that each student achieves this outcome, individual accountability is needed.

Realizing that much of the activity of a team happens when the team is meeting privately (without a professor), an effective and increasingly common way of addressing this tenet is to have team members provide feedback to each other and to the professor.

It is important to learn how to be an effective team member now, because most engineering work is done in teams, and studies show that most engineering graduates will have supervisory responsibility (at least project management) within five years of graduation. You may have worked in a team before, and if so you have probably noticed that some team members are more effective than others. Consider these three snapshots of interactions with engineering students. All three are true stories:

- Three team members approach the professor, concerned that they have not seen the fourth team member yet. The professor speaks with the student, who quickly becomes despondent, alerting the professor to a number of serious burdens the student is bearing. The professor alerts an advisor and the student gets needed help.
- A student, acting as the team spokesperson, tells the professor that one team member never comes to meetings. The professor speaks to the nonparticipating student, who expects to be contacted about meetings by smartphone, not email. The professor explains that the student's expectations are unrealistic.
- A student is insecure about being able to contribute during team activities. After the team's ratings of that student are in, the professor talks to the class about the importance of participating and the different ways students can contribute to a team. In the next evaluation, the student receives the highest rating on the team.

Peer evaluations are a useful way for team members to communicate to one another and to their professor about how the members of a team are performing. Reviewing and evaluating job performance is a marketable skill and is as useful to the employee seeking a job or a job advancement as it is to the supervisor.

ENGAGE

A critical component of peer evaluations is giving constructive feedback. The following article from BetterUp provides five ways teammates can provide constructive feedback. Have students read the following article, then ask: What tips did the article provide for giving constructive feedback? *Scheduling a 1-on-1 feedback session, keep a positive tone and delivery and be specific, focus on only a few items at a time, create a plan for improvement collaboratively, and follow up to recognize growth.*

URL

https://www.betterup.com/blog/constructive-feedback

TEACHING TIP

If you will be leveraging teams significantly in the course and you have a sizable class, it would be helpful to use team formation software like CATME.

Random assignment is typically not effective for long-term teams.

URL

https://info.catme.org/

Source: NASA

Underwater Training for a Hubble Space Telescope Servicing Mission. NASA engineers (on SCUBA) and astronauts (in modified space suits) take advantage of the effects of neutral buoyancy to practice replacing a Wide Field Camera in preparation for a Hubble Space Telescope (HST) servicing mission. Extensive teamwork is required in a 6.2-million gallon pool in Houston, TX, to develop, refine, and practice the procedures to be used on spacewalks in orbit. Engineers from Goddard Space Flight Center in Greenbelt, MD, and Johnson Space Center in Houston work together to perfect the procedures and hardware, including specialized astronaut tools. The team uses models representing the flight items for conducting end-to-end tasks in the neutral buoyancy facility. This unique example of engineering teamwork has resulted in five very successful HST servicing missions involving complex astronaut spacewalks. The challenge to the team is to develop the nominal and contingency procedures and tools on the ground (and in the water!) to be used for mission success with HST in orbit 300 miles up.

R. Werneth, ME

5.1.3 Focus on What Your Teammates Do Rather Than What You Think of Them

It is challenging to give a team member a single rating on their effectiveness as a teammate because some team members will be helpful to the team in some ways, but engage in some behaviors that hinder the team. Another difficulty is that each team member is likely to consider some ways of contributing more valuable than others, so the evaluation of a particular teammate will be overly influenced by that teammate's performance in certain areas. The only way to be fair is to focus on behaviors—what your teammates do—rather than opinions, such as how you feel about them. One way to focus on behaviors would be to ask you to take an inventory of what behaviors your teammates demonstrate and how often. The result would be that you might need to answer 50 or more questions about each member of the team. It is difficult to stay focused on answering accurately when completing such a long survey. A better way to focus on behaviors is by using sample behaviors to anchor each point of a rating scale. A peer evaluation instrument that is widely used in engineering education (and in other fields) is the Comprehensive Assessment of Team-Member Effectiveness (CATME, see *http://info.catme.org/*). CATME measures five different types of contributions to a team using such a behaviorally anchored rating scale. Each scale includes representative behaviors describing exceptionally good, acceptable, and deficient performance in each area. Each team should determine what the level of expected behavior should be. Recognizing that an individual team member may exhibit a combination of behaviors, the CATME instrument also includes in-between ratings. The five types of contributions are described below the associate behaviors.

Contributing to the Team's Work describes a team member's commitment to the effort, quality, and timeliness of completing the team's assigned tasks.

- A student who is exceptional at contributing to the team's work
  - Does more or higher-quality work than expected
  - Makes important contributions that improve the team's work
  - Helps to complete the work of teammates who are having difficulty
- A student who does an acceptable job at contributing to the team's work
  - Completes a fair share of the team's work with acceptable quality
  - Keeps commitments and completes assignments on time
  - Fills in for teammates when it is easy or important
- A student who is deficient at contributing to the team's work
  - Does not do a fair share of the team's work; delivers sloppy or incomplete work
  - Misses deadlines; is late, unprepared, or absent for team meetings
  - Does not assist teammates; quits if the work becomes difficult

Interacting with Teammates measures how a team member values and seeks contributions from other team members.

- A student who is exceptional at interacting with teammates
  - Asks for and shows an interest in teammates' ideas and contributions
  - Improves communication among teammates; provides encouragement or enthusiasm to the team
  - Asks teammates for feedback and uses their suggestions to improve
- A student who does an acceptable job at interacting with teammates
  - Listens to teammates and respects their contributions
  - Communicates clearly; shares information with teammates; participates fully in team activities
  - Respects and responds to feedback from teammates
- A student who is deficient at interacting with teammates
  - Interrupts, ignores, bosses, or makes fun of teammates
  - Takes actions that affect teammates without their input; does not share information
  - Complains, makes excuses, or does not interact with teammates; accepts no help or advice

Keeping the Team on Track describes how a team member monitors conditions that affect the team's progress and acts on that information as needed.

- A student who is exceptional at keeping the team on track
  - Watches conditions affecting the team and monitors the team's progress
  - Makes sure teammates are making appropriate progress
  - Gives teammates specific, timely, and constructive feedback

124 Chapter 5 Tools for Design

- A student who does an acceptable job at keeping the team on track
  - Notices changes that influence the team's success
  - Knows what everyone on the team should be doing and notices problems
  - Alerts teammates or suggests solutions when the team's success is threatened
- A student who is deficient at keeping the team on track
  - Is unaware of whether the team is meeting its goals
  - Does not pay attention to teammates' progress
  - Avoids discussing team problems, even when they are obvious

Expecting Quality is about voicing expectations that the team can and should do high-quality work.

- A student who is exceptional at expecting quality
  - Motivates the team to do excellent work
  - Cares that the team does outstanding work, even if there is no additional reward
  - Believes that the team can do excellent work
- A student who does an acceptable job at expecting quality
  - Encourages the team to do good work that meets all requirements
  - Wants the team to perform well enough to earn all available rewards
  - Believes that the team can fully meet its responsibilities
- A student who is deficient at expecting quality
  - Is satisfied even if the team does not meet assigned standards
  - Wants the team to avoid work, even if it hurts the team
  - Doubts that the team can meet its requirements

Having Relevant Knowledge, Skills, and Abilities accounts for both the talents a member brings to the team and those talents a member develops for the team's benefit.

- A student who has exceptional knowledge, skills, and abilities
  - Demonstrates the knowledge, skills, and abilities to do excellent work
  - Acquires new knowledge or skills to improve the team's performance
  - Is able to perform the role of any team member if necessary
- A student who has an acceptable level of knowledge, skills, and abilities
  - Has sufficient knowledge, skills, and abilities to contribute to the team's work
  - Acquires knowledge or skills needed to meet requirements
  - Is able to perform some of the tasks normally done by other team members
- A student who has deficient knowledge, skills, and abilities is
  - Missing basic qualifications needed to be a member of the team
  - Unable or unwilling to develop knowledge or skills to contribute to the team
  - Unable to perform any of the duties of other team members

Comprehension Check 5-1

Research shows that team performance can be enhanced if team members reflect on their own and their teammates' performance and give each other high-quality feedback. High-quality ratings are consistent with observed behavior, which may or may not be high ratings. We also know that rating quality (again, consistency with observed behavior) improves with practice. Guided practice in giving and receiving feedback and in practicing self- and peer evaluations using behavioral criteria will help you improve. Please take your time in evaluating the members of the following fictitious team.

Pat Friendly and very well-liked, makes working fun, and keeps everyone excited about working together. Relies on teammates to make sure everything is going okay. Pays attention to keeping the team upbeat but does not seem to notice if the team's work is getting done. Struggles to keep up with the rest of the team and often asks teammates for explanations. The team has to assign Pat the least difficult jobs because Pat does not have the skills to do more complex work. Offers ideas when able, but does not make suggestions that add anything unique or important to the final product. Always shows up for meetings, prepares beforehand, and does everything promised. Is confident that the team can do everything that is essential. Agrees that the team should meet all explicit task requirements.

Chris Okay as a person and does not interfere with the contributions of others but rubs teammates the wrong way by frequently griping about the work and making excuses for not following through on promises to the team. Chris has the brains and experience to make a unique and valuable contribution, but does not try. The fact that Chris is so smart frustrates some teammates who have to try hard to accomplish tasks that would be easy for Chris. Ignores assigned tasks or does a sloppy job because "Robin will redo the work anyway." Misses meetings or shows up without assigned work. Contributes very little during meetings. Was late to one meeting because "no one told me the meeting time." Missed another meeting because "the alarm clock did not go off." After missing meetings, he asks lots of questions to make sure that everyone is making progress and the team's work is being accomplished. Spends more time checking that everyone else is doing their work than getting the job done. Chris always seems sure that the team will do fine and says that the team should do good work that fully meets the standards for acceptable performance. In response to a teammate's question about Chris' failure to deliver a promised piece of work, Chris said, "Why should I bother? Robin won't let the team fail."

Robin Very bright. Has far greater knowledge of the subject than any of the other team members. Extremely skilled in problem solving. Robin has very high standards and wants the team's work to be impressive, but Robin worries whether the team's work will be good enough to stand out. Robin completes a big chunk of the team's work and takes on a lot of the really difficult work. Does the work that Chris leaves unfinished. The quality of Robin's work is consistently outstanding. Tends to just work out the solutions and discourages teammates' attempts to contribute. Reluctant to spend time explaining things to others. Does not like to explain "obvious" things. Is particularly impatient with Pat's questions and once told Pat "You are not smart enough to be on this team." Complains that Chris is a "lazy freeloader." Sometimes gets obsessed with grand plans and ignores new information that would call for changes. Does not pay attention to warning signs that the current plan might not be effective until the problems are obvious. Then handles the situation as a crisis and takes over without getting team input. Robin is reluctant to acknowledge or discuss problems in the team until they affect his work.

Terry Not nearly as bright as Robin, but works to develop enough knowledge and skills to do the assigned tasks. Terry can usually fill in for other team members if given specific directions, but does not understand most of the tasks that other team members normally perform. Does more grunt work than any of the other team members, but does not do as good a job as Robin and does not take on difficult tasks. Sometimes makes mistakes on the more complex work. Super responsible, spends a lot of time giving one-on-one help to Pat. Always on time to meetings. Often calls to remind everyone (especially Chris) about meetings and usually makes some nice comment about one of the teammate's strengths or a valuable contribution that the teammate has recently made. Terry is outgoing and highly supportive

ENGAGE

There are numerous examples of biomimetics/biomimicry. Vox has a video on biomimicry that could be useful to show. It primarily uses the example of the bullet train and the kingfisher.

The world is poorly designed. But copying nature helps.
https://www.youtube.com/watch?v=iMtXqTmfta0

What are the three main ways that engineers can mimic nature? *Engineers can mimic shape (form) [self-cleaning surfaces for paint on a car], process [ant communication used in software/algorithms], and ecosystem [circular economy, the idea that all "waste" is used in something else and not something that ends up in a landfill].*

INTEGRATE CONTENT AREAS

Have students research other examples of biomimicry and report their findings about how the biology in the application informs engineering design. A few examples are listed here: down feather jackets, termite mound cooling, shark skin antimicrobial film, shock absorbers from woodpeckers, Velcro, and adhesives from mussels.

ENGAGE

For a good laugh, you may consider showing your students some of the absurd patents that have been filed in the past. The website called Totally Absurd Patents keeps a catalog of them.

URL

www.totallyabsurd.com/inventionsarchive.htm

Comprehension Check 5-1 *(continued)*

of teammates when well-rested, but is sometimes too tired to get excited about teammates' ideas. Is not defensive when teammates offer feedback, but does not ask for teammates' suggestions, even when teammates' input could help Terry to do better work. Terry thinks that the team can do great work and encourages teammates to do their best. When Robin expresses doubts if the team can do superior work, Terry reassures everyone that the team is capable of outstanding work. When the team is headed in the wrong direction, Terry is quick to notice and say something, but usually does not suggest a way to fix it. Terry reviews the team's objectives and alerts the team to anything that comes up that would affect the team. Terry was reluctant to press the issue when Robin's plan ignored one of the guidelines specified for the project.

To test your ability to focus on individual behaviors, go to *https://www.catme.org/login/survey_instructions* and rate each team member on each type of contribution to the team. On the Scenario Results page, a green arrow indicates that your rating matches the expected rating. If your rating does not match the expert rating, the blue arrow shows your rating and the red arrow indicates the rating experts would have assigned. If you count one point for every level separating your rating from the expert rating on the five different types of contribution, a low score is best, indicating the greatest agreement with the expert ratings. You can mouse over the red arrows to read the rationales underlying the expert ratings.

5.2 Drawing Ideas from Prior Art and Nature

While brainstorming is well known, it is far from the only approach to generating ideas. The most common approach to solving a problem is to find out if it has been solved before. The biblical proverb, "there is nothing new under the sun" shows that it has long been known that all of our ideas have roots in our previous experience—even if we cannot explicitly identify where an idea came from. This approach can range from "phone a friend" to a comprehensive search of published material, including journals, patents, and product catalogs.

Contacting a colleague to ask if any similar problem has been solved previously is most effective for engineers who have large, diverse networks to draw on. An experienced engineer with a robust network may know just the person to go to when certain kinds of questions arise:

- *"Javon, I've got to put some electronics where they will be exposed to gamma radiation—what are my options for shielding?"*
- *"Isabela, I've run into a problem—how can I damp the vibration in this equipment so that the operator doesn't feel uncomfortable?*

prior art

biomimetics

A more formal study of the way a problem has been solved before is called research on **prior art**, particularly when discussing existing solutions in a legal sense related to a patent application. More generally, this sort of research will extend to related products. Printers, fax machines, and copiers have many common technologies, so an engineer designing a new copy machine would be expected to be familiar with recent advances in the other two technologies. A special subset of prior art is known as **biomimetics**—exploring how a particular problem has been solved in nature. Birds were studied for many years in the pursuit of a technological means of human flight, and more recently, the structure of the wings of an owl that permit silent flight have been studied in hopes of reducing the noise made by aircraft.

Another example of biomimicry is the design of the bullet train. Previously, a significant pressure change would occur whenever a train would enter a tunnel because of the air being pushed by the speeding train. The pressure would continue to build until the train left the tunnel, at which point the pressure would be released all at once. This would result in a high intensity sound, often called a sonic boom. While birdwatching, Dr. Eiji Nakatsu noticed that a particular type of bird, called a kingfisher, was able to quickly dip its beak into water in search of food without making a significant splash. From this observation, he drafted a concept for a bullet train matching the curvature of the kingfisher's beak (Figure 5-1). This insight allowed the air to pass more freely past the train, reducing the pressure buildup causing the sonic booms.

There are various systems of questions that encourage the exploration of a larger design space from a more diverse perspective. Examples are Edward de Bono's Six Thinking Hats and Alex Osborn's checklist, and more recently Tom Kelley's Ten Faces of Innovation.

Figure 5-1 Biomimicry: Kingfisher and bullet train.

5.3 Models of Brainstorming

brainstorming

Most high school graduates have heard the term **brainstorming**, and many have participated in a process called by that name. This term refers to one particular process by which ideas are generated. Idea generation follows three common rules: encourage a lot of ideas, encourage a wide variety of ideas, and do not criticize. The third rule is important if the first two are to be achieved. It has been said: "The best way to get a good idea is to get lots of ideas!" Brainstorming may be useful if individuals are stuck and hearing the ideas of others helps get them thinking differently, but individuals thinking separately and then pooling their ideas typically outperform the results of brainstorming as a group. Sometimes, brainstorming fails because participants don't follow the rules well, but there are other processes that limit the productivity of brainstorming. In brainstorming, only one person can contribute at a time without interrupting, so ideas may be lost or idea generation might be slowed while participants wait their turn. Further, some participants will limit their participation out of laziness, introversion, or social pressure to avoid dominating discussion. More effective than brainstorming is generating ideas in a group where any idea is welcome, but critical reflection is permitted to spend more time on productive directions as long as the group doesn't run out of ideas. It's still important that everyone in the group feels safe to contribute, but critical reflection makes better use of everyone's time. Formal brainstorming may be useful in groups where criticism is likely to prevent some people from contributing.

Defining criteria before generating ideas seems as if it might threaten the "do not criticize" objective—proposed solutions could be steered to match the expectations established by the criteria and thus limit the range of ideas. While defining criteria first might constrain idea generation, generating solutions first might cause participants to develop an opinion of a "best solution" and then force their criteria choices to fit their preferred idea. The process of generating solutions and the process of generating criteria to evaluate those solutions are linked. The second interaction is more damaging to the quality of the design process, so we recommend identifying constraints and criteria first.

TEACHING TIP

The idea of generic brainstorming has been praised without much critique on its weaknesses, especially how the rule about not discounting wild ideas is applied. Thus, hybrid brainstorming, with an individual phase before decision-making with the team is preferred to preserve team member anonymity in producing ideas. It's suggested that hybrid brainstorming be used in the classroom as a general rule. The specific procedures such as 6-3-5 Brainwriting or the Six Thinking Hats is up to the team/group.

INTEGRATE CONTENT AREAS

As a quick combinatorics question, you can ask students how many ideas will be generated by a group of *n* people using the same rules: *The total number of ideas is 3*n^2. Each round gives 3*n new ideas; with n people, there are n rounds. Therefore, you get n papers with 3*n ideas. The product of n and 3n is then the total number of ideas.*

ELL

Students who struggle with verbal communication might be more comfortable with writing brainstorming ideas first, but the time limit could be an issue. Thus, you may suggest such teams do this process asynchronously with enough time that the student(s) can comfortably communicate their ideas.

5.3.1 Hybrid Brainstorming

Note that while brainstorming is widely characterized as a way to generate numerous ideas that are assumed to be better than what can be developed independently, the method is not without its detractors. Given the need for a safe space to generate off-the-way, potentially silly ideas, the potential issues quickly bubble to the surface. Because some individuals are more comfortable than others to share ideas, the potentially transformative and wild ideas may not be spoken. The reluctance to share ideas because the individual fears a negative reaction is called **evaluation apprehension**. Some team members may not participate because there is no intent to measure individual output, the only concern of brainstorming is the collective output—a long list of potential ideas. You may have had a group project in the past where the team received a good grade, but one team member didn't bother contributing to the final product at all (hopefully you weren't that team member). This phenomenon is called **free riding**. Finally, one team member may be the one driving the conversation, limiting opportunities for others to speak—a process called **production blocking**. Considering these issues, there are a few structured methods of brainstorming teams should consider.

evaluation apprehension

free riding

production blocking

hybrid brainstorming

To avoid the issues raised for traditional brainstorming, it is suggested that teams use a hybrid brainstorming model. **Hybrid brainstorming** involves two phases—one individual and one team. In the first phase, each team member generates a list of ideas individually with absolutely no communication with the team. After everyone has developed a list, anonymize them and begin grouping similar ideas into categories as a team. After grouping the ideas into themes, rank order the ideas individually. Come back together as a team and compare the rankings. From there, pursue the most frequently highly ranked ideas.

5.3.2 "6-3-5" Brainwriting

A similar model for brainstorming to sidestep the issues with typical brainstorming is called 6-3-5 Brainwriting. Developed by Bernd Rohrbach in 1968, the goal of 6-3-5 Brainwriting is to develop a large list of ideas in 30 minutes—108 ideas to be precise. Not only does it produce a lot of ideas, but the method is also simple to implement.

First, gather a team of six individuals who have a deep knowledge of the needs and requirements for the design; otherwise the quality of the ideas could be impacted. Although six is the traditional number, it's possible to do the exercise with four, five, or seven people. Too many people in the group would make the exercise a bit too unwieldy. After clarifying the problem statement for which the team will be brainstorming, a session looks like the following:

1. Everyone begins with a blank sheet of paper. This sheet could be a template with a 6×3 grid on it (6 rows and 3 columns). The number of rows can be the team size, but typically the number of rounds is 6.
2. Round 1 begins. Each person has 5 minutes to write down 3 ideas on their sheet.
3. After 5 minutes, everyone passes their sheet to the person on the left.
4. The next round begins. Each person has 5 minutes to write down 3 ideas on their sheet *that builds on the ideas listed so far.*
5. After 5 minutes, everyone passes their sheet to the person on the left.
6. Continue steps 4 and 5 until 6 rounds pass (or the number of rounds equal to your team size).
7. After the last round, the team then discusses the ideas developed on each of the sheets and decides which are the top ideas to pursue.

The key here is that there is *no verbal communication* at all during the 6-3-5 Brainwriting exercise. The only verbal communication comes after the final round of writing. 6-3-5 Brainwriting follows the hybrid brainstorming model, with an individual phase followed by a team phase. Screening the ideas can include the grouping performed in the generic hybrid brainstorming model in addition to the rank ordering of the ideas.

Note that there are several different ways teams can generate ideas, 6-3-5 is only one specific *structured* method of doing so. Just keep in mind that a hybrid model will help the team overcome the typical issues with brainstorming like free riding, evaluation apprehension, and production blocking.

5.3.3 The Six Thinking Hats

vertical thinking

lateral thinking

Problems require a combination of thinking patterns, two examples of which are vertical and lateral thinking. **Vertical thinking** involves using a linear, sequential process for solving problems. The focus is on beginning with a set of givens and existing knowledge, then narrowing down possibilities systematically until a final single solution is produced. On the other hand, **lateral thinking** begins a similar way, but expands out in different directions—not necessarily in a linear fashion—to generate multiple possible ideas. Although these thinking patterns have their place in the engineering design process, sometimes people can get bogged down in a particular mindset. Perhaps you've been in a group that is all positive, everyone is optimistic, and no ideas are being given any substantial critique. Maybe you've also been in a situation where no idea can quite get beyond the first few sentences before it's completely discarded. How do you focus the team on a specific mindset so potential solutions can be explored comprehensively?

six thinking hats

Enter the **six thinking hats**, created by Edward de Bono, which first appeared in his 1985 book, *Six Thinking Hats*. The premise of the six thinking hats is to engage the team in viewing a problem through six different perspectives. Each metaphorical hat is given a color to act as a shorthand for a specific thought process the wearer should use: blue, green, red, yellow, black, and white. During the brainstorming process, the team can "put on their blue hats" or "put on their red hats," which corresponds to cycling through the six perspectives in Table 5-1.

Table 5-1 The Six Thinking Hats and their Purposes

Hat Color	When wearing this hat, you...
Blue	will ask for summaries, results, and draw conclusions. You are most concerned about process. Remind the group about the problem you're trying to address. What are the benefits?
Green	will use lateral thinking to generate the most ideas possible. Explore the widest range of ideas possible without constraints.
Red	will express your feelings without needing to justify them. Collect opinions.
Yellow	will only look for the positives in the presented ideas. Consider all the possible benefits, no matter how infeasible they may feel.
Black	will be cautious and skeptical of ideas. Be critical and point out everything that could go wrong with the presented ideas. *It's easy to overuse this one, don't let it dominate the brainstorming.*
White	will be focused on the facts and given information. Consider the data you have and could retrieve. Scope out the problem.

Note that some colors have cultural implications that are not related to the thought pattern associated with the hat. For example, although the green hat is about lateral thinking and being as creative as possible, the color green can be associated with infidelity in Chinese culture—obviously not very positive. Ultimately, the actual colors of the hats are completely irrelevant. What matters is cycling through the thinking patterns the hats encapsulate.

In order to use the hats, there are several different orders that a team can try. One suggested order is **Blue > White > Green > Yellow > Red > Black**. Each hat is "worn" for about two minutes.

Hats can be reintroduced as needed by the team leader. The sessions will always begin and end with Blue because the Blue hat is focused on setting the stage for the problem that the other hats will be used to scope out solutions and make decisions. You are free to vary the hats as needed for the application.

DISCUSSION

After completing the online instrument, have students discuss their results in groups. You may also form hypothetical teams to have students predict how effective the team might function.

All Black Hats: *Will likely be unable to move forward on an idea, pessimism would overwhelm the team and stall progress.*

All White Hats: *Might not appropriately filter out relevant information or be sensitive to specific populations that could be impacted by the solution (i.e., Green Hat thinking).*

All Green Hats: *Would be able to generate a lot of ideas, but might be unable to narrow down their options effectively.*

TEACHING TIP

Emphasize that vetical and lateral thinking are *both* needed to come up with suitable solutions to a design problem. If only vertical thinking is performed, it's possible that students could get stuck in a narrow set of options that rely too heavily on a specific body of knowledge or existing/familiar designs. This is called *design fixation.*

TEACHING TIP

In order to use the hats effectively, it would be useful to model how a team would move through these hats in a brainstorming session. Consider giving the students a simple design challenge, such as a device that would improve the lives of high school students and walk through possible product ideas using each hat.

ENGAGE

Students can take an online quiz to get a score for which hat they most align with. One such quiz is given in the following URL:

Six Thinking Hats - Jensen Test
https://www.gotoquiz.com/six_thinking_hats_jensen_test

You may consider forming teams using the Six Thinking Hats as one of the factors. One of the guiding ideas of the team formation with the Six Thinking Hats is to avoid placing more than one Black hat on a single team.

ENGAGE

To differentiate Wrong Theory Protocol from other brainstorming techniques, ask the following: What is considered the optimal solution when using Wrong Theory Protocol? *Emphasize that Wrong Theory Protocol is about creating more empathetic designs; it places explicit priority on designs that are ethical and humane.*

TEACHING TIP

To get the solution of Wrong Theory Protocol, the team may generate some uncomfortable ideas during the initial ideation phase. Without proper safeguards in the class to avoid offensive ideas, Wrong Theory Protocol is likely inappropriate for an in-class activity. Be very careful about outlining ground rules for what is appropriate in classroom discussion before attempting a brainstorming session.

5.3.4 Wrong Theory Protocol

In 2020, Vanessa Svihla and Luke Kachelmeier presented a new ideation technique to help designers focus on creative and empathetic ideas. Although the supposed ideal in other idea-generating frameworks is on the quality of ideas, the Wrong Theory Protocol places empathy at the center of the design process. With empathy being the goal to optimize, engineers must understand the needs of diverse stakeholders, particularly those who are typically marginalized. The best solutions in the Wrong Theory Protocol are those which are ethical and humane.

Note that the humanity of a solution may seem to be the "best" solution in terms of quality or effectiveness. Consider the use of strong pesticides in agriculture. Although these chemicals are extremely effective in reducing the impact of pests on crop yields, they can linger on the food we eat and runoff waste can make its way into nearby water sources. Plastics are convenient for all sorts of applications, especially single use items like plastic bags and rings to hold drink cans in packs, but often find their way into nature where they wreak havoc on wildlife—including humans, as we've learned more about the effect of microplastics on our bodies. In March 2022, scientists found microplastic pollution in our blood for the first time. Engineers must consider the long-term effects of their designs, which requires the consistent use of ethical frameworks to weigh potential design alternatives.

Wrong Theory Protocol attempts to address the push and pull between empathy and quality by reversing the problem statement. In other words, instead of generating ideas for the best and more humane design, the prompt is to create the *worst possible design that harms and humiliates.* Vanessa Svihla and Luke Kachelmeier provide the example of a doghouse. The most inhumane doghouse would have a bed made of glass shards, a roof attached with a sprinkler to spray the dog consistently, and have rotating blades for walls. The session lasts about 60-75 minutes. Note that the whole session isn't about generating the worst possible design, only the first half. After half of the time expires, everyone shares their design ideas and must defend the assertion that it is the worst design. The second half of the session flips the problem statement back to generating beneficial ideas by building off of the terrible ideas in the first phase. Preliminary evidence suggests that Wrong Theory Protocol can support creative ideation and more empathetic ideas.

5.4 Organizing Your Ideas with Morphological Charts

morphological chart

Once engineers generate ideas for their design's various functions, those different sets of ideas need to be combined into concepts. To accomplish this task, a team can use a tool called a morphological chart. A **morphological chart** is a visual method for generating design concept ideas by splitting a potential design into necessary functions and methods of achieving those functions. In other words, the method involves creating a table with the functions down the rows and different methods of achieving those functions across the columns. A morphological chart is a natural next step after performing a functional decomposition. All the chart does is organize the ideas from the ideation tools already presented into a tabular format.

Although the different methods of achieving the functions can be described using words, coming up with a sketch to represent the method can be helpful. Translating a series of words into a concept versus a series of pictures might be more difficult for some individuals. For example, consider the design of a chair. Would it be more helpful to have the description of a seat that says "egg shaped with beveled edges and thin oval cutouts laterally" or a picture?

Creating concepts using a morphological chart is as simple as drawing a line down the functions intersecting different methods on the way to the bottom. The concept is the combination of those individual ideas. Table 5-2 shows an example of a morphological chart for a door stop. Drawing through "ratchet-style, folds down, folds up, and on door side" yields a foldable metal and rubber stopper, while "friction, placed on floor, pick up and store, and sleek profile," results in a standard door stop (Figure 5-2).

Table 5-2 Morphological Chart

Function	Method 1	Method 2	Method 3
Stops door	Friction	Magnetic catch on door	Ratchet-style
Engage door stop	Always engaged	Placed on floor	Folds down
Disengage door stop	User tugs on door	Pick up and store	Folds up
Does not obstruct user	Sleek profile	Inside door	On door side

Figure 5-2 Standard door stops.

5.5 Comparing Designs and Making Decisions

The criteria identified earlier guided our design and must also guide the process of choosing a solution from a list of possible solutions. Simply identifying criteria is not enough information for a decision—each proposed solution must be evaluated against those criteria. The first step will be to eliminate any solutions that do not meet the minimum requirements—those solutions are out-of-bounds and need not be considered further. Those unreasonable solutions may have been important in the process of generating ideas, because some features of those unreasonable solutions may have been incorporated in other solutions, or maybe because hearing those ideas helped others think differently, leading to better solutions—but we don't need to feel guilty about throwing them away now—they have served their purpose.

Once the minimum requirements have been satisfied, there are many ways of applying the remaining criteria to select a solution. Voting is a quick way to reduce a large number of choices to a smaller number of choices, because one of many voting processes (weighted voting, multivoting, ranking, and more) can quickly eliminate many noncontroversial options. In spite of the use of voting to select leaders in many countries, voting is not the best approach to make a final selection, because it is likely to disenfranchise a large fraction of voters, particularly in a two-party system.

decision matrices

Other approaches are typically used to further narrow the set of choices, such as making **decision matrices**. In this approach, you use a table for each criterion to summarize how each of the solutions compares with others. An example is shown in Table 5-3 for the criterion "safety." Among the table entries, 0 indicates that that option in that column is worse than the option in that row; 1 indicates that both options rank equally; and 2 indicates that that option in that column is better than the option in that row.

ENGAGE

It's suggested that students get in the habit of checking their ideas with existing products and carefully describing the similarities and differences between the generated concept and the existing product.

One potential in-class activity to tie back to the idea of using prior art is to have students conduct a simple patent search for the door stop example in Table 5-2.

TEACHING TIP

There are several ways to complete a decision matrix. We present likely the most simple technique for decision analysis.

TEACHING TIP

It's critical to note that the decision matrix does not provide the optimal solution—especially if the attributes are measured based on ratings. The results are influenced by the weights chosen by the engineers. The result should be interpreted as a promising concept that should be prototyped at some level.

INTEGRATING CONTENT AREAS

An algebra problem that integrates linear inequalities would be used to determine how much a weight needs to change in order to change the outcome of the analysis.

For example, in Table 5-4, one question could be: what does the weight on cost need to be in order for option 2 to have a higher score than option 1?

Let the weight of Cost be x. Then, we solve the inequality:

Score of Option 2 > Score of Option 1
$6x + 32 + 70 + 20 > 2x + 80 + 70 + 10$
$4x > 38$
$x > 9.5$

So the weight of Cost needs to be at least 9.5.

DISCUSSION

What would happen if instead of rating the options using a rubric, we rank order them where 1 is the worst and *n* is the best (for *n* options). *The distance between ranks next to one another is ignored. For example, the difference between option 2 and 3 could be 34, and between 3 and 4 could be 560; yet, when rank ordered, the difference in both cases would be 1.*

Table 5-3 Comparing Options

Safety	Option 1	Option 2	Option 3
Option 1		0	1
Option 2	0		0
Option 3	1	2	
Total	1	2	1

Since a high level of safety is preferred, "better" means "safer." In the example, the first column indicates that option 1 is less safe than option 2, but the same as option 3. The resulting totals indicate that option 2 ranks best in terms of safety. To complete the process, generate similar tables for each of the criteria being considered and then sum the totals to identify the best solution.

weighted decision matrix

A disadvantage of the decision matrices approach is that all criteria have equal weight, whereas some criteria are likely to be more important than others. An alternative approach is to use a **weighted decision matrix**, shown in Table 5-4. In this approach, each option is scored against each of the criteria.

Table 5-4 Options with a Weighted Decision Matrix

	Weights	Option 1	Option 2	Option 3
Cost	2	2	6	10
Safety	8	10	4	6
Weight	10	7	7	2
Wow	5	2	4	6
Totals		4 + 80 + 70 + 10 = 164	12 + 32 + 70 + 20 = 134	20 + 48 + 20 + 30 = 118

In Table 5-4, the weights are in the first column and each option has been assigned a score from 0 to 10, indicating how well that option meets each criterion. This approach may be inconsistent in that a 7 for one rater may be different from a 7 for another rater, so it can help to better define the scale. For example, the options may be scored on the scale shown in Table 5-5 as to how well the option fits each criterion.

Table 5-5 Sample Scoring Rubric

Score	Meaning
0	Not satisfactory
1	Barely applicable
2	Fairly good
3	Good
4	Very good; Ideal

An important aspect of such scoring methods is that they are not a strictly mathematical determination, but rather provide input to the decision-making process. Based on the scores that result, judgment is required to ensure that all criteria have been properly considered and that there is agreement on the final decision.

5.6 Redesigning a Kitchen Gadget or Product: An Example of a Blender and Food Processor

Cooking is a source of joy and culture for people around the world. As we've progressed technologically, the rudimentary tools used in years past have been replaced with more automated machines that make cooking faster, safer, and more precise. Consider a chimichurri sauce, which is often found in Argentinian and Uruguayan cuisine. The most basic ingredients include parsley, garlic, and oregano combined with olive oil and red wine vinegar. The home cook who does not have the time to delicately chop up the parsley, garlic, and oregano can turn to a food processor. If you are not familiar with the food processor, it essentially is a miniature blender that is designed for mostly solid foods that would normally be chopped or sliced into small pieces—but might be tedious to do so by hand. If you want to make a delicious smoothie with mint chocolate chip ice cream, then use a blender. If you have a bunch of tomatoes that you want to chop up into a bunch of chunks, then use a food processor. Let's explore these two products and foreshadow a conceptual design for a kitchen product you'll be embarking on in this section.

We can delineate these two products using engineering design terminology. Consider a situation where we were redesigning a blender to be more like a food processor. In this case, we are trying to make a chimichurri sauce, but the blender doesn't quite work without some liquid to allow the ingredients to move around and reach the smaller blade at the bottom. The liquid is needed because the blender works using a vortex; by creating a force known as *shear*, the liquid is sucked down from the blender's top and thrown back into the sides in a process called *cavitation*. Ultimately, this process breaks up the food into smaller and smaller bits. Because we need some circulation, we add some olive oil and red wine vinegar as our liquid (perhaps a bit more than suggested), but the blender essentially liquifies the solid ingredients that come into contact with the blades and is leaving ingredients closer to the sides mostly untouched. We don't want the parsley, garlic, and oregano completely mixed into a liquid; we only want them chopped up into finer pieces. Therefore, one of our criteria for the blender redesign is: *the product finely chops dry ingredients.*

But what do we mean by *finely chops*? How small should the pieces be? To clarify our criterion, and to identify and define our engineering problem, we would create design specifications that quantify our criteria such that we can test if our solution is meeting the requirements. A culinary rule of thumb for a fine chop is one-fourth of an inch. Therefore, our design specification would be *the product finely chops dry ingredients to approximately 1/4" in length.* The design specifications allow us to be more precise and test how well our design functions—beyond fuzzy evaluations like "the prototype chopped the ingredients into somewhat small pieces but not small enough."

What's limiting our design? Recall that criteria let us know that our design is successful; more specifically, it must meet all the criteria to function properly. On the other hand, constraints are restrictions that limit exactly what we can do. Constraints often take the form of time, resources, and money, but that simple list is not exhaustive. Perhaps this redesigned blender needs to fit in with an existing product line offered by a major kitchen appliance manufacturer you're working for, which could involve certain aesthetic features like being made with a specific material or set of colors. For example, the firm may require that the plastic Acrylonitrile Butadiene Styrene (ABS) be used because they use it in their other product line and don't want to order different material from another supplier. Note that ABS is a food-grade plastic, resilient to impacts and abrasions, which is probably good given the product involves a spinning blade and solid ingredients flying around in the main container. Therefore, a constraint here could be: *the product must be made using Acrylonitrile Butadiene Styrene (ABS).*

Next, we should identify our inputs, outputs, and functions; in other words, what needs to go into the product, what are we expecting to come out, and what functions are necessary to transform the inputs into the outputs. The main input-output relationship we would like is for dry ingredients to be transformed into chopped dry ingredients. The function that transforms them is "chop ingredients." We would also expect this device to be powered by something, so a source of power should be an input. The power is used to engage whatever the chopping mechanism is, so we can include a function called "engage chopper." Figure 5-3 shows how the inputs, outputs, and functions would relate using a simple functional block diagram.

TEACHING TIP

The introductory part of this chapter provides a concise review of ideas from Chapters 4 and 5. You may consider walking through this scenario in class should you choose to use the project in this chapter to model the expected deliverables.

TEACHING TIP

This is a good time to reiterate to students the difference between criteria, constraints, and specifications.

TEACHING TIP

You can point to Table 5-6 as an example of a morphological chart that is not symmetric. Not every function/subsystem needs an equal number of methods.

ENGAGE

Have students create other concepts beyond what's listed. What's the most unique concept they can come up with in groups?

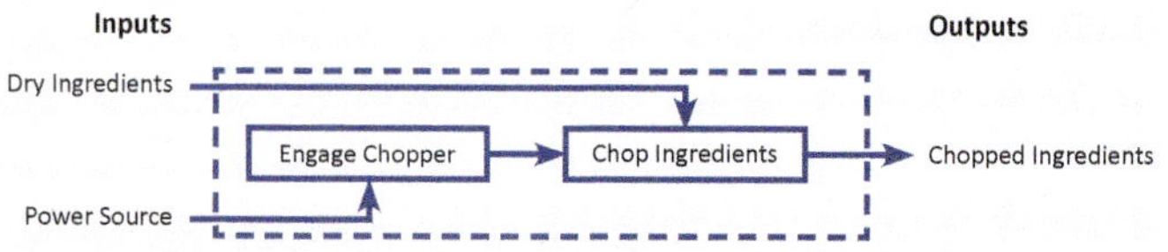

Figure 5-3 Block diagram.

But how will we accomplish the function "chop ingredients" and "engage chopper?" To generate concepts for how this product might work, we can use a morphological chart. Recall that a morphological chart is a tool that specifies a design's primary functions and different ways in which those can be accomplished. These charts can also include different qualities of subsystems, such as materials and power sources. Table 5-6 shows a morphological chart with the two functions and three methods of accomplishing each. By combining different methods together, we can generate concept ideas.

Table 5-6 Simple Morphological Chart for a Design Which Outlines Different Product Ideas to Chop Dry Ingredients

Function/Subsystem	Method 1	Method 2	Method 3	Method 4
Chop ingredients	Pressure	Shear	Friction	
Chopping instrument	Fixed blade	Removable blade	Rounded object	
Container	Pitcher	Cylinder	Bowl	None
Engage chopper	Motorized	Manual		

For example, combining "pressure," "removable blade, "none," and "manual" is just chopping the ingredients ourselves by hand—which we're trying to avoid. What happens if we combine "friction," "rounded object," "bowl," and "manual?" Those who have some experience in the kitchen might recognize this combination of methods as a mortar and pestle! A mortar and pestle is a set of two Stone Age time period tools that consist of a stone bowl called the mortar, and another stone that is shaped such that it is grippable, called the pestle. The user crushes the ingredient by applying force and grinding them until the desired consistency is reached. In fact, using a mortar and pestle is a traditional Argentinian method for making chimichurri. But still, we're looking for a solution that can automate our chopping process without turning our ingredients into a complete liquid. As we've established before, the blender operates by generating a shearing force that causes cavitation. Because we most often want to use liquids in a blender, having a pitcher container would be helpful to dispense the output. Therefore, a blender would combine "shear," "blade," "pitcher," and "motorized."

We can continue the process of combining different methods to create different concepts, and some may require a bit more imagination than others. For example, what famous "As Seen On TV" product promoted by pitchman Vince Offer is the combination of "pressure," "fixed blade," "cylinder," and "manual?" If you're struggling to develop concepts, consider revisiting the different subsystems and functions identified in your functional decomposition. Perhaps certain items need to be broken down further to uncover discrete differences that can be drawn among methods for performing required functions.

Intellectual Property (IP)

intellectual property (IP)

When you design an object, invent something, create written work or art, or create a symbol, this can be considered intellectual property. **Intellectual property (IP)** is owned by someone or some entity; it can be owned by you, your company, or your school. Often, these creations are part of a job or a result of a school assignment, and usually, the creation is essentially minor. If you were to invent something big or create a work of art that could be sold, or anything you consider to be valuable, especially outside of school or work, there are ways to protect your IP. You can apply for a *patent*, which is a formal acknowledgement that you created this invention. A *copyright* is a method of protecting a written work (look at the first few pages of any book to see an example), a work of art, a technical drawing, etc. A *trademark* protects a specific business idea from being sold by another business.

TEACHING TIP

We will provide suggestions to increase and decrease the complexity of projects throughout this text. Each differentiated instruction tip will be appended in the neighborhood of the appropriate step of the design process or prompt.

DIFFERENTIATED INSTRUCTION

You can adjust this prompt to several issues in the kitchen if you'd like to give students more freedom to define the problem. Encourage them to explore different topics such as: (1) common injuries in the kitchen, (2) laborious cooking techniques that can be improved, (3) ways to make the kitchen "smarter" (i.e., integrate with the Internet of Things), (4) designing better kitchens for older adults. You will be able to use the same prompts.

TEACHING TIP

We wrote this project as a "bare essentials" kind of activity to get students into the habit of following the design process in its early stages, considering how essential they are. Thus, it is a *conceptual design.* If you have standard crafting materials in class, you could allow them to develop rough prototypes.

DESIGN ACTIVITY

A Conceptual Design for Kitchen Gadgets

Problem Statement

From the simplest to the most involved dishes, making something from scratch in the kitchen can require several tools. Although a knife can be used to perform a variety of functions, including chopping, peeling, and separating different items, there are tools that have been made that specifically address niche needs in the kitchen. For example, consider our example of chopping the dry ingredients for the chimichurri. There are several products that exist specifically to perform the action of chopping that are motorized or manually operated. The purpose of this design activity is to allow you to practice some of the design thinking tools that help engineering teams generate ideas for a potential solution to a problem. Here, we will focus on a specific scenario in the kitchen where a home chef needs your help.

Gabriel has always been a home cook and loves to prepare large meals for his family and neighbors. However, after breaking his wrist in an accident playing a game of tennis with his friends, he's noticed that his dexterity has worsened. Therefore, kitchen tasks involving grip and twisting have become somewhat painful and more difficult to accomplish. Your task is to come up with a conceptual design for a kitchen product or appliance to help Gabriel create a favorite recipe of his, zucchini ravioli.

To create zucchini ravioli (in simple terms), he peels thin slices of zucchini, salts them, and pats them down with a paper towel. He mixes the filling with a few different kinds of cheese and spinach, which he places in the center of two zucchini slices and folds up into a little package. The repetitive nature of the folding process is particularly troublesome for Gabriel.

Materials

- Device connected to the Internet
- Writing utensils (different colors would be helpful)
- Paper (can be 8½ × 11 or larger, such as poster size)

Identify and Define Your Engineering Problem

Understanding Who Your Design Is For

To ensure your solution is appropriate, you need to understand the needs of the user. Conduct initial research on common issues related to dexterity like Gabriel is experiencing. You may consult online resources or someone you may know who has dexterity issues. Record your findings in your engineering notebook.

DIFFERENTIATED INSTRUCTION

You may consider making the initial research phase more comprehensive by requiring that students conduct an interview with someone they know who has issues with dexterity. If you elect to have them research broader topics, such as designing better kitchens for older adults, then they have the ability to interview any older adult they know. Laborious cooking techniques would also lend itself to interviewing parents, guardians, or older siblings.

Develop Criteria, Constraints, and Design Specifications

Now that you've taken time to understand your user's needs, next you must translate those needs into criteria, constraints, and design specifications. These will be used to guide the development of your solution and ensure it is successful. While defining constraints and criteria, be sure to think of factors such as performance, time and financial constraints, **ergonomics**, safety, and what state-of-the-art solutions may already exist. Larger designs also require looking at environmental, social, ethical, political, regulatory, and legal factors. In your design notebook, address the following prompts:

1. What are the criteria you identified for this design?
2. What are the constraints limiting the possibilities of this design?
3. How will you know if your design is successful? Translate your criteria and constraints into measurable design specifications.
4. Consult with another team of students in your class and compare your criteria and constraints. What differences and similarities do you notice?

Be sure to follow good practices when working in your engineering notebook:

1. Enter your work sequentially and chronologically.
2. Make an accurate and complete reflection of the progress being recorded.
3. Include sketches, drawings, and pictures where appropriate.
4. Use all pages; no pages should be removed. There should be no loose entries or pages.
5. Date and number each page.
6. Identify and line out (cross out) any unused spaces.
7. Keep and cross out any mistakes in the notebook. Do not erase or obliterate errors.
8. Include test data and calculations.

Functional Decomposition

With our criteria, constraints, and design specifications outlined, now we can decompose our potential solution into functions that it must accomplish and use these problem-solving techniques to to successfully develop our technological solutions. Using your answers from the previous section, record the following in your design notebook:

1. What are the inputs, outputs, and functions involved in the potential design solution?
2. What subsystems are necessary to generate the desired output?
3. Develop a functional block diagram relating the inputs, outputs, and functions.
4. Compare your functional block diagram to a different team of students than you consulted in the "Develop Criteria, Constraints, and Design Specifications" step. What differences and similarities do you notice?

TEACHING TIP

Considering this project should serve as a model for how students should engage in the engineering design process, ensure that they have measurable design specifications. They don't have to put in specific numbers at this stage, only a blank or variable where the quantity would go.

DIFFERENTIATED INSTRUCTION

For more advanced students, you should expect them to provide reasonable estimates for their design specifications, based on their research and physics knowledge.

TEACHING TIP

Consider collecting all of the criteria and constraints in a shared workspace, like a Google or Microsoft Form. Then, display the results to the class and have a group discussion.

TEACHING TIP

Push students to come up with as many ideas as possible. Have them aim for 100 as an aspirational number. You may require a list of these ideas as a deliverable.

TEACHING TIP

This is a good time to reinforce the idea of using prior art and biomimetics. You may emphasize that students find examples in nature as well, to guide their concept generation, if appropriate. This could serve as a challenge prompt.

DIFFERENTIATED INSTRUCTION

Students often do not link explicitly back to their criteria/constraints when deciding on attributes for a decision matrix. For students who need additional help here, consider having them create a table of the criteria/constraints/specifications and simplify them into attributes that can be assigned weights in the decision matrix.

Along a similar line, students frequently neglect to justify the weights they chose. As an additional prompt, have them conduct a *sensitivity analysis* by changing their weights to slightly value different attributes over another to see how their results change.

DIFFERENTIATED INSTRUCTION

To ensure a variety of assessment evidence, you may choose to let students pick their mode of communication. Or, you may elect to only check their design notebook, have them write a report, or make a presentation. It's not recommended that students be assessed using all models for each project.

TEACHING TIP

Challenge students to think critically about new things that they must measure as they consider how they would test their prototype.

You may also ask if they identified additional stakeholders beyond Gabriel.

Generating Potential Concepts for a Prototype

Now that you have considered the different functions and subsystems needed for your design to meet its criteria and constraints, we can turn to generating potential concepts. In your design notebook, address the following prompts:

1. Create a morphological chart using the functions and subsystems identified in your functional decomposition. Identify at least three methods per function or subsystem.
2. Once your morphological chart is complete, identify five potential concepts using the methods you specified.
3. From the five concepts, create a sketch and a description that capture how the concept would work. Next, search online to see if that concept has been implemented in an existing kitchen product. If there is an existing product, paste the image into your design notebook and explain how your concept is similar and different. If you cannot locate an existing product, justify why you believe your concept is unique.

Selecting a Concept

After generating a series of design alternatives, the concepts from the morphological chart, we now must pick a concept to move forward with. One tool for making a viable prototype selection is a decision matrix. In your design notebook, answer the following:

1. Using your criteria and design specifications, set up a decision matrix to begin weighing the viability of the five concepts.
2. Determine weights to place on the different criteria.
3. Calculate the "winning" design using your weights.
4. Summarize your findings.

Testing the Prototype

Because we are sticking to a conceptual design, we will assume you developed a prototype based on the concept selected from your decision matrix. How do we know it works? In your design notebook, address the following prompts:

1. Develop a test plan to measure whether your design achieves the criteria you specified at the beginning of the challenge.
2. Did you revise or add design specifications during your test plan development?
3. What are potential failure modes of the prototype? Of the final design?

Report Your Solution

Finally, it is time to report out to the class. Your teacher may elect to have you present a PowerPoint summarizing your processes from each stage of the design process outlined here. You may also write a short report covering the same material. In your design notebook, create a bulleted list of the main decisions, designs, and activities you made across the different stages of this project. If you are making a PowerPoint presentation, create an outline for how you will present the information.

Critical Design Review

When the design is complete, the design is reviewed one final time. The Critical Design Review is the final check on the design. It happens after the definition and analysis of the criteria, constraints, and simulations. The Critical Design Review wraps up the analysis, and indicates that the prototype development and testing are complete. The U.S. military describes a CDR this way, "The CDR establishes the initial product baseline. A successful CDR is predicated on the determination that the subsystem requirements, subsystem detailed designs, results of peer reviews, and plans for test and evaluation form a satisfactory basis for proceeding into system implementation and integration."

In this case, your team should ensure that all criteria and constraints are satisfied, and that the prototype functioned as intended. Your team should discuss ergonomics and safety, and examine whether there are any environmental, social, political, or ethical concerns.

IN-CLASS ACTIVITIES

Chapter 5

ICA 5-1

With your team, compose a plan to build the longest bridge possible using the K'Nex™ pieces provided by your instructor. The longest part of your bridge will be defined as the longest stretch of K'Nex pieces that are not touching another surface (such as a table, floor, chair, etc.).

During the planning phase (15 minutes), your team will only be allowed to connect eight K'Nex pieces together at any one moment. As soon as your team has finished the planning phase, disconnect all K'Nex pieces and place them in the provided container. The container will be shaken before you begin, so do not bother attempting to order the pieces in any way.

During the building phase (60 seconds), the restriction on the number of connected pieces goes away, but your team will not be allowed to talk. Your instructor will say GO, and then STOP, after 60 seconds has elapsed. At the end of the building phase, your team must step away from the bridge and remove all hands and other body parts from the K'Nex bridge. If any pieces fall after time is called, you are not allowed to stabilize the structure in any way.

ICA 5-2

With your team, come up with a plan to build the tallest tower possible using the K'Nex™ pieces provided by your instructor. The tallest portion of your tower will be defined as the tallest stretch of K'Nex pieces that are not touching another surface (such as a table, floor, chair, etc.).

During the planning phase (15 minutes), your team will only be allowed to connect eight K'Nex pieces together at any one moment. As soon as your team has finished the planning phase, disconnect all K'Nex pieces and place them in the provided container. The container will be shaken before you begin, so do not bother attempting to order the pieces in any way.

During the building phase (60 seconds), the restriction on the number of connected pieces goes away, but your team will not be allowed to talk. Your instructor will say GO, and then STOP after 60 seconds has elapsed. At the end of the building phase, your team must step away from the tower and remove all hands and other body parts from the K'Nex tower. If any pieces fall after time is called, you are not allowed to stabilize the structure in any way.

ICA 5-3

Following the rules for brainstorming (encourage a lot of ideas, encourage a wide variety of ideas, and do not criticize), develop ideas for the following with your team.

(a) A better kitty litter box
(b) A new computer interface device
(c) A new kind of personal transportation device
(d) Reducing noise pollution
(e) Reducing light pollution
(f) A new board game
(g) A squirrel-proof bird feeder
(h) A no-kill mole trap
(i) A tub toy for children of 4 years or younger
(j) A jelly bean dispenser
(k) A new musical instrument
(l) A self-cleaning bird bath
(m) A new smartphone application

CHAPTER 5 ASSESSMENT QUESTIONS

1. Which of the following statements about design is true?
   (a) The best designs are the result of a singular creative genius.
   (b) The goal of any designer is to find the single best solution.
   (c) Designs must satisfy many objectives.
   (d) The best design will always satisfy all of the objectives.
   (e) None of the above statements is true.

2. Which of the following statements about design is NOT true?
   (a) Design is an iterative process.
   (b) Customer feedback is part of the design process.
   (c) Designs must result in tangible products.
   (d) Failure is an important part of the design process.

3. Which of the following do NOT describe a possible design criteria?
   (a) Design criteria should be specific.
   (b) Design criteria may include a must-have feature in the design.
   (c) Design criteria may include a should-have feature in the design.
   (d) Design criteria may include preferences.
   (e) Design criteria may include potential solutions.

4. Which of the following is NOT a good criterion for a new vehicle?
   (a) The car should be a silver color.
   (b) The car should be spacious.
   (c) The car should have at least 15 cubic feet of cargo space.
   (d) The car must have a backup camera.

5. What is the best way to choose a final design?
   (a) Let each engineer argue for the merits of their design ideas.
   (b) Use the design criteria to compare designs.
   (c) Build prototypes of each design idea and test them.
   (d) Ask a customer focus group to choose the best design.

6. Which of the following is NOT a good way to build teamwork?
   (a) Focus on personal relationships.
   (b) Set rules for team behavior.
   (c) Agree on the purpose of the team.
   (d) Clearly define a means of measuring team progress.

7. What should be the first consideration when creating a project timeline?
   (a) Determine the project's due date.
   (b) Determine which tasks depend on other tasks.
   (c) Create a matrix of responsibilities for each team member.
   (d) Elect a team leader.

8. Which of these groups of people are teams, and which are groups? Why?
   (a) A rugby team
   (b) Four people moving a couch from one room to another
   (c) Twenty people watching a movie at a movie theater
   (d) Four people who rotate driving duty on a long road trip
   (e) Three TV announcers broadcasting a WNBA game

9. What are major characteristics of a team leader?

10. List three things you can do to be an exceptionally successful team member.

11. List two things you can do to be an effective team leader.

12. Search for an example of an engineering design that has been affected by something in nature—an example of biomimicry. What characteristic of something found in nature led to an improvement in an engineering design?

13. *Your instructor may assign you to a team for this problem.* Your team is tasked with designing a new method to serve food in the cafeteria of their school. Use a shared document (such as a Google doc) to work collaboratively to:
    (a) Define criteria and constraints of your design.
    (b) Brainstorm possible ideas. Record your ideas in the document.

14. *Your instructor may assign you to a team for this problem.* Your team has been tasked with addressing a problem at a local park. The park needs a method for people to park their bikes at different locations of the park. Use hybrid brainstorming to generate a list of possible ideas and possible issues as your team addresses this problem.

15. *Your instructor may assign you to a team for this problem.* Your team has been tasked with brainstorming answers to the following question: "Can we predict potential changes caused by the advancement of technology?" Use hybrid brainstorming and record your final results in a shared document (such as a Google doc).

16. **THE GREAT EGG DROP**

    This is a chance to use problem-solving techniques to solve a problem. You may have seen a "drop the egg" contest, where a device is built to keep the egg from breaking when dropped from a certain height. This challenge is a variation of the egg-drop challenge.

    You cannot protect the egg in any way but are allowed to design something for it to land on. You will be allowed three drops per team and will be assigned at random the heights from which you will drop the egg. The egg must freefall after release. If the egg breaks open, the height will be taken as zero. If the eggshell cracks, the height from which you dropped it will be divided by 2. If you drop the egg and miss the catching apparatus, that is your tough luck, and a zero height will be used—suggesting that you need to devise a way to always hit the target. Use a decision matrix to determine the best possible approach.

    Your grade will be determined by the following equation:

    Ranking = (height from which egg is dropped)/(weight of catching apparatus)

    After demonstrations, the average ranking number for each team will be calculated and the value truncated to an integer. The teams will then be ranked from highest to lowest value. The heights and weights will be measured in class. The actual grade corresponding to your class ranking will be determined by your instructor. Be sure to maintain a paper or digital portfolio using the engineering documentation process.

17. **TREE HEIGHT**

    We have been contacted by a power company to conduct a study of tree height and interference with high-voltage wires. Among other requirements, the company is looking for a quick, easy, and inexpensive method to measure the height of a tree.

    Your project is to develop different methods of measuring the height of a tree. As a test case, use a tree designated by your instructor. You may *not* climb the tree as one of the methods! In addition, you may not harm the tree or leave any trace of your project behind.

    To sell your methods to the customer, you must create a poster. The poster should showcase your measurement methods, including instructions on how to conduct the experiment, any important calculations, graphs or photos, and your resulting measurements. The poster will be graded on neatness, organization, spelling and grammar, formatting, and strength of conclusions. You may use any piece of poster board commercially available, or a "science fair" board. The poster can be handwritten, or typed and attached, or… here is a chance to use your creativity! If you present more than one method, you must indicate and justify your "best" choice.

Your grade will be determined as follows:

- Method 1: 30 points
- Method 2: 20 points
- Method 3: 10 points
- Presentation Board: 40 points

18. A two-liter soft-drink container, nearly full of water and open to the atmosphere, is placed on the floor. Where could you locate an orifice on the side of the bottle so that the jet of water that squirts out will have the maximum range? Keep the bottle filled by continuously pouring water in the container as the tests are conducted. Justify your answer with theory by discussion rather than equations.

19. Your team is tasked with designing a device to store and charge smartphones in a school library. Your goal is to present a prototype for your design in written and verbal form. For each step, use collaborative, Web-based tools such as Google Docs.

    (a) Your team should brainstorm to decide on criteria and constraints that apply to your solution.

    (b) From there, use one of the brainstorming strategies from this chapter to generate ideas.

    (c) Use a decision matrix to select one design you intend to prototype, and a morphological chart to determine how to implement the different functions your device will exhibit.

    (d) Using common materials such as cardboard, glue, tape, and so on, design and build a non-functional prototype of the device.

    (e) Develop a short report describing your design. Include your decision matrix and morphological chart and a picture of your final prototype. Include documentation of the decisions, designs, and activities during each phase.

    (f) Develop a presentation to show your design in 90 seconds or less. Be sure to show your physical prototype.

    (g) Prepare a memo to your teacher describing how you dealt with any conflict in your team and how your team came to a consensus.

20. Which of the six thinking hats typically represents how you engage as a team member? Which is the most different from your normal behavior? What do you have to be sure to do to serve your team in the role which is the most different from your typical behavior?

21. Your friend is ready to buy a car, but they cannot make a decision. You say "Fortunately for you, I just studied decision matrices in school! Let's do one for your car!" Your friend says "I am most concerned about gas mileage, and that's, oh, about 70% of my decision. After that, number of cupholders is about 5%, gross vehicle weight is about 10%, and cost is about 15% of the total." After you marvel at how much your friend has considered these

criteria, you assemble a decision matrix in your quest to help. You find the following choices on your local car lot:

- The Jeep Wrangler: 22 mpg, 4 cupholders, $42,000, with a weight of 5,100 pounds
- The Chevrolet Colorado: 2 cupholders, $27,700, 26 mpg, and 4,700 pounds
- The Toyota Prius: 3,200 pounds, 53 mpg, 6 cupholders, $30,000
- A 2013 Ford Focus: 27 mpg, $13,495, 3,200 pounds, 4 cupholders.

Use a decision matrix to advise your friend on which car they should buy according to their criteria.

22. You have been tasked to form a team to design a new amusement park ride. Why would an interdisciplinary team prove to be valuable?
    (a) Identify the goals of the project.
    (b) List the objectives and requirements that you would expect.
    (c) List five different disciplines that could contribute, and briefly describe why each would be a valuable addition.
    (d) Establish and evaluate potential constraints in each of these areas: health, safety, social, environmental, ethical, political, regulatory, and legal.

23. When designing a new kitchen device to help create a zucchini ravioli, what possible constraints or criteria may be defined that relate to the following?
    (a) Performance
    (b) Financial constraints
    (c) Safety
    (d) Ergonomics
    (e) Social aspects
    (f) Environmental considerations

24. Given your design of the kitchen gadget to help make zucchini ravioli, write a concise statement addressing your predictions of the following:
    (a) Expected performance
    (b) Failure modes of the device
    (c) Reliability of the final product

25. Given your design of a kitchen gadget to make zucchini ravioli, describe each term as it could affect the implementation of your design:
    (a) Materials
    (b) Personnel
    (c) Resources
    (d) Funding
    (e) Manufacturability
    (f) Feasibility

OVERVIEW

The need for precise communication in the preparation of functional documents distinguishes technical drawing from the expressive drawing of the visual arts. Artistic drawings are subjectively interpreted; they can have multiple meanings. Technical drawings are understood to have one intended meaning.

OBJECTIVES

- Set up, create, and modify drawings
- Store and retrieve geometry
- Demonstrate and use appropriate line types in engineering drawings
- Draw two-dimensional, single-view objects
- Create multi-view working drawings using orthographic projection
- Dimension objects using current American National Standards Institute (ANSI) standards
- Draw single-line two-dimensional pictorial representations
- Create working drawings that include section views
- Develop and present ideas using sketching and computer-aided design and drafting (CADD)
- Draw conclusions using industry standard visualization techniques and media

CHAPTER 6

Engineering Graphics and Design Tools

TOPICS

Engineering designs start as images in the mind's eye of an engineer. Design tools have become a platform for communicating and integrating the design process from invention through to a finished, manufactured assembly. The computer has made it possible to create an integrated design that has been tested through simulation to ensure that all parts will fit together and behave as intended. Software for **computer-aided design**, or **CAD**, includes a variety of additional tools related to simulated stress-testing, assembly, and tolerancing. Today, manufacturing and engineering design are connected through Design for Manufacturing (DFM) concepts and Finite Element Analysis (FEA). The end result is more than just a visual representation, but a virtual part or assembly ready to be manufactured and put into service.

computer-aided design

6.1 Engineering Graphics

Visualizing an image in your mind is a familiar experience. Visual thinking is a foundation of engineering. The image can be realized at many different levels of abstraction: size, weight, how it deforms under stress, or how it dissipates heat, for example. The versatility of today's technology now permits a single person to tie all the component parts together to build a device from a collection of visual images. The images must be conveyed to other engineers, designers, fabricators, and test labs. Those images must be constructed in such a way that they are readily recognizable, consistent, and in a readable format. This ensures that the visual ideas are clearly and unambiguously conveyed to others. **Engineering graphics** is a specific way of presenting images of parts or assemblies. Despite the growing connectedness of computer design tools, there is still a visual satisfaction in a graphical design implementation.

engineering graphics

In this chapter, we will be covering details on making a completed engineering drawing that describes exactly what a part does and what it looks like. It is useful to visualize the end result that you have in mind before you begin. Figure 6-1 is an example of an engineering drawing that provides a context for some of the elements we will be discussing.

ENGAGE

Define the term *technical drawing.* *(Technical drawing is essential for communicating ideas in industry and engineering. To make the drawings easier to understand, people use familiar symbols, perspectives, units of measurement, notation systems, visual styles, and page layout. Together, such conventions constitute a visual language, and help to ensure that the drawing is unambiguous and relatively easy to understand. These drafting conventions are condensed into internationally accepted standards and specifications that transcend the barrier of language, making technical drawings a universal means of communicating complex mechanical concepts.)*

ENGAGE

Explain to your students that engineering drawings are considered "legal documents" in a court of law. An engineering drawing is a legal document (that is, a legal instrument) because it communicates all the needed information about "what is wanted" to the people who will expend resources turning the idea into reality. It is thus a part of a contract; the purchase order and the drawing together, as well as any ancillary documents (engineering change orders [ECOs], called-out specs), constitute the contract. Thus, if the resulting product is wrong, the worker or manufacturer are protected from liability as long as they have faithfully executed the instructions conveyed by the drawing. If those instructions were wrong, it is the fault of the engineer.

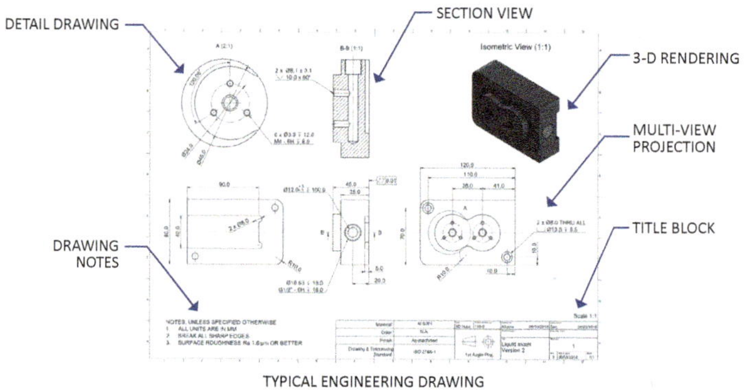

Figure 6-1 Engineering drawing showing typical design elements.

We will be covering the following topics:

- **3-D projections**—For rendering a realistic-looking version of the finished piece
- **Multi-view projections**—Used to show the overall object usually from three orthogonal views
- **Detail drawings**—A drawing focused on a single part or to provide extra detail where needed for clarity
- **Assembly drawings**—A completed assembly that shows what the finished part looks like, sometime sectioned to show interior views
- **Layout**—The overall arrangement of the visual elements and drawing technical information
- **Line types**—How to correctly represent specific features depending on how they are to be viewed in the context of the drawing
- **Drawing standards**—Conventions for layout, drawing sizes, and differences between ISO and ANSI representations
- **Dimensioning standards**—Conventions for correctly delineating dimensions so they are unambiguous and not redundant

Engineering drawings, whether done using a pencil and paper or a computer, start with a blank page or screen. The image in the engineer's mind must be built within the digital world for representation on the computer screen. The creative nature of this activity is similar to that of an artist. Sometimes, designs begin with a simple pencil and paper sketch (Figure 6-2).

Figure 6-2 A simple architectural sketch.

6.2 Projections, Perspective, and Point of View

The first authentic record of engineering graphics dates to 2130 B.C. based on a statue in the Louvre, Paris. The statue is of an engineer and governor of a small city-state in Babylon. At the base of the statue are measuring scales and scribing instruments along with a plan of a fortress engraved on a stone tablet.

Except for the use of pen and paper rather than stone tablets, it was not until printed books appeared around 1450 that the techniques of graphics advanced. Around the same time, **pictorial-perspective drawing** was being developed by the Italian artist Paolo Uccello. This type of drawing presents an object as it would look to the human eye or in a photograph. One of Uccello's paintings, shown in Figure 6-3, illustrates an early version of this type of drawing. The essential characteristic of a perspective drawing is that parallel lines converge at a point in the distance, just as parallel railroad tracks seem to converge in the distance. Copper-plate engravings permitted the production of finely detailed technical drawings in large numbers using pictorial perspective. The pictorial-perspective drawings were crucial to the advancement of technology through the Renaissance and until the beginning of the Industrial Revolution.

pictorial-perspective drawing

Figure 6-3 An example of early pictorial perspective.

Technical drawings, limited by the medium of pen and paper (or vellum) gave rise to two-dimensional drawings, also called 2-D drawings. A **2-D drawing** is a compact way to depict all of the detail of an object. A common example of a 2-D drawing that you would likely find is a blueprint, as shown in Figure 6-4.

2-D drawing

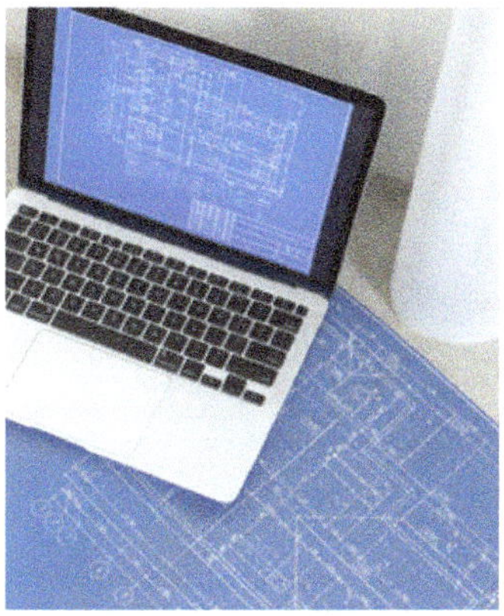

Figure 6-4 A blueprint is one kind of 2-D drawing.

Architects commonly used blueprints to plan and show how a new building should be laid out. But these drawings could not convey adequately the details of the construction of an object. One solution to this problem was the use of the **exploded view** developed in the 15th century and perfected by Leonardo da Vinci. The exploded view of an assembly of individual parts shows the parts spread out along a common axis, as shown for the water pump in Figure 6-5. The exploded view reveals details of the individual parts, along with showing the order in which they are assembled.

exploded view

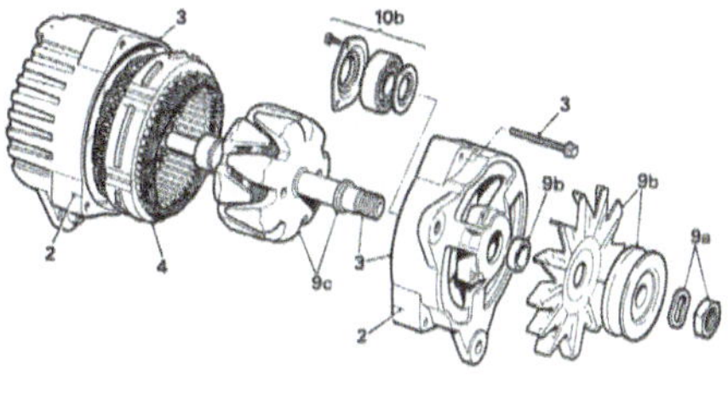

Figure 6-5 Assembled and exploded view of a water pump.

The Industrial Revolution brought with it the need to more closely tie together the concept of

ENGAGE

Visit the PaoloUccello.org website with your students.

URL

http://www.paolouccello.org/

Study the paintings in the online gallery and search for examples of "perspective" in his work. St. George and the Dragon c. 1456, Miracle of the Desecrated Host (Scene 1) 1465–69, and many others will clearly show perspective lines and vanishing points in the painting.

ENGAGE

Define the term *plane of projection.* *(Answer: A projection plane, or plane of projection, is a type of view in which graphical projections from an object intersect. Projection planes are used often in descriptive geometry and graphical representation. A picture plane in perspective drawing is a type of projection plane. With perspective drawing, the lines of sight between an object and a picture plane return to a vanishing point and are not parallel. With parallel projection, the lines of sight from the object to the projection plane are parallel.)*

ENGAGE

List the following types of drawings from the simplest to the most detailed: orthographic drawing, sketch, perspective drawing, and isometric drawing.

(Answer: Sketch, isometric drawing, perspective drawing, orthographic drawing.)

ENGAGE

Visit the Boeing Consulting Vocational Study Information website with your students. This is an online reference for engineering graphics.

URL

http://www.boeingconsult.com/tafe/dwg/dwg1/2d-iso.htm

a design with the final manufactured product using technical drawing. The perspective drawing of a simple object in Figure 6-6a shows pictorially what the object looks like. However, it is difficult to accurately represent dimensions and other details in a perspective drawing. By contrast, **orthographic projections**, developed in 1528 by German artist Albrecht Dürer, accomplish this quite well. An orthographic projection typically shows three views of an object. Each view shows a different side of the object (front, top, and side). An example of an orthographic projection is shown in Figure 6-6b. Orthographic projections are typically easy to draw, and the lengths and angles in orthographic projections can be accurately represented. As a result, orthographic drawings can convey more information than a perspective drawing. But their interpretation takes more effort than a pictorial perspective, as is evident from Figure 6-6. French philosopher and mathematician René Descartes laid the foundation for the mathematical principles of projections by connecting geometry to algebra in the 17th century. Much later, Gaspard Monge, a French mathematician, "invented" the mathematical principles of projection known as **descriptive geometry**. These principles form the basis of engineering graphics today. Because those principles were thought to be of such strategic importance, they remained military secrets until 1795. By the 19th century, orthographic projections were used almost universally in mechanical drawing, and they are still the basis for many engineering drawings today.

orthographic projections

descriptive geometry

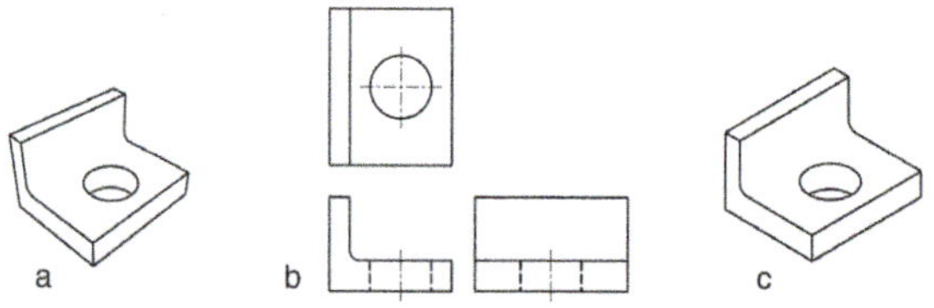

Figure 6-6 (a) Perspective projection; (b) Orthographic projection; (c) Isometric projection.

The **isometric projection** (Figure 6-6c) is used more often today than the pictorial-perspective projection. Historically, it was used for centuries by engravers. Then, in the early 19th century, William Farish, an English mathematician, formalized the isometric projection and introduced it to engineers. The isometric projection simplifies the pictorial perspective. In an isometric projection, parallel lines remain parallel, rather than converging to a point in the distance. In addition, the three coordinate axes have the same **metric** (hence the name *iso-metric*). This means that a length unit (say, a centimeter) along one axis has identical length in the drawing along the other two axes. Keeping parallel lines parallel and retaining the same metric along all axes in an isometric projection distorts the appearance of the object slightly. But the distortion in an isometric projection is negligible for objects of limited depth. For situations in which the depth of the object is large, such as an architectural view down a long hallway, the pictorial-perspective is preferable. The advantage of the isometric projection, though, is that it is much easier to draw than a pictorial-perspective projection.

isometric projection

metric

6.2.1 3-D Projections

The goal in engineering graphics is to represent an image of a physical object (either real or virtual) so that the object represented by the image can be conveyed to other persons or shared in the manufacturing process. Objects can be shown as **3-D projections** or **multiview projections**. Figure 6-7 shows the handle of a pizza cutter both ways.

3-D projections

multiview projections

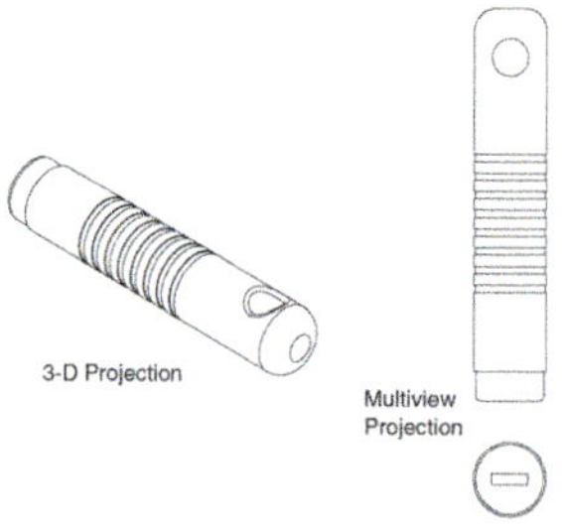

Figure 6-7 3-D projection and multiview projection of a handle.

The 3-D projection clearly suggests the three-dimensional character of the handle, even though it is displayed on a two-dimensional medium (the page). 3-D projections are useful in that they provide an image that is similar to the image in the designer's mind. But 3-D projections are often weak in providing adequate details of the object, and there is often some distortion of the object. For instance, the circular hole through the right end of the handle becomes an ellipse in an isometric 3-D projection.

Multiview projections are used to overcome the weaknesses of 3-D projections. Multiview projections (such as orthographic projections) are a collection of flat 2-D drawings of the different sides of an object. For instance, the side and bottom end of the pizza cutter handle are shown in the multiview projection in Figure 6-7. Because there are two views, it is quite easy to depict the details of the object. In addition, taken together, multiview projections provide a more accurate representation of the object than the 3-D projection—a circular hole appears as a circle in a multiview projection. On the other hand, multiview projections require substantial interpretation, and the overall shape of an object is often not obvious upon first glance. Consequently, the combination of the overall image provided by 3-D projections and the details provided by multiview projections yields a representation of an object that is best. The shape of the object is immediately evident from the 3-D projection, and the detail needed for an accurate description of the object is available from the multiview projection.

Two types of 3-D projections are available in most CAD software: isometric and perspective. Three projections of a house are shown in Figure 6-8; these 3-D projections represent all three dimensions of the house in a planar image. Each type of 3-D projection has its advantages and disadvantages.

The isometric projection has a standard orientation that makes it the typical projection used in CAD. In an isometric projection, the width and depth dimensions are sketched at 30° above the horizontal as shown in Figure 6-9. This results in the three construction line parallel projection angles at the upper front corner of the cube being equal to 120° in the plane of the paper. Lines along the edges of the cube, known as **construction lines** (lines which help us draw, but are removed in the finished drawing), are parallel, so this type of projection is also known as a **parallel projection**. Notice that the three sides of the cube are equal in length in the drawing, leading to the term isometric (equal measure). Isometric drawings work quite well for objects of limited depth. However, an isometric drawing distorts the object when the depth is significant. In this case, a pictorial-perspective drawing is better.

construction lines

parallel projection

A pictorial-perspective, or simply **perspective**, projection is drawn so that parallel lines converge in the distance, unlike isometric projections, where parallel lines remain parallel. As shown in Figure 6-10, the projectors converge toward two points, known as vanishing points. This results in **foreshortening**—lines of an object are changed to produce

perspective

foreshortening

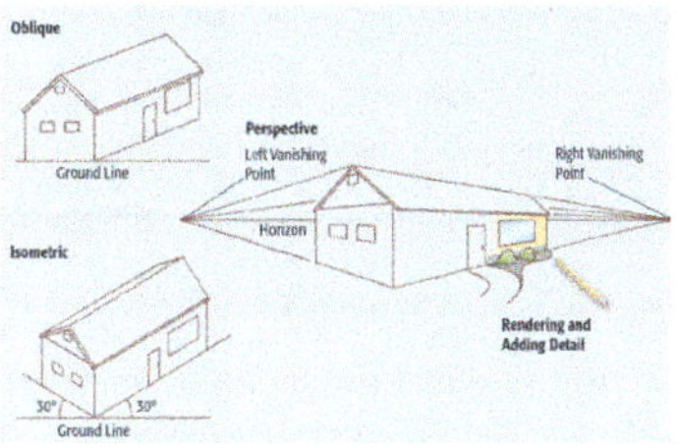

Figure 6-8 Isometric, oblique, and perspective projections.

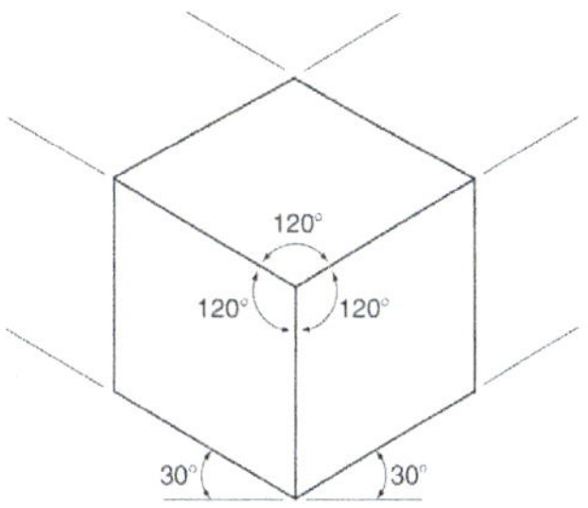

Figure 6-9 Isometric projection of a cube showing parallel projectors.

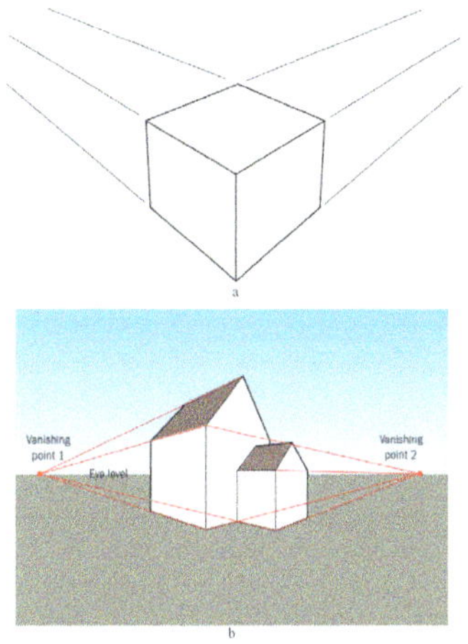

Figure 6-10 Pictorial-perspective projection of (a) a cube and (b) a building showing converging projectors.

ENGAGE

Visit the ArtyFactory.com website with your students. Navigate to "perspective drawing" and choose the linear and aerial perspective page. This reference can illustrate the concepts of perspective from eye level, low eye level, and high eye level.

URL

http://www.artyfactory.com/perspective_drawing/perspective_index.html

ENGAGE

Visit the National Council of Teachers of Mathematics *Illuminations* website with your students. Navigate to "Isometric Drawing Tool" and allow the students time to explore the tool and create isometric shapes.

This tool is particularly helpful if you plan to incorporate other spatial visualization practices in your course.

URL

http://illuminations.nctm.org/activity.aspx?id=4182

ENGAGE

Visit the Clan MacCAD YouTube website with your students. Ask your students to watch the video entitled *Third Angle Projection vs. First Angle Projection 3D animation Part 1*

URL

https://www.youtube.com/watch?v=bk2E8P33Ztc

ENGAGE

Visit the Clan MacCAD YouTube website with your students. Ask your students to watch the video entitled *Third Angle Projection vs. First Angle Projection 3D animation Part 2.*

URL

https://www.youtube.com/watch?v=yGjVnXgUpQM

ENGAGE

Visit the PracticalStudent.com website with your students. Define "The Glass Box" method of developing orthographic projections.

URL

http://www.practicalstudent.com/subjects/tg/projectionsystems/oplesson1.shtml

(Answer: "The Glass Box" is the name sometimes given to a theoretical see-through box that any object can fit inside. It is a very useful tool when it comes to trying to explain how orthographic projection works.

Ask your students to compare "first angle projection" and "third angle projection.")

the illusion of the part extending into the third dimension. Perspective projections can also be drawn with a single vanishing point, which is useful for showing interiors of buildings, or with three vanishing points, which can display perspective for very large objects. Perspective projections are most useful in providing a realistic image of an object when the object spans a long distance, such as the view of a bridge or aircraft from one end or of a tall building viewed from near its base. Generally, small manufactured objects are adequately represented by isometric or oblique projections as shown in Figure 6-9.

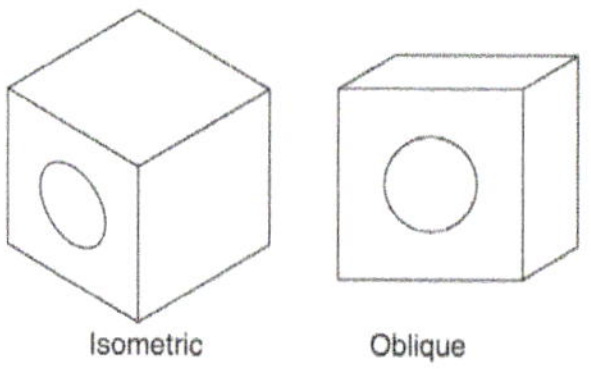

Figure 6-11 Isometric and oblique projections of a cube with a hole.

oblique projection

Two types of pictorial representations used frequently for 3-D rendering drawings are the isometric and oblique views. The **oblique projection** places the principal face of the object parallel to the plane of the paper with the axes in the plane of the paper perpendicular to one another. The axis into the paper is at an arbitrary angle with respect to the horizontal. Figure 6-11 compares an isometric projection of a cube with a hole in it with an oblique projection of the same object. The advantage of the oblique projection is that details on the front face of the object retain their true shape. For instance, the circle on the front face is circular in the oblique projection, while it is elliptical in the isometric projection.

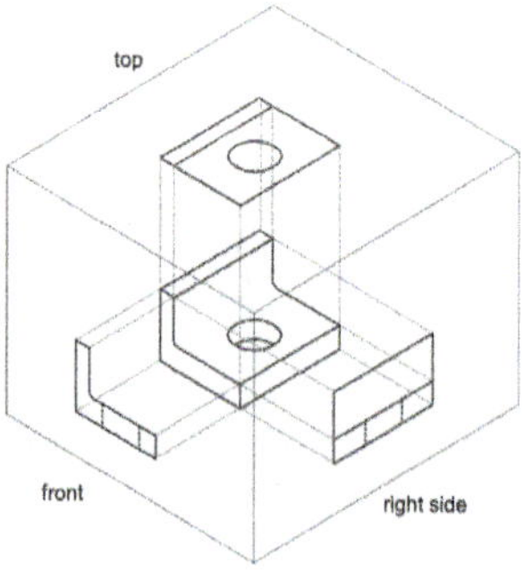

Figure 6-12 Sides of an object can be projected onto an imaginary glass box surrounding the object.

6.2.2 Multiview Projections

The standard multiview projection in engineering graphics is the orthographic projection. Although 3-D projections provide a readily identifiable visual image of an object, multiview projections are ideal for showing the details of an object. Dimensions can be shown easily, and most features remain undistorted in multiview projections.

An orthographic projection is most easily thought of as a collection of flat 2-D views of different sides of an object—front, top, side, and so forth. For instance, two orthographic projections could be used for a coffee mug. The front view would show the sidewall of the mug along with the loop forming the handle. The top view would show what one would see looking down into the mug—a circular rim of the mug, the bottom of the inside of the mug, and the top of the handle that sticks out of the side of the mug. Dimensions of the mug could easily be added to the projections of each side of the mug to create an engineering drawing.

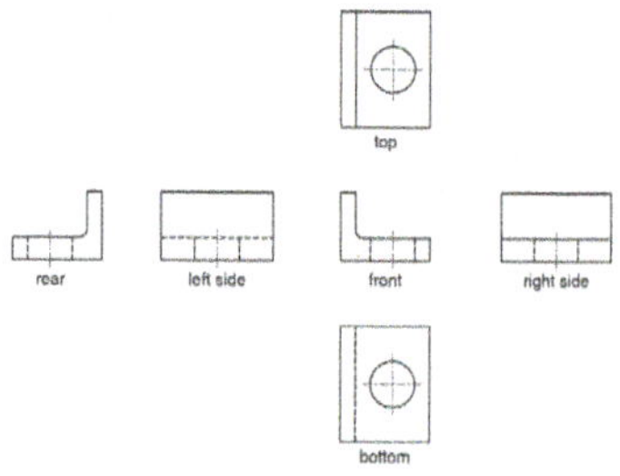

Figure 6-13 Third-angle projection resulting from unfolding the imaginary glass box surrounding the object.

One way of looking at multiview projections is to imagine a glass box surrounding the object as shown in Figure 6-12. The image of each side of the object can be projected onto the wall of the glass box. Now an observer on the outside of the box can see each side of the object as projected onto each of the six walls of the box. Solid lines show the edges evident in the projection, and dashed lines show lines that are hidden by the object. Now imagine unfolding the glass box as if each of the edges of the glass box were a hinge, so that the front view is in the middle. Now the unfolded glass box represents all six sides of the object in a single plane as shown in Figure 6-13. The dimensions of the object remain the same in all views. For example, the horizontal dimension (the width of the object) in the front view is identical to that same dimension in the top view and the bottom view. The views also remain aligned so that the bottom edge in the front view is even with the bottom edge in the right-side view and so on. In addition, the same edges in adjacent views are closest together. For instance, the same edge of the object is at the left side of the front view and the right side of the left-side view. This edge in the front view is closest to the same edge in the left-side view. See Figure 6-14 for an example using a toilet.

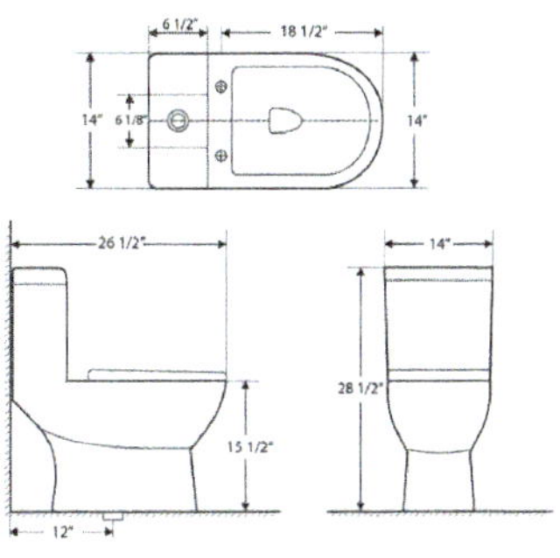

Figure 6-14 Orthographic views show only two dimensions at a time.

In many cases, three views are needed to represent an object accurately, although in some cases only two views are necessary, and in other cases more than three views are needed to show complex features of the object. It is helpful to select the side of the object that is most descriptive of the object as the front view. Sometimes this may place an object so that what is normally thought of as the front of the object is not shown in the front view of the multiview projection. For example, what is usually described as being the side of a car should be chosen as the front view, because this view is probably THE most descriptive and easily recognizable as a car. A view of the front of a car is not as descriptive or as obvious as a view of the side of the car. Furthermore, the object should be properly oriented in the front view. For instance, a car should be shown with its wheels downward in their normal operating position for the front view. The other views that are shown in addition to the front view should be views that best represent features of the object. Normally the minimum number of views necessary to accurately represent the object is used. The standard practice is to use the front, top, and right-side views. But the choice of which views to use depends on the object and which details need to be shown most clearly.

auxiliary view

A problem that can occur in orthographic projections is that one of the faces of the object is at an angle to the orthographic planes that form the imaginary glass box. An example is the object shown in Figure 6-15. The circular hole with a keyway (the rectangular cutout on one side of the hole) that is perpendicular to the angled face appears in both the top view and the right-side view. However, it is distorted in both of these views, because it is on an angled plane of the object. This requires an **auxiliary view** to avoid this distortion. In this case, a view of the object is drawn so that the angled face is parallel to the auxiliary view plane. The view is based on the viewer looking at the object along a line of sight that is perpendicular to the angled face. Seen from this direction, the circular hole in the auxiliary view appears as a circle without any distortion. As suggested by the projection lines (dash-dot lines) in Figure 6-15, the auxiliary view is projected from the front view in the same way as the top and right-side views are projected. Thus, its position with respect to the front view depends on the orientation of the angled face. It is normal practice not to include hidden lines or other features that are not directly related to the angled surface in the auxiliary view.

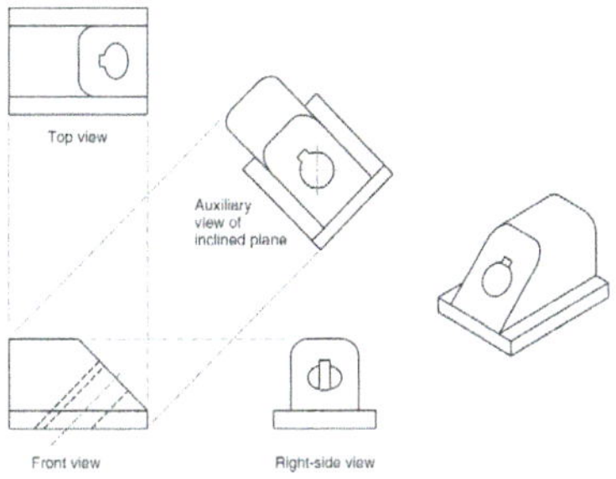

Figure 6-15 Auxiliary view of an inclined face.

6.2.3 Detail Drawings

detail drawings

Detail drawings document the detailed design of individual components using orthographic views. The detail drawing is the representation of a design that is specific enough that all of the information necessary for the manufacture of the part is provided. As a result, the detail drawing includes the necessary views, dimensions, and specifications required for

ENGAGE

Visit the Cognifront Products YouTube website with your students. Ask your students to watch the video entitled *Introduction to Auxiliary Views.*

URL

https://www.youtube.com/watch?v=vZbrcAGOB_o

ENGAGE

Visit the Clan MacCAD YouTube website with your students. Ask your students to watch the video entitled *Section Views in Engineering Drawing & CAD – The Basics.*

URL

https://www.youtube.com/watch?v=doyJeJSwiLQ

ENGAGE

Visit the Design1atMonash YouTube website with your students. Ask your students to watch the video entitled *Assembly Drawings 1.*

URL

https://www.youtube.com/watch?v=RGr2vzch-SI

manufacturing the part. Figure 6-16 shows an example of a detail drawing of a part of a pizza cutter. The detail drawing includes fully dimensioned orthographic views, notation of the material that the part is to be made from, information on the acceptable tolerances for the dimensions, and a title block that records important information about the drawing. Often, an isometric projection is included in the detail drawing to further clarify the shape of the part. Detail drawings provide sufficient detail so that the part can be manufactured based on the drawing alone.

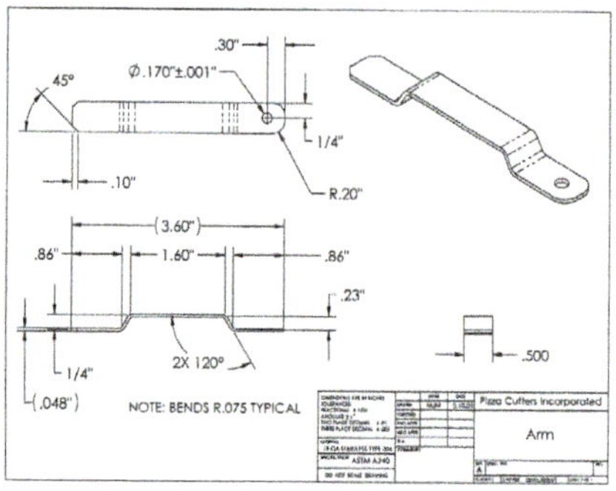

Figure 6-16 Detail drawing of a sheet metal part for a pizza cutter.

6.2.4 Assembly Drawings

assembly drawings

Assembly drawings show how the components of a design fit together. Dimensions and other details are usually omitted in assembly drawings to enhance clarity. Several styles of assembly drawings are commonly used. Sometimes the assembly drawing is just an isometric projection of the fully assembled device. But an exploded isometric projection is often helpful to show how the individual parts are assembled, as shown in Figure 6-17 for a pizza cutter. In some cases, a sectioned assembly, or cut-away view, shows how complicated devices are assembled. A cutting plane passes through the assembly, and part of the device is removed to show the interior of the assembly. Numbers or letters can be assigned to individual parts of the assembly on the drawing and keyed to a parts list.

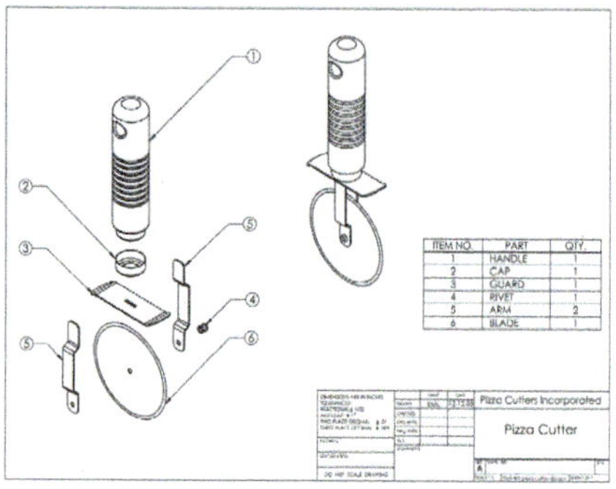

Figure 6-17 Assembly drawing of a pizza cutter.

parts list

bill of materials

Finally, a **parts list**, or **bill of materials**, must be included with a set of working drawings. The parts list includes the part name, identification number, material, number required in the assembly, and other information (such as a catalog number for standard parts like threaded fasteners). An example is shown in Figure 6-18 for a pizza cutter. The parts list is used to ensure that all parts are ordered or manufactured and brought to the central assembly point.

ITEM	QTY.	PART NO.	DESCRIPTION
1	1	52806	HANDLE
2	1	52825	CAP
3	1	42886	GUARD
4	1	97511	RIVET
5	2	55654	ARM
6	1	56483	BLADE

Figure 6-18 A bill of materials for a pizza cutter.

Taken together, the detail drawings of each individual part, the assembly drawing, and the bill of materials provide a complete set of working drawings for the manufacture of a part.

6.3 Freehand Sketching

ideation

An integral part of the creative design process is **ideation**, the generation of concepts or ideas to solve a design problem. Often, freehand sketching can be used to quickly explore and communicate concepts that come about in the mind's eye. The process of sketching can solidify and fill out rough concepts. Furthermore, sketching captures the ideas in a permanent form that can be used to communicate the concept to others. In this way, sketches often act as stepping stones to refine and detail the original concept or generate new ideas. Many great design ideas are first sketched on a blank sheet of paper, in a lab notebook, or even on an electronic tablet with a stylus.

Freehand sketching requires few tools—just a pencil and paper. One of the advancements in CAD as a design tool is that with the use of an electronic sketch pad, often the CAD program can interpret a sketch directly into an orthographic representation. Usually, the CAD program will require a little bit of help in scaling the three orthogonal views, but there is a lot of flexibility in these programs. This can also speed up the production of a 3-D printed scale model and the process of iterating a design.

Three types of pictorial sketches are used frequently in freehand sketching: oblique, isometric, and orthographic. The process of freehand sketching is very similar for all three types of projections.

There are three main steps:

- First, construction lines (light or very thin lines) are used to create the rough primary view and the part "envelope"—the largest imagined box that would contain the part.
- Second, the primary view is "blocked in," or drawn in darkly with a thicker pen or more pressure with a pencil to emphasize all the visible edges in the primary view.
- Third, construction lines are used to project or extend the visible edges of the part and then those lines are blocked in. Keep in mind that construction lines guide the proportions of the finished sketch.

These three steps are followed regardless of the particular view, but there are differences, as discussed previously, depending on whether the view is oblique, isometric, or orthographic. The differences are explained below.

The isometric projection tilts the part so that no surface of the part is in the plane of the paper, and the orthographic imitates the way in which the part would be portrayed as the typical CAD format on a screen. One advantage of the orthographic representation is that it may ultimately be easier to build the part in the CAD software, since it is a common output format.

6.3.1 Oblique Sketching

The oblique projection places the principal face of the object parallel to the plane of the paper. This face is the basis for the primary view. The advantage of the oblique projection is that details on the front face of the object retain their true shape. This often makes oblique freehand sketching easier than isometric sketching, where no plane is parallel to the paper. The disadvantage of the oblique projection is that it does not appear as "photorealistic" as an isometric projection.

An oblique freehand sketch is easy, since it begins with a two-dimensional representation of the front face of the object. Figure 6-19 shows the steps involved in quickly sketching a part with a circular hole.

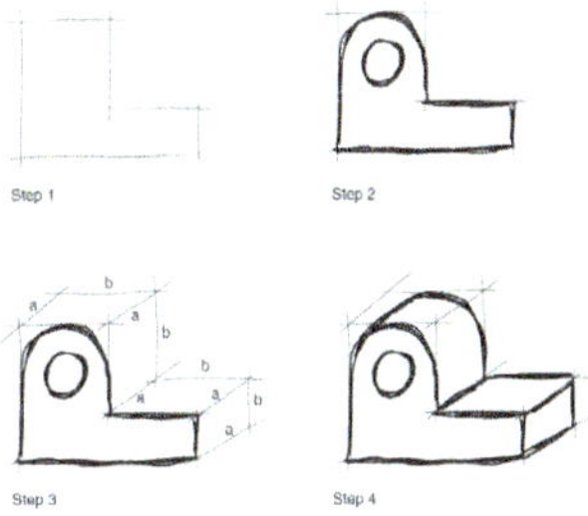

Figure 6-19 Creating a simple oblique freehand sketch.

ENGAGE

Visit the Massachusetts Institute of Technology 2.009 Engineering Processes website with your students. Ask your students to watch the videos listed under *Sketching Tutorials*.

URL

http://web.mit.edu/2.009/www/resources/sketching/index.html

Have your students practice some of the basic techniques using the tutorials on perspective, or download the videos yourself to develop practice problems.

DIFFERENTIATED INSTRUCTION

Spatial visualization is a key skill for all disciplines of engineering. Building this skill is associated with a greater likelihood of continuing and graduating in an engineering major. You can assess your student's spatial visualization skills using the Revised Purdue Spatial Visualization Test (Revised PSVT:R). Those with lower scores can be directed with resources to practice sketching, mental cuts, and mental rotations.

URL

https://www.silc.northwestern.edu/revised-purdue-spatial-visualization-test-revised-psvtr-visualization-of-rotations/

ENGAGE

Visit the DBA Design YouTube website with your students. Ask your students to watch the video entitled *Using Oblique and Isometric techniques.*

URL

https://www.youtube.com/watch?v=w503xvjt7mA

Ask the students to sketch their initials after they finish the video.

For oblique sketches, all of the receding construction lines must be parallel to each other and are usually at an angle of 30° to 45° with respect to the horizontal. The receding lines end at the appropriate depth for the object. The final sketch, while rough and lacking detail, clearly shows the design intent for the part.

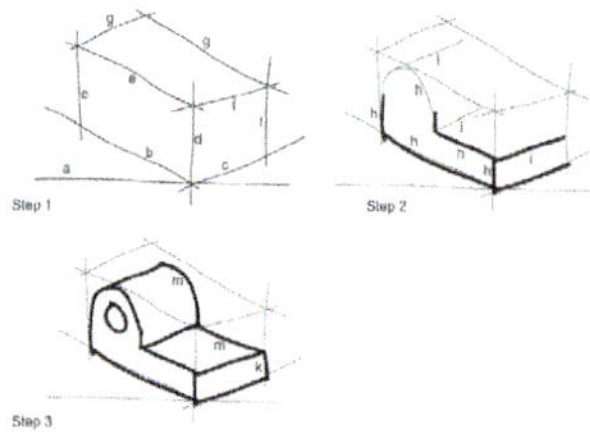

Figure 6-20 Creating a simple isometric freehand sketch.

6.3.2 Isometric Sketching

Isometric freehand sketches are somewhat more difficult to master than oblique sketches because no face is in the plane of the paper for an isometric projection. The steps to construct a simple freehand isometric sketch are shown in Figure 6-20.

In the isometric sketch, the first construction line is a horizontal line. A vertical construction line is drawn to intersect this horizontal line and then two construction lines, each 30° above the horizontal line (lines b and c in Figure 6-20) are drawn. The rest of the lines, e, f, and g, from above are used to create an enclosing box for the primary view. Using those construction lines as guidelines ensures that features that exist in planes that are parallel to each other must be kept parallel.

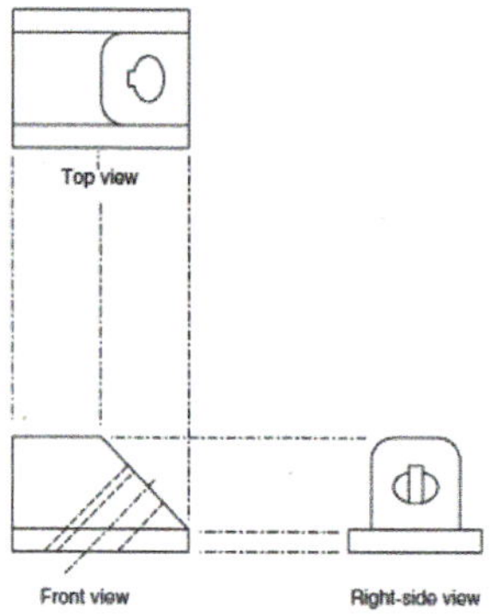

Figure 6-21 Creating a simple orthographic sketch.

6.3.3 Sketching Orthographic Views

The orthographic view (Figure 6-21) is what we typically think of when we think of a traditional multiview representation. It starts out similarly to sketching an oblique view. In the orthographic version of a part, you start with one face that is in the plane of the paper and what would typically be the primary view. This view should be the "natural" position of the part when it is in use and a view that minimizes the number of hidden lines. This is usually placed in the lower left side of the drawing. This will leave room for a top view above the primary view and a side view to the right of the primary view.

Next, move to the top view. You do this by carrying the construction lines of all edges and features (like a hole drilled in one face) that will need to be represented, upward to frame the top view. Then, using those witness lines as a guide, block out the edges and features using a darker or heavier line. Be sure to match the dimensions of the actual part as closely as you can by eye for the scale that you are using in the drawing. Follow a similar process on the side view. Carry the construction lines to the right, keeping them at right angles to the lines between the primary and the top view. Then, block in the finished object by using heavier lines to outline the features of that view, again paying attention to the scaling.

An advantage of this type of sketching is that it mimics the finished view that will eventually be produced by the CAD program. This helps in planning the layout and thinking about the dimensioning requirements. It also helps develop the skill of visualizing objects in three dimensions.

The choice of whether to use an oblique, an isometric, or an orthographic projection is often arbitrary. Because the oblique projection is easier to sketch, it is sometimes preferred. On the other hand, an isometric projection provides a more photorealistic image of the object, while an orthographic projection matches what the finished CAD drawing will look like.

6.3.4 Sketching Auxiliary Views

line of sight

When one of the faces of the part is at an angle to the orthographic planes, it may be necessary to sketch an auxiliary view so that the features on the angled face are not distorted. The view is based on the observer looking at the object along a **line of sight** that is perpendicular to the angled face. Sketching an auxiliary view is a lot like sketching a standard orthographic view. Consider the object shown in Figure 6-22, having an angled face with a circular hole and keyway (rectangular cutout). Only the auxiliary view shows the circle without any distortion.

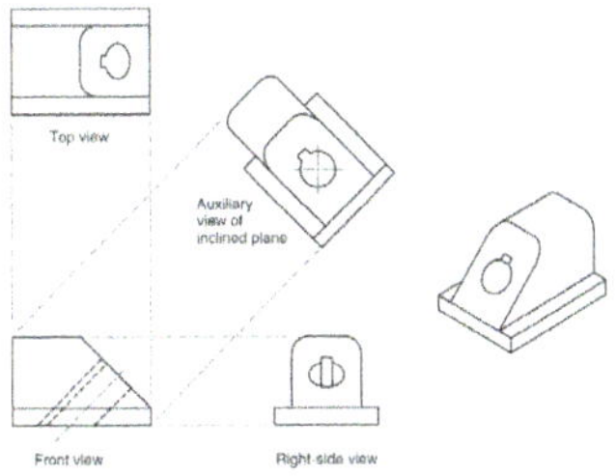

Figure 6-22 Auxiliary view of an inclined face.

Details of sketching an auxiliary view vary, depending on the nature of the object to be sketched. However, Figure 6-23 shows the steps that would usually be used for the object shown in Figure 6-22, as an example.

Follow the procedure very much like the one for orthographic projection. The primary difference is that the construction lines will be placed so that they are perpendicular to the primary face of the feature that you are trying to highlight with this auxiliary view. Pay attention to proportion and make sure that the auxiliary view shows the "true view" of the feature that it is representing.

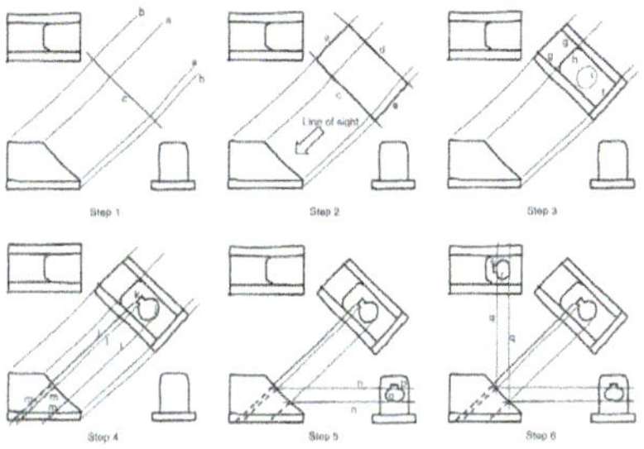

Figure 6-23 Creating an auxiliary view in a freehand sketch.

Notice how the auxiliary view is used to more accurately sketch the features (hole and keyway) that are on the inclined surface when they are projected to the other views after the auxiliary sketch has been created. The technique for sketching auxiliary views depends on the object being sketched. If the inclined surface were on the left edge of the front view instead of the right edge, the auxiliary view would be sketched to the left and above the front view instead of to the right. Likewise, the examples shown in this chapter only demonstrate the typical steps that can be used to generate freehand sketches.

Professional Success: Will Freehand Sketching Ever Become Obsolete?

CAD has almost totally eliminated pencil-and-paper drawings. So, what is the future of freehand sketching? Pencil-and-paper (or tablet and stylus) sketches are quick, efficient, easily modified, and easily conveyed to others. So, this will remain the technique of choice for early design brainstorming sessions.

A logical progression is to more closely couple the sketching process to a CAD model through machine learning. Sketches are not geometrically perfect, however a tablet could be taught that a round object is either a hole or a tube or a shaft, depending on selections from a menu. Even the Microsoft PowerPoint program has built-in alignment tools for geometric shapes. Though it's not designed for a CAD interface, some of those same techniques could be translated to a CAD format. Freehand sketching does develop a feel for a physical part and a tablet that assists in generating oblique or isometric sketches as they are drawn is a logical enhancement.

ENGAGE

Visit the Engineering Essentials website with your students. Ask your students to watch the video entitled *Auxiliary view*.

URL

http://engineeringessentials.com/ege5/files/ege/adv/adv_page1_ex2.htm

ENGAGE

Visit the Wikipedia page on 3D Projection.

URL

https://en.wikipedia.org/wiki/3D_projection

Ask the students to compare and contrast the different types of projection drawings.

ENGAGE

What does the acronym CAD stand for?
(CAD stands for Computer-Aided Design).

QUESTION

Describe a CAD system in your own words.
(Answer: A CAD workstation has many parts. It includes a computer, keyboard, display screen, drawing tablet, and plotter. CAD software offers a variety of drawing tools that can be used either from the keyboard and mouse or from the drawing tablet.)

QUESTION

What equipment or tools does a designer need to create a drawing using CAD?
(Answer: A designer needs a CAD workstation that includes a computer, keyboard, display screen, drawing tablet, plotter, and CAD software.)

6.4 CAD—Computer-Aided Design

The introduction of the computer revolutionized engineering graphics. In the first iteration of using computers for graphics, it was viewed as a high technology replacement for a pencil and paper—basically computer drafting. As more capabilities were added, it gradually evolved to be more of a design and manufacturing tool. Other acronyms were added; Computer-Aided Engineering (CAE), Computer-Aided Manufacturing (CAM), and other terms were used to encompass a broader range of computer-related design tools.

Several different models for representing a part's geometry have been used in CAD, including 2-D models and 3-D wireframe and surface models. However, solid modeling is the current state of the art in CAD. Solid modeling has the inherent advantage of more accurately representing the design intent of the part that is modeled.

model

A CAD **model** is a virtual part that exists not only as an image, but as a full-blown engineered part, ready to be produced. The model is an engineering document of record. It contains all of the design information, including the geometry, dimensions, tolerances, materials, and manufacturing information. A CAD model replaces the paper blueprints and engineering drawings used a few decades ago.

2-D model

The simplest model used in CAD is a **2-D model**. It is essentially the computer graphics equivalent of an orthographic projection created using a pencil and paper, as shown in Figure 6-24. A 2-D model represents a three-dimensional object with several views, each showing one side of the object projected onto a plane. On the computer, a 2-D model appears as lines and curves on a flat surface, and it is up to the engineer or designer to interpret several views to create a mind's eye image of the three-dimensional object. The CAD software electronically stores the 2-D model as several views but does not inherently "know" that the views can be connected to one another to form a three-dimensional object.

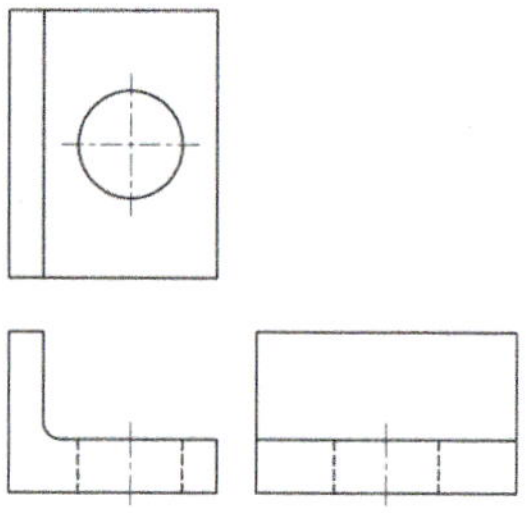

Figure 6-24 2-D model of an object.

2 ½-D model

The term **2 ½-D model** has been introduced as a way to represent a family of parts. It is not a commonly used term because all real-world objects inherently are 3-D models. Some people are of the mind that the ½ dimension represents a variation of a 2-D part into a third dimension. This works when the extra dimension is an extrusion of a two-dimensional shape. In this case, the CAD software electronically stores the cross-sectional shape of the object and the depth that it is **extruded** is the ½ dimension. It is a way of describing objects that have a constant cross section but could be of varying lengths.

extruded

For instance, the simple shape shown in Figure 6-25 is the 2 ½-D model of the object shown using a 2-D model in Figure 6-24 without the hole. The object, including the hole, cannot be represented as a 2 ½-D model because the hole results in a change in the cross section. A 2 ½-D model could be used to represent a shape such as an I-beam or C-channel.

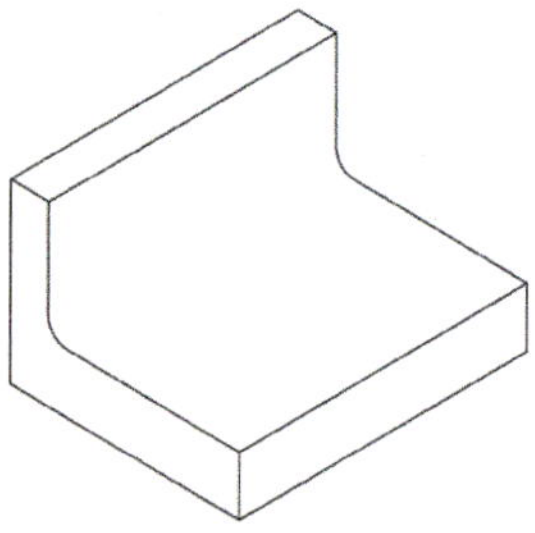

Figure 6-25 2 ½-D model of an object formed by extruding a 2-D shape.

3-D model

A **3-D model** is the most general model used in CAD software. In this case, the CAD software electronically stores the entire three-dimensional shape of the object. Of course, a two-dimensional drawing representing the orthographic views of the object can be automatically created from the 3-D model. Most CAD software is based on 3-D models as the starting point for the design. This is analogous to the way that a physical part is machined from a raw initial shape.

QUESTION

How has the introduction of CAD affected the fields of drawing and design?
Answer: (CAD has made great changes in drawing and design. Clothing designers use CAD to create patterns and images for new fabrics and clothing. Engineers often use two-dimensional working drawings to label the different parts of a machine and indicate how it is assembled. Landscape architects use CAD to show the arrangement of trees, shrubs, walkways, and lighting in an outdoor area. Urban planners show how new streets and traffic lights will affect the flow of traffic through a city. Other CAD-designed products include cars, aircraft, furniture, rugs and other textiles, and electronic games.)

wireframe model

The simplest 3-D model is a **wireframe model**. Figure 6-26a shows a wireframe representation of the object that we have been using as an example. The problem inherent in a wireframe representation is quite clear in Figure 6-26a—it is difficult to interpret the drawing because all of the edges are visible. Figure 6-26b shows the object with the hidden lines shown as dashed lines to indicate that they are behind a surface. 3-D models not based on wireframes are able to deal with hidden lines more effectively and offer other advantages. In a wireframe model, only edges of the object are represented. The CAD software needs to store and retrieve the locations of the vertices and information about which vertices are to be connected to each other by lines. A circle or curved section can be represented by a series of closely spaced vertices with linear edges connecting them, or as an arc defined by a center, radius, and end points. Before more powerful computers were available, the low storage requirements of a wireframe model made it quite popular, however these issues have largely gone away, and wireframes have become much less common.

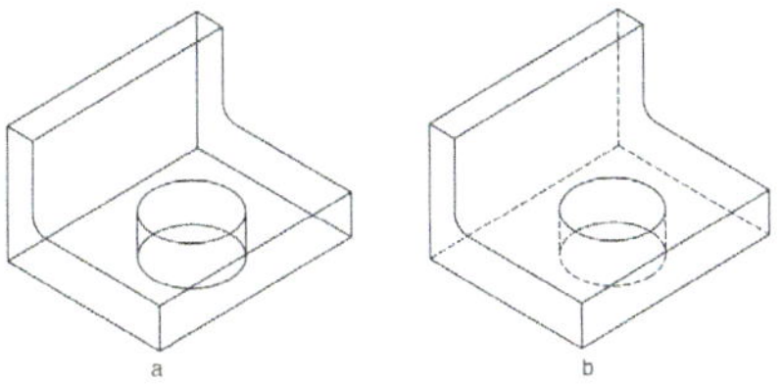

Figure 6-26 (a) 3-D wireframe model; (b) 3-D wireframe model with hidden lines shown as dashed lines.

surface model

A 3-D **surface model** defines the object in terms of surfaces such as plates (flat) and shells (curved) in addition to edges. This makes it easy to determine whether a line or surface is hidden, because the surfaces are defined. Using a surface model also permits the construction of smoothly curved surfaces, such as a circular hole, rather than using many line segments to form a curved line as with some wireframe models. It is also possible to smooth one surface into another surface to provide a sculpted shape. Figure 6-27a shows the simple object with hidden lines removed altogether. Although a surface model can be displayed to look like a wireframe, either with or without hidden lines, it can also be displayed by assigning different degrees of shading to the surfaces. A virtual light source is assumed to be near the object to provide a three-dimensional lighting effect as shown in Figure 6-27b. This is known as **shading** or **rendering**. Although a variety of shading techniques are available, all depend on determining how much light from a virtual light source strikes each portion of the virtual object. The use of shading provides a very realistic image of the object, which permits a much more vivid communication of the nature of the object. However, the problem that remains with a surface model is that the interior of the object remains undefined. The surface model just represents the shell of the object.

shading

rendering

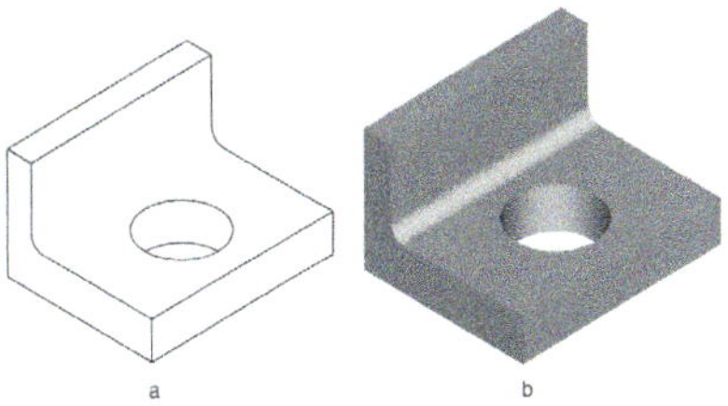

Figure 6-27 (a) 3-D wireframe model with hidden lines removed. (b) 3-D model with shading.

6.5 Solid Modeling and Design for Manufacturing

solid modeling

All the issues about ease of viewing, whether a model is 2 ½-D or 3-D, or how smoothly an object is rendered, go away with **solid modeling**. This is the most sophisticated and common method of representing a manufactured object. Unlike wireframe or surface models, a solid model is a virtual representation of an object. It is endowed with all the characteristics of the final, manufactured product. It has all the mechanical characteristics, volume, and density, as well as surfaces and edges.

constructive solid geometry

primitive modeling

The first attempt at solid modeling was a technique known as **constructive solid geometry** or **primitive modeling**, which is based on the combination of geometric primitives such as right rectangular prisms (blocks), right triangular prisms (wedges), spheres, cones, and cylinders. Each primitive could be scaled to the desired size, translated to the desired position, and rotated to the desired orientation. Then one

ENGAGE

Discuss the advantages of CAD drawings compared to those that are hand-drawn. Advantages of CAD drawings include the following:

1. CAD saves time by combining design and drafting into a single process. Architectural or mechanical drawings can be created much faster with CAD than with hand drawing.
2. Information stored in the computer can help prevent errors in the design.
3. Because designs can be revised onscreen, much less time is needed to change and improve existing designs.
4. Designs are more accurate. When changes must be made to a design, they can be made more consistently from one design to the next.
5. CAD systems can provide a list of the parts that are needed to make a product, as well as the cost of each.
6. CAD systems can evaluate how well a design will work and point out potential problems before construction or manufacturing begins.

ENGAGE

Ask your students to complete an online image search with their browser. Students should search *wireframe, surface*, and *solid* 3-D models and find images that depict the similarities and differences among the three types of CAD models used in most software programs. Ask the students which models are more difficult to create, which ones will require more memory on the device, and which give the most accurate depiction of the product being displayed.

ENGAGE

Visit the *AutoDesk* Knowledge Network website with your students. Navigate to the AutoCAD pages on Modeling 3D Objects.

URL

https://knowledge.autodesk.com/support/autocad/getting-started/caas/CloudHelp/cloudhelp/2015/ENU/AutoCAD-Core/files/GUID-9DACE807-BC9D-4357-B47E-C6199F6AF1A2-htm.html

Explain to your students that most CAD programs have the ability to produce solid, surface, and mesh objects, which offer different functionality that, when used together, create a powerful suite of 3-D modeling tools. For example, you can convert a primitive solid to a mesh to take advantage of mesh creasing and smoothing.

NOTE: Instructors can research 3-D modeling in whatever software package they prefer; introducing the students to modeling using the class software will be beneficial.

primitive could be added to or subtracted from other primitives to make up a complex object. It is easy to see how two blocks minus a cylinder (hole), as shown in Figure 6-28, could represent an object similar to the object in Figure 6-27.

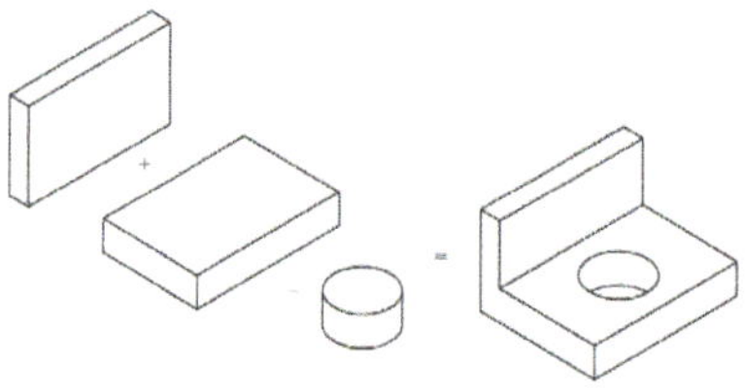

Figure 6-28 Combining geometric primitives to represent an object.

constraint-based solid modeling

Constraint-based solid modeling overcomes the weakness of constructive solid geometry modeling by making the modeling process more intuitive. Instead of piecing together geometric primitives, the constraint-based modeling process begins with the creation of a 2-D sketch of the profile for the cross section of the part. The initial sketch need not be particularly accurate; it only needs to reflect the basic geometry of the part's cross-sectional shape. Details of the cross section are added later. The next step is to constrain the 2-D sketch by adding enough dimensions and parameters to completely define the shape and size of the 2-D profile. The name constraint-based modeling arises because the shape of the initial 2-D sketch is "constrained" by adding dimensions to the sketch. Finally, a three-dimensional object is created by **revolving** or extruding the 2-D sketched profile. Figure 6-29 shows the result of revolving a simple L-shaped cross section by 270° about an axis and extruding the same L-shaped cross section along an axis. In either case, these solid bodies form the basic geometric solid shapes of the part. Other features can be subsequently added to modify the basic solid shape.

revolving

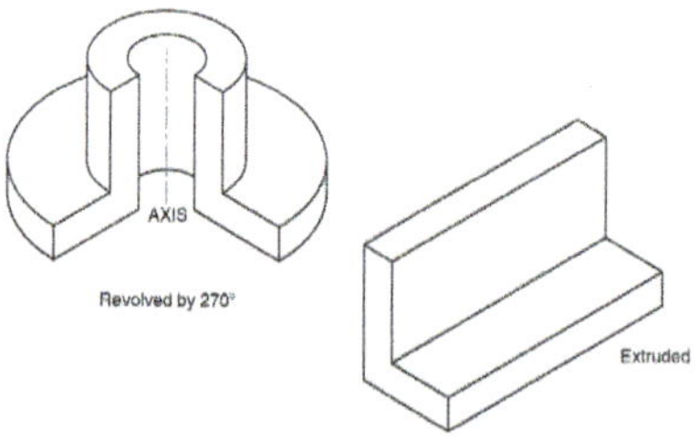

Figure 6-29 Revolving by 270° and extruding an L-shaped cross section.

Once the solid model is generated, all the surfaces are automatically defined, so it is possible to shade it in the same way as a surface model is shaded. It is also easy to generate 2-D orthographic views of the object. This is a major advancement from other modeling schemes. With solid modeling, the two-dimensional drawings are produced as views of the three-dimensional virtual model of the object.

In traditional CAD, the three-dimensional model is derived from the two-dimensional drawings. One might look at solid modeling as the sculpting of a virtual solid volume of material. Because the volume of the object is properly represented in a solid model, it is possible to slice through the object and show a view of the object that displays the interior detail. Once several solid objects have been created, they can be assembled in a virtual environment to ensure that they fit together properly and to visualize the assembled product.

6.5.1 The Nature of Solid Modeling

feature-based modeling

features

Feature-based modeling attempts to make the modeling process more efficient by creating and modifying geometric **features** of a solid model in a way that represents how geometries are created using common manufacturing processes. Features in a part have a direct analogy to geometries that can be manufactured or machined. A base feature is a solid model that is roughly the size and shape of the part that is to be modeled. The base feature is the 3-D solid created by revolving or extruding a cross section, such as those shown in Figure 6-29. It can be thought of as the initial work block. All subsequent features reference the base feature either directly or indirectly. Additional features shape or refine the base feature. Examples of additional features include holes or cuts in the initial work block.

This analogy between feature-based modeling and manufacturing processes is demonstrated in Figure 6-30 for making a handle for a pizza cutter. Beginning at the top of the figure, we follow the steps that an engineer would use to create a solid model, or virtual part, on the left and the steps that a machinist would take to create the same physical part on the right. The engineer, using solid modeling software, begins by creating a two-dimensional profile, or **cross section**, of a part, in this case a circle (shown in isometric projection). The analogous step by a machinist is to choose a circular bar stock of material with the correct diameter. Next the engineer extrudes, or stretches, the circular cross section along the axis perpendicular to the plane of the circle to create a three-dimensional base feature (a cylinder in this case). The equivalent action by a machinist is to cut off a length of bar stock to create an initial work block. Now the engineer adds features by cutting away material on the left end to reduce the diameter and by rounding the right end of the cylinder. The machinist performs similar operations on a lathe to remove material from the cylinder. Next, the engineer creates a circular cut to form a hole through the cylinder on the right end. The machinist drills a hole in the right end of the cylinder. Finally, the engineer creates a pattern of groove cuts around the handle. Likewise, the machinist cuts a series of grooves using a lathe. In similar fashion, a geometric shape could be added to the base feature in the solid model, analogous to a machinist welding a piece of metal to the work block. Feature-based techniques give the engineer the ability to easily create and modify common manufactured features. As a result, planning the manufacture of a part is facilitated by the correspondence between the features and the processes required to make them.

cross section

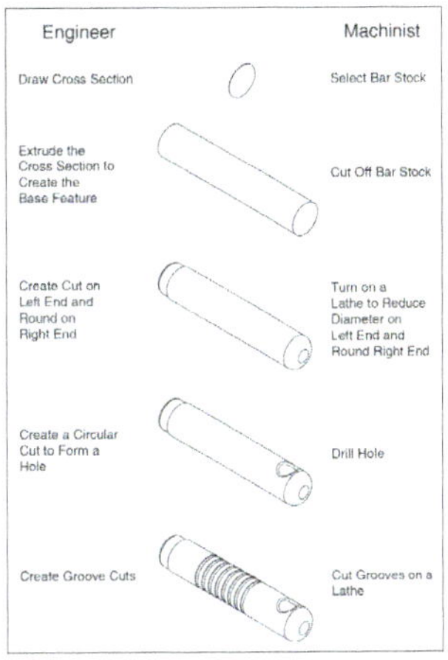

Figure 6-30 The analogy between feature-based modeling done by an engineer or a designer and common manufacturing processes done by a machinist.

Constraint-based modeling permits the engineer or designer to incorporate "intelligence" into the design. Often, this is referred to as design intent. Unlike traditional CAD software, the initial sketch of a two-dimensional profile in constraint-based solid modeling does not need to be created with a great deal of accuracy. It just needs to represent the basic geometry of the cross section. The exact size and shape of the profile is defined through assigning enough parameters to fully "constrain" it.

dimensional constraints

There are two primary types of constraints. **Dimensional constraints** are used to specify distances between items in a solid model. For example, a hole is prescribed by the location of its center, typically in terms of horizontal and vertical distances from features of a part, such as edges or other holes. In addition, the radius and depth of the hole need to be specified. Given these four-dimensional constraints, the hole size, depth, and location are fully defined.

geometric constraints

Instead of specifying numerical dimensions, **geometric constraints** define positional relationships between entities in the model in terms of the geometry. For instance, a pipe can be formed from a solid circular rod by creating a hole that is concentric with the circumference of the rod. The radius and depth of the hole require dimensional constraints, but the position of the hole does not. Instead, the position can be prescribed purely by a geometric constraint, specifically that the hole is concentric with the rod. Often, geometric constraints are particularly useful in creating assemblies of individual parts, as demonstrated in Section 6.5.3.

parametric

Another aspect of solid modeling is that the model is **parametric**. This means that parameters of the model may be modified to change the geometry of the model. A dimension is a simple example of a parameter. When a dimension is changed, the geometry of the part is updated. Thus, the *parameter drives the geometry*. For example, the position of the hole could be specified parametrically, so that the center of the hole is located at a position that is one-half of the total length of the part. If the total length was specified as 80 mm,

ENGAGE

Visit the Dassault Systemes website with your students. Navigate to the "SolidWorks Help" section and search for *feature-based models.*

URL

https://help.solidworks.com/2012/english/solidworks/acadhelp/feature-based_models.htm

Explain to your students that most CAD programs offer similar means for producing 3-D models. "Feature-based" modeling typically requires the creation of a "base." This feature is the basis on which you create the other features. The base feature can be an extrusion, a revolve, a sweep, a loft, thickening of a surface, or a sheet metal flange. However, most base features are extrusions. Dassault provides a list of typical features and a brief explanation of their uses.

NOTE: Instructors can research 3-D modeling in whatever software package they prefer; introducing the students to modeling using the class software will be beneficial.

ENGAGE

Visit the SpaceClaim (now a part of ANSYS) YouTube website with your students. Ask your students to watch the video *Direct versus Feature based modeling: An Overview.*

URL

https://www.youtube.com/watch?v=yTzlPjk93Ds

ENGAGE

Visit the CCAD Inc – AutoCAD Specialists YouTube website with your students. Ask your students to watch the video *Applying simple constraints in AutoCAD.*

URL

https://www.youtube.com/watch?v=mD-ulMWVH1g

NOTE: Instructors can research geometric constraints in whatever software package they prefer; introducing the students to constraints using the class software will be beneficial.

ENGAGE

Visit the Siemens Product Lifecycle Management software website with your students. Ask your students to read the article entitled *The 10 Golden Rules of Solid Modeling.*

URL

https://blogs.sw.siemens.com/solidedge/the-10-golden-rules-of-solid-modeling/

NOTE: Instructors can research history-based modeling in whatever software package they prefer; introducing the students to the rules for modeling using the class software will be beneficial.

then the hole would be 40 mm from either side. However, if the length was changed to 100 mm, then the hole would automatically be positioned at half of this distance, or 50 mm from either side. Thus, no matter what the length of the part is, the hole stays in the middle of the part. The power of this approach is that when one dimension is modified, all linked dimensions are updated according to specified mathematical relations, instead of having to update all related dimensions individually.

history-based modeling

parent-child relation

The last aspect of solid modeling is that the order in which parts are created is critical. This is known as **history-based modeling**. For example, a hole cannot be created before a solid volume of material in which the hole occurs has been modeled. If the solid volume is deleted, then the hole is deleted with it. This is known as a **parent–child relation**. The child (hole) cannot exist without the parent (solid volume) existing first. Parent–child relations are critical to maintaining design intent in a part. Most solid modeling software recognizes that if you delete a feature with a hole in it, you do not want the hole to remain floating around without being attached to a feature.

6.5.2 From Virtual to Manufactured

This analogy of building the virtual part while mimicking the actual manufacturing process has other benefits. It forces the designer to start thinking about the order of operations. If a particular machine, say a lathe, is used extensively for several manufacturing steps, the features in the part should be designed to minimize the times that the part would have to be removed and then re-chucked in the lathe. Cross drilling a hole on a rounded surface may require a special tool to prevent accidentally drilling an off-center hole. So, the process of designing the part inherently contains the steps, tools, and fixtures as well as the sequence of operations that will produce the most economical part.

This same concept can be carried to a finished assembly. Once all the individual parts have been created, it is now possible to assemble them virtually. As we saw in our pizza cutter, the original design of the hole through the guard could not be assembled because it did not provide enough clearance for the arm. Assembling a part virtually and thinking through each step is part of the concept of Design for Manufacturability (DFM). Two of the main goals of DFM are to make a part easier and less expensive to produce. It can also reduce weight, reduce scrap, and allow for partial or full automation of the manufacturing process.

6.5.3 Virtual Assembly

One of the great advantages of solid modeling software is that it allows engineers or designers to assemble modeled parts to create a virtual assembly. This helps the designer work out problems with part interactions early in the design phase, before manufacturing has begun. Often it is possible to check automatically for clearances or interferences between parts. A virtual assembly also permits the engineer or designer to get an idea of what the assembled product looks like before it is fabricated. In the case of the pizza cutter, the assembly begins by bringing two parts of the pizza cutter into the assembly window of the CAD software, in this case the rivet and one arm, as shown in Figure 6-31.

Constraints are then used to orient the two parts properly. For example, a geometric constraint is used to make the shaft of the rivet and the cylindrical surface of the hole concentric with one another, as shown in Figure 6-32.

Next, the surface of the arm that the arrow points to in Figure 6-32 must be constrained so that it coincides with the underside of the head of the rivet that the arrow points to in Figure 6-33.

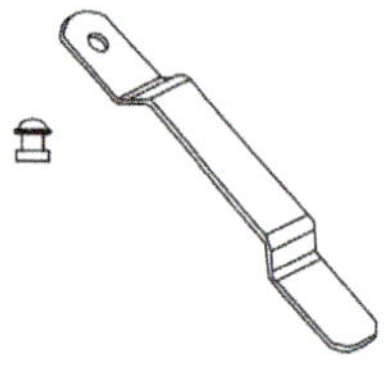

Figure 6-31 The pizza cutter assembly begins as two parts, the rivet and an arm, oriented randomly in space.

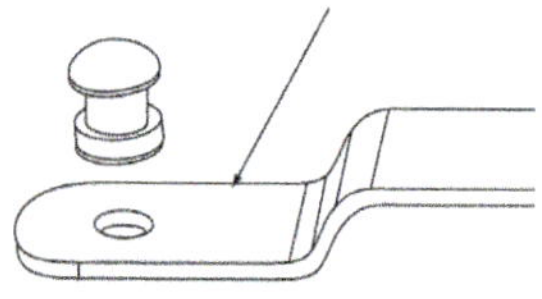

Figure 6-32 The shaft of the rivet is constrained to be concentric with the hole in the arm.

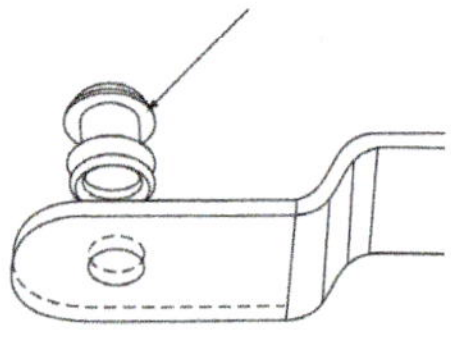

Figure 6-33 The underside of the head of the rivet (see arrow) is constrained to be coincident with the upper plane of the arm (indicated by the arrow in Figure 6-32).

The blade can be added by making the hole in the blade concentric with the shaft of the rivet and by making the two surfaces with arrows in Figure 6-34 coincident. Likewise, the hole in the second arm is concentric to the shaft of the rivet, and the surfaces indicated in Figure 6-35 are made coincident. At this point, the two arms can still rotate with respect to one another; hence, a geometric constraint is added so that the surfaces indicated in Figure 6-36 are parallel. Similar geometric constraints can be used to assemble the remaining parts to create the finished assembly, shown in Figure 6-37.

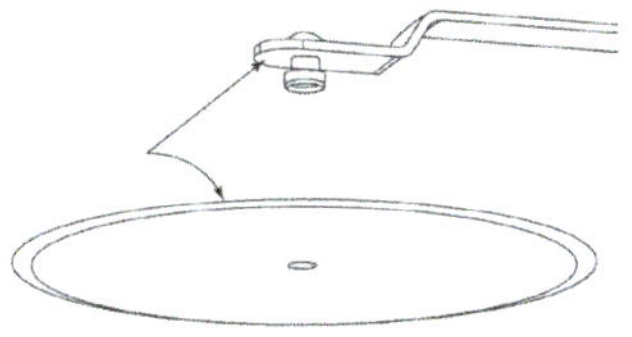

Figure 6-34 The lower surface of the arm and the upper surface of the blade (indicated by arrows) are constrained to be coincident.

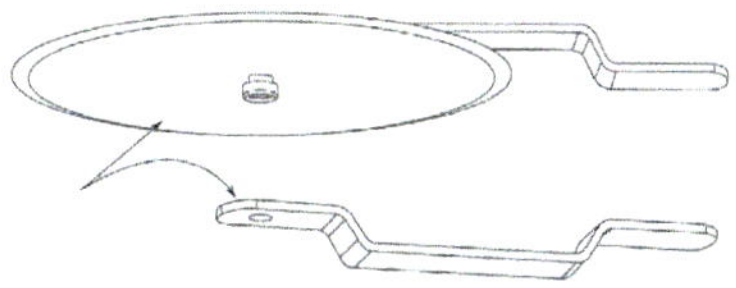

Figure 6-35 The lower surface of the blade and the upper surface of the second arm (indicated by arrows) are constrained to be coincident.

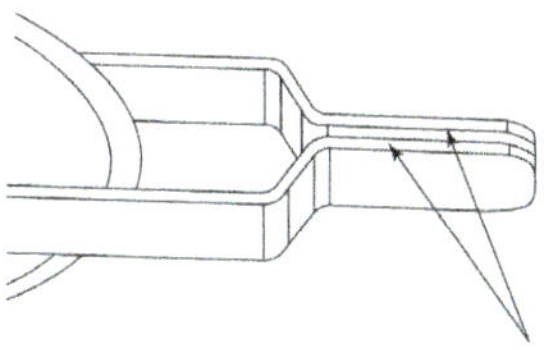

Figure 6-36 The sides of the two arms (indicated by arrows) are constrained to be parallel so that the arms are aligned with each other.

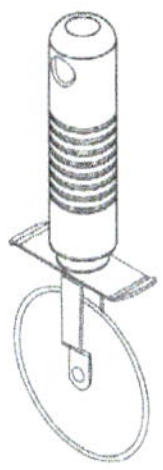

Figure 6-37 The assembled pizza cutter.

interference

Once the pizza cutter is assembled, it is possible to check for **interference** between parts. In this case, there was an error in the metal guard that is at the end of the handle to prevent the user's hand from slipping down the handle onto the sharp blade. Notice in Figure 6-38 that the rectangular assembly that holds the blade cannot fit through the circular hole in the guard. The solid modeling software will detect this problem, which can be fixed by returning to the CAD file of the guard and changing the hole feature from a circular hole to a rectangular hole. Because of the associative nature of the CAD software, this modification in the part will be automatically reflected in the assembly, resulting in the modified guard in Figure 6-39, shown in an exploded view.

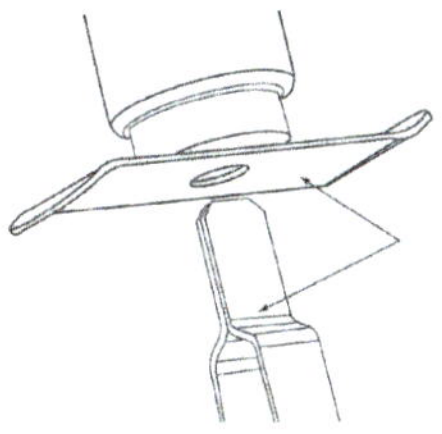

Figure 6-38 The rectangular ends of the arms cannot fit through the circular hole of the guard, a problem that is automatically detected by the CAD software as interference between the parts.

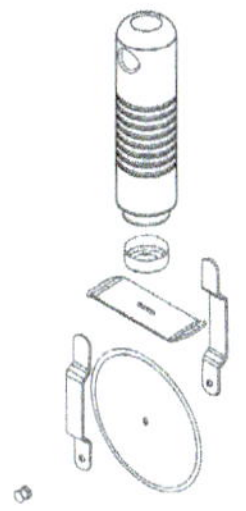

Figure 6-39 An exploded view of the pizza cutter showing the guard with a rectangular hole instead of a circular hole.

ENGAGE

Visit the Hagerman & Company, Inc. YouTube website with your students. Ask your students to watch the video *Virtual Parts and Assemblies.*

URL

https://www.youtube.com/watch?v=k0IuJnV39gM

NOTE: Instructors can research the virtual assembly of parts in whatever software package they prefer; introducing the students to assemblies using the class software will be beneficial.

ENGAGE

Visit the American National Standards Institute (ANSI) website with your students. Ask your students to read the "Standards Are Everywhere!" page.

URL

https://ansi.org/education/activities/standards-student-programs/k-12-students

Enlarge any of the posters provided and use the visuals to explain ANSI standards to your students. Standards are everywhere! Can you spot them?

ENGAGE

Visit the Wisc–Online educational website, maintained by Wisconsin's Technical Colleges, with your students. Ask your students to watch the video entitled *Basic Types of Dimensioning Used in Engineering Drawings.*

URL

https://www.wisc-online.com/learn/technical/engineering/eng18704/basic-types-of-dimensioning-used-inengineeri

Ask your students to use the interactive menu to see both *Unidirectional* and *Aligned* dimensions, along with other basic types of dimensions.

The steps that have been shown in creating the handle and the pizza cutter assembly are merely examples of how solid-modeling CAD software can be used in the design of parts. There are many alternative methods of accomplishing the same thing, such as changing the order in which the features are added to the base feature or changing the order in which the parts are assembled. Furthermore, many details have been left out, particularly with regard to differences in how various CAD software packages accomplish the creation of parts and assemblies. However, the steps used to create the handle and the pizza cutter assembly demonstrate some key characteristics of solid modeling—starting with a base feature and adding other features to it; using constraints, both dimensional and geometric, to incorporate "design intent"; and using the associative capability of the software to permit changes to ripple through associated objects such as the assembly.

6.6 Drawing Standards

Not only is the accurate and clear depiction of details of a part necessary in an engineering drawing, but it is also necessary that the drawing conform to commonly accepted standards or conventions. There are two reasons for the existence of standards and conventions. First, using standard symbols and projections ensures a clear interpretation of the drawing by the viewer. For example, the symbol Ø is associated with a means to indicate the diameter of a circle. Without this convention, it would be more difficult to unambiguously represent circular features on a drawing.

unidirectional

aligned

Most CAD programs automatically use a standardized presentation of drawings, usually based on standards from the American National Standards Institute (ANSI; see *https://www.ansi.org*) in the United States and the International Standards Organization (ISO; see *https://www.iso.org*) in the rest of the world. Figure 6-40 shows the dimensions of a part in one view using both the ANSI and ISO standards. Notice that the numbers for the dimensions in the ANSI standard are **unidirectional**, having the same orientation. The values for the dimensions in the ISO drawing are **aligned** with the dimension line. Even with the use of CAD software to ensure standard presentation in drawings, it is up to the engineer to implement drawing standards in some cases. An example is in the placement of dimensions in a drawing. The user controls where dimensions appear on the drawing. Furthermore, even in cases where the CAD software automatically implements drawing standards, it is still necessary to understand what the conventions mean so that they are properly interpreted.

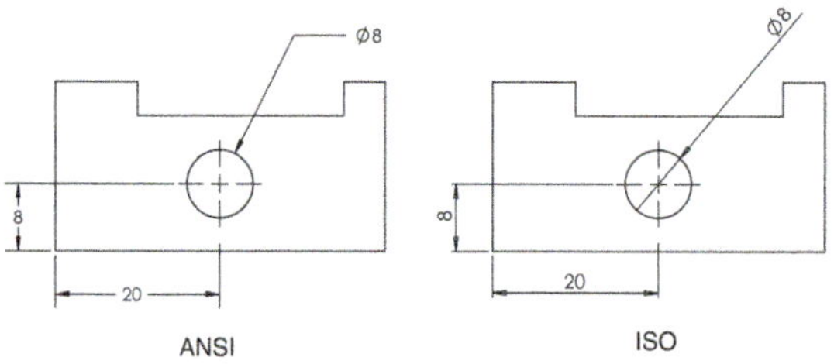

Figure 6-40 Drawings of the same object using the ANSI and ISO standards.

There is some variability among different CAD software packages in how various details are shown. For instance, the length of centerlines, the type of line used for hidden lines, and the shape of arrowheads may differ from one CAD software package to another. An example is the gray color sometimes used for hidden lines. Although hidden lines are generally shown as dashed lines in final drawings, they may be shown as gray or other colors while the part is being modeled.

6.6.1 Sheet Layout

Engineering drawings are created on standard-size sheets that are designated by the code indicated in Table 6-1. Engineering drawings are almost always done in landscape orientation so that the long side of the drawing is horizontal. The choice of drawing size depends upon the complexity of the object depicted in the drawing. The drawing should be sized so that the projections of the part, dimensions, and notes all fall within the borders of the drawing with adequate spacing, so that the drawing is not cluttered. The drawing should be large enough so that all details are readily evident and readable. As a result, regardless of the physical size of the part, simple parts are usually drawn on smaller sheets because it is not necessary to show much detail. Complex parts are usually drawn on larger sheets so that there is room to show adequate detail. When using CAD software, it is necessary to consider the printed drawing size in addition to the drawing size on the computer screen. On the computer screen, it is possible to zoom in to read detailed items, but this cannot be done for a printed drawing. Consequently, the sheet size should be chosen carefully.

Table 6-1 Standard Sheet Sizes

ANSI	ISO
A - 8.50" × 11.00"	A4 - 210 mm × 297 mm
B - 11.50" × 17.00"	A3 - 297 mm × 420 mm
C - 17.00" × 22.00"	A2 - 420 mm × 594 mm
D - 22.00" × 34.00"	A1 - 594 mm × 841 mm
E - 34.00" × 44.00"	A0 - 841 mm × 1189 mm

title block

The **title block** on a drawing contains important information about the drawing. Normally the title block of a drawing is in the lower right corner of the drawing. ANSI standard title blocks can be used, but often individual companies use their own standard title block. The title block, such as the one shown in Figure 6-41, is used primarily for drawing control within a company. The title block includes information regarding the part depicted in the drawing, such as its name and part number; the person who created the drawing; persons who checked or approved the drawing; dates the drawing was created, revised, checked, and approved; the drawing number; and the name of the company. This information allows the company to track the drawing within the company if questions arise about the design.

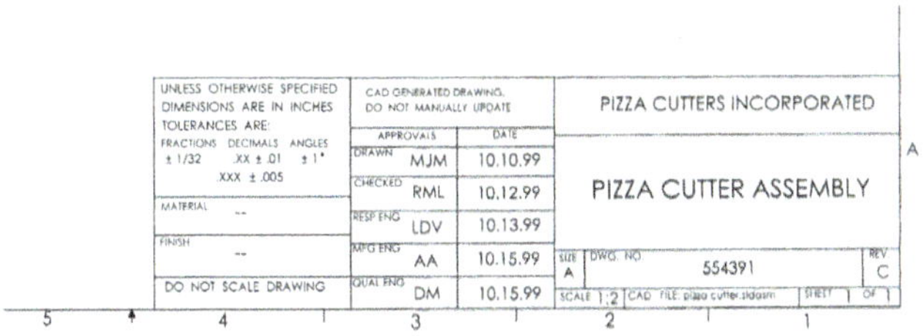

Figure 6-41 A typical title block.

ENGAGE

Visit CADstudio's CADforum website with your students. Navigate to the "CAD/BIM Blocks" section.

URL

http://www.cadforum.cz/catalog_en/?q=free+title+blocks

Explain to your students that all CAD programs utilize title blocks to display their drawings with the appropriate information on them. Title blocks come in many forms. Some are built into the software package you are using; many more are available for free by downloading them and adding them to your software's title block folder. All of them are constrained to ANSI standards.

NOTE: Instructors can research title blocks in whatever software package they prefer; introducing the students to title blocks using the class software will be beneficial. Creating a school-specific title block for all the students to use is also an excellent exercise to introduce the concept.

ENGAGE

Visit the Wisc–Online educational website, maintained by Wisconsin's Technical Colleges, with your students. Ask your students to watch the video entitled *Basic Types of Lines Used in Engineering Drawings.*

URL

https://www.wisc-online.com/learn/technical/engineering/eng16004/basic-types-of-lines-used-in-engineering-draw

Ask your students to use the interactive menu to see examples of all the standard types of lines used in engineering drawings.

Other information can appear in the title block to provide details about the manufacture of the part, such as the material for the part, the general tolerances, the surface finish specifications, and general instructions about manufacturing the part (such as "remove all burrs"). Many times, though, this information is shown as notes on the drawing rather than in the title block.

scale

Finally, the title block provides information about the drawing itself, such as the units used in dimensioning (typically inches or millimeters) and the scale of the drawing. The **scale** of the drawing indicates the ratio of the size of the drawing to the size of the actual object. It can be thought of as the degree to which the object is enlarged or shrunk in the drawing. Scales on drawings are denoted in several ways. For instance, a drawing is created such that 1 inch on the drawing represents 2 inches on the physical object. This can be reported in the Scale box in the title block as HALF SCALE, HALF SIZE, 1 = 2, 1:2, or 1/2" = 1" . In the numerical representations, the left number is the length of the drawing, and the right number is the length of the actual object. A part that is drawn twice as large as its physical size would be denoted as DOUBLE, 2 × (denoting "2 times"), 2 = 1, or 2:1. In some cases, such as architectural drawings, the scale can be quite large. For instance, 1/4" = 1' - 0"corresponds to a scale of 1:48, where " denotes inches and ' denotes feet. Common scales used in engineering drawing are 1 = 1, 1 = 2, 1 = 4, 1 = 8, and 1 = 10 for English units, and 1:1, 1:2, 1:5, 1:10, 1:20, 1:50, and 1:100 for metric units. A 1:1 scale is often specified as FULL SCALE or FULL SIZE.

6.6.2 Lines

The types of lines used in engineering drawing are standardized. Different line types have different meanings and understanding these different types of lines makes interpreting drawings easier. The line types vary in thickness and in the length and number of dashes in dashed lines. Most CAD software automatically produces line types that correspond to the way the line is used in the drawing. Figure 6-42 shows some of the more commonly used line types:

- **Visible lines** (thick, solid lines) represent the outline of the object that can be seen in the current view.
- **Hidden lines** (thin, dashed lines) represent features that are hidden behind surfaces in the current view.
- **Centerlines** or **symmetry lines** (thin, long-dash/short-dash lines) mark axes of rotationally symmetric parts or features.
- **Dimension lines** (thin lines with arrowheads at each end) indicate sizes in the drawing.
- **Extension lines** or **witness lines** (thin lines) extend from the object to the dimension line to indicate which feature is associated with the dimension.
- **Leader lines** (thin, solid lines terminated with arrowheads) are used to indicate a feature with which a dimension or note is associated.

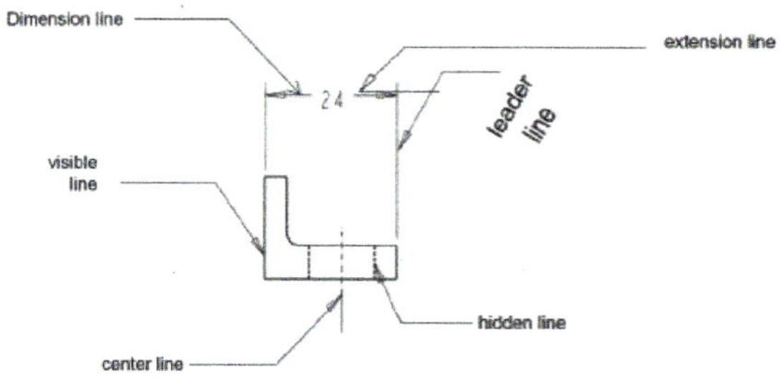

Figure 6-42 Types of lines used in drawings.

The arrowheads for dimension lines, leader lines, and other types of lines may be filled or not filled, depending on the CAD software.

precedence of lines

It is common in technical drawings that two lines in a particular view coincide. When this occurs, a convention known as the **precedence of lines** dictates which line is shown—visible lines have precedence over hidden lines, which have precedence over centerlines. An example is shown in Figure 6-43. In the top view, line A is shown as a solid visible line, since the visible line that forms the top left edge has precedence over the dashed hidden line for the side of the hole. Line B of the top view is shown as a dashed hidden line, since it has precedence over the centerline of the hole. Likewise, line C in the side view is shown as a visible line, since the edge on the right side has precedence over the centerline of the hole. Visible lines, hidden lines, and centerlines all have precedence over extension lines.

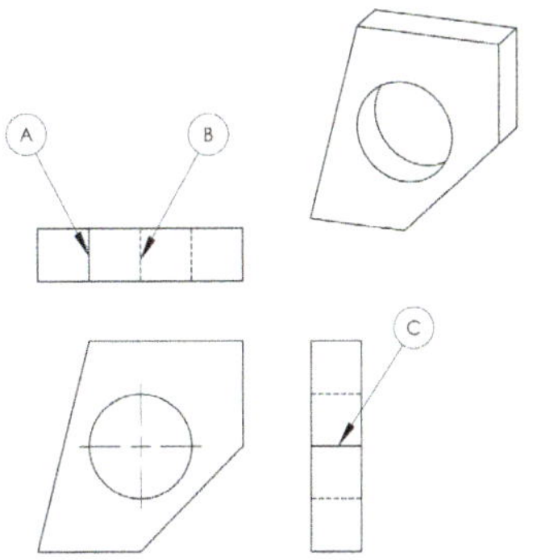

Figure 6-43 An example showing the precedence of lines.

Conventions also exist for the intersection of lines in a drawing. For instance, an extension line is always drawn so that there is a slight, visible gap between its end and the outline of the object, as shown in Figure 6-42. Similarly, center marks appearing at the center of a circle are drawn so that the short dashes of each line intersect to form a small cross, as shown in Figure 6-44. Fortunately, CAD software automatically takes care of most of the details related to lines, as well as their length and position. The engineer or designer only needs to know how to properly interpret the various line types.

6.6.3 Dimension Conventions

overdimensioned

Engineering drawings are usually drawn to scale, but it is still necessary to specify numerical dimensions for convenience and to ensure accuracy. In general, sufficient dimensions must be provided to define the geometry of the part precisely, but redundant dimensions should be avoided. Furthermore, dimensions on the drawing should reflect the way the part is made or the critical dimensions of the part. For example, Figure 6-44 shows three drawings of a plate with two holes. Figure 6-44a is **overdimensioned**, having one redundant dimension. It is clear that, given the distances from the left edge to the left hole, the distance from the right edge to the right hole, and the overall width of the plate, the distance between the holes has to be 6 mm. Thus, the 6-mm dimension could be omitted. However, the dimensions that are shown in the drawing often suggest which dimensions are most important. If the 6-mm dimension in Figure 6-44a were omitted, it would suggest that the distance between each hole and the nearest edge of the plate is more important than the distance between the holes. But suppose, instead, that the distance between the holes is critical. This might be necessary if the holes in the plate are supposed to align with another part. Then the dimensions shown in Figure 6-44b would clearly indicate that the distance between the holes is the most important dimension by omitting the dimension from the right hole to the right edge. Figure 6-44c indicates that the distance from the left edge to each hole, not the distance between the holes, is critical. The engineer or designer must consider the optimal way to display dimensions to clearly convey which dimensions are most important.

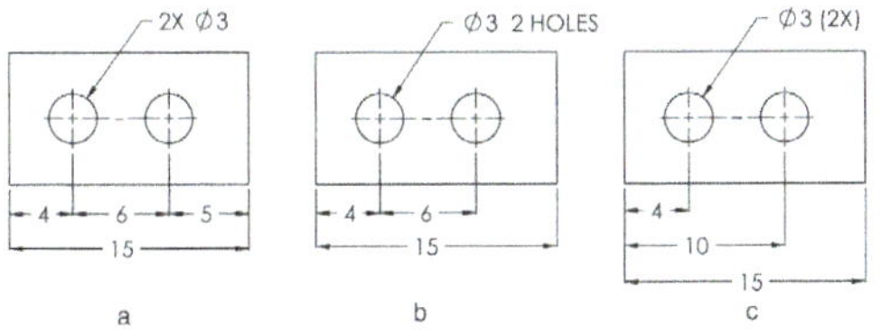

Figure 6-44 (a) An overdimensioned drawing; (b) A dimensioning scheme that emphasizes the 6-mm distance between the holes; (c) A dimensioning scheme that emphasizes the distance from the left edge to each hole.

TEACHING TIP

When dealing with precedence of lines, the order should clear. A visible edge or line would overwrite the dashed line. Hidden edges are more important to note than a centerline, so hidden lines take precedence over those centerlines.

ENGAGE

Visit the Wisc–Online educational website, maintained by Wisconsin's Technical Colleges, with your students. Navigate to the Technical section and choose "Engineering." watch the video entitled *Print Reading: An Alphabet of Lines in Print Reading.*

URL

https://www.wisc-online.com/learn/technical/machine-tool/mtl17903/print-reading-an-alphabet-of-lines-in-print-r

Ask your students to use the interactive menu to see examples of the alphabet of lines in print reading.

ENGAGE

Visit the Dassault Systemes website with your students. Navigate to the SolidWorks Dimension Leaders PropertyManager section and search for Witness/Leader Display.

URL

http://help.solidworks.com/2012/english/solidworks/sldworks/hidd_dve_dimension_leaders.htm

Explain to your students that dimensioning circles, arcs, and holes in your designs can be done with your CAD software. Have them differentiate between linear and radial dimensions and then navigate to the "Hole Callouts" section to see the proper way to display callouts for common machining processes.

URL

http://help.solidworks.com/2012/english/solidworks/sldworks/HIDD_Options_Callout.htm

NOTE: Instructors can research specific dimensioning techniques (like arcs and holes) in whatever software package they prefer; introducing the students to dimensioning using the class software will be beneficial.

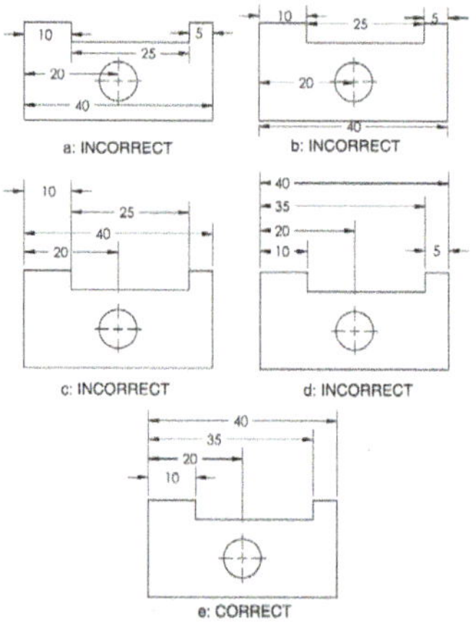

Figure 6-45 (a) Placing dimensions on the object is incorrect; (b) Placing dimensions so that dimension lines align with visible lines of the object or are very close to the object is incorrect; (c) Placing dimensions in random order is incorrect; (d) Overdimensioning is incorrect; (e) A properly dimensioned drawing.

The position of the dimensions is also important. Figure 6-45 shows several examples of incorrect ways of dimensioning a drawing. It is best to avoid having dimensions inside the boundary of the object, as is the case in Figure 6-45a. Dimensions should be kept outside the boundaries of the object whenever possible, unless clarity is improved in some way by having them inside the boundary. Dimension lines should not be too close to the boundaries of the object, as is the case with the 40-mm dimension in Figure 6-45b. A typical spacing is 10 mm (3/8 inch) for the distance of the first dimension from the object and 6 mm (1/4 inch) for the spacing between subsequent dimensions. It is also difficult to interpret the drawing when the dimension line coincides with a visible line of the object, as is the case in Figure 6-45b for the 20-mm dimension (the dimension line coincides with the center mark line) and the 25-mm dimension (the dimension line looks like a continuation of the upper boundary of the part). In Figure 6-45c, the dimensions are in a random order with some smaller dimensions farther from the boundary of the object (such as the 10-mm dimension) than larger dimensions (such as the 40-mm dimension). This results in dimension lines crossing extension lines, which should be avoided. It is often helpful to stack the dimensions in order of increasing length as shown in Figure 6-45d. However, there are three problems in this case. First, placing the dimension numbers right above one another is somewhat difficult to read. Second, there is a redundant dimension—the 5-mm dimension is unnecessary. Third, the 10-mm dimension is so small that there is no dimension line, and only the arrowheads appear. This is readily corrected by placing the arrowheads outside of the extension lines instead of inside, as shown in Figure 6-45e. In addition, in this case the dimension numbers are staggered, or offset, by a small amount to make them more easily readable. All the dimensioning conventions are aimed at making the drawing as easily readable as possible so that the information on the drawing is conveyed clearly to the reader.

contour dimensioning

It is natural for a person reading the drawings to look for the dimensions of a feature in the view where the feature occurs in its most characteristic shape and where it is visible (as opposed to hidden). This is known as **contour dimensioning**. For example, the location and size of a hole should be dimensioned in the view where the hole appears as a circle. Likewise, it is best to show the overall dimensions of the object in a view that is most descriptive of the object. If possible, both the horizontal and vertical location of a feature should be dimensioned in the same view. The upper part of Figure 6-46 shows an object properly dimensioned in millimeters to show the relation of the dimensions to the features in the most obvious manner. In the lower part of Figure 6-46, the dimensioning has several problems. The overall vertical dimensions of the L-shaped profile, the 8- and 35-mm dimensions, are shown on the right view where it is not clear that the object is L-shaped. The diameter of the hole is shown in a view where the hole is shown only as a pair of parallel hidden lines, not as a circle. And the vertical location of the hole is shown in one view, while the horizontal location is shown in another view. Finally, the vertical location of the hole is shown as a distance to its edge (the 7-mm dimension). Generally, the locations of holes should be dimensioned to the center of the hole.

The conventions for millimeter and inch dimensioning are different. For millimeter dimensioning, dimensions less than 1 mm have a zero preceding the decimal point, and integer dimensions may or may not include a decimal point and following zeros. Thus, legitimate millimeter dimensions are 0.8, 6, 8.00, and 1.5. Inch dimensions do not include a zero before the decimal point for values less than 1 inch. Zeros to the right of the decimal point are usually included. Legitimate inch dimensions are .800, 6.00, 8.00, and 1.500. For both millimeter and inch dimensions, the number of digits to the right of the decimal point may be used to indicate the tolerance (acceptable variation). It is standard practice to omit the units for dimensions. However, it is wise

ENGAGE

Visit the NextWave Technology YouTube website with your students. Ask your students to watch the video entitled *dimensioning in SketchUp.* SketchUp is a 3-D modeling computer program for a wide range of drawing applications such as architectural, interior design, civil and mechanical engineering, film, and video game design—and is available in a freeware version. (www.sketchup.com/)

URL

https://www.youtube.com/watch?v=wbuJkJkUp_8

NOTE: Instructors can research dimensioning in whatever software package they prefer; introducing the students to dimensioning using the class software will be beneficial.

to note the units in the title block or as a note on the drawing. Circles and arcs are dimensioned using special symbols and rules. A radius is denoted using a leader line with an arrow pointing at the arc, as shown in the upper portion of Figure 6-46. Sometimes a small cross is placed at the center of the radius. The position of the center should be dimensioned unless the radius is for a rounded edge, in which case the arc is positioned by virtue of being tangent at its ends to the sides forming the corner that is rounded, as shown in Figure 6-46. The dimension for a radius begins with a capital letter R to indicate radius. Full circles are dimensioned using the diameter with a leader line having an arrow pointing at the circle, as shown in Figures 6-44 and 6-46. If space permits, the leader line can extend across a diameter of the circle with arrows at each end of the diameter. The dimension for a diameter begins with the Greek letter phi (Ø) to indicate diameter. The location of the center of the circle must also be dimensioned.

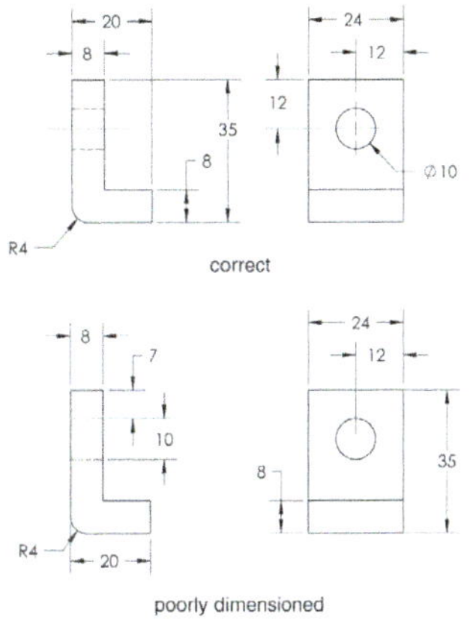

Figure 6-46 Properly and improperly dimensioned drawings.

Sometimes maintaining clarity makes it preferable to dimension several concentric circles from a side view, where the circle does not appear as a circle, thus violating the rule to dimension a feature in its most characteristic shape. This also permits the depth of the feature to be indicated along with the diameter in the same view of the object. Figure 6-47 shows how concentric circles can be dimensioned from a side view. The diameter dimensions to the left use Ø to indicate diameter. The dimensions above the object indicate the length of the features corresponding to particular diameters along the axis.

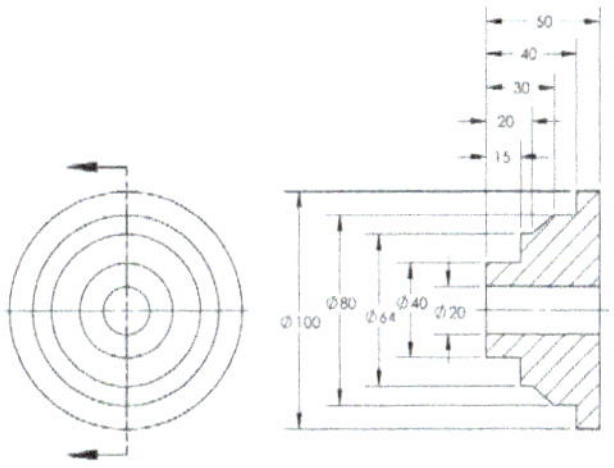

Figure 6-47 Proper dimensioning of concentric circles.

Angles are dimensioned as decimals or in degrees and minutes (60 minutes is one degree), so that 32.5 and 32°30' are equivalent, where ' indicates minutes. The dimension line for an angle is drawn as an arc with its center at the apex of the angle, as shown in Figure 6-48a. Here, a combination of angles and length dimensions are used to fully dimension the object.

Note that the arrows are "inside" for the 76° dimension and "outside" for the 45° dimension. Either type can be used, depending on the space available and readability. Figure 6-48b shows how the same shape can be dimensioned without using any angles. In some ways, this figure is less clear than Figure 6-48a, because some extension lines cross and there are more dimension lines. At first glance, it may appear that

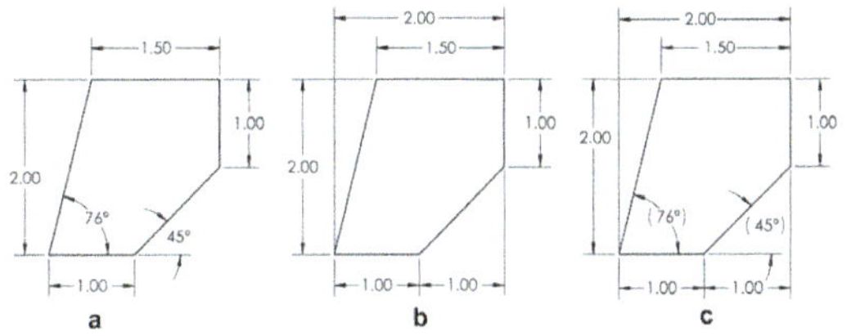

Figure 6-48 (a) Properly dimensioned object using angles; (b) Properly dimensioned object without using angles; (c) Angles are in parentheses to indicate that they are reference dimensions.

ENGAGE

Ask your students to visit the Wikipedia page on "Geometric dimensioning and tolerancing."

URL

https://en.wikipedia.org/wiki/Geometric_dimensioning_and_tolerancing

Navigate to the section on "Dimensioning and tolerancing philosophy." This is an excellent resource for the basic rules of dimensioning according to ANSI standards.

ENGAGE

Visit the Cimquest YouTube website with your students. Ask them to watch the video *SolidWorks Reference Dimensions.*

URL

https://www.youtube.com/watch?v=Db9nDcbPUj8

reference dimension

Figure 6-48b is overdimensioned, since the 2-inch dimension on the top is simply the sum of the two 1-inch dimensions on the bottom. However, strictly speaking, all these dimensions are necessary. If the lower right 1-inch dimension were omitted, the upper-right corner of the object would not necessarily be constrained to be 90°. In practice, when lines are drawn at right angles, a 90° angle is usually implied, so the lower right 1-inch dimension could be omitted without creating any ambiguity. Figure 6-48c shows the length dimensions in addition to the angles. While this drawing is overdimensioned, the display of the angles may aid in clarifying the dimensions for the reader of the drawing. Whether to include extra dimensions depends on what critical aspects of the dimensions need to be indicated on the drawing. If it is necessary that the angle in Figure 6-48c be exactly 45°, then this angle should be specified on the drawing and the 1-inch horizontal dimension should be in parentheses to indicate it is a **reference dimension** provided for convenience to the reader. On the other hand, if the 1-inch dimension is critical, it should be indicated on the drawing and the angle could be provided as a reference dimension for convenience, as shown in Figure 6-48c. Reference dimensions are provided only for information to help in interpreting the drawing. They are not intended to be measured or to be used to govern the manufacture of the part. Generally, it is preferable to show the dimensions in the clearest form possible. For this object, Figure 6-48a is superior to the other dimensioning schemes by using the fewest dimensions that overlap the least.

Multiple features that are identical need not be individually dimensioned. For example, the diameter of the two 3-mm holes in Figure 6-44a are dimensioned with a leader line to one hole with the dimension 2 × Ø3. The 2 × indicates "two times." Alternative dimensions are Ø3 (2 ×), Ø3 2 PLACES, or Ø3 2 HOLES.

There are several other "rules" for dimensioning shown here, in no particular order:

- Dimension lines should be outside of the outline of a part whenever possible.
- Dimension lines should not cross one another.
- Dimensions should be indicated on a view that shows the true length of the feature. This is particularly important when dimensioning features that are on a surface that is at an angle to the plane of the orthographic view.
- Each feature should be dimensioned only once. Do not duplicate the same dimension in different views.
- Dimension lines should be aligned and grouped when possible, to promote clarity and uniform appearance. See, for example, the placement of the horizontal dimensions in Figure 6-45e.
- The numerical dimension and arrows should be placed between the extension lines where space permits. If there is space only for the dimension, but not the arrows, place the arrows outside of the extension lines. When the space is too small for either the arrows or the numerical value, place both outside of the extension lines.
- Place the dimension no closer than about 10 mm (3/8 inch) from the object's outline.
- Dimensions should be placed in clear spaces, as close as possible to the feature they describe.
- When dimensions are nested (such as the horizontal dimensions in Figure 6-45e), the smallest dimensions should be closest to the object.
- Avoid crowding dimensions. Leave at least 6 mm (1/4 inch) between parallel dimension lines.
- Extension lines may cross visible lines of the object.
- Dimensions that apply to two adjacent views should be placed between the views unless clarity is enhanced by placing them elsewhere. The dimension should be attached to only one view. Extension lines should not connect two views.
- Dimension lines and extension lines should not cross, if possible. Extension lines may cross other extension lines.
- A centerline may be extended to serve as an extension line, in which case it is still drawn as a centerline.

- Centerlines should not extend from one view to another view.
- Leader lines are usually sloped at about 30°, 45°, or 60° and are never horizontal or vertical.
- Numerical values for dimensions should be approximately centered between arrowheads unless several dimensions are nested, as in Figure 6-45e. In this case, dimensions should be staggered.
- When a rough, noncritical dimension (such as a round) is indicated on a drawing, add the note TYP to the dimension to indicate that the dimension is "typical" or approximate.
- Sometimes it is helpful to include a reference dimension that is redundant or otherwise unnecessary, but is included for the convenience of the person reading the drawing. Such dimensions are usually in parentheses.

6.7 Assembly Drawings

An assembly drawing shows all the components of a design either assembled or in an exploded view. Many times, assembly drawings include sections. Most dimensions are omitted in assembly drawings, although sometimes an overall height or length may be indicated. Individual parts are not dimensioned, but some dimensions of the assembled mechanism may be included. Hidden lines are seldom necessary in assembly drawings, although they can be used where they clarify the design. Leader lines attached to a ballooned letter or detail number, as shown in Figure 6-49, reference the parts of the assembly. The leader lines should not cross and nearby leader lines should be approximately parallel. Sometimes parts are labeled by name rather than number. The parts list may be on the assembly drawing (usually on the right side or at the bottom) or it may be a separate sheet. The assembly drawing may also include machining or assembly information in the form of notes on the drawing. Often, assembly drawings include assembly sections. These are typically orthographic or pictorial section views of parts as put together in an assembly. The assembly cross section on the left side of Figure 6-50 shows the interior structure of the parts of a pizza cutter and how they fit together. The detail view in the upper right of Figure 6-50 shows the cross section through the rivet to indicate how the parts are assembled.

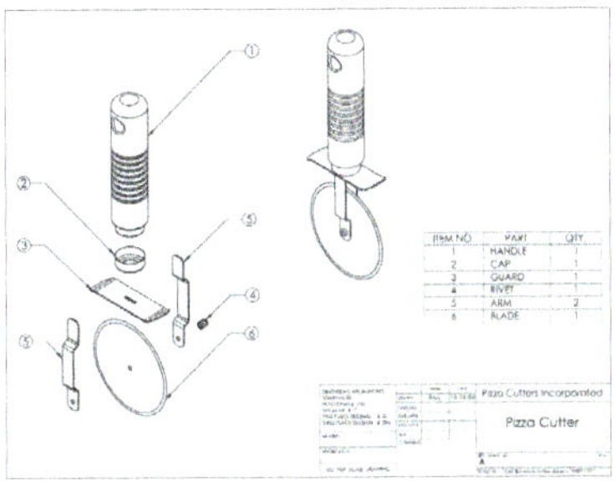

Figure 6-49 Assembly drawing of a pizza cutter.

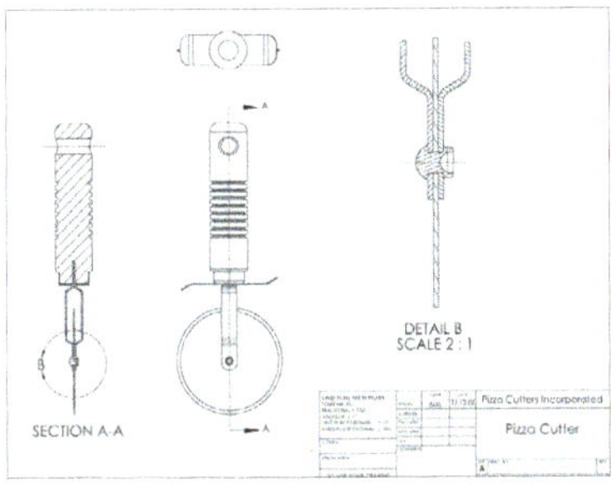

Figure 6-50 Detail view of a small section of an assembly.

6.8 Integration with Other Design Tools

Early on, we referred to CAD software as an ecosystem, with the growth of increasingly inexpensive memory and cloud storage for drawings. Related add-on tools are readily available for use with your CAD system. Here are a few examples:

Stress Modeling—This allows the designer to virtually create a high degree of force or torque on a virtual part and see color-coded stress intervals displayed on the part. It gives the designer the opportunity to see if a design variation might be possible that can reduce the stresses on the part, increasing its operational life. Related testing programs are available that can help determine heat flow, pressure-related stresses (in a pipe or pressure vessel), resonant frequencies, and fluid dynamics (for example, a building in a high wind).

ENGAGE

Visit the Xype YouTube website with your students. Ask them to watch the video *SolidWorks Tips – Assembly Drawings.*

URL

https://www.youtube.com/watch?v=FieRlqEFl04

TEACHING TIP

One of the ways engineers accomplish this is a technique called *finite element analysis*, which has been built into 3-D modeling software to allow the engineer to test the design's feasibility. You can show your students an example of FEA using the following video.

URL

https://www.youtube.com/watch?v=NpUWuATcy14

172 Chapter 6 Engineering Graphics and Design Tools

Assembly Instructions—Using individual parts and sub-assemblies can allow a manufacturing engineer to turn a series of drawings or 3-D renderings into assembly instruction for manufacturing personnel. This can include the tooling to be used to help with assembly, maximum torque to be applied to bolts, and so forth.

Integrating with Construction—In the architectural world, CAD programs are used for design, however the scaling is much larger than typically manufactured parts (such as windows, air conditioning equipment, and escalators). Using Building Information Management (BIM) software allows architects and general contractors to specify and include CAD drawings of these pieces of equipment in their overall design of their buildings in the form of drawing "outlines." This way they can access the detailed information if they need to, but for purposes of "assembling" their building, they can treat these elements as just blocks to be placed in the right places in their building.

Generative Design—It is possible to build a version of a part by specifying a range of constraints rather than specific constraints. Let's say that you have a part, and it requires two bolt holes—one at either end—and must have a ¼" threaded hole in the middle. It could easily be made by cutting a piece of flat stock, drilling the two holes at either end, and then drilling and tapping the hole in the middle. If, however, you are trying to minimize the weight and it could be made of any variety of materials, then you can have the CAD program generate a family of solutions by generating several versions that cover the range of design possibilities. A computer can quickly generate multiple designs, giving the designer several options that all meet the allowable range of constraints. This design, for example might end up looking like a flat, two-bladed propellor with a large threaded hub in the middle. It would likely be more easily printed with a 3-D printer, than using conventional machining tools.

The ecosystem of CAD and CAM software continues to expand. The general trend is to make the design portion more integrated into the manufacturing process. Rather than treat CAD as a discrete step in a process, everything is being positioned to be part of a continuous, seamless process. We are still a long way from that point, but it's clear that is where the whole process of creating new products is headed.

CHAPTER 6 ASSESSMENT QUESTIONS

1. How does freehand sketching differ from technical drawing?
2. Design a poster pointing out three safety hazards in your laboratory or classroom. Include sketches of each as part of the poster.
3. Explain how "paint" or "draw" programs such as Adobe Illustrator, Microsoft Paint, or drawing tools in Microsoft Word differ from CAD engineering graphics software.
4. Create the 2-D Orthographic Projection for a) through i). Include proper centerlines and hidden lines.

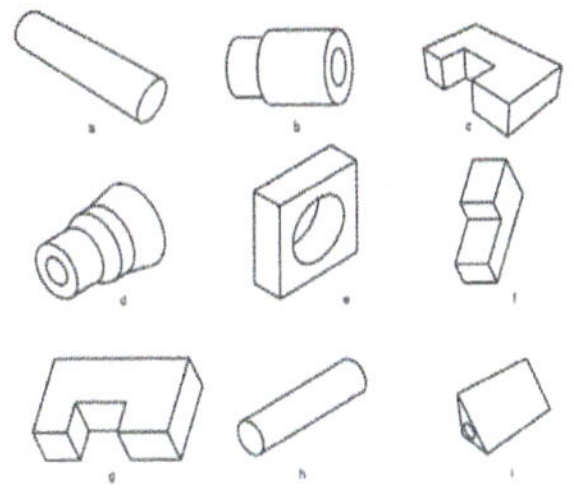

5. The items shown are 2 inches wide, 1.5 inches high, and 1 inch deep. All holes are through holes. (The objects are shown as oblique projections.)
   (a) Create freehand orthographic sketches of the objects in the figure below. Include proper hidden lines and centerlines..
   (b) Create freehand isometric sketches of the objects.
   (c) For the following diagrams, draw a 2-D drawing of the face facing you. (There is no need to draw in any hidden lines.)

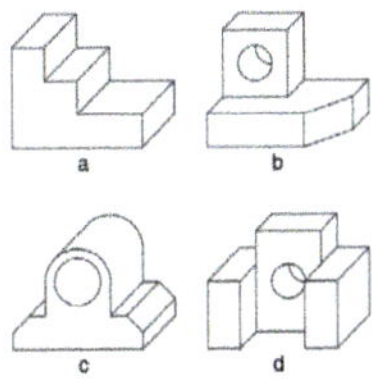

6. Draw the orthographic projection of the following objects (a-c). Ensure all hidden lines are accounted for.

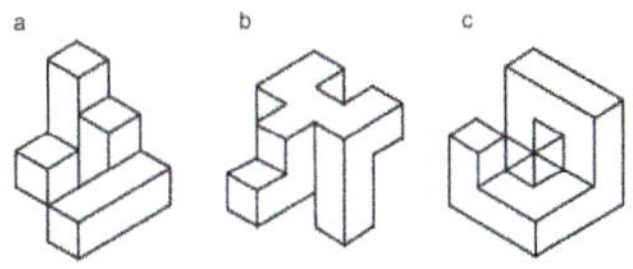

174 Chapter 6 Engineering Graphics and Design Tools

7. Use a CAD package to draw each figure from questions 5 and 6.

8. Look around your classroom. With a partner, create a blueprint of your specific room.

9. Create a freehand orthographic sketch of a coffee mug. Only two views are necessary to fully represent the mug.

10. The wireframe model shown in the figure is a small cube centered within a large cube. The nearest corners of the two cubes are connected. For instance, the lower left front corners of both cubes are connected with a line. This wireframe model is ambiguous; that is, it could represent several different solid bodies. Freehand sketch or trace the figure. Then use dashed lines for hidden lines to show at least two different solid-model configurations that are possible for this wireframe model.

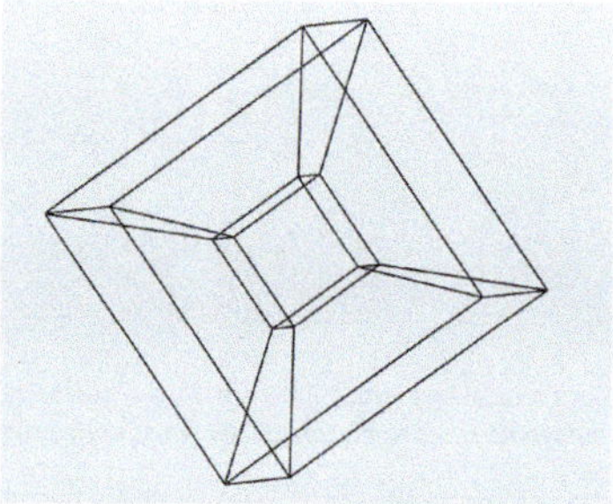

11. Consider the following objects. Sketch what the base feature would look like. List what features would be added to model the object. The type of base feature to be used (extrude, revolve, or both) is noted.
    (a) Hexagonal cross section of a wooden pencil that is sharpened to a point (do not include eraser)—extrude.
    (b) Coffee mug without the handle—revolve.
    (c) Nail—extrude and revolve.
    (d) Push pin—revolve.

12. Three orthographic views and an isometric projection of an object that has overall dimensions of 4 * 4 * 4 inches are to be shown in a drawing. Determine the best ANSI sheet size for an object of this size at the scales indicated below
    (a) Half size.
    (b) 2:1.
    (c) 1 = 1.

13. The drawing shown below has many errors in the dimensioning. Sketch or draw with CAD the orthographic views and show the proper placement of the dimensions.

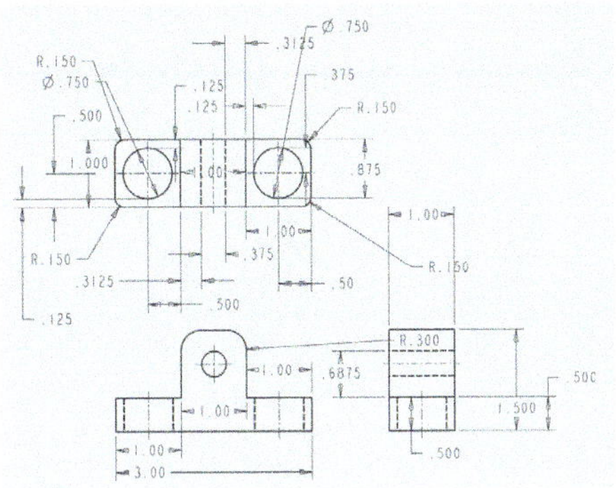

14. The figure below shows the dimensioned front view and several different corresponding right-side views, each with a horizontal width of 1 inch. Using CAD, draw three orthographic views (front, top, and right side) as well as the auxiliary view of the inclined surface for the corresponding right-side view.

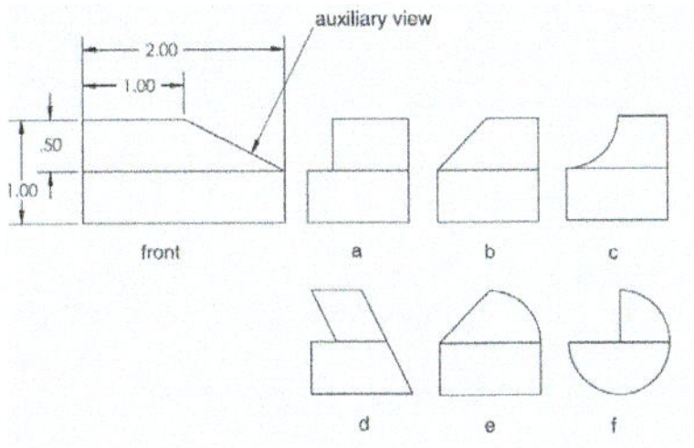

15. Research:
   (a) Sketch a half-door-hinge.
   (b) Draw a half-door-hinge using a CAD program. Be sure to properly dimension your drawing.
   (c) Store and close the image. Wait at least an hour, then open the image to review your work.

TEACHING TIP

You may chose to emphasize this exercise as "storing" and "retrieving" geometry.

16. Find a smartphone:

    (a) Sketch the smartphone.

    (b) Draw the smartphone using a CAD program.

17. Assume we are producing a CAD drawing and the following line types are used: visible, hidden, center, dimension, extension, and leader. Explain how each of the lines help in the creation and interpretation of the drawing.

18. Given the following orthographic projection, describe the object if you were to build it using the visualization. How would you detail this object to someone who does not have the projection?

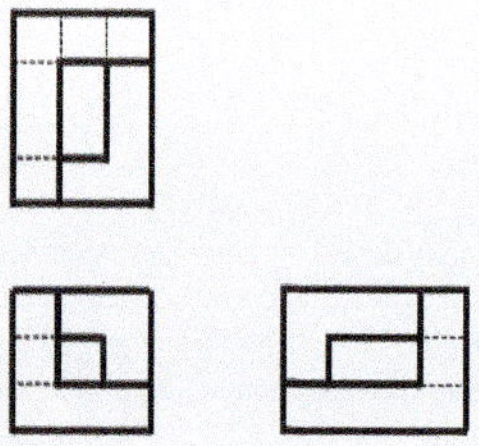

OVERVIEW

In order for students to be prepared to conduct analyses to solve engineering problems, a few ideas from physics need to be reviewed. These include basic SI units, conversions, and dimensions. Next, we introduce common quantities used in engineering calculation, such as force, weight, pressure, and density. The project at the end of the chapter serves an application of forces to the analysis of trusses, which is a critical concept for civil engineers—structural engineers, in particular.

CHAPTER 7

Essential Physics Ideas

TOPICS

In order to get started with the nitty gritty of engineering design, we need to familiarize ourselves with typical physical phenomena across engineering disciplines.

As aspiring engineers, you must learn to distinguish among many terms that laymen tend to use interchangeably. You must also understand the technical meaning of terms that are misunderstood by those untrained in science or engineering. One term that is often misunderstood is "dimension." To most people, a dimension refers to a straight line (length, one dimension), a flat surface (area, two dimensions), or a solid object (volume, three dimensions). Some folks might include time as a fourth dimension. The number of concepts classified as dimensions is far broader.

dimension
unit

A **dimension** is a measurable physical idea; it generally consists solely of a word description with no numbers. A **unit** allows us to quantify a dimension—to state a number describing how much of that dimension exists in a specific situation. Units are defined by convention and related to an accepted standard.

- Length is a dimension. There are many units for length, such as mile, foot, meter, light-year, and fathom.
- Time is a dimension. There are many units for time, such as second, minute, hour, day, fortnight, year, and century.
- Temperature is a dimension. There are many units for temperature, such as Celsius, Fahrenheit, and kelvin.

The dimensions of length, time, and temperature are familiar to us, but in reality, we do not often use these words since they are fairly vague.

We do not say . . .	We do say . . .
It is really a long **length** to Lumberton.	Lumberton is about 175 **miles** away.
Bake the cake for a **time**.	Bake the cake for 35 **minutes**.
Set the oven to a high **temperature**.	Set the oven to 450 **degrees Fahrenheit**.

The difference between the left and the right columns is that the statements on the left refer to dimensions and those on the right refer to established standards or units.

OBJECTIVES

- Build and test designs against design specifications, evaluate the results of those tests, and present their analyses
- Work in teams to design and build a project related to Civil Engineering (Tower Building)
- Write a reflection of the project

OBJECTIVES

- Use consistent units for all measurements and computations
- Use clear and concise written, verbal, and visual communication techniques
- Use collaborative tools such as desktop or web-based applications to share and develop information
- Use problem-solving techniques to develop technological solutions such as product, process, or system
- Apply the design process, including decision matrices, as a team participant
- Perform different roles within the project as a team member
- Formulate decisions using collaborative strategies such as decision and design matrices and conflict resolution
- Maintain an engineering notebook for the project per established conventions
- Develop and test the model for the project
- Demonstrate communication skills by preparing and presenting the project, including building consensus setback resolution and decision matrices
- Test and evaluate proposed solutions using engineering methods such as creating models, prototypes, mock-ups, or simulations or performing critical design review, statistical analysis, or experiments
- Predict performance, failure modes, and reliability of a design solution
- Apply structured techniques such as a decision tree, design matrix, or cost-benefit analysis to select and justify a preferred solution to a problem
- Prepare a project report that clearly documents the designs, decisions, and activities during each phase of the engineering design process
- Take minutes of a team meeting
- Translate design requirements into a design solution
- Use brainstorming methods to identify solutions to a design problem
- Recognize and demonstrate that there are many possible successful designs and that a design process does not always result in a single best design

ENGAGE

Units like the kilogram can be based off physical references. In fact, they are liable to change if the scientific community deems it necessary. For example, in 2018, the block defining the kilogram was discarded after it was realized that the kilogram could be based on unchanging quantities that have been measured. Consider sharing the following NPR article with your students.

Say Au Revoir To That Hunk Of Metal In France That Has Defined The Kilogram

https://www.npr.org/2018/11/13/666310991/say-au-revoir-to-that-hunk-of-metal-in-france-that-has-defined-the-kilogram

QUESTION

Define the terms *dimensions* and *units. We usually consider quantities like mass, length, time, and perhaps charge and temperature, as fundamental dimensions. Units give the magnitude of some dimension relative to an arbitrary standard. For example, when we say that a person is six feet tall, we mean that person is six times as long as an object whose length is defined to be one foot. Without units, a number is at best meaningless and at worst misleading to the reader.*

QUESTION

Define the *International System of Units,* commonly known as the metric system. *The International System of Units—in French, Le Système International d'Unités (SI) is the modern form of the metric system, and is the most widely used system of measurement. It comprises a coherent system of units of measurement built on seven base units.*

7.1 The Metric System

The SI system (Le Système International d'Unités), commonly known as the metric system, is the standard set of units for most of the world. Originally developed by French scientists under King Louis XVI, the SI system was finalized by the international scientific community as the standard unit system in 1971. This system defines seven base units, from which all others are derived. Table 7-1 shows the seven base units and their corresponding fundamental dimensions.

NOTE:

Within this text, dimensions are shown in braces { } and units in brackets [ ].

Table 7-1 Fundamental Dimensions and Base Units

Dimension	Symbol	Unit	Symbol
Mass	M	kilogram	kg
Length	L	meter	m
Time	T	second	s
Temperature	Θ	kelvin	K
Amount of substance	N	mole	mol
Light intensity	J	candela	cd
Electric current	I	ampere	A

7.1.1 SI Prefixes

SI prefix

The SI system is based upon multiples of 10. By using an **SI prefix** when reporting numbers, we avoid long strings of zeros. For example, instead of saying, "The distance to Atlanta is 198,000 meters," we would say, "The distance to Atlanta is 198 kilometers."

For a list of SI prefixes, refer to Table 7-2. Note that the abbreviations for all SI prefixes from deci- to kilo- are lowercase, whereas from Mega- to Yotta- are uppercase.

Table 7-2 SI Prefixes (Example: 1 millimeter [mm] = 1 x 10^{-3} meters [m])

Numbers Less than One			Numbers Greater than One		
Power of 10	Prefix	Abbreviation	Power of 10	Prefix	Abbreviation
10^{-1}	deci-	d	10^{1}	deca-	da
10^{-2}	centi-	c	10^{2}	hecto-	h
10^{-3}	milli-	m	10^{3}	kilo-	k
10^{-6}	micro-	μ	10^{6}	Mega-	M
10^{-9}	nano-	n	10^{9}	Giga-	G
10^{-12}	pico-	p	10^{12}	Tera-	T
10^{-15}	femto-	f	10^{15}	Peta-	P
10^{-18}	atto-	a	10^{18}	Exa-	E
10^{-21}	zepto-	z	10^{21}	Zetta-	Z
10^{-24}	yocto-	y	10^{24}	Yotta-	Y

Determining the appropriate SI prefix to use becomes simple when the number is placed in engineering notation: just examine the exponent. As a reminder, scientific and engineering notation are defined as follows:

scientific notation

Scientific notation is typically expressed in the form #.### × 10^N, where the single digit to the left of the decimal point is the most significant nonzero digit of the value being represented. N is an integer, and multiplying by 10^N serves to locate the true position of the decimal point.

engineering notation

Engineering notation is expressed in the form ###.### × 10^M, where M is an integer multiple of 3, and the number of digits to the left of the decimal point is one, two, or three as needed to yield a power of 10 that is indeed a multiple of 3.

Example 7-1

Express the following values using scientific notation, engineering notation, and the correct SI prefix.

Standard	Scientific	Engineering	With Prefix
(a) 43,480,000 m	4.348×10^7 m	43.48×10^6 m	43.48 Mm
(b) 0.0000003060 V	3.060×10^{-7} V	306.0×10^{-9} V	306.0 nV
(c) 9,860,000,000 J	9.86×10^9 J	9.86×10^9 J	9.86 GJ
(d) 0.0351 s	3.51×10^{-2} s	35.1×10^{-3} s	35.1 ms

Note that the numeric values of the mantissa are the same in the last two columns, and the exponent in engineering notation specifies the metric prefix.

Comprehension Check 7-1

Express the following values using an appropriate SI prefix such that there are only one, two, or three digits shown to the left of the decimal.

Standard	With Prefix
(a) 3100 joules [J]	
(b) 26,510,000 watts [W]	
(c) 459,000 seconds [s]	
(d) 0.00000032 grams [g]	

7.1.2 Official SI Rules

When reporting units using the SI system, follow these official rules:

- **If a unit abbreviation appears as a capital letter, it has been named after a person; all other abbreviations appear as lowercase letters.** For example, the abbreviation "N" stands for "newton," the SI unit of force named after Isaac Newton.

  Correct: The book weighs 5 N. **Incorrect**: The book weighs 5 n.

  Correct: The rod is 5 m long. **Incorrect**: The rod is 5 M long.

  The one exception to this rule is the volumetric unit of liter. The abbreviation is shown as L, since a lowercase l can be confused with both the number 1 and the uppercase letter I.

TEACHING TIP

When describing scientific notation, it may be helpful to reiterate that one of the advantages of rewriting numbers in this fashion is that very big and very small numbers can be represented and manipulated in a compact fashion. You may choose to demonstrate this fact by multiplying two numbers in scientific notation.

ENGAGE

Visit the Mars Climate Orbiter website, created by the National Aeronautics and Space Administration (NASA), with your students.

URL
https://mars.nasa.gov/mars-exploration/missions/mars-climate-orbiter/

According to the website, what was the reason the orbiter was "lost"? *The failure board's first report identifies eight contributing factors that led directly or indirectly to the loss of the spacecraft. "The 'root cause' of the loss of the spacecraft was the failed translation of English units into metric units in a segment of ground-based, navigation-related mission software."*

- **Abbreviations of units are not shown as plural.**
  Correct: 10 centimeters = 10 cm **Incorrect**: 10 centimeters ≠ 10 cms
- **Abbreviations are not shown with a period unless they appear at the end of a sentence.**
  Correct: The rod is 5 mm long. **Incorrect**: The rod is 5 mm. long.
- **Abbreviations are written in upright Roman type (m, k, L) to distinguish them from mathematical variables (*m, k, l*), which are indicated by italics.**
- **One space separates the number and abbreviation, except with the degree symbol referring to an angle.**
  Correct: 5 mm or 5° **Incorrect**: 5mm or 5 °
- **Spaces or commas may be used to group digits by threes.**
  Correct: 1 000 000 or 1,000,000 **Incorrect**: 100,0000
- **Abbreviations for derived units formed by multiple units are joined by a space or interpunct (the center dot).** Care must be taken to avoid confusing SI prefixes with units.
  Correct: kg m or kg · m **Incorrect**: kgm or kg * m
  This is particularly important when confusion might arise. For example, "ms" stands for millisecond, but "m s" stands for meter second. In cases like this, using a center dot (m · s) is preferable since it is less likely to be misunderstood.
- **Abbreviations for derived units formed by dividing units are joined by a virgule (the "slash" /) or shown with a negative exponent.** Care must be taken to appropriately display the entire denominator.
  Correct: $N/(m\ s^2)$ or $N\ m^{-1}\ s^{-2}$ **Incorrect**: $N/m\ s^2$ or $N/m/s^2$
- **Do not combine prefixes to form a compound prefix. Use a single prefix.**
  Correct: picojoules (pJ) **Incorrect**: millinanojoules (mnJ)
  Correct: gigaseconds (Gs) **Incorrect**: kilomegaseconds (kMs)

Comprehension Check 7-2

Indicate if the following units are correctly expressed according to the official SI rules. If the unit is incorrectly displayed, show the correction.

(a) Reading this sentence took 5 Secs.
(b) The average person's pupils are 60mms. apart.
(c) One gallon is the same as 380 microkiloliters.

7.2 Other Unit Systems

Prior to the adoption of the SI unit system by the scientific community, several other systems of units were used and are still used today, particularly in the United States. Great Britain officially converted to metric in 1965, but it is still common there to see nonmetric units used in communications for the general public.

It is important to know how to convert between all unit systems. Table 7-3 compares three systems. The system listed as AES (American Engineering System) is in common use by the general public in the United States. The USCS (United States Customary System) is commonly called "English" units.

Table 7-3 Comparison of Unit Systems, with Corresponding Abbreviations

Dimension	SI (MKS)	AES	USCS
Length	meter [m]	foot [ft]	foot [ft]
Mass	kilogram [kg]	pound-mass [lb_m]	slug
Time	second [s]	second [s]	second [s]
Temperature	Celsius [°C]	Fahrenheit [°F]	Fahrenheit [°F]
	kelvin [K]	Rankine [°R]	Rankine [°R]

Accepted Non-SI Units

The units in Table 7-4 are not technically in the SI system, but due to their common usage, are acceptable for use in combination with the base SI units.

NOTE:

1 liter does not equal 1 cubic meter!

Table 7-4 Acceptable Non-SI Units

Unit	Equivalent SI	Unit	Equivalent SI
astronomical unit [AU]	1 AU = 1.4959787×10^{11} m	day [d]	1 d = 86,400 s
atomic mass unit [amu]	1 amu = $1.6605402 \times 10^{-24}$ g	hour [h]	1 h = 3600 s
electronvolt [eV]	1 eV = $1.6021773 \times 10^{-19}$ J	minute [min]	1 min = 60 s
liter [L]	1 L = 0.001 m^3	year [yr]	1 yr = 3.16×10^7 s
		degree [°]	1° = 0.0175 rad or 1 rad = 57.3°

7.3 Unit Conversions

LENGTH:

1 m = 3.28 ft
1 km = 0.621 mi
1 ft = 12 in
1 in = 2.54 cm
1 mi = 5280 ft
1 yd = 3 ft

We use conversion factors to translate from one set of units to another. Let us examine the conversions found for the dimension of length, as shown in the margin, beginning with the conversion: 1 meter [m] = 3.28 feet [ft]. By dividing both sides of this equation by 3.28 feet, we obtain

$$\frac{1\ \text{m}}{3.28\ \text{ft}} = 1$$

DISCUSSION

What is the proper procedure for converting units?

Unit conversion is a multi-step process that involves multiplication or division by a numerical factor, selection of the correct number of significant digits, and rounding.

DISCUSSION

The United States uses another system of units. What other unit systems are acceptable for the general public?

Certain units are not part of the International System of Units; that is, they are outside the SI, but are important and widely used.

(See examples in Table 7-4.)

ENGAGE

Visit Convert-Me.com with your students for length conversions.

URL
https://www.convert-me.com/en/convert/length/

Ask the students to type a "1" in the mile text box and then to click "Convert Me." How many kilometers are there in a mile? *(1.609)* How many meters? *(1,609)* How many furlongs to a U.S. surveyor? *(8)* How many nautical miles to a ship's navigator? *(0.869)*

or in other words, "There is 1 meter per 3.28 feet." If we divide both sides of the original expression by 1 meter, we obtain

$$1 = \frac{3.28 \text{ ft}}{1 \text{ m}}$$

or in other words, "In every 3.28 feet there is 1 meter." The number 1 is dimensionless, a pure number. *We can multiply any expression by 1 without changing the expression.* We use this rule to convert between unit systems.

For example, on a trip we note that the distance to Atlanta is 123 miles [mi]. How many kilometers [km] is it to Atlanta? From the conversion table, we can find that 1 kilometer [km] = 0.621 miles [mi], or

$$1 = \frac{1 \text{ km}}{0.621 \text{ mi}}$$

By multiplying the original quantity of 123 miles by 1, we can say

$$(123 \text{ mi})(1) = (123 \text{ mi})\left(\frac{1 \text{ km}}{0.621 \text{ mi}}\right) = 198 \text{ km}$$

Note that we could have multiplied by the following relationship:

$$1 = \frac{0.621 \text{ mi}}{1 \text{ km}}$$

We would still have multiplied the original answer by 1, but the units would not cancel, and we would be left with an awkward, meaningless answer.

$$(123 \text{ mi})(1) = (123 \text{ mi})\left(\frac{0.621 \text{ mi}}{1 \text{ km}}\right) = 76 \frac{\text{mi}^2}{\text{km}}$$

We want to use the ratio that will make the unit we want to eliminate cancel from the original expression.

As a second example, we are designing a reactor system using 2-inch [in] diameter plastic pipe. The design office in Germany would like the pipe specifications in units of centimeters [cm]. From the conversion table, we find that 2.54 centimeter [cm] = 1 inch [in], or

$$1 = \frac{2.54 \text{ cm}}{1 \text{ in}}$$

By multiplying the original quantity of 2 inches by 1, we can say

$$(2 \text{ in})(1) = (2 \text{ in})\left(\frac{2.54 \text{ cm}}{1 \text{ in}}\right) = 5 \text{ cm}$$

In a final example, suppose a car travels at 40 miles per hour [mi/h or mph]. Stated in words, "A car traveling at a rate of 40 mph will take 1 hour to travel 40 miles if the velocity remains constant." By simple arithmetic this means that the car will travel 80 miles in 2 hours or 120 miles in 3 hours. In general,

$$\text{Distance} = (\text{velocity})(\text{time elapsed at that velocity})$$

Suppose the car is traveling at 40 mph for 6 minutes. How far does the car travel? Simple calculation shows

$$\text{Distance} = (40)(6) = 240 \text{ miles}$$

TIME:

1 d = 24 h
1 h = 60 min
1 min = 60 s
1 yr = 365 d

Without considering units, the preceding example implies that if we drive our car at 40 mph, we can cover the distance from Charlotte, North Carolina, to Atlanta, Georgia, in 6 minutes! What is wrong? Note that the velocity is given in miles per hour, and the time is given in minutes. We need to apply the conversion factor that 1 hour [h] = 60 minutes [min]. If the equation is written with consistent units attached, we get

$$\text{Distance} = \left(\frac{40 \text{ mi}}{\text{h}}\right)\left(6 \text{ min} \left| \frac{1 \text{ h}}{60 \text{ min}}\right.\right) = 4 \text{ mi}$$

IMPORTANT CONCEPT:

Be sure to *always* include units in your calculations and your final answer!

It seems more reasonable to say, "traveling at a rate of 40 miles per hour for a time period of 6 minutes will allow us to go 4 miles."

To convert between any set of units, the following method demonstrated in Examples 7-2 to 7-8 is very helpful. This procedure is easy to use but take care to avoid mistakes. If you use one of the conversion factors incorrectly, say, with 3 in the numerator instead of the denominator, your answer will be in error by a factor of 9.

Unit Conversion Procedure

1. Write the value and unit to be converted.
2. Write the conversion formula between the given unit and the desired unit.
3. Make a fraction, equal to 1, of the conversion formula in Step 2, such that the originial unit in Step 1 is located so the orginial unit will cancel.
4. Multiply the term from Step 1 by the fraction devleoped in Step 3.
5. Cancel units, perform mathematical calculations, and express the answer in "reasonable" terms.

Example 7-2

Convert the length 40 yards [yd] into units of feet [ft].

Method	Steps
(1) Term to be converted	40 yd
(2) Conversion formula	1 yd = 3 ft
(3) Make a fraction (equal to one)	$\frac{3 \text{ ft}}{1 \text{ yd}}$
(4) Multiply	$40 \text{ yd} \left\| \frac{3 \text{ ft}}{1 \text{ yd}}\right.$
(5) Cancel, calculate, be reasonable	120 ft

ENGAGE

Students can follow a simple 4-step procedure to convert units properly:

1. *What do you have?*
   Starting Units
2. *What do you want?*
   Desired Units
3. *Identify conversion factor(s).*
   Ratio or Linking Units
4. *Cancel units where you can and do the math.*

ENGAGE

Visit the National Institute of Standards and Technology (NIST) website with your students. Navigate to the page on unit conversion.

URL
http://www.nist.gov/pml/wmd/metric/unit-conversion.cfm

What does the NIST say about unit conversions?

Unit conversion is a multi-step process that involves multiplication or division by a numerical factor, selection of the correct number of significant digits, and rounding.

TEACHING TIP

Getting students used to the idea that multiplying by a conversion factor is like multiplying by 1 can be a conceptual challenge. Writing out the explicit step of multiplying by 1 can help make the connection more obvious.

Example 7-3

Convert the time 456,000 seconds [s] into units of minutes [min].

Method	Steps
(1) Term to be converted	456,000 s
(2) Conversion formula	1 min = 60 s
(3) Make a fraction (equal to one)	$\frac{1\text{ min}}{60\text{ s}}$
(4) Multiply	$\frac{456{,}000\text{ s}}{} \Big\vert \frac{1\text{ min}}{60\text{ s}}$
(5) Cancel, calculate, be reasonable	7600 min

Comprehension Check 7-3

The highest mountain in the world is Mount Everest in Nepal. The peak of Mount Everest is 29,029 feet [ft] above sea level. Convert the height into units of miles [mi].

Comprehension Check 7-4

To be considered a full-time employee, companies in the United States required you work more than 30 hours [h] in a week. Convert the time into units of minutes [min].

7.4 Conversions Involving Multiple Steps

LENGTH:

1 m = 3.28 ft
1 km = 0.621 mi
1 in = 2.54 cm
1 ft = 12 in
1 mi = 5280 ft
1 mi = 1.609 km
1 yd = 3 ft

Sometimes, more than one conversion factor is needed in a single quantity. We can multiply by several conversion factors, each one of which is the same as multiplying by 1. For example, suppose we determined that the distance to Atlanta is 123 miles [mi]. How many yards [yd] is it to Atlanta? From the conversion table, we do not have a direct conversion between miles and yards, but we see that both can be related to feet. We can find that 5280 feet [ft] = 1 mile [mi], or

$$1 = \frac{5280\text{ ft}}{1\text{ mi}}$$

We can also find that 1 yard [yd] = 3 feet [ft], or

$$1 = \frac{1\text{ yd}}{3\text{ ft}}$$

By multiplying the original quantity of 123 miles by 1 using the first set of conversion factors, we can say:

$$(123\text{ mi})(1) = (123\text{ mi})\left(\frac{5280\text{ ft}}{1\text{ mi}}\right) = 649{,}440\text{ ft}$$

If we multiply by 1 again, using the second set of conversion factors and applying reasonableness:

$$(649{,}440\text{ ft})(1) = (649{,}440\text{ ft})\left(\frac{1\text{ yd}}{3\text{ ft}}\right) = 216{,}000\text{ yd}$$

This is usually shown as a single step:

$$(123\text{ mi})\left(\frac{5280\text{ ft}}{1\text{ mi}}\right)\left(\frac{1\text{ yd}}{3\text{ ft}}\right) = 216{,}000\text{ yd}$$

Example 7-4

Convert the length 40 yards [yd] into units of millimeters [mm].

Method	Steps
(1) Term to be converted	40 yd
(2) Conversion formula	1 yd = 3 ft 1 ft = 12 in 1 in = 2.54 cm 1 cm = 10 mm
(3) Make fractions (equal to one)	$\frac{3\ \text{ft}}{1\ \text{yd}}\ \frac{12\ \text{in}}{1\ \text{ft}}\ \frac{2.54\ \text{cm}}{1\ \text{in}}\ \frac{10\ \text{mm}}{1\ \text{cm}}$
(4) Multiply	$\frac{40\ \text{yd}}{} \Big\vert \frac{3\ \text{ft}}{1\ \text{yd}} \Big\vert \frac{12\ \text{in}}{1\ \text{ft}} \Big\vert \frac{2.54\ \text{cm}}{1\ \text{in}} \Big\vert \frac{10\ \text{mm}}{1\ \text{cm}}$
(5) Cancel, calculate, be reasonable	37,000 mm

Here, there are several other viable solution paths. For example, we could have converted yards to feet, then feet to meters, then meters to millimeters.

Example 7-5

Convert 55 miles per hour [mi/h or mph] to units of meters per second [m/s].

NOTE:

We have two unit groups to convert here, miles to meters, and hours to seconds.

Method	Steps
(1) Term to be converted	55 mph
(2) Conversion formula	1 km = 0.621 mi 1 km = 1000 m 1 h = 60 min 1 min = 60 s
(3) Make fractions (equal to one) (4) Multiply	$\frac{55\ \text{mi}}{\text{h}} \Big\vert \frac{1\ \text{km}}{0.621\ \text{mi}} \Big\vert \frac{1000\ \text{m}}{1\ \text{km}} \Big\vert \frac{1\ \text{h}}{60\ \text{min}} \Big\vert \frac{1\ \text{min}}{60\ \text{s}}$
(5) Cancel, calculate, be reasonable	24.6 m/s

Example 7-6

Convert the power of 3,780,000 kilowatts [kW] into units of gigawatts [GW].

NOTE:

Treat SI prefixes as units.

Method	Steps
(1) Term to be converted	3,780,000 kW
(2) Conversion formula	$1\ \text{kW} = 1 \times 10^3\ \text{W}$ $1\ \text{GW} = 1 \times 10^9\ \text{W}$
(3) Make a fraction (equal to one)	$\frac{1 \times 10^3\ \text{W}}{1\ \text{kW}}\ \ \frac{1\ \text{GW}}{1 \times 10^9\ \text{W}}$
(4) Multiply	$\frac{3{,}780{,}000\ \text{kW}}{} \Big\vert \frac{1 \times 10^3\ \text{W}}{1\ \text{kW}} \Big\vert \frac{1\ \text{GW}}{1 \times 10^9\ \text{W}}$
(5) Cancel, calculate, be reasonable	3.78 GW

QUESTION

Define the acronym *MPH,* and give some history behind its origins.

Miles per hour is a common unit of speed in the Imperial system of units used in the UK and the USA. MPH first came into common usage with the regular stage coaches in the 18th century, because running the stages on time required a timetable. In the early 19th century, with the advent of the railroads, speeds became significantly higher. At one time, many people thought that all the passengers would asphyxiate if the trains went over 20 mph. This myth was quickly dispelled as trains were soon traveling at 50 mph and above and the passengers were surviving the experience.

186 Chapter 7 Essential Physics Ideas

Example 7-7

Convert the volume of 40 gallons [gal] into units of cubic feet [ft^3].

By examining the "Volume" box in the conversion table, we see that the following facts are available for use:

$$1\ L = 0.264\ gal \quad \text{and} \quad 1\ L = 0.0353\ ft^3$$

By the transitive property, if a = b and a = c, then b = c. Therefore, we can directly write

$$0.264\ gal = 0.0353\ ft^3$$

VOLUME:

1 gal = 3.785 L
1 gal = 4 qt
1 L = 1000 cm^3
1 L = 0.264 gal
1 L = 0.0353 ft^3
1 L = 33.8 fl oz
1 mL = 1 cm^3
1 m^3 = 1000 L
1 pt = 16 fl oz
1 qt = 2 pt

Method	Steps
(1) Term to be converted	40 gal
(2) Conversion formula	0.264 gal = 0.0353 ft^3
(3) Make fractions (equal to one)	$\frac{0.0353\ ft^3}{0.264\ gal}$
(4) Multiply	$\frac{40\ gal}{} \Big\vert \frac{0.0353\ ft^3}{0.264\ gal}$
(5) Cancel, calculate, be reasonable	5.3 ft^3

This picture shows 5-gallon water bottles made from polycarbonate. Millions of these bottles are made each year around the world to transport clean water to remote locations.

The use of polycarbonate to contain products for consumption has raised safety concerns because bisphenol A is leached from the plastic into the stored liquid. In July 2012, the U.S. Food and Drug Administration banned the use of BPA in bottles and cups used by infants and small children.

Photo courtesy of E. Stephan

RULES OF THUMB

1 quart = 1 liter	1 cubic foot = 7.5 gallons
1 cubic meter = 250 gallons	1 cubic meter = 5.55-gallon drums
1 cup = 250 milliliters	1 golf ball = 1 cubic inch

One frequently needs to convert a value that has some unit or units raised to a power, for example, converting a volume given in cubic feet to cubic meters. It is critical in this case that the power involved be applied to the *entire* conversion factor, both the numerical values and the units.

Example 7-8

Convert 35 cubic inches [in^3] to cubic centimeters [cm^3].

NOTE:

When raising a quantity to a power, be sure to apply the power to both the value and the units.

Method	Steps
(1) Term to be converted	35 in^3
(2) Conversion formula	1 in = 2.54 cm
(3) Make fractions (equal to one)	$\frac{(2.54\ cm)^3}{(1\ in)^3}$
(4) Multiply	$\frac{35\ in^3}{} \left| \frac{(2.54)^3\ cm^3}{1\ in^3} \right.$
(5) Cancel, calculate, be reasonable	574 cm^3

Note that in some cases, a unit that is raised to a power is being converted to another unit that has been defined to have the same dimension as the one raised to a power. This is difficult to say in words, but a couple of examples should clarify it.

If one is converting square meters [m^2] to acres, the conversion factor is *not* squared, since the conversion provided is already in terms of length squared: 1 acre = 4047 m^2.

If one is converting cubic feet [ft^3] to liters [L], the conversion factor is *not* cubed, since the conversion provided is already in terms of length cubed: 1 L = 0.0353 ft^3.

Comprehension Check 7-5

In January 2008, *Scientific American* reported that physicists Peter Sutter and Eli Sutter of Brookhaven National Laboratories made a pipette to measure droplets in units of a zeptoliter [zL]. Previously, the smallest unit of measure in a pipette was an attoliter [aL]. Convert the measurement of 5 zeptoliters [zL] into units of picoliters [pL].

Comprehension Check 7-6

Officially, a hurricane is a tropical storm with sustained winds of at least 74 miles per hour [mi/h or mph]. Convert this speed into units of kilometers per minute [km/min].

Comprehension Check 7-7

Many toilets in commercial establishments have a value printed on them stating the amount of water consumed per flush. For example, a label of 2 Lpf indicates the consumption of 2 liters per flush. If a toilet is rated at 3 Lpf, how many flushes are required to consume 20 gallons [gal] of water?

ENGAGE

Visit Wikipedia with your students. Perform a search for "dimensionless quantities." Observe the list. Explain to your students that a *dimensionless quantity* is any number or defined expression which is either unitless or has composite dimensions which cancel out to unity. These are found in numerous fields such as mathematics, physics, engineering, and economics. In contrast, quantities containing dimensions include length, time, mass, luminosity, and electric charge, as well as products of these fundamental quantities, such as volume, speed, and voltage.

7.5 Derived Dimensions and Units

derived dimensions

With only the seven base dimensions in the metric system, all measurable things in the known universe can be expressed by various combinations of these concepts, called **derived dimensions**. As simple examples, area is length squared, volume is length cubed, and velocity is length divided by time.

As we explore more complex parameters, the dimensions become more complex. For example, the concept of force is derived from Newton's second law, which states that force is equal to mass times acceleration. Force is then used to define more complex dimensions such as pressure, which is force acting over an area, or work, which is force acting over a distance.

derived unit

Sometimes, the derived dimensions become quite complicated. For example, electrical resistance is mass times length squared divided by both time cubed and current squared. Particularly in the more complicated cases like this, a **derived unit** is defined to avoid having to say things like "The resistance is 15 kilogram-meter squared divided by second cubed ampere squared." It is much easier to say "The resistance is 15 ohms," where the derived unit "ohm" equals one (kg m^2)/(s^3 A^2).

Quantity	Fractional Notation	Exponential Notation
Velocity	$\frac{L}{T}$	$L^1 T^{-1}$
Acceleration	$\frac{L}{T^2}$	$L^1 T^{-2}$

One way to determine the dimensions of a quantity, such as volume, is to examine the common units used to express the quantity. While volume can be expressed in gallons, it can also be expressed as cubic feet or cubic meters. The units of cubic meters express volume in a manner easily transferred to dimensions. The units of gallons and of cubic feet and of cubic meters are dimensionally equal to length cubed.

Currently, there are officially 22 named derived units in the SI system. All are named after famous scientists or engineers who are deceased (see Table 7-5). It is worth noting that numerous common derived dimensions do not have a corresponding named derived SI unit. For example, there is no named derived SI unit for the derived dimension velocity as there is for force (newton) or electrical resistance (ohm).

NOTE:

One letter can represent several quantities in various engineering disciplines. For example, the letter "*P*" can indicate pressure, power, or vertical load on a beam. It is important to examine and determine the nomenclature in terms of the context of the problem presented.

Table 7-5 Common Derived Units in the SI System

Dimension	SI Unit	Base SI Units	Derived from
Force (*F*)	newton [N]	$1\text{ N} = 1\frac{\text{kg m}}{\text{s}^2}$	$F = ma$ Force = mass times acceleration
Energy (*E*)	joule [J]	$1\text{ J} = 1\text{ N m} = 1\frac{\text{kg m}^2}{\text{s}^2}$	$E = Fd$ Energy = force times distance
Power (*P*)	watt [W]	$1\text{ W} = 1\frac{\text{J}}{\text{s}} = 1\frac{\text{kg m}^2}{\text{s}^3}$	$P = E/t$ Power = energy per time
Pressure (*P*)	pascal [Pa]	$1\text{ Pa} = 1\frac{\text{N}}{\text{m}^2} = 1\frac{\text{kg}}{\text{m s}^2}$	$P = F/A$ Pressure = force per area
Voltage (*V*)	volt [V]	$1\text{ V} = 1\frac{\text{W}}{\text{A}} = 1\frac{\text{kg m}^2}{\text{s}^3\text{ A}}$	$V = P/I$ Voltage = power per current

7.6 Force

When you push a grocery cart, it moves. If you keep pushing, it keeps moving. The longer you push, the faster it goes; the velocity increases over time, meaning that it accelerates. If you push a full grocery cart that has a high mass, it does not speed up as much, meaning it accelerates less than a cart with low mass. Simply put, the acceleration (a) of a body depends on the force (F) exerted on it and its mass (m). This is a simple form of Newton's second law of motion and is usually written as $F = ma$ (see Table 7-6).

IMPORTANT CONCEPT:

F = ma
SI unit of force = newton [N]

Table 7-6 Dimensions of Force

Quantity	Common Units	Exponents						
		M	L	T	Θ	N	J	I
Force	N	1	1	−2	0	0	0	0

newton

The SI unit of force, the **newton** [N], is defined as the force required to accelerate a mass of one kilogram at a rate of one meter per second squared (see Table 7-7). It is named for Sir Isaac Newton (1643–1727). Newton's *Principia* is considered one of the world's greatest scientific writings, explaining the law of universal gravitation and the three laws of motion. Newton also developed the law of conservation of momentum, the law of cooling, and the reflecting telescope. He shares credit for the development of calculus with Gottfried Leibniz.

In the SI system, mass, length, and time are base units and force is a derived unit; force is found from combining mass, length, and time using Newton's second law. The SI system is called *coherent*, because the derived unit is set at one by combining base units. The AES system is considered non-coherent as it uses units that do not work together in the same fashion as the SI units do. There are two uses of the term "pound" in the AES system, which occurred in common usage long before Newton discovered gravity. To distinguish mass in pounds and force in pounds, the unit of mass is given as pound-mass (lb_m) and the unit of force is given as pound-force (lb_f). One pound-force is the amount of force needed to accelerate one pound-mass at a rate of 32.2 feet per second squared. Since this relationship is not easy to remember or use in conversions, we will stick with SI units for problem solving, following the procedure discussed in the chapter on fundamental dimensions.

NOTE:

In general, a "pound" can be used as a unit of mass or force. For distinction, the following convention is used:
– pound-mass [lb_m]
– pound-force [lb_f]

Table 7-7 Definitions of Force Units in Various Unit Systems

Unit System	Mass	Acceleration	Force	
SI	1 kg	1 m/s^2	1 N = 1 kg m/s^2	coherent
AES	1 lb_m	32.2 ft/s^2	1 lb_f = 32.2 lb_m ft/s^2	non-coherent

Example 7-9

A professional archer is designing a new longbow with a full draw weight of 63 pound-force [lb_f]. The draw weight is the amount of force needed to hold the bowstring at a given amount of draw, or the distance the string has been pulled back from the rest position. What is the full draw weight of this bow in units of newtons [N]?

Convert to Base SI Units	Steps
(1) Convert term	63 lb_f
(2) Apply conversion formula	1 N = 0.225 lb_f
(3) Make a fraction (4) Multiply	$\frac{63\ lb_f}{} \Big\vert \frac{1\ N}{0.225\ lb_f}$
(5) Cancel, calculate, be reasonable	280 N

Example 7-10

NOTE:

The term **thrust** is used to express a force used to push an object very quickly, such as a rocket or an airplane.

A ship is being designed to use an engine that runs continuously, providing a small but constant acceleration of 0.06 meters per second squared [m/s^2]. If the ship has a mass of 30,000 kilograms [kg], what is the thrust (force) provided by the engine in units of pound-force [lb_f]?

Calculate Required Parameters	Steps
(1) Determine appropriate equation	$F = m\,a$
(2) Insert known quantities	$F = \frac{30{,}000\ kg}{} \Big\vert \frac{0.06\ m}{s^2}$
(3) Calculate	$F = 1800\ \frac{kg\ m}{s^2}$

This is apparently our final answer, but the units are puzzling. If the unit of force is the newton, and if this is a valid equation, then our final result for force should be newtons. If we consider the dimensions of force

Quantity	Common Units	Exponents M	L	T
Force	N	1	1	−2

A unit of force has dimensions $F\ \{=\}\ ML/T^2$*, which in terms of base SI units would be* $F\ [=]\ kg\ m/s^2$*. As this term occurs frequently, it is given the special name "newton." Anytime we see the term* [$kg\ m/s^2$]*, we know we are dealing with a force equal to a newton.*

IMPORTANT CONCEPT:

1 N = 1 (kg m) / s^2

(3) Calculate	$F = \frac{1800\ kg\ m}{s^2} \Big\vert \frac{1\ N}{1\ \frac{kg\ m}{s^2}} = 1800\ N$

Convert to Desired Units	Steps
(1) Convert term	1800 N
(2) Apply conversion formula	1 N = 0.225 lb_f
(3) Make a fraction (4) Multiply	$\frac{1800\ N}{} \Big\vert \frac{0.225\ lb_f}{1\ N}$
(5) Cancel, calculate, be reasonable	405 lb_f

Comprehension Check 7-8

The engine on a spacecraft nearing Mars can provide a thrust of 15,000 newtons [N]. If the spacecraft has a mass of 750 kilograms [kg], what is the acceleration of the spacecraft in miles per hour squared [mi/h^2]?

7.7 Weight

mass
weight
gravity

The **mass** of an object is a fundamental dimension. Mass is a quantitative measure of how much of an object there is, or in other words, how much matter it contains. The **weight** (w) of an object is a force equal to the mass of the object (m) times the acceleration of **gravity** (g). This is a special application of Newton's second law, where acceleration is equal to gravity.

IMPORTANT CONCEPT:

Weight is a FORCE
w = mg
SI unit of weight = newton [N]

While mass is independent of location in the universe, weight is dependent upon both mass and gravity (Table 7-8).

On the Earth, gravity is approximately 9.8 meters per second squared [m/s^2]. On the moon, gravity is approximately 1.6 meters per second squared [m/s^2]. A one-kilogram [kg] object acted on by Earth's gravity would have a weight of 9.8 newtons [N], but on the moon it would have a weight of 1.6 newtons [N]. Unless otherwise stated, assume all examples take place on the Earth.

Table 7-8 Dimensions of Weight

Quantity	Common Units	Exponents						
		M	L	T	Θ	N	J	I
Weight	N	1	1	–2	0	0	0	0

192 Chapter 7 Essential Physics Ideas

IMPORTANT CONCEPT:

Mass of an object: A quantitative measure of how much of an object there is.

Weight of an object: A quantitative measure of the force exerted on the object due to gravity.

DEVILISH DERIVATION

Newton's law of universal gravitation states

$$F = G\frac{m_1 m_2}{r^2}$$

where:

G is universal gravitational constant $G = 6.673 \times 10^{-11}(\mathrm{N\,m^2})/\mathrm{kg^2}$
m is the mass
r is the distance between the centers of mass of two bodies

On the Earth, the distance between the center of a body and the center of the Earth is approximately the radius of the Earth, r_e.

The mass of one of the bodies can be defined as is the mass of the Earth. Rewrite the equation:

$$F = m\left[G\frac{m_e}{r_e^2}\right]$$

The quantity in square brackets is a constant (call it "g"). We call the force "the weight (w) of the body." So,

$$w = mg$$

This is the common equation that relates weight and mass. The value for g on Earth is calculated to be 9.8 meters per second squared [m/s²], or 32.2 feet per second squared [ft/s²].

Example 7-11

What is the weight of a 225-kilogram [kg] bag of birdseed in units of newtons [N]?

Calculate Required Parameters	Steps
(1) Determine appropriate equation	$w = mg$
(2) Insert known quantities	$w = \frac{225\ \mathrm{kg}}{} \Big\vert \frac{9.8\ \mathrm{m}}{\mathrm{s}^2}$
(3) Calculate, be reasonable	$w = 2205\ \frac{\mathrm{kg\ m}}{\mathrm{s}^2} \Big\vert \frac{1\ \mathrm{N}}{1\frac{\mathrm{kg\ m}}{\mathrm{s}^2}} = 2205\ \mathrm{N}$

Comprehension Check 7-9

The mass of the human brain is 1360 grams [g]. State the weight of the human brain in units of newtons [N] on the Earth.

Comprehension Check 7-10

The mass of the human brain is 1360 grams [g]. State the weight of the human brain in units of newtons [N] on the Earth's moon. The gravity on the Earth's moon is 1.6 meters per second squared [m/s²].

7.8 Density

density

IMPORTANT CONCEPT:

$\rho = m/V$

$\gamma = w/V$

Density (ρ, Greek letter rho) is the mass of an object (m) divided by the volume the object occupies (V). Density should not be confused with weight—think of the old riddle: which weighs more, a pound of feathers or a pound of bricks? The answer is they both weigh the same amount, one pound, but the density of each is different. The bricks have a higher density than the feathers, since the same mass takes up less space.

specific weight

Specific weight (γ, Greek letter gamma) is the weight of an object (w) divided by the volume the object occupies (V). Table 7-9 compares density and specific weight in terms of fundamental dimensions.

Table 7-9 Dimensions of Density and Specific Weight

Quantity	Common Units	Exponents						
		M	L	T	Θ	N	J	I
Density	kg/m^3	1	−3	0	0	0	0	0
Specific weight	N/m^3	1	−2	−2	0	0	0	0

Example 7-12

NOTE:

Upon conversion from units of grams per cubic centimeter to pound-mass per cubic foot, the answer should be ≈ 60 times larger.

The density of sugar is 1.61 grams per cubic centimeter [g/cm^3]. What is the density of sugar in units of pound-mass per cubic foot [lb_m/ft^3]?

Convert to Desired Units	Steps
(1) Term to be converted	$1.61\ g/cm^3$
(2) Conversion formula (3) Make fractions (4) Multiply	$1.61\ \frac{g}{cm^3} \Big\vert \frac{2.205\ lb_m}{1000\ g} \Big\vert \frac{1000\ cm^3}{0.0353\ ft^3}$
(5) Cancel, calculate, be reasonable	$101\ lb_m/ft^3$

Example 7-13

NOTE:

Upon conversion from units of grams per cubic centimeter to kilograms per cubic meter, the answer should be = 1000 times larger.

The density of a biofuel blend is 0.72 grams per cubic centimeter [g/cm^3]. What is the density of the biofuel in units of kilograms per cubic meter [kg/m^3]?

Convert to Desired Units	Steps
(1) Term to be converted	$0.72\ g/cm^3$
(2) Conversion formula (3) Make fractions (4) Multiply	$0.72\ \frac{g}{cm^3} \Big\vert \frac{1\ kg}{1000\ g} \Big\vert \frac{100^3\ cm^3}{1\ m^3}$
(5) Cancel, calculate, be reasonable	$720\ kg/m^3$

ENGAGE

Why do we use 55 gallon drums? Have students explore their origin and how many industries use this standard drum size.

Origin:
https://www.skolnik.com/blog/oils-long-history-with-the-55-gallon-steel-drum

More information:
https://www.iqsdirectory.com/articles/55-gallon-drum.html

194 Chapter 7 Essential Physics Ideas

A vast array of valuable compounds can be formed by microbial cultures. Oil produced by the fungi Pythium irregulare can be extracted and used for biofuels or pharmaceutical compounds. Biosystems engineers culture the microorganism, design the bioreactor, and extract the valuable compounds using sustainable, ecoprocessing techniques.

Photo courtesy of C. Drapcho

Example 7-14

What is the weight of water, in units of pound-force [lb_f], in a 55-gallon [gal] drum completely full? Assume the density of water to be 1 gram per cubic centimeter [g/cm^3]. Ignore the weight of the drum.

Convert to Base SI Units	Steps	
(1) Term to be converted	55 gal	1 g/cm^3
(2) Conversion formula (3) Make fractions (4) Multiply	$\frac{55\ gal}{} \Big\vert \frac{1\ L}{0.264\ gal} \Big\vert \frac{1\ m^3}{1000\ L}$	$\frac{1\ g}{cm^3} \Big\vert \frac{1\ kg}{1000\ g} \Big\vert \frac{100^3\ cm^3}{1\ m^3}$
(5) Cancel, calculate	0.208 m^3	1000 kg/m^3

Calculate Required Parameters	Steps
(1) Determine appropriate equation	$w = m\,g$
(2) Insert known quantities	$w = \frac{m}{} \Big\vert \frac{9.8\ m}{s^2}$
(1) Determine appropriate equation	$m = \rho V$
(2) Insert known quantities	$m = \frac{1000\ kg}{m^3} \Big\vert \frac{0.208\ m^3}{}$
(3) Calculate, be reasonable	$m = 208\ kg$
(2) Insert known quantities	$w = \frac{208\ kg}{} \Big\vert \frac{9.8\ m}{s^2}$
(3) Calculate	$w = 2038\ \frac{kg\ m}{s^2} \Big\vert \frac{1\ N}{\frac{1\ kg\ m}{s^2}} = 2038\ N$

NOTE:

Here, we must first determine mass, then use this to determine weight. We repeat the process to Calculate Required Parameters twice.

Convert to Desired Units	Steps
(1) Term to be converted	2038 N
(2) Conversion formula (3) Make a fraction (4) Multiply	$\frac{2038\ N}{} \Big\vert \frac{0.225\ lb_f}{1\ N}$
(5) Cancel, calculate, be reasonable	460 lb_f

Specific Gravity

specific gravity

IMPORTANT CONCEPT:

$SG = \frac{\rho_{object}}{\rho_{water}}$

In technical literature, density is rarely given; instead, the **specific gravity** is reported. The specific gravity (SG) of an object is a dimensionless ratio of the density of the object to the density of water, shown in Table 7-10 in terms of fundamental dimensions. It is convenient to list density in this fashion so *any* unit system may be applied by our choice of the units of the density of water. The specific gravities of several common substances are listed in Table 7-11.

Table 7-10 Dimensions of Specific Gravity

Quantity	Common Units	Exponents						
		M	L	T	Θ	N	J	I
Specific gravity	–	0	0	0	0	0	0	0

Table 7-11 Specific Gravity Values for Common Substances

IMPORTANT CONCEPT:

Density of water
= 1 g/cm³
= 1 kg/L
= 1000 kg/m³
= 62.4 lbₘ/ft³
= 1.94 slug/ft³

Liquids	SG	Solids	SG
Acetone	0.785	Aluminum	2.70
Gasoline	0.739	Baking soda	2.16
Glycerin	1.26	Copper	8.96
Mercury	13.6	Gold	19.3
Olive oil	0.703	Iron	7.87
Propane	0.806	Lead	11.4
Sea water	1.03	Polyvinyl chloride (PVC)	1.38
Water	1.00	Silicon	2.33

IMPORTANT CONCEPT:

Specific Gravity Limits
0.5 < Solids < 23
Liquids ~1
*exceptions:
iodine, 4.93
mercury, 13.6
0.0001 < Gases < 0.001

When calculating or considering specific gravities, it is helpful to keep in mind the range of values that you are likely to have.

The densest naturally occurring elements at normal temperature and pressure are osmium and iridium, both with a specific gravity close to 22.6. *The densest substances that a normal person is likely to encounter are platinum (SG = 21.5) and gold (SG = 19.3).* Thus, if you calculate a specific gravity to be higher than 23, you have almost certainly made an error.

Most liquids are similar to water, with a specific gravity around 1. One notable exception is mercury, with a specific gravity of 13.6.

On the lower end of the scale, *the specific gravity of air is about 0.001*, whereas hydrogen has a specific gravity of slightly less than 0.0001. Therefore, if you get a specific gravity value less than about 10^{-4}, you need to check your work very carefully.

Example 7-15

NOTE:

Since we want the final units in kg/m^3, we choose a density of water value = $1000\ kg/m^3$

The specific gravity of butane is 0.599. What is the density of butane in units of kilograms per cubic meter [kg/m^3]?

Calculate Required Parameters	Steps
(1) Determine appropriate equation	$\rho_{object} = (SG)(\rho_{water})$
(2) Insert known quantities	$\rho_{object} = (0.599)\left(1000\ \frac{kg}{m^3}\right)$
(3) Calculate, be reasonable	$\rho_{object} = 599\ \frac{kg}{m^3}$

Example 7-16

Mercury has a specific gravity of 13.6. What is the density of mercury in units of slugs per liter [slug/L]?

NOTE:

Since we want the final units in slug/L, we choose a density of water value = $1.94\ slug/ft^3$, then convert ft^3 to L.

Calculate Required Parameters	Steps
(1) Determine appropriate equation	$\rho_{object} = (SG)(\rho_{water})$
(2) Insert known quantities	$\rho_{object} = (13.6)\left(1.94\ \frac{slug}{ft^3}\right)$
(3) Calculate	$\rho_{object} = 26.384\ \frac{slug}{ft^3}$

Convert to Desired Units	Steps
(1) Term to be converted	$26.384\ slug/ft^3$
(2) Conversion formula (3) Make a fraction (4) Multiply	$\frac{26.384\ slug}{ft^3} \Big\vert \frac{0.0353\ ft^3}{1\ L}$
(5) Cancel, calculate, be reasonable	0.931 slug/L

Example 7-17

The specific weight of carbon tetrachloride is 15.6 kilonewtons per cubic meter [kN/m^3]. What is the mass of a solid rectangle of brass, with a volume of 1.5 cubic meters [m^3], in units of kilograms [kg]?

NOTE:

Unless otherwise specified, assume all questions refer to Earth, where gravity is 9.8m/s^2

Convert to Desired Units	Steps
(1) Term to be converted	$15.6\ kN/m^3$
(2) Conversion formula (3) Make a fraction (4) Multiply	$\frac{15.6\ kN}{m^3}\Big\vert\frac{1000\ N}{1\ kN}$
(5) Cancel, calculate, be reasonable	$15600\ N/m^3$

Calculate Required Parameters	Steps
(1) Determine appropriate equation	$\gamma = \frac{w}{V}$ so $w = \gamma V$
(2) Insert known quantities	$w = (15600\ N/m^3)(1.5\ m^3)$
(3) Calculate	$w = 23400\ N$
(1) Determine appropriate equation	$w = m g$ so $m = \frac{w}{g}$
(2) Insert known quantities	$m = \frac{23400\ N}{}\Big\vert\frac{s^2}{9.8\ m}\Big\vert\frac{1\ kg\ m}{N\ s^2}$
(3) Calculate	$m = 2387.75\ kg$

Comprehension Check 7-11

Convert 50 grams per cubic centimeter [g/cm^3] into units of pound-mass per cubic foot [lb_m/ft^3].

Comprehension Check 7-12

A 75-gram [g] cylindrical rod is measured to be 10 centimeters [cm] long and 2.5 centimeters [cm] in diameter. What is the specific gravity of the material?

Comprehension Check 7-13

A 75-gram [g] cylindrical rod is measured to be 10 centimeters [cm] long and 2.5 centimeters [cm] in diameter. What is the specific weight of the material, in units of pounds-force per cubic foot [lb_f/ft^3]?

ENGAGE

If you'd like to go over the history of temperature, you can show the following video.

Fahrenheit to Celsius: History of the thermometer

https://www.youtube.com/watch?v=rARnTlPax8E&t=26s

What is the issue with Fahrenheit, Celsius, and other classical thermometers when using them in calculations? Why is Kelvin preferred? *Because Kelvin is an absolute measurement of temperature, whereas Fahrenheit and Celsius are both relative measures. You can further elaborate that Kelvin has a direct relationship with kinetic energy, making the measure compatible with equations describing physical laws.*

7.9 Temperature

temperature

Temperature was originally conceived as a description of energy: heat (thermal energy) flows spontaneously from "hot" to "cold." But how hot is "hot"? The thermometer was devised as a way to measure the "hotness" of an object. As an object gets warmer, it usually expands. In a thermometer, a temperature is a level of hotness that corresponds to the length of the liquid in the tube. As the liquid gets warmer, it expands and moves up the tube. To give temperature a quantitative meaning, numerous temperature scales have been developed.

Fahrenheit
Celsius

Many scientists, including Isaac Newton, have proposed temperature scales. Two scales were originally developed at about the same time—**Fahrenheit** [°F] and **Celsius** [°C]—and have become widely accepted by the general public. Gabriel Fahrenheit (1686–1736), a German physicist and engineer, developed the Fahrenheit scale in 1708. Anders Celsius (1701–1744), a Swedish astronomer, developed the Celsius scale in 1742. The properties of each scale appear in Table 7-12.

You may wonder why the Celsius scale seems so reasonable, and the Fahrenheit scale so random. Actually, Mr. Fahrenheit was just as reasonable as Mr. Celsius. Mr. Celsius set the freezing point of water to be 0 and the boiling point to be 100. Mr. Fahrenheit took as 0 a freezing mixture of salt and ice, and as 100 body temperature. With this scale, it just so happens that the freezing and boiling points of water work out to be strange numbers.

NOTE:

Human body
~37°C = 98°F
Room temperature
~21°C = 70°F

Table 7-12 Properties of Water

Scale	Freezing Point	Boiling Point	Divisions Between Freezing and Boiling
Fahrenheit [°F]	32	212	180
Celsius [°C]	0	100	100
Kelvin [K]	273	373	100
Rankine [°R]	492	672	180

Some units can cause confusion depending on how they are used in calculation. One of those is temperature. One reason for this is that we use temperature in two different ways: **(1) reporting an actual temperature value and (2) discussing the way a change in temperature affects a material property.** To clarify, we resort to examples.

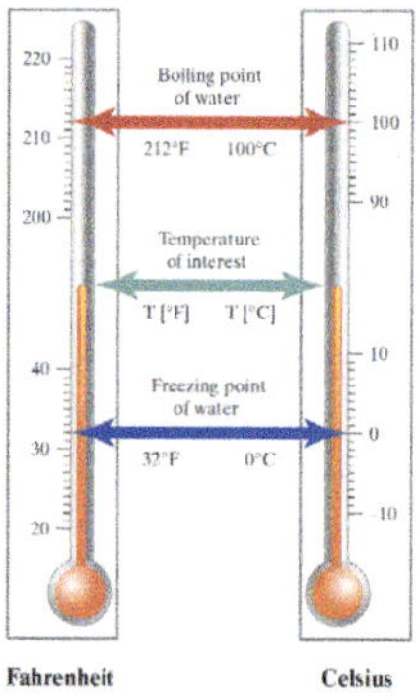

7.9.1 Calculating Temperature Values

When an actual temperature reading is reported, such as "the temperature in this room is 70°F," how do we determine this reading in another temperature scale? The scales have different zero points, so they cannot be determined using a single conversion factor but require a formula. Most of you are familiar with the formula to calculate between Fahrenheit and Celsius, but this equation is cumbersome to remember.

$$T[°F] = \frac{9}{5}\,T[°C] + 32$$

IMPORTANT CONCEPT:

When actual temperature readings are calculated:

$$\frac{T[°F] - 32}{180} = \frac{T[°C] - 0}{100}$$

$$T[K] = T[°C] + 273$$

$$T[°R] = T[°F] + 460$$

Let us imagine we have two thermometers, one with the Fahrenheit scale and the other with the Celsius scale. We set two thermometers side by side so that the freezing point and the boiling point of water are at the same location on both thermometers. We are interested in the relationship between these two scales. From this figure we see that the fraction of the distance from the freezing point to the boiling point in both scales is the same. This means that we can write

$$\frac{T[°F] - 32}{212 - 32} = \frac{T[°C] - 0}{100 - 0}$$

This relationship is really all we need to know to relate a temperature in degrees Fahrenheit to degrees Celsius. You can easily do the algebra to calculate from Fahrenheit to Celsius, or vice versa. By remembering this form, you do not have to remember if the value is 9/5 or 5/9, or to add or subtract 32. This formula is determined by the method of interpolation.

kelvin

degrees Rankine

There are numerous other temperature scales, and two are worth mentioning: **kelvin** [K] and **degrees Rankine** [°R]. The kelvin scale is named for First Baron William Thomson Kelvin (1824–1907), an English mathematician and physicist. Kelvin first proposed the idea of "infinite cold," or absolute zero, in 1848, using the Celsius scale for comparison. The Rankine scale is named for William J. M. Rankine (1820–1872), a Scottish engineer and physicist, who proposed an analogy to the kelvin scale, using the Fahrenheit scale for comparison. Both men made significant contributions to the field of thermodynamics.

IMPORTANT CONCEPT:

Absolute temperature scales are never negative.

The kelvin and Rankine scales are "absolute," which means that at absolute zero, the temperature at which molecules have minimum possible motion, the temperature is zero. *Absolute temperature scales therefore have no negative values.*

In the kelvin scale, the degree sign is not used; it is simply referred to as "kelvin," not "degrees kelvin." It is the base SI unit for temperature and the most frequently used temperature unit in the scientific community.

Example 7-18

The hottest temperature in the United States ever recorded by the National Weather Service, 56.7 degrees Celsius [°C], occurred in Death Valley, California, on July 10, 1913. State this value in units of degrees Fahrenheit [°F].

Calculate Required Parameters	Steps
(1) Determine appropriate equation	$\frac{T[°F] - 32}{212 - 32} = \frac{T[°C] - 0}{100 - 0}$
(2) Insert known quantities	$\frac{T[°F] - 32}{180} = \frac{56.7 - 0}{100}$
(3) Calculate, be reasonable	$T = 134$°F

Comprehension Check 7-14

The temperature of dry ice is −109.3 degrees Fahrenheit [°F]. State this temperature in units of kelvins [K].

IMPORTANT CONCEPT:

When properties that contain temperature are converted using

$$\frac{1\,°\text{C}}{1.8\,°\text{F}} \quad \frac{1\,\text{K}}{1\,°\text{C}} \quad \frac{1\,°\text{R}}{1\,°\text{F}}$$

we read the units under consideration as "per degree Fahrenheit," with the clue being the word "per."

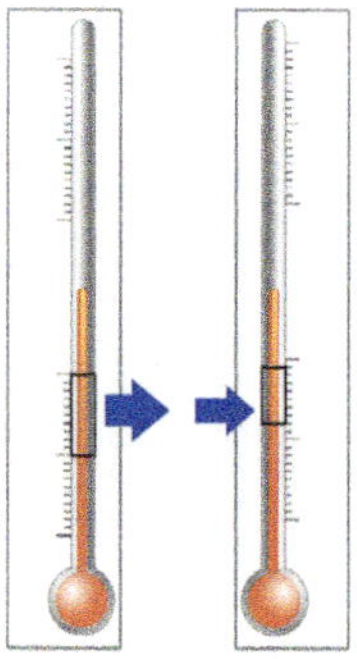

Fahrenheit Celsius

7.9.2 Conversions Involving Temperature Within a Material Property

When considering how a change in temperature affects a material property, we use a scalar conversion factor. In general, we encounter this in sets of units relating to the property of the material; for example, the units of the thermal conductivity are given by W/(m K), which is read as "watts per meter kelvin." When this is the case, we are referring to the size of the degree, not the actual temperature.

To find this relationship, remember that between the freezing point and the boiling point of pure water, the Celsius scale contains 100 divisions, whereas the Fahrenheit scale contains 180 divisions. The conversion factor between Celsius and Fahrenheit is $100°\text{C} \equiv 180°\text{F}$, or $1°\text{C} \equiv 1.8°\text{F}$.

A note of clarification about the term "per" when reading the sentence: "Gravity on earth is commonly assumed to be 9.8 meters per second squared," there is often little confusion in translating the words to symbols: $g = 9.8\ \text{m/s}^2$. For a more complex unit, however, this can present a challenge. For example, the sentence "The thermal conductivity of aluminum is 237 calories per hour meter degree Celsius," can be confusing because it can be interpreted as:

$$k = 237\,\frac{\text{cal}}{\text{h m °C}} \quad \text{or} \quad k = 237\,\frac{\text{cal}}{\text{h}}(\text{m °C})$$

$$\text{or} \quad k = 237\,\frac{\text{cal}}{\text{h m}}\,°\text{C}$$

Officially, according to SI rules, *when writing out unit names anything following the word "per" appears in the denominator of the expression.* This implies the first example listed is correct.

Example 7-19

The specific heat (C_p) is a measure of the ability of an object to store heat. Specific heat is a material property, and values are available in technical literature. The specific heat of copper is $0.385\ \text{J/(g °C)}$, which is read as "joules per gram degree Celsius." Convert this to units of $\text{J/(lb}_m\ °\text{F)}$, which reads "joules per pound-mass degree Fahrenheit."

NOTE:

Specific heat [J/(g K)]
Air = 1.012
Aluminum = 0.897
Copper = 0.385
Helium = 5.1932
Lead = 0.127
Water = 4.184

Convert to Desired Units	Steps
(1) Term to be converted	$0.385\,\frac{\text{J}}{\text{g °C}}$
(2) Conversion formula	
(3) Make a fraction (4) Multiply	$\frac{0.385\ \text{J}}{\text{g °C}}\left\lvert\frac{1000\ \text{g}}{2.205\ \text{lb}_m}\right\rvert\frac{1\ °\text{C}}{1.8\ °\text{F}}$
(5) Cancel, calculate, be reasonable	$97\,\frac{\text{J}}{\text{lb}_m\ °\text{F}}$

Comprehension Check 7-15

The specific heat of copper is 0.09 British thermal units per pound-mass degree Fahrenheit [$\text{BTU/(lb}_m\ °\text{F)}$]. Convert into units of British thermal units per kilogram kelvin [BTU/(kg K)].

7.10 Energy 201

7.10 Energy

energy

Energy is an abstract quantity with several definitions, depending on the form of energy being discussed. You may be familiar with some of the following types of energy.

7.10.1 Types of Energy

work · potential energy · kinetic energy · energy of translational motion · thermal energy

- **Work** (W) is energy expended by exertion of a force (F) over a change in distance (Δx). As an example, you exert a force on (push) a heavy desk so that it slides across the floor.
- **Potential energy** (PE) is a form of work done by moving a weight (w) over a change in a vertical distance (ΔH). Recall that weight is mass (m) times gravity (g). Note that this is a special case of the work equation, where force is weight and distance is height.
- **Kinetic energy** (KE) is a form of energy possessed by an object in motion. If a constant force is exerted on a body, then by $F = ma$, we see that the body experiences a constant acceleration, meaning the velocity increases linearly with time. Since the velocity increases as long as the force is maintained, work is being done on the object. Another way of saying this is that the object upon which the force is applied acquires kinetic energy by a change in velocity, also called **energy of translational motion**. For a nonrotating body moving with some velocity (v) the kinetic energy can be calculated by $KE_T = (^1/_2)mv^2$, assuming the body was initially at rest ($v_i = 0$).
- **Thermal energy** or heat (Q) is energy associated with a change in temperature (ΔT). It is a function of the mass of the object (m) and the specific heat (C_p), which is a property of the material.

IMPORTANT CONCEPT:

Work
$W = F\Delta x$
Potential Energy
$PE = mg\Delta H$
Kinetic Energy
$KE_T = \frac{1}{2}m(v_f^2 - v_i^2)$
Thermal Energy
$Q = mC_p\Delta T$

Where Does $KE = ½mv^2$ Come From?

If a constant force is applied to a body,

- That body will have a constant acceleration (remember $F = m\,a$).
- Its average velocity is the average of its initial and final values.

Average velocity	$v = (v_f + v_i)/2$
Distance traveled	$d = v\,t$
Work done	$W = F\,d$
	$= (m\,a)d$
	$= m\,a\,v\,t$

Acceleration is the change in velocity over time, or $a = (v_f - v_i)/t$
Substituting for a and v in the work equation

$$W = m\,a\,v\,t = m[(v_f - v_i)/t]\,[(v_f + v_i)/2][t]$$
$$W = \left(\tfrac{1}{2}\right)m\left(v_f^2 - v_i^2\right)$$

This is given the name *kinetic energy*, or the energy of motion. This expression is for the *translation* of a body only.

ENGAGE

Visit the InStepNanopower.com website with your students. Read about the first practical footwear embedded energy harvester and watch the video about the product.

URL

http://instepnanopower.com/

Ask your students how this simple concept could be "scaled-up" to create larger energy harvesting systems. *(Examples: Speed bumps on a road or floors that harvest energy with every step someone takes on them.)*

QUESTION

Name three *renewable* and three *nonrenewable* sources of energy. *(Answer: Renewable energy sources include human and animal muscle power, water, wind, geothermal energy, and solar energy Nonrenewable energy sources include oil, gas, coal, and nuclear fission.)*

QUESTION

Give five examples of *synthetic* materials. *(Answer: Nylon, rayon, polyurethane, synthetic diamonds, PVC, neoprene.)*

QUESTION

What are the seven main sources of energy?

Answer:

- *Human and animal muscle power—from food*
- *Solar energy—from the sun*
- *Wind energy—from the airflow across Earth's surface*
- *Gravitational energy—from tides and falling water*
- *Geothermal energy—from heat inside the Earth*
- *Chemical energy—from fossil fuels (coal, oil, gas) and wood*
- *Nuclear energy—from the conversion of radioactive matter into energy*

ENGAGE

Explain the difference between *potential* energy and *kinetic* energy. *Energy that is moving is called kinetic energy. Energy that is stored is called potential energy.*

7.10.2 Calories and BTUs and Joules—Oh My!

joule

ENERGY:

SI unit of energy = joule [J]

The SI unit of energy is **joule**, defined as one newton of force acting over a distance of one meter (Table 7-13). The unit is named after James Joule (1818–1889), an English physicist responsible for several theories involving energy, including the definition of the mechanical equivalent of heat and Joule's law, which describes the amount of electrical energy converted to heat by a resistor (an electrical component) when an electric current flows through it. In some mechanical systems, work is described in units of foot pound-force [ft lb_f].

British thermal unit

NOTE:

1 BTU is ≈ 1000 times as big as 1 joule.

1 calorie is ≈ 4 times as big as 1 joule.

For energy in the form of heat, units are typically reported as British thermal units and calories instead of joules. A **British thermal unit** [BTU] is the amount of heat required to raise the temperature of one pound-mass of water by one degree Fahrenheit. A calorie [cal] is the amount of heat required to raise the temperature of one gram of water by one degree Celsius.

Table 7-13 Dimensions of Energy

Quantity	Common Units	Exponents						
		M	L	T	Θ	N	J	I
Energy	J	1	2	−2	0	0	0	0

Example 7-20

A 50-kilogram [kg] load is raised vertically a distance of 5 meters [m] by an electric motor. How much energy in units of joules [J] was required to raise the load?

First, we must determine the type of energy. The parameters we are discussing include mass (kilograms) and height (meters). Examining the energy formulas given above, the equation for potential energy fits. Also, the words "load is raised vertically a distance" fit with our understanding of potential energy.

Calculate Required Parameters	Steps
(1) Determine appropriate equation	$PE = m\,g\,\Delta H$
(2) Insert known quantities	$PE = \dfrac{50\ \text{kg}}{} \bigg\vert \dfrac{9.8\ \text{m}}{\text{s}^2} \bigg\vert \dfrac{5\ \text{m}}{}$
(3) Calculate, be reasonable	$PE = 2450\ \dfrac{\text{kg m}^2}{\text{s}^2}$

This is apparently our final answer, but the units are puzzling. If the units of energy are joules and if this is a valid equation, then our final result for energy should be joules. If we consider the dimensions of energy:

Quantity	Common Units	Exponents		
		M	L	T
Energy	J	1	2	−2

A unit of energy has dimensions $E \{=\} M\ L^2/T^2$, which in terms of base SI units would be $E\ [=]\ kg\ m^2/s^2$. As this term occurs frequently, it is given the special name "joule." Anytime we see this term ($kg\ m^2/s^2$), we know we are dealing with an energy, equal to a joule.

(3) Calculate, be reasonable	$PE = 2450\ \dfrac{\text{kg m}^2}{\text{s}^2} \bigg\vert \dfrac{1\ \text{J}}{1\ \frac{\text{kg m}^2}{\text{s}^2}} = 2450\ \text{J}$

Example 7-21

In the morning, you like to drink your coffee at a temperature of exactly 70 degrees Celsius [°C]. The mass of the coffee in your mug is 470 grams [g]. To make your coffee, you had to raise the temperature of the water by 30 degrees Celsius [°C]. How much energy in units of British thermal units [BTU] did it take to heat your coffee? The specific heat of water is 4.18 joules per gram degree Celsius [J/(g °C)].

First, you must determine the type of energy we are using. The parameters discussed include mass, temperature, and specific heat. Examining the energy formulas given above, the equation for thermal energy fits. Also, the words "How much energy . . . did it take to heat your coffee" fit with an understanding of thermal energy.

Calculate Required Parameters	Steps
(1) Determine appropriate equation	$Q = m\,C_p \Delta T$
(2) Insert known quantities	$Q = \frac{470\ \text{g}}{} \Big\vert \frac{4.18\ \text{J}}{\text{g}\ °\text{C}} \Big\vert \frac{30\ °\text{C}}{}$
(3) Calculate	$Q = 59{,}370$ J

Convert to Desired Units	Steps
(1) Term to be converted	59,370 J
(2) Conversion formula (3) Make a fraction (4) Multiply	$\frac{59{,}370\ \text{J}}{} \Big\vert \frac{9.48 \times 10^{-4}\ \text{BTU}}{1\ \text{J}}$
(5) Cancel, calculate, be reasonable	56 BTU

Comprehension Check 7-16

You push an automobile with a constant force of 20 pound-force [lb_f] until 1500 joules [J] of energy has been added to the car. How far did the car travel in units of meters [m] during this time? You may assume that frictional losses are negligible.

Comprehension Check 7-17

One gram [g] of material A is heated until the temperature rises by 10 kelvins [K]. If the same amount of heat is applied to one gram [g] of material B, what is the temperature rise of material B in units of kelvins [K]?

The specific heat (C_p) of material A = 4 joules per gram kelvin [J/(g K)]

The specific heat (C_p) of material B = 2 joules per gram kelvin [J/(g K)]

7.11 Power

power
watt
horsepower

Power is defined as energy per time (Table 7-14). The SI unit of power is **watt**, named after James Watt (1736–1819), a Scottish mathematician and engineer whose improvements to the steam engine were important to the Industrial Revolution. He is responsible for the definition of **horsepower** [hp], a unit of power originally used to quantify how the steam engine could replace the work done by a horse.

ENGAGE

Visit the Bozeman.com website with your students. Watch the video entitled *Energy, Work & Power.*

URL

https://www.bozemanscience.com/energy-work-power

QUESTION

How much power can the human body produce? *(Answer: A trained cyclist can produce about 400 watts of mechanical power for an hour or more, but adults of good average fitness average between 50 and 150 watts for an hour of vigorous exercise. A healthy well-fed laborer over the course of an 8-hour work shift can sustain an average output of about 75 watts. The yield of electric power is decreased by the inefficiency of the human-powered generator, since all real generators will incur some losses during the energy conversion process.)*

URL

https://en.wikipedia.org/wiki/Human_power

NOTE:
Power is the RATE at which energy is delivered over time.

IMPORTANT CONCEPT:
$P = E/t$
SI unit of power = watt [W]

Table 7-14 Dimensions of Power

Quantity	Common Units	Exponents						
		M	L	T	Θ	N	J	I
Power	W	1	2	−3	0	0	0	0

To help understand the relationship between energy and power, imagine the following. Your 1000-kilogram car has run out of gas on a level road. There is a gas station not far ahead, so you decide to push the car to the gas station. Assume that you intend to accelerate the car up to a speed of one meter per second (about 2.2 miles per hour), and then continue pushing at that speed until you reach the station. Ask yourself the following questions:

- Can I accelerate the car to one meter per second in one minute?
- On the other hand, can I accelerate it to one meter per second in one second?

Most of you would probably answer "yes" to the first and "no" to the second, but why? Well, personal experience! But that is not really an explanation. Since the change in kinetic energy is the same in each case, to accelerate the car in one second, your body would have to generate energy at a rate 60 times greater than the rate required if you accelerated it in one minute. The key word is *rate*, or how much energy your body can produce per second. If you do the calculations, you will find that for the one-minute scenario, your body would have to produce about 1/90 horsepower, which seems quite reasonable. On the other hand, if you try to accomplish the same acceleration in one second, you would need to generate 2/3 horsepower. Are you two-thirds as powerful as a horse?

As another example, assume that you attend a class on the third floor of the engineering building. When you are on time, you take 2 minutes to climb to the third floor. On the other hand, when you are late for class, you run up the three flights in 30 seconds. In which case do you do the most work (expend the most energy)? In which case do you generate the most power?

Example 7-22

A 50-kilogram [kg] load is raised vertically a distance of 5 meters [m] by an electric motor in 60 seconds [s]. How much power in units of watts [W] does the motor use, assuming no energy is lost in the process?

This problem was started in Example 7-20. The energy used by the system was found to be 2450 joules [J], the analysis of which is not repeated here.

Calculate Required Parameters	Steps
(1) Determine appropriate equation	$P = \frac{E}{t}$
(2) Insert known quantities	$P = \frac{2450\ \text{J}}{60\ \text{s}} \left| \frac{1\ \text{W}}{1\ \frac{\text{J}}{\text{s}}} \right.$
(3) Calculate, be reasonable	$P = 41\ \text{W}$

Note that since power = energy/time, energy = power × time. We pay the electric company for energy calculated this way as kilowatt-hours [kW h]. If power is constant, we can obtain the total energy involved simply by multiplying the power by the duration of time that power is applied. If power is *not* constant, we would usually use calculus to determine the total energy, but that solution is beyond the scope of this book.

Comprehension Check 7-18

A motor with a power of 100 watts [W] is connected to a flywheel. How long, in units of hours [h], must the motor operate to transfer 300,000 joules [J] to the flywheel?

7.12 Efficiency

efficiency

IMPORTANT CONCEPT:

Efficiency is always less than 100%.

input

output

Efficiency (η, Greek letter eta) is a measure of how much energy or power is lost in a process. In a perfect world, efficiency would always be 100%. All energy put into a process would be recovered and used to accomplish the desired task. We know that this can never happen, *so efficiency is always less than 100%.* If a machine operates at 75% efficiency, 25% of the energy put into the process is lost.

The use of the terms "input" and "output" requires some explanation. The **input** is the quantity of energy or power supplied to the mechanism from some source that the mechanism requires to operate. The **output** is the amount of energy or power that is actually applied to a task by the mechanism. Note that the rated power or power listed on the device label refers to the input power—the power needed to operate the device. In an ideal, 100% efficient system, the input and output would be equivalent. In an inefficient system (the real world), the input is equivalent to the sum of the output and the power or energy lost.

IMPORTANT CONCEPT:

$\eta = P_{out} / P_{in}$

$P_{in} = P_{out} + P_{lost}$

P_{in} = quantity required by mechanism to operate

P_{out} = quantity actually applied to task

P_{lost} = quantity wasted during the application

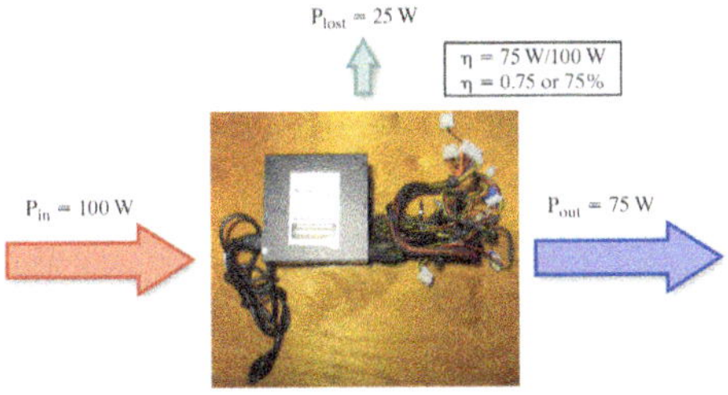

Photo courtesy of E Stephan

Orders of Magnitude

Table 7-15 gives you an idea of orders of magnitude of power and energy as related to real-world objects and phenomena. All values are approximate and, in most cases, have been rounded to only one significant figure. Thus, if you actually do the calculations from power to energy, you will find discrepancies. A few things of possible interest:

- U.S. power consumption (all types) is one-fifth of the total world power consumption.
- The Tsar Bomba generated 1.5% of the power of the sun, but only lasted 40 nanoseconds.
- Total human power consumption on the planet is about 0.01% (1/10,000) of the total power received from the sun.

ENGAGE

Visit the U.S. Department of Energy's website. Explore the Zero Energy Ready Home link and take the "DOE Tour of Zero."

URL
http://energy.gov/eere/buildings/doe-tour-zero

Ask your students what elements are common to the homes highlighted on the pages. *(Examples: High-performance insulation system, draft protection, high-efficiency comfort system, high-efficiency appliances, and advanced lighting technology.)*

INTEGRATE CONTENT AREAS

Orders of magnitude can be tied to the concept of a logarithm. The logarithm is the inverse to exponentiation and allows us to make sense of quantities that span large orders of magnitude. A common example is the loudness of sound, which is measured in decibels (dB). The following URL provides a list of noise levels for your students to peruse.

URL
https://music.eku.edu/sites/music.eku.edu/files/ekuhealthandsafety.pdf

You may also introduce *logarithmic plotting* if deemed applicable

ENGAGE

Visit the U.S. Department of Energy's Solar Decathlon website. Navigate to the Challenges link and read the guidelines for the latest challenge.

URL
https://www.solardecathlon.gov/

Ask your students who is eligible *(college teams)*, how long of a commitment it requires *(2 years)*, and what the basic requirements are for the sun-powered energy-efficient homes:

- *Is affordable, attractive, and easy to live in*
- *Maintains comfortable and healthy indoor environmental conditions*
- *Supplies energy for cooking, cleaning, entertainment, and commuting*
- *Provides adequate hot water*
- *Produces as much or more energy than it consumes*

Table 7-15 Order of Magnitude for Power and Energy Comparison

Power	"Device"	Energy per Hour	Energy per Year
10 fW	Minimum reception power for cell phone	40 pJ	300 nJ
	Single human cell	4 nJ	30 mJ
10 mW	DVD laser	40 J	300 kJ
500 mW	Cell phone microprocessor	2 kJ	15 MJ
750 W	Power per square meter bright sunshine	3 MJ	25 GJ
20 kW	Average U.S. home	80 MJ	600 GJ
100 kW	Typical automobile	400 MJ	3 TJ
150 MW	Boeing 747 jet	500 GJ	5 PJ
1 GW	Large commercial nuclear reactor	4 TJ	30 PJ
20 GW	Three Gorges hydroelectric dam (China)	80 TJ	600 PJ
4 TW	U.S. total power consumption	15 PJ	100 EJ
20 TW	Total human power consumption	80 PJ	600 EJ
100 TW	Average hurricane	400 PJ	(NA)
200 PW	Total power received on the Earth from sun	1 ZJ	6 YJ
400 YW	Total power of the sun	10^6 YJ	10^9 YJ

Example 7-23

A standard incandescent light bulb has an efficiency of about 5%; thus, $\eta = 0.05$. An incandescent bulb works by heating a wire (the filament) inside the bulb to such a high temperature that it glows white.

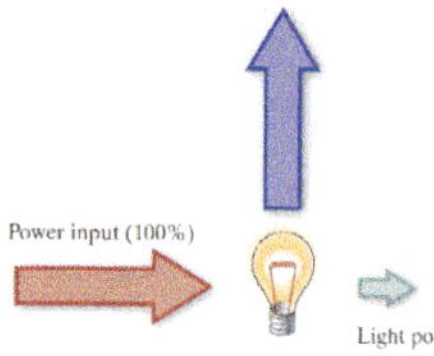

If a 100-watt [W] bulb is turned on for 15 minutes [min], how much energy, in units of joules [J], is "lost" as heat during the 15-minute [min] period?

Calculate Required Parameters	Steps
(1) Determine appropriate equation	$E = P\,t$
(2) Insert known quantities	$E = (P)(15\ \text{min})$
(1) Determine appropriate equation	$P_{in} = P_{out}/\eta$ $P_{lost} = P_{in} - P_{out}$
(2) Insert known quantities	$100\ \text{W} = P_{out}/0.05$
(3) Calculate	$P_{out} = 5\ \text{W}$ $P_{lost} = 100\ \text{W} - 5\ \text{W} = 95\ \text{W}$
(2) Insert known quantities	$E = (95\ \text{W})(15\ \text{min})$
(3) Calculate, be reasonable	$E = 95\ \text{W} \left\lvert \frac{15\ \text{min}}{} \right\rvert \frac{60\ \text{s}}{1\ \text{min}} \left\lvert \frac{1\ \text{J}}{(1\ \text{W s})} \right. = 85{,}500\ \text{J}$

Example 7-24

Over the past few decades, the efficiency of solar cells has risen from about 10% to the most recent technologies achieving about 40% conversion of solar energy to electricity. The losses are due to several factors, including reflectance and resistive losses, among others.

Assume you have an array of solar cells mounted on your roof with an efficiency of 28%. If the array is delivering 750 watts [W] of electricity to your home, how much solar power, in units of watts [W], is falling on the photoelectric cells?

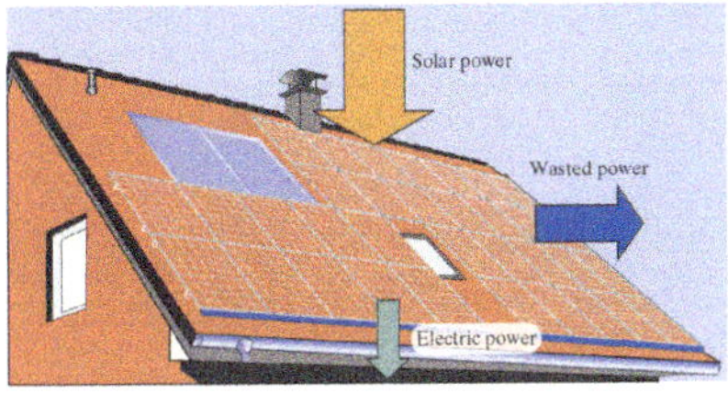

NOTE:

How much power is wasted by the array of solar cells? If the array received 2680 W and delivered 750 W, then the difference is 2680- 750 = 1930 W; thus, 1930 W are wasted.

Calculate Required Parameters	Steps
(1) Determine appropriate equation	$P_{in} = P_{out}/\eta$
(2) Insert known quantities	$P_{in} = 750\ W/0.28$
(3) Calculate, be reasonable	$P_{in} = 2680\ W$

Example 7-25

If your microwave takes 2 minutes [min] to heat your coffee in Example 8-23, how many watts [W] of input power does your microwave require, assuming that it is 80% efficient? Remember that our answer was 59,370 joules [J], before we converted the final answer to units of British thermal units [BTU].

Calculate Required Parameters	Steps
(1) Determine appropriate equation	$P_{in} = P_{out}/\eta$
(2) Insert known quantities	$P_{in} = P_{out}/0.8$
(1) Determine appropriate equation	$P_{out} = E/t$
(2) Insert known quantities	$P_{out} = \frac{59{,}370\ J}{2\ min} \left\vert \frac{1\ min}{60\ s} \right\vert \frac{1\ W\ s}{1\ J}$
(3) Calculate	$P_{out} = 494.75\ W$
(2) Insert known quantities	$P_{in} = 494.75\ W/0.8$
(3) Calculate, be reasonable	$P_{in} = 618.4\ W$

Comprehension Check 7-19

A motor with an input power of 100 watts [W] is connected to a flywheel. How long, in units of hours [h], must the motor operate to transfer 300,000 joules [J] to the flywheel, assuming the process is 80% efficient?

ENGAGE

Visit the U.S. Department of Energy's Office of Energy Efficiency & Renewable Energy website. Ask students to read the Federal Agency Use of Renewable Electric Energy page.

URL
https://www.energy.gov/eere/femp/federal-agency-use-renewable-electric-energy

TEACHING TIP

This section is designed to address the Precision Exam standards. You may elect to skip this section if it's outside the scope of your class. The project itself will still be applicable.

TEACHING TIP

To emphasize the different ways trusses can be put together, it would be worth showing the class a series of bridge pictures and having them sketch the pattern of the truss for themselves.

Comprehension Check 7-20

If a 50-kilogram [kg] load was raised 5 meters [m] in 50 seconds [s], determine the minimum rated wattage [W] of the motor needed to accomplish this task, assuming the motor is 80% efficient.

Example 7-26

A simple two-stage machine is shown in the diagram that follows. Initially, an electric motor receives power from the power grid, accessed by being plugged into a standard electrical wall socket. The power received by the motor from the wall socket is the "input" power or the power the motor uses or requires (Point A).

The spinning drive shaft on the motor can then be used to power other devices, such as a hoist or a vacuum cleaner or a DVD drive; the power available from the spinning shaft is the "output" power of the motor (Point C). In the process of making the drive shaft spin, however, some of the input power is lost (Point B) because of friction.

The power available from the spinning shaft of the motor (Point C) is then used to operate some device: the hoist or vacuum cleaner or DVD drive (Point E). In other words, the output power of the motor is the input power to the device it drives. This device will have its own efficiency, thus wasting some of the power supplied to it by the motor (Point D).

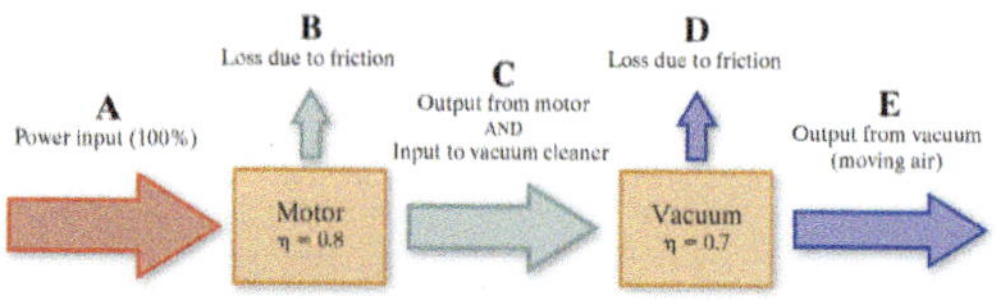

NOTE:

The overall efficiency of two linked devices is the product of the two efficiencies.

Assume that a motor with an efficiency of 80% is used to power a vacuum cleaner that has an efficiency of 70%. If the input power to the motor (Point A) is one-half horsepower [hp], what is the output power of the vacuum (Point E)?

The output from the motor (Point C) is $P_C = \eta P_A = 0.8\ (0.5\ hp) = 0.4\ hp$. This is the input to the vacuum, so the output from the vacuum (Point E) is

$$P_E = \eta P_C = 0.7(0.4\ hp) = 0.28\ hp$$

What is the overall efficiency of this machine?

The input power (Point A) is 0.5 hp, and the output power (Point E) is 0.28 hp; thus, the efficiency of this linked system is:

$$\eta = P_A/P_E = 0.28\ hp/0.5\ hp = 0.56, \text{ or an efficiency of } 56\%$$

7.13 Using Forces in Trusses

truss
members

A common structure you might see across contexts like roofing and on bridges is a truss. A **truss** is composed of long slender objects called **members**, which could be made of several different types of materials like woods and metals. These structures are used to provide support for a load, such as vehicles traveling over the Ohio River or the shingles on the roof of your school. As you might imagine, the members pinned together that form the truss have been studied extensively to understand how we might best arrange them geometrically and which materials to use to piece together the structure. Here, we will discuss simple trusses, which are composed of triangular elements that are pinned together at their joints.

In order to analyze a truss, we'll need to leverage what we know about forces. There are two assumptions that need to be made. *First, all forces are applied at the joints, so we ignore the weight of the individual*

tension
compression
tensile force
compressive force

members. We can make this assumption because it is most often the case that the weight of the members pales in comparison to load that is being supported at the joints. *Second, smooth pins are used to attach the members to one another*. In practice, you'll often see the members welded or bolted to a common surface, called a gusset plate. Accordingly, the members can be **tension** or **compression**. For a member that is tension, there is a **tensile force** that is pulling the member in such a way that it would elongate. On the other hand, members not being pulled will be compressed by a **compressive force**.

When analyzing a truss (Figure 7-1), it is crucial to note which forces are tensile and which are compressive. For instance, if a member is going to be compressed, it is wise to ensure it is not too thin. Otherwise, the member could easily buckle under the load.

Figure 7-1 Example of a truss.

7.13.1 Resolving Forces

method of joints

In order to determine the forces within the members composing a truss, one method is called the **method of joints**. In the method of joints, we turn our attention to the pins holding the members together. Based on our first assumption, that is where the loads are. Because the truss isn't moving, it is said to be in equilibrium. Using Newton's second law, a particle of mass m that is acted upon by a force F will accelerate at a rate of a in the same direction of the force. This law is often written as:

$$F = ma$$

Because the truss is in equilibrium, the truss isn't accelerating. Therefore, the total force is zero. This must mean that the members in the truss are arranged in such a way that all the forces cancel one another out. This is often stated using the following equation:

$$\sum F = 0$$

where the Σ (the Greek letter sigma) stands for repeated addition, which could be translated into the phrase, "the sum of." Thus, the mathematical sentence above would read, *the sum of the forces (in the truss) is equal to zero*. In these simple trusses, we'll be working in two dimensions. Because the trusses are assembled in two dimensions, the forces can point in any direction—360 degrees. To properly analyze the truss, we'll need to decompose forces into two components, a horizontal component and a vertical component. To obtain their associated values, we need some simple trigonometry.

resultant force
sine
cosine

Consider a force that is 100 Newtons [N] that points 60 degrees above the horizontal. This force is equivalent to two forces that are pulling entirely in the horizontal and vertical directions. The force that is the combination of the two purely horizontal and vertical forces is called the **resultant force**. In order to find those horizontal and vertical forces, we use the trigonometric functions **sine** and **cosine**. These mathematical functions are used to model periodic phenomena like sound and light waves but are needed here because of their properties relating to triangles. If we draw in an imaginary line from the tip of the vector down to the horizontal, you'll see we just created a triangle. The horizontal and the line we drew are the legs of the triangle; the horizontal is the leg *adjacent* to the 60-degree angle and the new line is the leg *opposite* of the angle. The force itself is the hypotenuse of the triangle.

Sine is the function associated with the leg *opposite* of the angle and cosine is associated with the leg *adjacent* to the angle. Here, we use the Greek letter theta θ to represent the angle. Using sine and cosine, we can write the formula for the x component of the force (F_x) which acts horizontally and the y component of

TEACHING TIP

It is important to distinguish whether a member is in compression or tension. So, emphasize that all forces in a truss must be accompanied by a T for Tension or C for Compression when solving truss problems.

TEACHING TIP

One of the classes that nearly all engineers take is called Statics. The essence of the class involves quantifying forces acting upon solid objects which are in *equilibrium*. The fundamental equation describing the sums of forces as zero describes a specific state, where the object is not accelerating. This very simple physical relationship allows us to determine all of the forces acting upon the object, often using basic trigonometry and algebra.

ENGAGE

What is the y component of a 500 N force with an angle of 180 degrees from the horizontal? *There is no y component because sin(180 deg) = 0. Therefore, the y component is 500sin(180 deg) = 0.*

INTEGRATE CONTENT AREAS

At this point, it might be beneficial to review the Unit Circle with students to ensure they are properly drawing the triangle associated with a given angle measurement from the x axis.

You can use the following resultant force calculator to provide a few examples of decomposing a force into its x and y components.

URL
https://calculator.academy/resultant-force-calculator/

ENGAGE

To emphasize how changing the point of rotation impacts the value of the moment, consider placing a point, X above the force B. The distance above B doesn't matter.

Now A is 2m away from X and A is pushing the beam counterclockwise. So now the moment from A about X is NEGATIVE 200 N-m (because it is now rotating in the opposite direction).

The moment from B about X is ZERO because the force lies in a straight line with X vertically or horizontally. You may consider having students hold a textbook between their hands and have a partner apply force vertically above and horizontally to the side of their hands (standing in for a pin connection). Have them apply a force outside of those straight lines and note how the book obviously rotates.

The force C is now 3m away from X and is pushing the beam in the same direction as it would have in the O case. The moment from C about X is POSITIVE 300 N-m.

the force (F_y) which acts vertically.

$$F_x = F\cos(\theta),\qquad F_y = F\sin(\theta)$$

In the context of our example, this means the force can be decomposed into horizontal and vertical forces by plugging in the magnitude of the force and the angle into the appropriate trigonometric function:

$$F_x = (100)\cos(60°) = 50\,[N],\qquad F_y = (100)\sin(60°) = 50\sqrt{3} = 86.6\,[N]$$

If we want the resultant force back, all we do is apply the Pythagorean Theorem. The triangle we made to decompose the main force into its horizontal and vertical components is a right triangle, after all.

$$F = \sqrt{F_x^2 + F_y^2}$$

For example, to verify it works (after rounding):

$$F = \sqrt{(50)^2 + (86.6)^2} = \sqrt{2500 + 7499.56} = \sqrt{9999.59} \sim 100\,[N]$$

Because there are two directions we need to care about when doing the analyses, the equation that represents equilibrium should be split into the horizontal and vertical directions as well.

$$\sum F_x = 0,\qquad \sum F_y = 0$$

These equations are at the heart of the method of joints.

7.13.2 Moments

In our free body diagram, we should identify places where there is a tendency for the object to rotate with respect to a certain point—especially if we need to solve for it. A moment is always with respect to *something*, most often a point or axis. When you turn a doorknob, for example, you are creating a moment by exerting a force in a twisting motion that creates torque. The moment is with respect to the center of the doorknob.

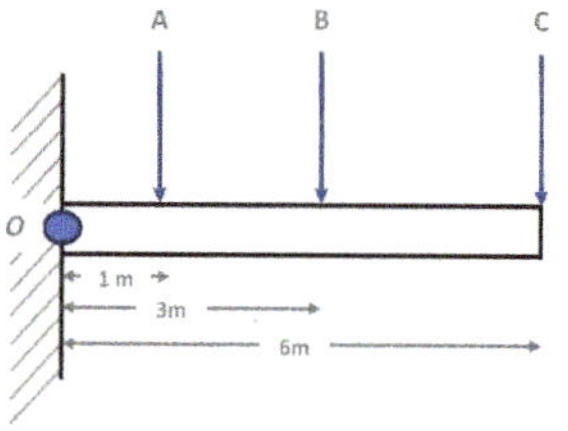

Figure 7-2 Example of moments at various distances.

A moment is simply the force multiplied by the *perpendicular* distance to that point. It is not necessarily the shortest distance to the point. For example, consider the three forces labeled A, B, and C in Figure 7-2. In order to determine the moment about the beams connection to the wall (a point labeled O), all we need is the distance from the wall to the force d and the magnitude of the force F. Therefore, the moment M would be:

$$M = Fd$$

INTEGRATE CONTENT AREAS

For a more complicated method of joint problems, it is necessary to use ideas from linear algebra to solve the system of equations. When the sum of the forces is not enough, taking a moment or two in strategic locations can provide the necessary constraints to find a unique solution to the problem. You may choose to discuss systems of equations. For example:

- If the number of variables is more than the number of equations, then the system is under-determined (no solutions or infinitely many).
- If there are more equations than variables, then the system is over-determined (almost always no solution).

We're looking for a case where the number of variables equals the number of equations. Any new equation we find using forces or moments should contain redundant information—meaning after we get an equal number of equations and unknowns, the new equation should confirm what we already knew about the unknown values. It's easy to forget about moments when solving truss problems, so remind students that they should consider how a moment around a convenient point may give them the additional equation they need.

In our example, say our forces were 100 N each. As we get further and further away from *O*, the magnitude of the moment increases. For A, the moment is 100 N-m (Newton meters). At B, the moment is 300 N-m, three times as much as the moment at A. Likewise the moment caused by C is twice as big as B. This observation is the same concept as why it is much easier to open a pickle jar when you use your whole arm instead of just your wrist—you get more leverage, or more precisely, torque.

Just like the two force component equations, because the truss is in equilibrium, there should also be no rotations occurring. Thus, we have one more equation at our disposal.

$$\sum M_p = 0$$

7.13.3 Free Body Diagramming

free body diagram

moments

reactions

To uncover all the forces in a physics problem like solving a truss, it is incredibly valuable to draw what is called a **free body diagram**. A free body diagram is a representation of a single object or subsystem composed of multiple objects that is used to examine the different forces acting on it. These forces could be external, like a pushing force on an object or heavy load sitting on the object, or internal, such as tension in a rope. The free body diagram could also include **moments**, which are measures of the tendency for an object to rotate around a specific point. Finally, there are **reactions**, which are the forces and moments that an object exerts on its supports. For example, in two dimensions, a beam attached to a fixed support, like a wall, would have two reaction forces (both horizontally and vertically) and one moment. Why? Because the fixed support is resisting movement in the horizontal and vertical directions in addition to any sort of rotations. On the other hand, something like smooth pin or hinge would only have two reaction forces, horizontal and vertical, because the hinge would resist movement in those directions but would allow rotations.

Once all the forces, reactions, and moments are identified, take note of their distances from one another. Label the points these different forces, reactions, and moments occupy so they can be referenced in the analysis. That's *all* we need. The purpose of the free body diagram is to strip away all the unnecessary detail. Although you may be able to recreate a beautiful sketch of the object in question, it will simply be noise as you try to dig through the picture for the relevant quantities that you need to solve the problem.

normal force

For example, consider the scenario in Figure 7-3. We have a beam connected to a rigid surface, a wall, that has three different things happening across it. First, we have a force being applied 1m from the wall. We also have an object with some mass *m*, which means we could find its weight, W, using the equation W = mg, where g is the gravitational constant. Whenever an object is resting on a surface, there is another force called the **normal force**, which is the force the surface the object is resting on exerts back to avoid the two solids passing through one another. The normal force is perpendicular to the surface. A similar common force is tension, which we've already mentioned in the context of the truss members but will occur in objects like ropes and strings when an object exerts a force on it, like the weight of an object.

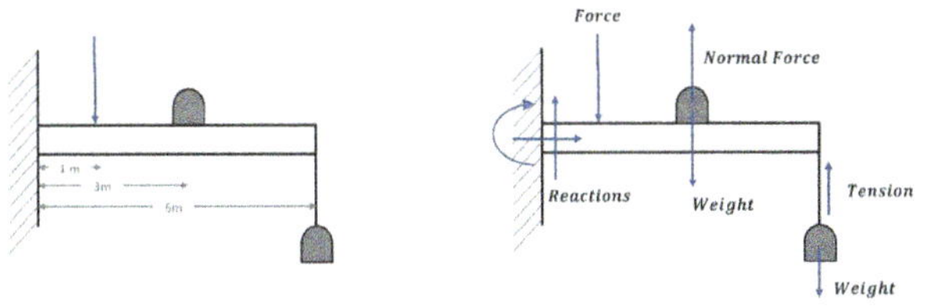

Figure 7-3 A beam connected to a wall subjected to a force, a weight, with another weight hanging off its end.

TEACHING TIP

Moments are always around a point or axis, so the sum of the moments must also be with respect to a point or axis. Be sure to emphasize this in your notation when solving problems on the board.

TEACHING TIP

If students get stuck when solving trusses, ensure they have the reactions appropriately represented in the free body diagram.

TEACHING TIP

These are not the only possible reactions, but are presented here to demonstrate how varied the potential reactions are. A good rule of thumb is to determine how the support cannot move. If it can't move a particular way, there's a force and/or moment resisting that motion.

TEACHING TIP

A big conceptual hurdle that students have issues with is establishing conventions. Emphasize to the class that as long as you are consistent with whatever coordinate system you're using, you'll come to the same conclusion. In this case, it's helpful to note the positive direction (which could be placed right by the initial equation). You could demonstrate the same equation by reversing the positive direction to the left and resolve the problem to demonstrate this point.

You will also notice the presence of reactions where the beam connects to the wall. These appear because they are the forces resisting the other forces present in the scenario. Depending on how the truss is attached to a surface, there may be one, two, or three reactions. There are several ways for truss members to be attached to a surface; a selection of these attachments is shown in Figure 7-4. In (A), we have a smooth pin which resists motion in the horizontal and vertical directions but could allow rotations. In (B), we have a rocker, which is often installed on long bridges to allow for metal to expand during hotter seasons. Because it allows horizontal motion, there is no reaction along that axis, however it will have a reaction in the direction *in which it contacts the surface*. To juxtapose the horizontal rocker, the situation in (C) is a smooth rod making direct contact with a surface at an angle of θ. Just like the rocker, it will have a single reaction that is perpendicular to the point of contact. Finally, and perhaps the most complicated, is the connection to a fixed support (D). It will have two reactions and a moment.

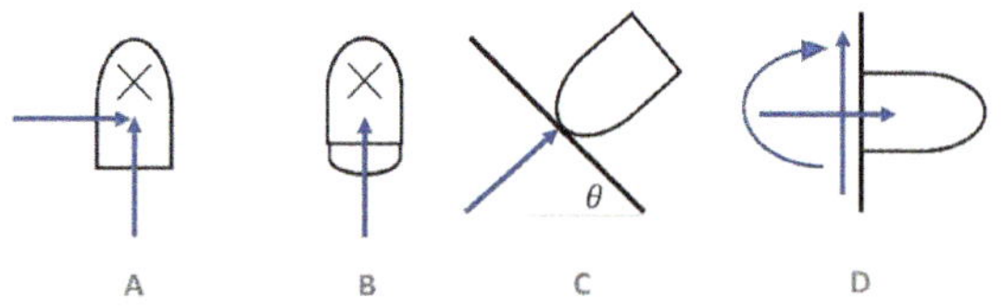

Figure 7-4 Four examples of the types of connections in trusses.

When creating a free body diagram for a truss, we will be focusing on the joints—where the members connect to one another. When we solve a truss, we will use information we find at other joints, but the method we'll outline here doesn't concern itself with what's happening at the other joints. Consider the situation in Figure 7-5. There are three members connected at a single joint labeled A. Therefore, there are three internal forces we want to draw, F_{AB}, F_{AC}, and F_D. We don't want to include any forces at B, C, or D because we will model them when we do the joint they are associated with. There are also two marked angles θ_{CD} and θ_{DB}. We will also carry these over to our free body diagram. That's it! All we need are the drawn forces and angles.

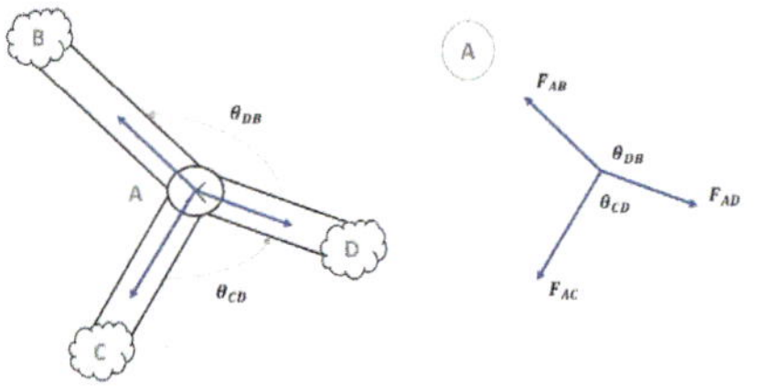

Figure 7-5 Constructing a free body diagram for a joint in a truss

7.13.4 Solving a Simple Truss Using the Method of Joints

We can examine a truss by creating a free body diagram of each joint that has *at least* one known force and *at most* two unknown forces. Why the restrictions? It's mathematical. If we have too many unknowns, there's no way to say anything definitive about an undetermined force other than expressing it in terms of forces we may or may not know. When you draw the unknown forces, draw them facing out from the pin—meaning they are in tension. The math will let us know the actual direction. If we crunch the numbers and the result is positive, then the force is indeed a tensile force. If it is negative, that's okay; all that means is that the force is compressive.

When you place the *xy* axis on the free body diagram, you're free to choose its orientation. Therefore, it is wise to pick the orientation that is easiest to decompose the forces into their horizontal and vertical pieces. From there, you can apply the sine and cosine functions to split the resultant forces into their x and y components. Let's look at the following simple truss in Figure 7-6.

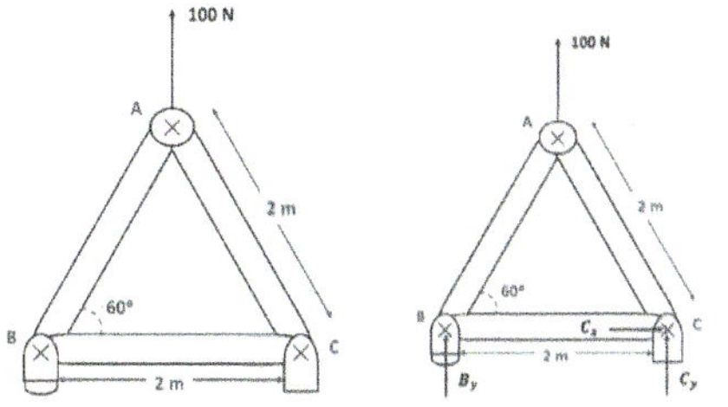

Figure 7-6 A simple truss and its support reactions.

This truss is in the form of an equilateral triangle pinned together at the top with a 100 Newton force pulling straight upward at the point A. At points B and C, the truss is connected to the ground via a rocker and smooth pin. The rocker allows the truss to rock left and right, which means there's only one vertical reaction. The smooth pin resists vertical and horizontal movement, which results in two support reactions each. When solving a truss, it is not necessary to find all the support reactions. However, it is necessary to find at least some, in order to reduce the number of unknowns at a joint. First, let's take care of the support reactions. We have five possible equations; we could add all of the forces in the x direction, all of the forces in the y direction, or find the moment about the points A, B, and C. With four unknowns and five equations, that's more than we need. Look at the forces in the x direction. Whenever we sum forces, we specify the positive direction, which is most often to the right.

$$\overset{+}{\rightarrow}\sum F_x = 0$$

The only support reaction in the x direction is C_x. This means the first equation is:

$$C_x = 0$$

Let's keep going and see what else we can learn. What about the sum of the forces in the y direction?

$$+\uparrow\sum F_y = 0$$

There are three forces in this case, the known 100 Newton force, and the two support reactions B_y and C_y. Everything is pointing in the positive direction so we can add them all up:

$$100 + B_y + C_y = 0$$
$$B_y + C_y = -100$$

Take a moment here to make a prediction. What do you think these support reactions should be, given the geometry? Next, we'll use the moment equations starting at B. Why? Because we'll only be considering the support reactions at C. Because we'll set our center of rotation at B, the forces at B can't contribute! Remember that moments are based on the perpendicular distance from the center of rotation; this means that 100 N force is 1m away from B and C_y is 2m away. Note that C_x lies on the same line as B, so it is 0m away from B perpendicularly. We'll assume that counterclockwise rotation is positive, so both the 100 N and C_y forces would cause counterclockwise rotation. Putting all of that together:

TEACHING TIP

After drawing the free body diagram, the next step is to determine the support reactions. These form the basis for iteratively solving the equations at each joint.

TEACHING TIP

Just like identifying the positive direction for the force calculations, noting positive rotation is also important to emphasize.

ENGAGE

When trying to pick a point for a moment, you should consider the joint with the most forces acting upon it to form any necessary equation to solve for the reactions. Why does that rule of thumb help you solve for unknown forces? *Because the forces at that joint will not contribute at all to the moment. They are 0 m away from the rotation point. Therefore, if you don't know three forces at a joint but need to know one force at another joint, you can effectively ignore the three forces you don't know.*

TEACHING TIP

It is helpful to make it abundantly clear that all we care about in the method of joints are the forces acting at the joint itself. When looking at joint A for instance, we only need to consider the internal forces and reactions at joint A. This greatly simplifies the analysis.

Be careful to mark the angle measurements when creating a free body diagram at a joint.

$$\overset{+}{\circlearrowleft}\sum M_B = 0$$

$$(100 \times 1) + (C_y \times 2) + (C_x \times 0) = 0$$

$$100 + 2C_y = 0$$

$$2C_y = -100$$

$$C_y = -50\ N$$

Okay, so the support reaction at C that points vertically is 50 N but pointing down. Looking back at our truss, that makes sense. The 100 N force is pulling up, so the support is resisting that force at C. Remember your prediction? Now that we know C_y, we can determine B_y from an equation we already have:

$$B_y + C_y = -100$$

$$B_y + (-50) = -100$$

$$B_y = -50\ N$$

So, B_y is the exact same. All that is left is to determine the support reactions in the horizontal direction. What might those be? If there is a force pulling straight up, would we expect any resistance horizontally at the pins? No, we wouldn't. But we can verify that using one more moment equation. Let's pick A.

$$\overset{+}{\circlearrowleft}\sum M_C = 0$$

$$(100 \times 1) + (C_y \times 2) + (C_x \times 0) = 0$$

This is enough information to begin with the method of joints. Remember that each joint should have *at least* one known force and *at most* two unknown forces. Looking at our picture, that's exactly what we have at each joint. Let's begin with the method at joint A (Figure 7-7). *Whenever we draw an unknown force, draw it facing out from the pin.* That configuration lets us immediately know that the member is in tension if the force is positive and in compression if the force is negative.

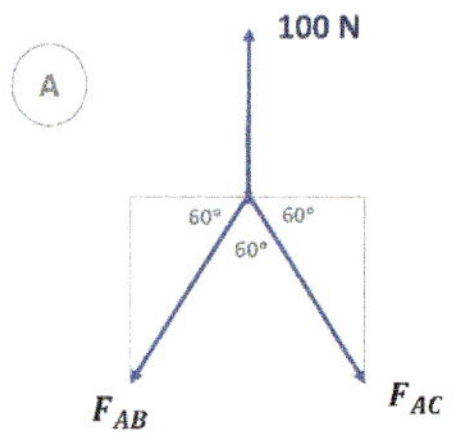

Figure 7-7 Free body diagram of joint A.

At joint A, we have the known 100 N force and the two internal forces for members connecting A to B and A to C. We call those forces F_{AB} and F_{AC}, respectively. We have two unknown forces, so we only need two equations. Let's sum the forces in the x direction. The x component of F_{AB} will point to the left, which means we need to append a negative sign.

$$\overset{+}{\rightarrow}\sum F_x = 0$$

$$-F_{AB}\cos(60°) + F_{AC}\cos(60°) = 0$$

The value of cos(60°) is 1/2, but it's common to both terms. That means we can divide it out and find that:

$$F_{AB} = F_{AC}$$

Next, we'll look at the vertical forces:

$$\overset{+}{\uparrow}\sum F_y = 0$$

$$100 - F_{AB}\sin(60°) - F_{AC}\sin(60°) = 0$$

Rearranging the values, factoring out the sin(60°), and using the fact that $F_{AB}=F_{AC}$:

$$\sin(60°)\,(F_{AB} + F_{AC}) = 100$$
$$\sin(60°)\,(F_{AB} + F_{AB}) = 100$$
$$2\sin(60°)\,F_{AB} = 100$$

Dividing both sides by 2sin(60°), we find that the internal forces for the top two members are:

$$F_{AB} = F_{AC} = \frac{100}{2\sin(60°)} = \frac{100}{\sqrt{3}} = \mathbf{57.7\ N\ (Tension)}$$

Because the forces are positive, they are in tension. This makes sense because of the scenario, a force pulling the members upward. Also, we'll keep the force's values in the form $\frac{100}{\sqrt{3}}$ to avoid rounding errors in the next step (Figure 7-8).

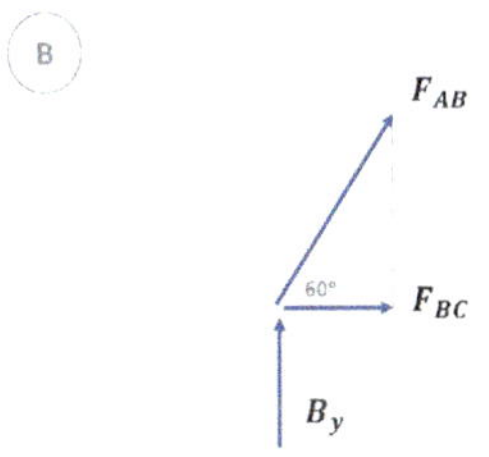

Figure 7-8 Free body diagram of joint B.

We only have one more internal force to determine, F_{BC}. Notice here that F_{AB} is pointing outward from joint B. That's because we just determined that the member from A to B is in tension. Therefore, the force at the linked joint should reflect that finding. In this situation, we have one unknown (F_{BC}) and two knowns (B_y and F_{AB}). In this case, we only need one equation, the sum of the forces in the x direction.

TEACHING TIP

As students become more comfortable with trigonometry, they won't need to rely as much on the x and y direction force equations.

DIFFERENTIATED INSTRUCTION

The method of joints is just one method of solving a truss. For more advanced students, you may consider introducing the other popular analytical approach called the *method of sections.*

- Take a picture of a bridge where a truss is prominently visible and recreate the pattern on the board.
- Draw a wavy line through several of the members to cut the truss in half, avoiding cutting more than three.
- Draw in forces just like the method of joints for one side of the truss.
- Ask students why this approach is a valid way to determine the internal forces of the truss. *Because the truss is in equilibrium, therefore, each point in the truss should also be in equilibrium.*
- Why would the method of sections be preferable to the method of joints? *You can solve for unknown forces without needing to iteratively solve in the case of method of joints.*
- How many members can you cut through at once? Why? *It depends on how many members you have already solved. As you cut through more unknown members, you'll need an additional equation to solve for the unknowns. Therefore, you should limit cutting through three unknowns (i.e., forces in the x-direction, forces in the y-direction, and the sum of moments).*

$$\overset{+}{\rightarrow}\sum F_x = 0$$

$$F_{AB}\cos(60°) + F_{BC} = 0$$

$$F_{BC} = -F_{AB}\cos(60°) = -\left(\frac{100}{\sqrt{3}}\right)\left(\frac{1}{2}\right) = -\frac{50}{\sqrt{3}} = \mathbf{28.9\ N\ (Compression)}$$

This time the force is negative, so the member is in compression. At this point, we're finished!

DESIGN ACTIVITY

Tower Building

Problem Statement

Let's put the idea of trusses into action. A tower truss is a freestanding structure that can be used as a column to support a variety of objects. You're an engineer in the entertainment industry, so the company you work for is often contracted to design elaborate sets for live musical acts. In fact, most recently, your company was approached to help the administrators at a local music festival build a stage that can support a series of speakers and lights for the various acts to use while they are performing. They believe the weight of the speakers to be between 15 and 25 kg and the lights to be between 10 and 15 kg per unit. They estimate that they would have roughly 16 speakers (8 on each side facing the audience) and 40 lights hanging from the truss above. They expect the speakers to hang approximately 6 to 8 feet above the main dancing area. Your job is to design the appropriate truss structure to handle those loads. The local music festival organizers communicate to you that they have a relatively tight budget, so it would be helpful to optimize the design as much as possible from a cost perspective.

Materials (Per Team)

- Device connected to the Internet
- Writing utensils (different colors would be helpful)
- Paper (can be 8½ × 11 or larger, such as poster size)
- 1 Protractor
- Rulers/tape measure
- 30 craft sticks
- 30 pipe cleaners
- 30 metal paper clips
- Roll of tape
- Weights that can hang from the craft sticks and pipe cleaners

Understanding Who Your Design Is For

To ensure your solution is appropriate, you need to understand the needs of the user. Conduct initial research on stage truss design and the various weights of stage lights and speakers. Record your findings in your engineering notebook and consider the following questions:

1. What is the range in speaker and stage light weight that you found online? What is the total load that the truss will have to sustain?
2. Where are speakers usually placed on trusses for concerts like what was described by the customer?
3. What is the range in height for tower trusses in the case of concert venues?

TEACHING TIP

Consider collecting all of the criteria and constraints in a shared workspace, like a Google Form or Microsoft Form. Then, display the results to the class and have a group discussion.

TEACHING TIP

Consider collecting all of the values found by the students in a shared workspace, like a Google Form or Microsoft Form. Then, display the results to the class and have a group discussion.

DIFFERENTIATED INSTRUCTION

The exact materials you have in your classroom might differ from what is suggested here. If you'd like to adjust the difficulty of this project, you can introduce further constraints on the project by...

- Limiting the amount of materials by introducing a cost to use each item. For example, craft sticks are sturdier than pipe cleaners, so the craft sticks should cost more. These can be real costs or costs you determine on your own. Depending on the cost system you devise, you can set a constraint on price from the festival organizers.
- Increase the expected weight that the tower needs to hold. You can use weights of increasing size, such as 20 g, 50 g, or 100 g. If the weights are too light, you can increase the capacity by placing textbooks on top of the tower.

TEACHING TIP

Students should be expected to assign reasonable values to design specifications at this stage.

DIFFERENTIATED INSTRUCTION

If you skipped the previous section, your students can ignore steps 3-4 in "Generating Potential Concepts..."

DIFFERENTIATED INSTRUCTION

You may consider having students work out the method of joints on the Pratt, Warren, and Gambrel trusses as part of the research phase with sample weights from the problem statement.

TEACHING TIP

Emphasize here that using the method of joints is a form of modeling to test different designs. This will be crucial for their team project.

Develop Criteria, Constraints, and Design Specifications

Now that you've taken time to understand your user's needs, next you must translate those needs into criteria, constraints, and design specifications. These will be used to guide the development of your solution and ensure it is successful. In your design notebook, address the following prompts:

1. What are the criteria you identified for this design?
2. What are the constraints limiting the possibilities of this design?
3. How will you know if your design is successful? Translate your criteria and constraints into measurable design specifications.
4. Consult with another team of students in your class and compare your criteria and constraints. What differences and similarities do you notice?
5. Develop a test plan to measure whether your design achieves the criteria you specified at the beginning of the challenge.
6. Did you revise or add design specifications during your test plan development?

Generating Potential Concepts for a Prototype

Now it is time to generate some potential concepts for your tower truss. We'll also perform some calculations to ensure the design is appropriate. In your design notebook, address the following:

1. Create a sketch for three concepts and a description that captures how the concepts would work.
2. Search online to see if that concept has been implemented by researching existing trusses for the top of the tower, such as a Pratt Truss, Warren Truss, and Gambrel Truss.
3. Use the method of joints to create a mathematical model of your three concepts. Find all internal forces in the members using the information you located in the "Understanding Who Your Design Is For" section, such as the total load from the speakers and lights.
4. What do you notice about the differences in the internal forces between your designs?
5. How can you test if your tower will be able to withstand the applied forces from the speakers and lights? Which criteria apply to this consideration?

Selecting a Concept

After generating a series of design alternatives, we now must pick a concept to move forward with. In your design notebook, address the following:

1. Using your criteria and design specifications, set up a decision matrix to begin weighing the viability of the three concepts.
2. Determine weights to place on the different criteria.
3. Calculate the "winning" design using your weights.
4. Summarize your findings.

Building a Prototype

Use the materials given to you to develop a prototype solution given the concepts you examined in the previous step.

Testing the Prototype

Because we are sticking to a conceptual design, we will assume you developed a prototype based on the concept selected from your decision matrix. How do we know it works? In your design notebook, address the following prompts:

1. How did the prototype perform?
2. Interpret the results: What worked and what didn't? What do you need to do to revise your prototype to meet the specifications?
3. Pair up with another team in the class. Compare your prototypes to one another, noting the similarities and differences in each prototype's performance.
4. Perform the method of joints on their prototype to verify their solution. Compare your answers and discuss any discrepancies.

Report Your Solution

Finally, it is time to report out to the class. Your teacher may elect to have you present a PowerPoint summarizing your processes from each stage of the design process outlined here. You may also write a short report covering the same material. In your design notebook, create a bulleted list of the main decisions you made across the different stages of this project. If you are making a PowerPoint, create an outline for how you will present the information. Next, reflect on the following questions:

1. What was the objective of this project?
2. What worked?
3. What didn't work and why didn't it work?
4. How did the design compare with the best and worst performers?
5. What would you do differently?
6. Was the objective accomplished?

TEACHING TIP

Because the students are developing a scale model of the tower with different materials, they'll need to consider a scaling factor for the dimensions and their maximum weight. Encourage them to think about how their prototype would scale given those constraints.

DIFFERENTIATED INSTRUCTION

If you skipped the section on the method of joints, your students can ignore step 4 in "Testing the Prototype."

DIFFERENTIATED INSTRUCTION

To ensure a variety of assessment evidence, you may chose to let students pick their mode of communication. Or, you may elect to only check their design notebook, have them write a report, or make a presentation. It's not recommended that students be assessed using all modes for each project.

IN-CLASS ACTIVITIES
Chapter 7

ICA 7-1

Express the following values using scientific notation, engineering notation, and an appropriate SI prefix such that there are only one, two, or three digits shown to the left of the decimal.

Standard	Scientific	Engineering	With Prefix
(a) 389,589,000 joules [J]			
(b) 0.0000000008 pascals [Pa]			

ICA 7-2

Express the following values using scientific notation, engineering notation, and an appropriate SI prefix such that there are only one, two, or three digits shown to the left of the decimal.

Standard	Scientific	Engineering	With Prefix
(a) 0.0698 meters [m]			
(b) 501,000,000,000 grams [g]			

ICA 7-3

Complete the following table.

	inches	feet	yards	meters	miles
Abbreviation	[in]				
(a)		90			
(b)					2

ICA 7-4

A category F5 tornado can have wind speeds of 300 miles per hour [mph]. What is this velocity in units of meters per second [m/s]?

ICA 7-5

Some species of bamboo have been measured to grow at a rate of up to 4 inches per hour [in/h]. Express this growth rate in units of micrometers per second [μm/s].

ICA 7-6

Which of the following is the longest distance?

(a) 26.4 miles [mi]

(b) 40 kilometers [km]

(c) 2500 yards [yd]

(d) 100,000 feet [ft]

ICA 7-7

Complete the following table.

	cubic inch	fluid ounces	gallon	liter	cubic foot
Abbreviation	[in^3]				
(a)					3
(b)				5	

ICA 7-8

Which of the following is the largest volume?

(a) 50 gallons [gal]

(b) 100 liters [L]

(c) 1.5 cubic meters [m^3]

(d) 2.5 cubic feet [ft^3]

ICA 7-9

Which of the following is the fastest speed?

(a) 50 centimeters per second [cm/s]

(b) 2.5 kilometers per hour [km/h]

(c) 1 mile per hour [mi/h or mph]

(d) 125 feet per minute [ft/min]

ICA 7-10

When shipping freight around the world, most companies use a standardized set of containers to make transportation and handling easier. The 40-foot [ft] long rectangular container is the most popular container worldwide. If the container is 2.4 meters [m] wide and has an enclosed volume of 2385 cubic feet [ft^3], what is the height of the container in units of inches [in]?

ICA 7-11

In NCAA basketball, the center circle diameter which encompasses the free throw line is 3.66 meters [m]. What is the area of the center circle, in units of square feet [ft^2]?

ICA 7-12

In many engineering uses, the value of "g" the acceleration due to gravity, is taken as a constant. However, g is actually dependent upon the distance from the center of the Earth. A more accurate expression for g is:

$$g = g_0 \left(\frac{R_e}{R_e + A} \right)^2$$

Here, g_0 is the acceleration of gravity at the surface of the Earth, A is the altitude above the Earth's surface, and R_e is the radius of the Earth, approximately 6380 kilometers [km]. Assume g_0 = 9.8 meters per second squared [m/s^2]. What is the value of g at an altitude of 20 miles [mi] in units of meters per second squared [m/s^2]?

ICA 7-13

Complete the following table to determine mass or weight, assuming the object is on Earth.

	pound-mass	kilogram	newton	pound-force
Abbreviation		[kg]		
(a)	10			
(b)			700	

ICA 7-14

Rank the following from largest mass to smallest mass.

Gravity on the Earth's moon = 1.6 meters per second squared [m/s^2]

Gravity on Mars = 3.71 meters per second squared [m/s^2]

(a) 1-kilogram [kg] object on Earth

(b) 2.75-pound-mass [lb_m] object on the Earth's moon

(c) 1.25-pound-force [lb_f] object on the Earth's moon

(d) 15-newton [N] object on Earth

(e) 15-newton [N] object on Mars

ICA 7-15

A typical American football weighs 14.5 ounces [oz]. If the ball is taken to the Earth's moon, what is the mass in units of grams [g] of the ball on the moon?

Gravity on the Earth's moon = 1.6 meters per second squared [m/s^2].

ICA 7-16

The average American consumes 3 pound-force [lb_f] of sugar each week. If the specific gravity of sugar is 1.59, determine the volume of sugar in liters [L] consumed by the average American in one year.

ICA 7-17

The inside of an airplane is kept at 75 degrees Fahrenheit [°F], or what most people would consider room temperature. What is the difference in temperature between the outside of the plane at cruising altitude (–60 degrees Celsius [°C]) and the inside of the plane at room temperature, in units of kelvins [K]?

ICA 7-18

In July of 2020, NASA launched a new space probe to Mars called *Perseverance*. *Perseverance* landed on Mars in February of 2021. When it travels through the Martian atmosphere during descent, the outside temperature of the heat shield is predicted to reach 3800 degrees Fahrenheit [°F]. What is this temperature in units of kelvin [K]?

ICA 7-19

The National Weather Service announces a Wind Chill Advisory when wind chills between –10 degrees Fahrenheit [°F] and –24 degrees Fahrenheit [°F] are expected. Which of the following temperature readings would indicate the need for a Wind Chill Advisory? Select all that apply.

(a) –5 kelvin [K]

(b) –26 degrees Celsius [°C]

(c) 244 kelvin [K]

(d) 430 degrees Rankine [°R]

(e) –268 degrees Celsius [°C]

ICA 7-20

Which of the following requires the expenditure of more work?

(a) Lifting a 100-newton [N] weight a height of 4 meters [m].

(b) Exerting a force of 50 pound-force [lb_f] on a sofa to slide it 30 feet [ft] across a room.

ICA 7-21

Which object—A, B, or C—has the most potential energy when held at the distance *H* above the surface of the ground?

(a) Object A: mass = 1 kilogram [kg]	height = 3 meters [m]
(b) Object B: mass = 1 slug	height = 3 feet [ft]
(c) Object C: mass = 1 gram [g]	height = 1 centimeter [cm]

224 Chapter 7 Essential Physics Ideas

ICA 7-22

It requires 5650 joules [J] of energy to lift an object 15 feet [ft] upward on Mars. If NASA is simulating this action on Earth by using 5650 joules [J] to lift the same object, what is true about how high the object can be lifted on Earth? Gravity on Mars = 3.71 meters per second squared [m/s^2].

(a) The height the object can be lifted is GREATER on Earth than on Mars.

(b) The height the object can be lifted is LESS on Earth than on Mars.

(c) The height the object can be lifted is the SAME on Earth as on Mars.

ICA 7-23

A 10-gram [g] rubber ball is released from a height of 6 meters [m] above a flat surface on the Earth's moon. Gravitational acceleration on the Earth's moon is 1.62 meters per second squared [m/s^2]. Assume that no energy is lost from frictional drag. What is the velocity, in units of meters per second [m/s], of the rubber ball the instant before it strikes the flat surface?

ICA 7-24

If a 10-kilogram [kg] rotating solid cylinder moves at a velocity (v), it has a kinetic energy of 36 joules [J]. Determine the velocity the object is moving in units of meters per second [m/s] if the kinetic energy is given by $KE = \frac{1}{2} mv^2 + \frac{1}{4} mv^2$.

CHAPTER 7 ASSESSMENT QUESTIONS

1. You are listening to an audiobook that has 10 hours [h] and 53 minutes [min] of listening time. If the audiobook was made using a hard copy version of a novel that is 387 pages long, what is the average speed the narrator used to record the audiobook in units of seconds per page [s/page]?

2. The last time Mars was closest to the Earth occurred on October 6, 2020. On this date, Mars was 38.6 million miles [mi] from the Earth. The next time Mars will be this close to the Earth will be September of 2035. Express the distance in units of meters [m] using an appropriate metric prefix so there are one, two, or three digits to the left of the decimal point.

3. One of the National Academy of Engineering Grand Challenges for Engineering is **Engineer the Tools of Scientific Discovery**. This challenge includes developing the equipment necessary to explore space. In January of 2020, NASA announced the TESS satellite had discovered an Earth-sized planet in a star's habitable zone—the area in a band around a star in which conditions may exist for liquid water to be present. The planet, called TOI 700 d, is located 101.4 light-years away from the Earth in the Dorado constellation. One light-year is 9.46×10^{12} kilometers [km]. How far is the planet TOI 700 d from the Earth in units of meters [m]? Use an appropriate SI prefix to report your answer so there are one, two, or three digits to the left of the decimal point.

4. The AbioCor™ artificial heart pumps at a rate of 10 liters per minute [L/min]. Express this rate in units of gallons per second [gal/s].

5. On April 20, 2010, the *Deepwater Horizon* oil drilling operation exploded, causing the largest marine oil spill ever recorded. Many estimates have been given to describe the volume of oil lost per day. One estimate, just before the well was capped, places the escaping oil at 8400 cubic meters per day [m^3/d]. One barrel [bbl] of oil is 42 gallons [gal]. Convert this rate of escaping oil into barrels per hour [bbl/h].

6. One of the National Academy of Engineering Grand Challenges for Engineering is **Provide Access to Clean Water.** Only 5% of water is used for households—the majority is used for agriculture and industry. It is estimated that 528 gallons [gal] of water are required to produce food for one person for one day [d]. How many liters per year [L/yr] are required to feed one person?

7. We recently read that some cheap land is for sale on Mars for 0.05 U.S. dollars per square inch [\$/in^2]. Determine the total cost in dollars if we purchase 1.5 acres.

8. In 2019, the price of gasoline in Norway was 7.32 U.S. dollars per gallon [\$/gal]. The exchange rate was 1 Norwegian Krone [NK] to 0.11 U.S. dollars [\$]. Express this price in Krone per liter [NK/L].

226 Chapter 7 Essential Physics Ideas

9. Most human heart rates are about 75 beats per minute [beat/min]. An average pumping rate is about 70 milliliters per beat [mL/beat]. How many gallons of blood does your heart pump in one day [gal/d]?

10. You wish to calculate the area of the NCAA volleyball court by "pacing" the distance. Your partner has already determined the length of the court to be 18 meters [m]. You pace the width of the court, and determine it is 17 paces. Your "pace" is equal to 21 inches [in]. Determine the area of the volleyball court in units of square meters [m^2].

11. A good swimmer moves forward 5 feet with each stroke [ft/stroke]. A good sprinter can run 100 meters [m] in 12 seconds [s]. What is the equivalent speed of the sprinter in units of strokes per hour [strokes/h]?

12. The Earth is traveling around the sun at a speed of about 67,000 miles per hour [mph]. If it takes one year [yr] for the Earth to circle the Sun and we assume the Earth's orbit is circular, what is the radius of the orbit in units of meters [m]? Use an appropriate SI prefix to report your answer so there are one, two, or three digits to the left of the decimal point.

13. How large a surface area in units of square feet [ft^2] will 1 gallon [gal] of paint cover if we apply a coat of paint that is 0.1 centimeter [cm] thick?

14. A box has a volume of 10 gallons [gal]. If two sides of the box measure 2.4 meters [m] by 2.4 feet [ft], what is the length of the third side of the box in units of inches [in]?

15. A 1.25-gallon [gal] quantity of molten (liquefied) plastic is going to be used to make a 1-foot [ft] by 8-inch [in] plastic block. If 0.75 liters [L] of the plastic is lost in the manufacturing process, how thick is the final plastic block, in units of inches [in]?

16. One of the National Academy of Engineering Grand Challenges for Engineering is **Manage the Nitrogen Cycle**. It is important to control the agricultural impact on the nitrogen cycle as engineers attempt to provide a sustainable food supply. To grow an experimental crop to test a new fertilizer, you create a small "farm" with the soil at a depth of 8 inches [in]. You use 250 buckets of soil, and each bucket holds 5 gallons [gal] of soil. What is the area of your "farm" in units of square feet [ft^2]?

17. In many engineering uses, the value of "g," the acceleration due to gravity, is taken as a constant. However, g is actually dependent upon the distance from the center of the Earth. A more accurate expression for g is:

$$g = g_0 \left(\frac{R_e}{R_e + A} \right)^2$$

Here, g_0 is the acceleration of gravity at the surface of the Earth, A is the altitude above the Earth's surface, and R_e is the radius of the Earth, approximately 6380 kilometers [km]. Assume $g_0 = 9.8$ meters per second squared [m/s^2]. If the value of g is 9 meters per second squared [m/s^2], what is the altitude in units of miles [mi]?

Chapter 7 Assessment Questions 227

18. While working on Mars, you find an unlabeled liquid in the laboratory. When you combine the liquid with water in a 1-liter [L] container, the liquid floats on top of the water. Which of the following could be the mystery liquid? Select ALL possible answers. The acceleration due to gravity on Mars is 3.71 meters per second squared [m/s^2].
    (a) Density = 1.5 slugs per cubic foot [$slug/ft^3$]
    (b) Density = 850 kilograms per cubic meter [kg/m^3]
    (c) Specific Weight = 850 newtons per cubic meter [N/m^3]
    (d) Specific Gravity = 0.9
    (e) Density = 10 pound-mass per gallon [lb_m/gal]
    (f) Specific Weight = 10 pound-force per gallon [lb_f/gal]
    (g) Specific Gravity = 1.9

19. At room temperature (70 degrees Fahrenheit [°F]), the specific gravity of pure water is 0.998. In your freezer, the temperature is 0 degrees Fahrenheit [°F]. At this temperature, the water has changed to ice; the specific gravity of ice is 0.92. Assume a 2-liter [L] bottle is filled with 2 quarts [qt] of room temperature tap water and then frozen. Will the bottle be able to hold the ice? You must prove your answer by determining both the mass of the water in units of kilograms [kg] and the volume of the ice in units of liters [L].

20. In NCAA softball, the softball is a sphere with a circumference between 11.88 and 12.13 inches [in] and a mass between 177 and 198 grams [g]. You find a ball and determine the mass is 180 grams [g] and the specific gravity is 0.45. While this specific gravity seems low for a solid, the ball is not made of a completely solid material but is instead filled with a sponge-like substance. Determine the radius of the ball you found and report the value in units of inches [in]. Does the ball you found fit the circumference requirements to be used as a softball?

21. The shot put is a track-and-field event using a spherical, metal ball. Assume you manufacture a shot put ball from a solid metal material. The ball weighs 16 pound-force [lb_f] and has a radius of 60 millimeters [mm]. What is the specific gravity of material used to make the shot put?

22. Gatorade® was created in 1965 at the University of Florida to help student athletes stay healthy. One electrolyte Gatorade helps to replace in the body is potassium. The Recommended Daily Allowance (RDA) of potassium for males over the age of 19 is 3.4 grams per day [g/day]. A 0.25-liter [L] bottle of Gatorade Perform® contains 30 milligrams [mg] of potassium. If a baseball player wanted to drink enough Gatorade during a single day to meet the RDA for potassium (assuming he did not get any potassium from other foods), how much Gatorade would he need to consume in units of gallons [gal]?

23. One of the National Academy of Engineering Grand Challenges for Engineering is **Restore and Improve Urban Infrastructure**. Infrastructure includes bridges and roads, but it also includes water and sewer systems, railroads, and power grids. In 2016, the state of Washington opened the SR 520 "floating" bridge. The bridge is made of 77 floating concrete boxes anchored to the bed of the lake using 58 steel cables. Each cylindrical cable is 3 inches [in] in diameter, 1000 feet [ft] long, and made of steel. Assume the specific gravity of steel is 7.9. What is the weight of one cable in units of pound-force [lb_f]?

24. According to regulations, cross-country skiing races should be canceled if the temperature at the start of the race is warmer than 5 degrees Celsius [°C] or colder than –35 (negative 35) degrees Celsius [°C]. For the following start temperature conditions, under which conditions could the cross-country skiing race be held?

    Indicate TRUE if the temperature listed is between the acceptable limits to conduct the ski race. Indicate FALSE if the temperature listed is outside of the acceptable limits. Indicate NOT POSSIBLE if the temperature listed is not physically possible.

    (a) 40 degrees Fahrenheit [°F]

    (b) 490 degrees Rankine [°R]

    (c) 5 kelvin [K]

    (d) –8 kelvin [K]

    (e) –500 degrees Fahrenheit [°F]

25. A "normal" blood pressure has a gauge pressure of 120 millimeters of mercury [mm Hg] (systolic reading) over 80 millimeters of mercury [mm Hg] (diastolic reading). Convert 120 millimeters of mercury into units of pound-force per square inch [psi].

26. A ball is thrown vertically into the air with an initial kinetic energy of 2500 joules [J]. As the ball rises, it gradually loses kinetic energy as its potential energy increases. At the top of its flight, when its vertical speed goes to zero, all of the kinetic energy has been converted into potential energy. Assume that no energy is lost to frictional drag, etc. How high does the ball rise in units of meters [m] if it has a mass of 5 kilograms [kg]?

27. A 100-watt [W] motor (60% efficient) is used to raise a 100-kilogram [kg] load 5 meters [m] into the air. How long, in units of seconds [s], will it take the motor to accomplish this task?

28. A robotic rover on Mars finds a spherical rock with a diameter of 10 centimeters [cm]. The rover picks up the rock and lifts it 20 centimeters [cm] straight up. The rock has a specific gravity of 4.75. The gravitational acceleration on Mars is 3.71 meters per second squared [m/s^2]. If the robot's lifting arm has an efficiency of 40% and required 10 seconds [s] to raise the rock 20 centimeters [cm], how much power in units of watts [W] did the arm use?

29. Given the stationary beam shown here, what is the moment generated by force C if force A is 20N, force B is 150N and force C is 75N?

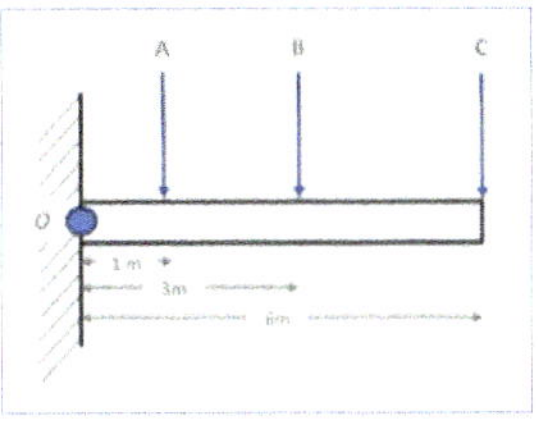

30. For the image shown below,

(a) What is the distance x if the beam is not in motion?

(b) What is the moment generated by force m_1?

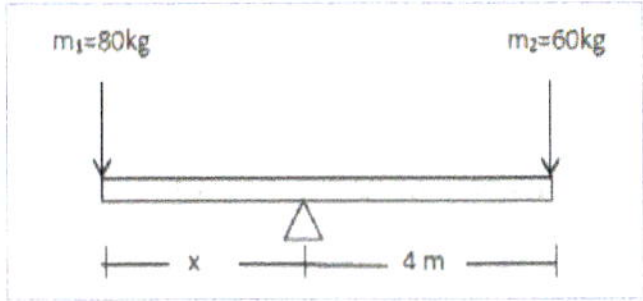

31. Draw a free body diagram of this system. You do not need to solve for any of the values.

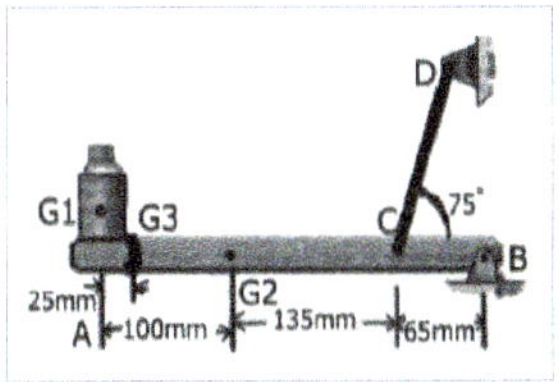

32. A 44,000 kg truck is on a 325-foot-long bridge. It is located 50 feet from one end. A 75 kg person stands on the bridge 200 feet from the other end. The bridge is in static equilibrium; in other words, it is not in motion. Find the forces supporting the bridge at each end, F_1 and F_2.

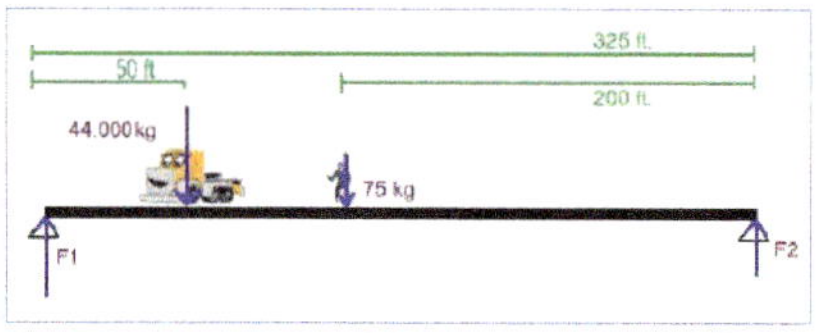

33. Find all of the internal forces in the following trusses (a-d).

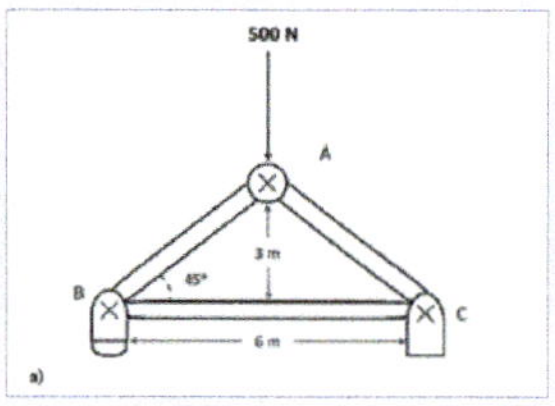

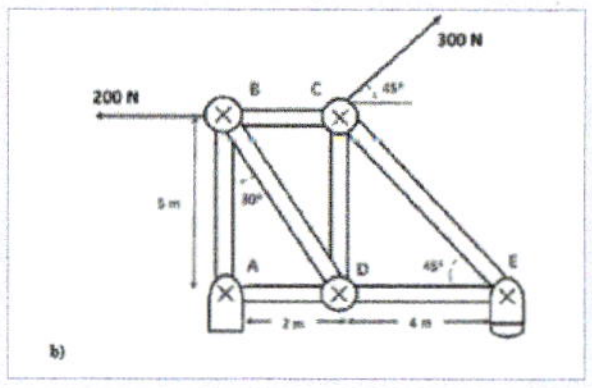

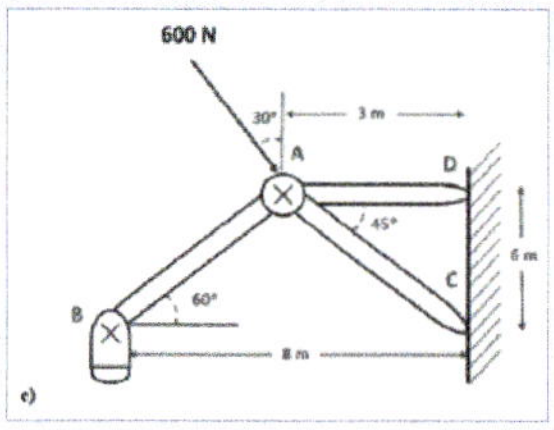

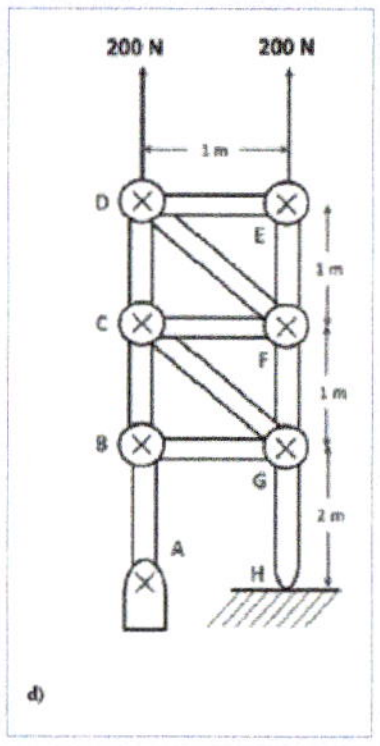

OVERVIEW

This chapter continues our dive into ideas from physics that inform the engineering design process. In particular, we focus on electrical and mechanical engineering for this unit. We will review foundational topics in electrical engineering and use a pinball design project to frame the combination of electrical and mechanical processes to create a functional product. We also discuss the principles of fluid flow, connecting it with water distribution systems.

CHAPTER 8

Leveraging Electrical and Mechanical Ideas

TOPICS

With an understanding of fundamental Physics ideas, we can dive more into specific areas of engineering. In Chapter 7, we examined how the basic ideas of force could be applied to solving a truss in civil engineering. In this case, we'll look at electrical and mechanical ideas that apply to systems, including electrical and physical components. The key to understanding emerging trends in electrical and mechanical engineering, or electromechanical systems where they intersect, is an understanding of these basic concepts. They apply to the development of smaller and more efficient phones, to power generation and distribution using wind or solar, and to fast and efficient transportation.

8.1 Electrical Concepts

The basic concepts of electricity and electrical devices are perhaps less familiar to most students than are many of the other physical phenomena covered previously in this text. This is due partly to lack of practical experience, and partly to the fact that in general these phenomena are themselves invisible, only their effects being perceptible. These effects range from receiving an electric shock to the almost magical performance of touch-screen devices. Table 8-1 summarizes the concepts discussed in this section, and Table 8-2 shows the representations of electrical devices in schematic drawings.

WARNING:

Electrical parameters can get confusing at times since V is used as both a variable name for voltage and for the unit volt, and C is used as a variable name for capacitance as well as the unit coulomb.

Table 8-1 Summary of Electrical Properties

Property	Symbol	Related Equations	Typical Units	Equivalent Units
Charge	Q	Coulomb's Law: $\lvert F\rvert = k_e \frac{\lvert Q_1 Q_2\rvert}{r^2}$	coulomb [C]	C = A s
Current	I	$Q = I\,t$	ampere [A]	A = C/s
Voltage	V		volt [V]	V = J/C
Resistance	R	Ohm's Law: $V = I\,R$	ohm [Ω]	Ω = V/A
Conductance	G	$G = 1/R$	siemens [S]	S = A/V
Electric power	P	$P = V\,I = I^2 R = V^2/R$	watt [W]	W = V A
Capacitance	C	$Q = C\,V$ $\quad E_C = \frac{1}{2} C\,V^2$	farad [F]	F = C/V
Inductance	L	$E_L = \frac{1}{2} L\,I^2$	henry [H]	H = V s/A

OBJECTIVES

- Use idealized equations that are fundamental to civil engineering: hydrostatic pressure from density and height; flow velocity in an ideal (frictionless) system using Bernoulli's Equation
- Describe how real-world factors change performance from the ideal: water tower height affects pressure driving force; pressure driving force affects flow rate from a pipe; fittings, bends, pipe length, and pipe diameter affect flow rate in a pipe
- Work in teams to design and build a project related to civil engineering and write a reflection
- Use clear and concise written, verbal, and visual communication techniques
- Use collaborative tools such as desktop or web-based applications to share and develop information
- Use problem-solving techniques to develop technological solutions such as product, process, or system
- Apply the design process, including decision matrices, as a team participant
- Perform different roles within the project as a team member
- Formulate decisions using collaborative strategies such as decision and design matrices and conflict resolution
- Maintain an engineering notebook for the project per established conventions
- Demonstrate communication skills by preparing and presenting the project, including building consensus setback resolution and decision matrices
- Predict performance, failure modes, and reliability of a design solution
- Apply structured techniques such as a decision tree, design matrix, or cost-benefit analysis to select and justify a preferred solution to a problem
- Prepare a project report that clearly documents the designs, decisions, and activities during each phase of the engineering design process
- Take minutes of a team meeting
- Translate design requirements into a design solution

OBJECTIVES

- Use brainstorming methods to identify solutions to a design problem
- Recognize and demonstrate that there are many possible successful designs and that a design process does not always result in a single best design

ENGAGE

Ask your students to list the different kinds of technology they use. Which are powered by electricity? Which are not? How does electricity compare with other sources of energy? Why is it used with so many types of technology?

TEACHING TIP

For a complete list of the electrical schematic symbols, direct your students to the RapidTables.com online reference tools. Navigate to electrical symbols and consult the table provided.

URL

http://www.rapidtables.com/electric/electrical_symbols.html

TEACHING TIP

Visit the Calculator.net resource page with your students. These tools are used to calculate the total resistance of a group of resistors in parallel or in series. A complex circuit can normally be divided into simple parallel and series circuits for the total resistance. Also included is a calculator to estimate the resistance of a conductor based on size and conductivity.

URL

http://www.calculator.net/resistor-calculator.html

Table 8-2 Schematic Symbols

Device	Schematic Symbol	Notes
Battery (Single Cell & Multi-cell)	V (single cell), V (multi-cell)	The battery terminal with the longer line next to it is always the positive.
Resistor	R	
Capacitor	C	For polarized capacitors, the positive terminal is the one with the straight line, not the curved line.
Inductor	L	The inductor symbol can also be drawn with curved bumps rather than loops.

8.1.1 Electric Charge

electric charge

Electrons and protons, as well as some subatomic particles, have a property known as **electric charge**. On a small scale, charge can be measured in terms of the elementary charge (e). The magnitude of the elementary charge on either an electron or a proton is 1, and by convention, the charge of a proton is called positive ($e = +1$) and that of an electron negative ($e = -1$).

A force acts on a charged particle when in the vicinity of another charged particle. If the charges are alike, both positive or both negative, the force is a repulsive force, and the charges tend to accelerate away from each other. If the charges are unlike, one positive and one negative, the force is attractive with the particles tending to accelerate toward each other.

charge

coulomb

The value of the elementary charge (e) is inconveniently small, so **charge** (Q) is generally quantified using the derived unit **coulomb** [C]. A charge of one coulomb represents the total charge on approximately 6.24×10^{18} protons. Another way to say this: the elementary charge of a single electron is $1/6.24 \times 10^{18} = 1.6 \times 10^{-19}$ C.

IMPORTANT CONCEPT:

SI unit of charge = coulomb [C]

$1\ e = 1.602 \times 10^{-19}$ C

The actual force exerted on a charged object varies with both the amount of charge on each object (Q_1 and Q_2) and the distance (r) between the charges. This relationship is defined by Coulomb's law, named after the French physicist Charles-Augustin de Coulomb (1736–1806) who first described and quantified the attractive and repulsive electrostatic force.

$$|F| = k_e \frac{|Q_1 Q_2|}{r^2}$$

In this equation, k_e is Coulomb's constant, and is approximately equal to 9×10^9 N m²/C². This version of Coulomb's law is a specific case for two charges only. In general, there are more than two charges in three-dimensional space, and they require three-dimensional vector representation. Such mathematics, however, are beyond the scope of this course.

TEACHING TIP

Consider a very simple circuit consisting of four light bulbs and one 6 V battery. If a wire joins the battery to one bulb, to the next bulb, to the next bulb, to the next bulb, then back to the battery, in one continuous loop, the bulbs are said to be in series. If each bulb is wired to the battery in a separate loop, the bulbs are said to be in parallel. If the four light bulbs are connected in series, there is the same current through all of them, and the voltage drop is 1.5 V across each bulb, which may not be sufficient to make them glow. If the light bulbs are connected in parallel, the currents through the light bulbs combine to form the current in the battery, while the voltage drop is across each bulb and they all glow.

8.1.2 Electric Current

Electric current is superficially analogous to a current of water. Just as a current of water is a movement of water molecules in a pipe or channel, electric current is a movement of electric charge in a wire or other solid material.

electric current
amperes

Electric current (I) is measured in **amperes** [A], named for Andre-Marie Ampere (1775–1836), the French physicist who is credited with discovering electromagnetism.

IMPORTANT CONCEPT:

SI unit of current = ampere [A]
1 A = 1 C / s

The derived unit coulomb is defined in terms of the ampere as one ampere second. This may be easier to understand by rephrasing this as a current of one ampere represents a movement of one coulomb of charge past any given point in the wire every second.

To put the magnitude of the ampere in context—if you receive an electric shock by sticking your finger in a light socket, you realize the sensation is rather unpleasant. In general, in domestic appliances, such a shock is typically about 0.005 amperes. This level of current is small enough that although unpleasant, your muscles will still respond to the commands from your brain, and you can release the live wire or pull your hand away. However, in circuits powering large appliances such as stoves or air conditioners, the current from a typical shock is roughly twice that value, or 0.01 amperes. This is very close to the current level that will overload your nervous system so that your muscles will no longer obey, and you become unable to let go. This is FAR more dangerous.

8.1.3 Voltage

To really understand voltage requires knowledge of the concept of the electric field. This is unfortunately a bit too complicated for the limited time and space we have here, so we will merely attempt to help you develop a feel for how voltage affects other electrical parameters. You will study electric fields in some depth in physics, typically the second physics course, and may learn even more in other courses, particularly if you choose to study electrical or computer engineering.

voltage

Voltage (V) is quantified using units of volts [V] and is a measure of how much work is required to move an electric charge in the vicinity of other electric charges. The unit of volt is named for Italian physicist Alessandro Volta (1745–1827), who possibly invented the first chemical battery, called a voltaic pile. A somewhat inaccurate explanation of voltage, although useful in understanding it, is that voltage is what pushes the charges around to create current. Some decades ago, voltage was commonly called electromotive force (EMF), but this has fallen out of favor for a variety of reasons, not least of which is that voltage is not a force, being dimensionally quite different.

IMPORTANT CONCEPT:

SI unit of voltage = volt [V]
1 V = 1 J / C

One volt is defined as one joule per coulomb. In other words, if one joule of energy is required to move one coulomb of charge from one place to another, the voltage between those two points is one volt.

IMPORTANT CONCEPT:

$W = Q\Delta V$

To move a solid object, work equals force times the distance through which that force moves an object: $W = F\Delta x$. Similarly, to move electric charge, work equals electric charge times the difference in voltage through which that charge moves: $W = Q\Delta V$.

DISCUSSION QUESTION

Volta's first scientific paper was written in 1765 about what we call *triboelectricity* today. Define the term. (*The triboelectric effect is a type of contact electrification in which certain materials become electrically charged after they come into frictional contact with a different material. Rubbing glass with fur, or a comb through the hair, can build up triboelectricity. Most everyday static electricity is triboelectric.)*

ENGAGE

André-Marie Ampère made the revolutionary discovery that a wire carrying electric current can attract or repel another wire next to it that's also carrying electric current. Visit the Famous Scientists – The Art of Genius website with your students; search for Ampère.

URL

http://www.famousscientists.org/andre-marie-ampere/

Ask your students what other discoveries Ampère made before he died in 1836. *(Ampère proposed the existence of a new particle, the electrodynamic molecule—a microscopic charged particle which we can think of as a prototype of the electron. Ampère discovered and named the element fluorine in 1810. Ampère proposed that the chemical elements—48 were known at that time—should be grouped according to their properties. He made a number of mistakes but successfully grouped:*

- *The alkali metals: sodium and potassium*
- *The alkali earth metals: magnesium, calcium, strontium, and barium*
- *The halogens: chlorine, fluorine and iodine)*

ENGAGE

Alessandro Volta was a physicist, a chemist, and a pioneer of electrical science. He is most famous for his invention of the electric battery. Visit the Famous Scientists—The Art of Genius website with your students; search for Volta.

URL

http://www.famousscientists.org/alessandro-volta/

ENGAGE

Visit the Smart Learning For All YouTube channel and search for Ohm's Law. Watch the video with your students to gain a better understanding of Ohm's Law.

URL

https://www.youtube.com/watch?v=iLzfe_HxrWI

8.1.4 Electrical Resistance

Resistance is a measure of how difficult it is to move charges through a material. In some substances, such as many metals, electrons can move quite easily. In other materials such as glass or air, considerable force, thus considerable voltage, is required to make electrons move therein.

resistance

Resistance (R) is quantified using units of ohms [Ω], where one ohm is defined as one volt per ampere. For example, if a 1-volt battery were connected to a device having a resistance of one ohm, one ampere of current would flow through it. The ohm is named for Georg Simon Ohm (1789–1854), the German physicist who first described the relationship linking voltage, current, and resistance.

IMPORTANT CONCEPT:

SI unit of resistance = ohm [Ω]

1 Ω = 1 V / A

Resistance relates the voltage across a device to the current through the device. **Take particular note of the choice of prepositions**—across and through. Understanding this choice will help you understand voltage and current.

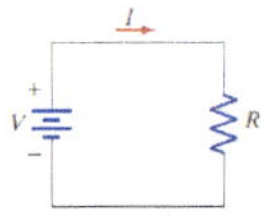

NOTE:

When denoting a current on a circuit diagram, an arrow is used to indicate the direction of the current.

Electric current is the movement of charge, typically electrons moving **THROUGH** a substance. Voltage is to some extent a measure of the force being exerted on the moving charges by forces at either end of the device. *This is where it can be a little confusing, particularly without using electric fields in the discussion.* However, imagine that on each side of a device is an accumulation of charge, each exerting a force on the electrons inside the device. Each of those forces might be a "push" or a "pull," and the total force on the electrons in the device is the difference in these forces. The difference in the forces from one side of the device to the other is referred to as the voltage **ACROSS** the device.

As an analogy, if you are trying to push a sofa across the room, but someone else is trying to push the sofa in the opposite direction with the same force, the net force is zero and the sofa does not move. If one person pulls and the other pushes, however, the sofa will move quicker than with either person alone.

Ohm's law

Resistance is related to current and voltage by **Ohm's law**. Note the following implications of Ohm's law:

- To maintain a specific current through a resistance requires a voltage proportional to the resistance. A larger resistance makes it harder to "push" the electrons through the device, thus a larger voltage is required.
- For a given voltage, current is inversely proportional to resistance. If the resistance increases, the voltage cannot "push" as many electrons through the device per second, so the current must decrease.

IMPORTANT CONCEPT:

Ohm's Law

$V = IR$

conductance

In some contexts, it is computationally simpler to use conductance instead of resistance. **Conductance** (G) is measured in siemens [S] and is simply the reciprocal of resistance. An older unit for conductance that you might find, particularly in older references, is the mho (ohm spelled backwards) and is represented by an upside-down omega [□]. The unit siemens is named for the German inventor Ernst Werner von Siemens (1816–1892) who, among other things, built the first electric elevator and founded the company known today as Siemens AG.

IMPORTANT CONCEPT:

$G = 1 / R$

SI unit of conductance = siemens or mho [S or □]

1 S = 1 / Ω = 1 A / V

Example 8-1

The voltage across a resistor is 15 volts [V], and the current through it is 6 milliamps [mA].
What is the value of the resistance in units of ohms [Ω]? Express the answer using an appropriate SI prefix such that there are only one, two, or three digits shown to the left of the decimal.

Calculate Desired Parameters	Steps
(1) Determine appropriate equation	$V = IR$ so $R = \frac{V}{I}$
(2) Insert known quantities	$R = \frac{15\ \text{V}}{6\ \text{mA}}\left\lvert\frac{1000\ \text{mA}}{1\ \text{A}}\right\rvert\frac{1\ \Omega\ \text{A}}{\text{V}}$
(2) Calculate, be reasonable	$R = 2500\ \Omega = 2.5\ \text{k}\Omega$

Comprehension Check 8-1

The current through a 12 kilo-ohm [kΩ] resistor is 25 microamps [μA]. What is the voltage [V] across the resistor?

8.1.5 Electric Power

electric power

Conceptually, **electric power** (P) is perhaps easiest to understand by examining the formula for gravitational potential energy. A mass has its potential energy increased by expending energy to lift it higher above the surface of the planet, since the mass of the object and the mass of the planet are mutually attracting each other. Similarly, forcing electrons closer to other electrons stores potential energy since they are mutually trying to repel each other. Recall our discussion of batteries. For each electron that is transferred to the negative terminal of the battery by the chemical reaction, a little bit of energy is "stored" in the battery. This is effectively electrical potential energy. The more electrons per second that are jammed together, the more energy per second is stored. Current is measured in charge (electrons) per second, power is proportional to current: $P \propto I$ or $P = XI$ where X is the proportionality constant.

IMPORTANT CONCEPT:
$P = VI$

Now think back to voltage. Voltage is a measure of how much energy is used to move a given amount of charge: one volt is one joule per coulomb. Therefore, voltage is the proportionality constant and $P = VI$.

Example 8-2

A semiconductor diode has 500 millivolts [mV] across it and 700 microamps [μA] of current through it. How much power is the diode absorbing, in units of watts [W]? Express the answer using an appropriate SI prefix such that there are only one, two, or three digits shown to the left of the decimal.

Calculate Desired Parameters	Steps
(1) Determine appropriate equation	$P = VI$
(2) Insert known quantities	$P = \left(\frac{500\ \text{mV}}{}\left\lvert\frac{1\ \text{V}}{1000\ \text{mV}}\right.\right)\left(\frac{700\ \mu\text{A}}{}\left\lvert\frac{1\ \text{A}}{1 \times 10^{6}\ \mu\text{A}}\right\rvert\frac{1\ \text{W}}{\text{V A}}\right)$
(3) Calculate, be reasonable	$P = 3.5 \times 10^{-4}\ \text{W} = 350\ \mu\text{W}$

For resistors, the electrical power absorbed is usually converted to heat, and we can use Ohm's law to replace either the voltage or the current in this power relationship:

$$P = VI = (IR)I = I^2R \quad \text{or} \quad P = VI = V(V/R) = V^2/R$$

ENGAGE

Visit Recoil Offgrid website with your students. Search for the article entitled *INFOGRAPHIC: Understanding the Grid.*

URL

https://www.offgridweb.com/preparation/infographic-understanding-the-power-grid/

Ask your students what new technologies are being introduced to our nation's grid, and what help they will provide. *("Microgrids" help distribute power but can be disconnected with no disruption to the larger grid. "Energy storage" allows the grid to store excess energy and distribute when needed. "Smart meters" enable better communications between the consumers and the utility companies.)*

ENGAGE

Visit the Florida State University's Molecular Expressions website with your students. Navigate to the Electricity & Magnetism Introduction page on capacitance.

URL

https://micro.magnet.fsu.edu/electromag/electricity/capacitance.html

Ask your students what the *Lyden Jar* was and how it was used to store an electric charge. *(It consisted of a stoppered glass jar filled with water with a nail piercing the stopper and dipped into the water. They connected the nail to an electrostatic charge. After disconnecting the nail from the source of the charge, they found that a shock could be felt by touching the nail. This demonstrated that the device had stored the charge.)*

IMPORTANT CONCEPT:

$P = VI = I^2R = V^2/R$

Note that these two relationships expressed in terms of resistance are only valid for resistors, not for other electrical components. However, it gives us a means to quickly calculate the power absorbed by a resistor when we know only the voltage or current, but not both.

Comprehension Check 8-2

A device contains a 1000-ohm [Ω] resistor with 120 volts [V] across it. What is the minimum power rating, in units of watts [W], of the device?

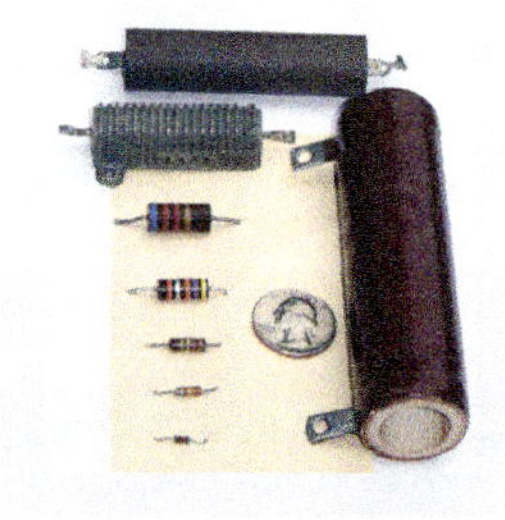

When specifying resistors, both the resistance and the wattage must be given. If you connect a resistor rated at 1 watt [W] in a circuit in which it will have to dissipate 100 W, it will literally burst into flames, or at least perform an imitation of popcorn by exploding! From bottom to top, the power rating of these resistors is ⅛ W, ¼ W, ½ W, 1 W, 2 W, 15 W, 25 W, and on the right, 50 W.

Photo courtesy of W. Park

8.1.6 Capacitance

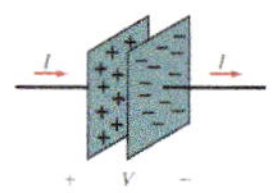

Another simple electrical device is the capacitor. The capacitor is formed by arranging two conducting, low resistance plates very close together, but separated by an insulator with extremely high resistance. Each plate has a wire connected to it. If a current is run into one of the plates of the capacitor, the charges accumulate on that plate since they cannot cross the insulating barrier to the other plate.

IMPORTANT CONCEPT:

$Q = CV$

$E = 1/2\ CV^2$

SI units of capacitance = farad [F]

$1\ F = 1\ C/V$

While accumulating negative charge on one plate, the negative electrons in the other plate are repelled, leaving behind an overall positive charge. This superficially gives the appearance that the current is going through the capacitor. However, the electrons entering one plate and leaving the other are different electrons, and a charge separation is accumulated on the plates of the capacitor.

capacitance

The voltage across the capacitor depends not only on the total charge stored, but also on the physical construction of the device, particularly the surface area of the plates. The charge (Q) stored in a capacitor is proportional to the voltage (V) across it. **Capacitance** (C) is the proportionality constant relating charge.

farads

Capacitance is measured in units of **farads** [F], where one farad equals one coulomb per volt. The unit of farad is named for Michael Faraday (1791–1867), an English scientist who made significant contributions to electromagnetism. He also made contributions to the field of chemistry, including Faraday's laws of electrolysis, discovering the chemical compound benzene, and inventing the rubber balloon.

NOTE:

One farad is a LARGE amount of capacitance. Most capacitors are much smaller than this, and are usually measured in microfarads [µF], nanofarads [nF], or even picofarads [pF].

Since capacitors can contain a separation of charge, they can store energy. The energy stored in a capacitor can be calculated by $E_C = \frac{1}{2}CV^2$. Note that the energy is proportional to the square of the voltage.

Example 8-3

A 0.01-microfarad [μF] capacitor is initially completely discharged (no stored charge, thus an initial voltage across it of zero). If a constant current of 4 milliamps [mA] charges the capacitor, how long will be required to change the voltage across it to 5 volts [V]? Express the answer in units of seconds [s], using an appropriate SI prefix such that there are only one, two, or three digits shown to the left of the decimal.

Calculate Desired Parameters	Steps
(1) Determine appropriate equation	$Q = C\,V$ and $Q = I\,t$ $t = \frac{C\,V}{I}$
(2) Insert known quantities	$t = \left(\frac{0.01\ \mu F}{} \middle| \frac{1\ F}{1 \times 10^{6}\ \mu F}\right)\left(\frac{5\ V}{}\right)\left(\frac{}{4\ mA} \middle| \frac{1000\ mA}{1\ A}\right)\frac{1\ C}{F\ V}\frac{1\ A\ s}{1\ C}$
(3) Calculate, be reasonable	$t = 1.25 \times 10^{-5}\ s = 12.5\ \mu s$

Comprehension Check 8-3

A constant current transfers 192 microcoloumbs [μC] of total charge onto one plate of a capacitor in 3 minutes [min].

(a) What is the current, in units of amperes [A]? Express the answer using an appropriate SI prefix such that there are only one, two, or three digits shown to the left of the decimal.

(b) If the capacitor has a capacitance of 6.8 microfarads [μF], what is the voltage in units of volts [V] across the capacitor?

When specifying capacitors, both the capacitance and the maximum voltage must be given. If you connect a capacitor rated at 25 V in a circuit in which it will have 200 volts [V] across it, the insulator between the plates will probably fail and the capacitor will be destroyed; it might even explode! The voltage ratings of the capacitors shown range from 25 V to 500 V. The big 10,000 microfarad [μF] capacitor on the left is rated at 50 V.

Photo courtesy of W. Park

IMPORTANT CONCEPT:

$V = L\,dI/dt$

$E = 1/2\,L\,V^2$

SI units of inductance = henry [H]

1 H = 1 (V s) / A

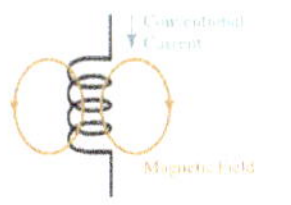

8.1.7 Inductance

If a current is moving through a wire, a magnetic field is generated surrounding that wire. If a wire is placed in a *changing* magnetic field, a current is induced in the wire. This is the physical basis for the final electrical device we will discuss, the inductor.

In its simplest form, an inductor is just a coil of wire. If a current flows through the coil, the magnetic field generated in each loop adds to the magnetic field generated in every other loop, generating a stronger magnetic field. This is the basis of the electromagnet. Industrial electromagnets can concentrate the magnetic field to a large enough value to pick up huge objects, like cars and trucks. However, in electrical engineering, inductors are more often used for their ability to store energy in the form of a magnetic field.

ENGAGE

Based on what they have learned about conductors and insulators, have students draw sketches—including specific atoms and electrons—comparing the movement of electrons in both.

ENGAGE

Joseph Henry was at the forefront of the great electromagnetic advances of the 1830s. He built the world's most powerful electromagnets and made practical breakthroughs that allowed Samuel Morse to invent the telegraph.

Visit the Famous Scientists – The Art of Genius website with your students; search for Henry.

URL

http://www.famousscientists.org/joseph-henry-2/

In 1830, Henry constructed the world's most powerful electromagnet. What did he call it? (*Albany Magnet*) How much more weight did it lift than its predecessor at 9 pounds? (*750 pounds*) By 1831, Henry's magnets could lift 2,000 pounds.

Note that once a magnetic field has been created in an inductor, if the source of the current that created the magnetic field is removed, the magnetic field will begin to collapse. However, a wire in a changing magnetic field will have a current induced in it, thus allowing the stored energy to be transferred to another place and form.

inductance
henrys

Inductance (L) is measured in units of **henrys** [H], named after Joseph Henry (1797–1878). Henry was an American scientist credited with discovering electromagnetic induction about the same time as Faraday. The voltage across an inductor is equal to the inductance of the device times the rate of change of current through the inductor. If the current in an inductor changes by one ampere per second, and a voltage is developed across the inductor of one volt, then the inductance is one henry.

NOTE:

Just as a pendulum transfers energy back and forth between kinetic and potential forms thus creating a physical oscillation, an inductor and a capacitor connected together can swap energy back and forth from a magnetic field to an electric field forming an electrical oscillation. For example, the Theremin, an electronic musical instrument that is played without touching it, relies on such oscillations.

The total energy stored in an inductor can be calculated by $E_L = \frac{1}{2} L I^2$. Note that the energy is proportional to the square of the current. You might note the similarity of this to both the formulae for energy stored in a capacitor $E_C = \frac{1}{2} C V^2$ and kinetic energy $KE = \frac{1}{2} m v^2$.

When specifying inductors, both the inductance and the maximum current must be given. If you connect an inductor rated at 5 milliamperes [mA] in a circuit which will have 2 amperes [A] through it, the wire will probably melt—a BAD idea! The inductor shown on the left is wound with many thousands of turns of extremely small wire and its current capacity is rated in microamperes. The one on the right has only ten turns of rather thick wire and can handle about an ampere.

Photo courtesy of W. Park

Example 8-4

The energy stored in a 50-millihenry [mH] inductor is 75 millijoules [mJ]. If all of this energy is transferred to a 15-microfarad [μF] capacitor, what is the voltage [V] across the capacitor?

Calculate Desired Parameters	Steps
(1) Determine appropriate equation	$E_L = E_C \quad E_L = \frac{1}{2} C V^2 \quad V = \sqrt{\frac{2 E_L}{C}}$
(2) Insert known quantities	$V = \sqrt{2 \frac{75\ \text{mJ}}{15\ \mu\text{F}} \left\| \frac{1 \times 10^6\ \mu\text{F}}{1\ \text{F}} \right\| \frac{1\ \text{J}}{1000\ \text{mJ}} \left\| \frac{1\ \text{V C}}{\text{J}} \right\| \frac{1\ \text{F V}}{\text{C}}}$
(3) Calculate, be reasonable	$V = 100\ \text{V}$

Comprehension Check 8-4

A 100-millihenry [mH] inductor has a current through it of 750 milliamperes [mA]. If the inductor discharges all of its energy into an initially discharged 5-microfarad [μF] capacitor, what is the voltage in units of volts [V] across the capacitor?

8.2 Applying the Concepts to Pinball

Figure 8-1 The playing area of a pinball machine.

When you walk into an arcade, certain games are essential. Perhaps one of the most recognizable arcade games is pinball. On its surface, the game seems straightforward. You begin by launching a ball using a spring within an enclosed space complete with bumpers, ramps, and flashing lights that you can only control with two flippers at the base of the machine (Figure 8-1). Between the flippers is a hole where the ball could easily slip through if you aren't quick enough with the flippers or the physics of the ball's trajectory don't cooperate and lead it to exit right outside the flippers' reach. You have three balls to get as high of a score as possible in most circumstances, but some machines have certain objectives like hitting a set of bumpers or switches to light up different elements of the board. Because you are attempting to hit specific targets, you need to take advantage of the physics related to collisions.

electromechanical system

A pinball machine is an example of an **electromechanical system**, which is a system that combines elements of electrical processes and mechanical processes. More broadly, these systems convert mechanical energy into electrical energy or electrical energy into mechanical energy. Although the term electromechanical systems may sound like a special classification, most of the systems you interact with daily that move are powered by some kind of electromechanical system. For example, electric motors convert electrical energy into mechanical energy by using current to drive a series of gears. Why is a pinball machine an electromechanical system? Consider how the score updates in classic pinball machines. When the score increments, that means the machine received an input from a switch impacted by the ball, which sent an electrical signal to the scoreboard to calculate and adjust the reel tracking the player's score. So mechanical energy was converted into electrical energy to send the signal, then back again to mechanical energy to adjust the score!

8.2.1 Components of The Pinball Machine

There are thousands of different designs of pinball machines, but there are a few common elements that most, if not all, machines use. The most common and most visible element is the ball. The ball inside the pinball machine is typically made from steel, and is 1 1/16" (or about 27 mm) in diameter. The pinball weighs 80 grams. This steel ball is the projectile that will careen through the machine, running into obstacles that, hopefully, add to the player's score.

Plunger

The ball is launched by pulling a spring-loaded plunger, powered by a spring that plunges the launcher into the ball, sending it flying into the field of play. The game designer can choose the strength of the spring to determine how much force will be used to launch the ball. When the launcher is pulled back, the plunger has some amount of potential energy. When it is released, the potential energy changes to kinetic energy. The kinetic energy of the plunger is transferred to the ball when it is released. The ball now shoots up the inclined ramp, overcoming friction and gravity, emerges at the top, and starts its path.

Figure 8-2 Pop bumpers, a bumper, and flippers.

Pop Bumpers

Pop bumpers appear throughout the field of play, and they often have a mechanical plate near the bottom (Figure 8-2). When the ball hits this plate, a mechanism at the top of the bumper is triggered to slam down onto the ball, launching it away from the pop bumper, and usually with a ringing bell noise (and some points added to the player's score).

DISCUSSION

As an example of a bare-bones, do-it-yourself pinball machine, show your class the following wooden pinball machine powered using an Arduino.

URL

https://www.youtube.com/watch?v=y2FeiB1704w

What kind of subsystems do your students recognize? This will help them during the functional decomposition step of their project.

Bumpers

Bumpers usually appear near the sides of the game and look like a surface on which the ball might innocently drift along. If the force of the ball is sufficient, a plunger will strike the bumper, sending the ball bouncing away.

Flippers

There are almost always two flippers at the bottom of the game. There may be an extra flipper or two higher in the playing field, but the main flippers are at the bottom of the playing field. They won't touch; there will be a gap that the ball can fit through, ending the turn.

There are other aspects designers may use, including additional ramps, spinners, pits where the ball might get stuck and launched, and the creative use of empty space.

8.2.2 Designing a Pinball Machine

Pinball machine designers will often start with a theme, whether that is a movie, television show, superhero, or rock band. The theme may not dictate every technical aspect but definitely will affect the decoration, lights, and other aesthetics. Therefore, in order to achieve the desired design, some mechanical and electrical processes need to be considered. For example, how will the lights be sequenced? Are they dependent upon pop bumpers, a switch, or flaps being struck by the ball once, twice, or thrice? These design choices will significantly impact how the pinball machine's inner workings will function.

spring constant

Let's examine the technical aspects of the initial ball launch. The main component of the plunger is a spring. The strength of the spring is measured by the **spring constant**. A higher spring constant makes it tougher to compress or elongate the spring. When the player pulls the plunger out, they compress a spring. The elastic potential energy in the spring is now

$$W = \bar{F}s$$

where $\bar{F}$ is the average force used to pull the plunger back (the force of compressing the spring), and s is the distance the spring is compressed.

The distance s is the displacement, $x_f - x_i$. The average force is $\frac{1}{2}(F_f + F_i)$. This means:

$$W = \left[\left(\frac{1}{2}F_f + F_i\right)\right]\left(s_f - s_i\right)$$

Hooke's Law tells us that the force needed to extend or compress a spring by a distance is proportional to that distance. In other words,

$$F = -kx$$

where x is the distance the spring is compressed, and k is the spring constant of the spring. In this case, the distance s (the distance the plunger is pulled) and the distance x (the distance the spring is compressed) are the same. Let's use Hooke's Law to rewrite this equation:

$$W = \left[-\frac{1}{2}\left(kx_f + kx_i\right)\right]\left(x_f - x_i\right)$$

or

$$W = \frac{1}{2}kx_i^2 - \frac{1}{2}kx_f^2$$

If we set the final position as 0, meaning the displacement is x_i, we can designate the energy in the spring as potential energy, and this gives us:

$$PE = \frac{1}{2}kx^2$$

This is the energy stored in the plunger. Once the player releases the plunger, this potential energy is converted to kinetic energy as the spring expands. Kinetic energy is expressed as:

$$KE = \frac{1}{2}mv^2$$

Where *m* is the mass of the ball, and *v* is the velocity of the ball at launch. Since all of the potential energy is transferred to the kinetic energy of the ball,

$$PE = KE. \ Therefore, \ \frac{1}{2}kx^2 = \frac{1}{2}mv^2, \ or \ kx^2 = mv^2$$

If the pinball is a typical 80 g steel ball, and if the value of *k* is given as 325 *N/m*, we can find an equation that will determine the velocity of the ball at launch based on the spring displacement, or how far back the player pulls.

$$v^2 = \frac{kx^2}{m} = \frac{325\frac{N}{m}x^2}{80g}$$

We should examine some units. PE and KE are energies, which are specified in Nm or J (Joules). Joules can use units of Nm, or can use $\frac{kg\,m^2}{s^2}$. The mass of the ball can be 80 g or 0.08 kg. Velocity can be in *m/s*. Therefore,

$$v^2 = \frac{kx^2}{m} = \frac{325\frac{N}{m}x^2}{0.08kg}$$

Let's see if our equation is reasonable. If a player pulls the plunger back 15 cm (or 0.15 m),

$$v^2 = 91.4\frac{m^2}{s^2}$$

or

$$v = 9.5\frac{m}{s}$$

which seems reasonable.

What if the player pulls back 20 cm? We can find the velocity of the ball as 12.74 *m/s*.

8.2.3 Design a Pinball Game Prototype

Problem Statement

The initial design of a pinball game often starts as a blank slate on, for example, a foam core board. Design elements are added to this prototype. Let's design a foam core, low-fidelity prototype of a pinball game. If foam core is not available, you can use the lid of a large box.

We won't be able to incorporate a full-scale electromechanical design, but our design can focus on an exciting game with an appropriate number of obstacles and working flippers.

The game must be adequately sized to be fun and allow the ball to travel around the board. The prototype must have simulated pop bumpers and bumpers, and must have (at least) 2 flippers on the bottom of the course with a gap just wide enough for the ball to pass through. You should also decide on a theme and incorporate that theme as much as possible. Remember to record minutes of each team meeting and record design decisions in your design notebook.

DIFFERENTIATED INSTRUCTION

You may choose to revisit this project during Chapter 9 to integrate and reinforce programming concepts. For example, when discussing conditional statements, you can have students create a flowchart for handling the scoring system.

TEACHING TIP

This list of materials is the minimum amount of supplies needed to create a prototype. However, you may allow students to add other materials to their prototypes based on what you have in your classroom.

DIFFERENTIATED INSTRUCTION

For this project, you may choose to emphasize creating a good testing plan. Therefore, you will want to cover sections 11.1 and 11.2, in Chapter 11 before starting this project. Have students carefully outline how they will measure the extent to which the prototype meets the stated criteria and constraints.

DIFFERENTIATED INSTRUCTION

If you have access to electronic supplies like breadboards, resistors, and a power supply, you can consider adding a criterion about having a light that illuminates when a bumper is hit.

TEACHING TIP

Students should recognize that there are two major subsystems needed to create a pinball machine: mechanical and electrical. Therefore, this realization should be represented in the functional decomposition. Don't let them forget the subsystem that accepts and dispenses change!

After they create their functional block diagram and share with their peers, you may consider walking through the following website that outlines the subsystems in a pinball machine.

URL

https://www.flippers.be/basics/101_how_pinball_machines_work.html

Materials

- One sheet of foam core, a box lid, or a shallow box
- Items to create pop bumper models: thread spools, small paper or plastic cups, thin cardboard tubes
- Rubber bands
- Dowel rods or pencils
- Craft sticks
- Tape, glue
- A ball (marble, high-bounce ball)

Develop Criteria, Constraints, and Design Specifications

Although we are developing our prototype of the pinball game for our own use, we need to specify our criteria, constraints, and design specifications. In your design notebook, address the following prompts:

1. What are the criteria you identified for this design?
2. What are the constraints limiting the possibilities of this design?
3. How will you know if your design is successful? Translate your criteria and constraints into measurable design specifications.
4. Consult with another team of students in your class and compare your criteria and constraints. What differences and similarities do you notice?
5. Develop a test plan to measure whether your design achieves the criteria you specified at the beginning of the challenge.
6. Did you revise or add design specifications during your test plan development?

Functional Decomposition

With our criteria, constraints, and design specifications outlined, now we can decompose our potential solution into functions that it must accomplish to be successful. Using your answers in the previous section, record the following in your design notebook.

1. What subsystems are necessary to generate the desired output?
2. Develop a functional block diagram relating the input, outputs, and functions.
3. Compare your functional block diagram to a different team of students than you consulted in the "Develop Criteria, Constraints, and Design Specifications" step. What differences and similarities do you notice?

Generating Potential Concepts for a Prototype

Now that you have considered the different functions and subsystems needed for your design to meet its criteria and constraints, we can turn to generating potential concepts. In your design notebook, address the following prompts.

1. Generate ideas for each of the functions and subsystems identified in the functional decomposition.
2. Organize the results into a morphological chart using the functions and subsystems identified in your functional decomposition.
3. Once your morphological chart is complete, identify five potential concepts using the methods you specified.
4. From the five concepts, create a sketch and a description that capture how the concept would work.

Selecting a Concept

After generating a series of design alternatives and the concepts from the morphological chart, we now must pick a concept to move forward with. One tool for making a viable prototype selection is a decision matrix. In your design notebook, answer the following:

1. Using your criteria and design specifications, set up a decision matrix to weigh the viability of the five concepts.
2. Calculate the "winning" design using your weights.
3. Summarize your findings.
4. Compare your selected concept to a different team of students than you consulted in the "Develop Criteria, Constraints, and Design Specifications" step. What differences and similarities do you notice? Discuss the process by which the other team arrived at the solution.

Developing and Testing the Prototype

Now that your team has chosen a design to move forward with, it is time to prototype the solution.

Testing the Prototype

Once you have a workable prototype, it's time to test whether the design meets the specifications outlined earlier in your process.

1. Implement the test plan with your prototype and record your results in your design notebook.
2. Interpret the results: What worked and what didn't? What do you need to do to revise your prototype to meet the specifications?
3. Your prototype currently should perform according to your specifications. If you were to develop the game further, what other functionality is needed to reach a product that is ready for broader distribution?

Report Your Solution

Finally, it is time to report out to the class. You may be asked to demonstrate your game, or your teacher may elect to have you present a PowerPoint summarizing your processes from each stage of the design process outlined here. You may also write a short report covering the same material. In your design notebook, create a bulleted list of the main decisions you made across the different stages of this project. If you are making a PowerPoint, create an outline for how you will present the information. Next, reflect on the following questions:

1. What was the objective of this project?
2. What worked?
3. What didn't work and why didn't it work?
4. How did the design compare with the best and worst performers?
5. What would you do differently?
6. Was the objective accomplished?

8.3 Properties of Materials

physical property

When considering the design of a physical product, it's critical to understand the physical, mechanical, and chemical properties of the materials we use to make it. A **physical property** of a material is something that can be observed or measured without altering the identity of the material. These properties can

DIFFERENTIATED INSTRUCTION

One of the deliverables for this project could be a detailed testing plan and results. Encourage students to take detailed notes on their results.

DIFFERENTIATED INSTRUCTION

To ensure a variety of assessment evidence, you may choose to let students pick their mode of communication. Or, you may elect to only check their design notebook, have them write a report, or make a presentation. It's not recommended that students be assessed using all modes for each project.

ENGAGE

What is the difference between a physical, chemical, and mechanical property? *(A physical property can be observed without doing anything to the material. A chemical property reveals how the material changes from one substance to another without changing the material's identity. A mechanical property involves applying an external force to the material.)*

Consider sharing the following link from ASM International, which lists several material properties. In fact, additional properties are mentioned like Electrical and Thermal.

URL

https://www.asminternational.org/documents/10192/1942078/typesofproperties.pdf

ENGAGE

What is stress and strain? *(Stress is the amount of force a material experiences per unit area and strain is the ratio of the change in the material's length.)*

ENGAGE

It is likely helpful to show students how a tensile test would be conducted. This short video from MTS Systems can be shown to the class to supplement your discussion.

URL

https://www.youtube.com/watch?v=lqR5cp2v1eQ

include density, color, and even melting point. We can also measure how well the material conducts heat and electricity, which are its thermal conductivity and electrical conductivity, respectively. The physical properties of materials can be extraordinarily helpful in narrowing down the selection of materials that are appropriate for a given criterion.

chemical property

phase

The key difference between a physical property and a **chemical property** is that the chemical property describes how the material changes from one substance to another. So, although we can apply heat to gold to determine its melting point is 1,948 degrees Fahrenheit or 1,064 degrees Celsius, we did *not* change the identity of the substance—it's still gold. That's because **phase**, which is the state of the material (liquid, gas, solid, or plasma), is a physical property of the material. We can observe the material's state directly. On the other hand, a material's ability to resist corrosion and its flammability are examples of chemical properties.

mechanical property

Although physical and chemical properties seem to cover the spectrum of what we would want to look for in a material, we still need to consider how the material performs under certain conditions. For example, when we are designing a product that needs to withstand a certain load such as a rung in a ladder, we would want to know which materials would be most suitable for those applications. Using a **mechanical property** of the material helps us make those decisions, which describes how the material behaves when subjected to an external force (or load). Mechanical properties of materials can include strength, stiffness, ductility, and hardness.

8.3.1 Engineering Stress and Strain

tensile test

Determining several mechanical properties of a material can be done using a **tensile test**, a common technique in material science. A tensile test involves placing a material in a machine that slowly applies tension, i.e., pulls the material, until it finally fails. An example of a tensile test machine is shown in Figure 8.3.

Figure 8-3 A machine used to conduct a tensile test.

stress

strain

When a force is applied to a material like what occurs during a tensile test, it will experience stress and strain. **Stress** (represented by the Greek letter sigma, σ) is defined to be the amount of force that material experiences per unit area (measured in pascals or Pa, but typically on the order of *mega*pascals or MPa), whereas **strain** (represented by the Greek letter epsilon, ε) is the ratio of the change in the material's length and its original length (this is unitless). A strain gauge or extensometer can be used to measure the elongation as the material is stretched. If you convert the applied force and the distances between marks on the strain gauge, the results will generalize to materials of differently sized cross-sections and sizes. To calculate stress, the measure of the internal forces experienced by the material, we divide the applied force (F) by the original cross-sectional area (A_0):

$$\sigma = \frac{F}{A_0}$$

To find the strain, the measure of the material's elongation because of the applied force, we divide the change in length (Δl) by the original length of the material (l_0):

$$\varepsilon = \frac{\Delta l}{l_0}$$

stress-strain curve

When we consider these values over the duration of the test, we can obtain a chart called the **stress-strain curve**. Pictured in Figure 8-4 is a generic stress-strain curve for a metal alloy.

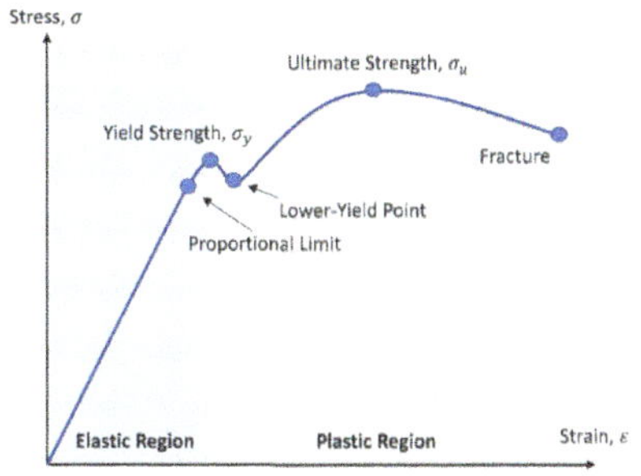

Figure 8-4 Idealized stress-strain curve for a generic metal alloy.

elastic region
Young Modulus
proportional limit
yield strength

There are two main regions on a stress-strain curve—an elastic region and a plastic region. When the material is initially subject to the load as part of the tensile test or any other externally applied force, there is a specific point that unless reached, the material will return to its original length. This is called the **elastic region**. Within the elastic region, one mechanical property can be measured, called the **Young (or Young's) Modulus**. Young's modulus is a measure of stiffness and is simply the slope of the line up to the **proportional limit**. The same principle governing how a spring functions (Hooke's Law) can be applied to materials in the elastic region; it is a linear relationship. The linear relationship breaks down at the proportional limit, but the material will remain elastic until the **yield strength** is reached. This is the point where the material will no longer return to its original length and will keep a 0.2% elongation even after the applied force is removed.

plastic region
ductility
lower-yield point
ultimate strength
necking

After the yield strength is reached, the **plastic region** is entered. At this point, any further deformation is permanent. The length of the plastic region can be used as a measure of **ductility**, which is a mechanical property of a material describing how well a material can sustain plastic deformation without failing. It is possible for the stress to decrease a bit after the yield strength is reached because there will be dislocations in the material's structure of the microscopic level. The lowest stress after the yield strength is called the **lower-yield point**. The material will continue to experience plastic deformation and reach its **ultimate strength**, the maximum stress that the material can withstand. When additional stress is experienced after its ultimate strength, the material will eventually fracture. Before failing, the material can experience **necking**, which is where stress is localized disproportionately in one smaller region. Necking will lead to a smaller and smaller cross-sectional area, like what's shown in Figure 8-5.

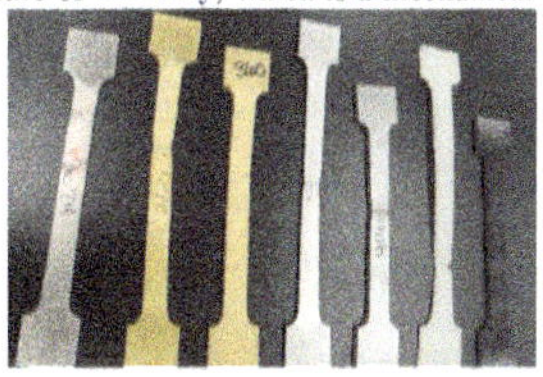

Figure 8-5 An example of necking in a material.

There are several repositories for finding these kinds of measurements for a material, such as the Material Properties Database and Estimation Tools website (matdat.com).

8.3.2 Testing Some Material Properties and Their Implications

Rockwell hardness test

Although the stress-strain curve can provide us with a wealth of information about a material, other tests are needed to determine other mechanical properties. For example, when measuring a material's hardness, there are a few tests that can be performed. The most common of which, the **Rockwell hardness test**, involves applying a 4 mm (or smaller) steel ball or diamond cone to a material's surface and measuring the depth

ENGAGE

Why does the stress decrease slightly before increasing to the ultimate strength? *(The material is experiencing dislocations at its molecular level because of the applied force.)*

TEACHING TIP

During this section, you may choose to recall the pinball project, specifically the spring used to fire the ball into the play area. When we choose our spring, we need to know how stiff it should be for optimal performance. This determination would be informed by the elasticity of the spring (i.e., the spring constant) as discussed here.

ENGAGE

Suppose an engineer needed a material that is resistant to stretching. How would you describe the slope of the elastic region? (*The slope of the line would be steeper. This can be reasoned out by realizing that strain captures how much the material stretches when it experiences an increasing applied force. Therefore, the value of stress can be high but strain should be low.)*

ENGAGE

The material properties discussed in this chapter are only a select few that engineers would consider during their design process. You can share the website MatWeb, which is a database of several material properties.

URL

https://www.matweb.com

Have students explore the data sheet for Aluminum 1060-H12. You may give them the URL directly or have them navigate the database themselves. What is its yield strength and ultimate strength?*(Its yield strength is 75.8 MPa and its ultimate strength is 82.7 MPa.)*

URL

https://www.matweb.com/search/DataSheet.aspx?MatGUID=0df660ac8b87434ca82af7681a6098e1

ENGAGE

Like the tensile test, you may want to show them the following video demonstrating the Charpy V-Notch test.

URL

https://www.youtube.com/watch?v=cvdvTtJPAwA

TEACHING TIP

One observation that can be made is that pressure and stress share the same formula.

Brinell hardness test

of the indentation. Another popular test, called the **Brinell hardness test**, is used for materials that have surfaces that are too rough or coarse. A 10 mm diameter steel ball is driven into the material's surface with a high load, often in the ballpark of 30 kN. The hardness number is found using a Brinell microscope or other optical system to measure two diameters of the impression made by the steel ball.

Charpy V-Notch test

To test a material's toughness, determining if it is brittle or ductile, a **Charpy V-Notch test** can be used. The test involves measuring how much energy it takes to fracture a material. The Charpy V-Notch test uses a pendulum that is dropped from a specified height that impacts a material that has been notched. The notch is a deliberate alteration made to the material, and it usually V or U shaped.

Thiele tube

Physical properties can also be tested through a variety of means. For example, melting point is classically determined using the **Thiele tube**, a special type of glassware that is designed to hold and heat Paraffin oil with a uniform temperature. The tube is heated using a heating element like a flame. A sample of the material (0.25-0.5 mL) is placed in a sealed capillary tube tied to a thermometer that will be immersed in the oil. The temperature is gradually increased until the material begins to melt.

safety factor

After taking relevant measurements, we use physical, chemical, and mechanical properties to inform material selections for our design. For example, when designing structures that will be subject to different forces, it's important to consider the **safety factor**—a measure of the degree to which the actual stress placed on the material is below the limiting stress. It is calculated by dividing the ultimate stress by the actual stress, so $FS = \sigma_u/\sigma_{actual}$. By definition, the factor of safety needs to be bigger than 1, otherwise the actual stress is beyond the ultimate stress and the material will fail. However, it's not good enough to be just above 1; engineers need to design to a higher factor of safety depending on the application. As the factor of safety increases, the safer the structure becomes. Increasing the factor of safety will often also increase the cost and impact the functionality of the structure, so there is a tradeoff that engineers must consider when addressing the established metrics. The factor of safety values can typically range from 1.5 to 10.

8.4 Pressure

pressure

pascal

Pressure is defined as force acting over an area, where the force is perpendicular to the area. In SI units, a **pascal** [Pa] is the unit of pressure, defined as one newton of force acting on an area of one square meter. The unit pascal is named after Blaise Pascal (1623–1662), a French mathematician and physicist who made great contributions to the study of fluids, pressure, and vacuums. His contributions with Pierre de Fermat on the theory of probability were the groundwork for calculus.

IMPORTANT CONCEPT:

$P = F/A$

SI unit of pressure = pascal [Pa]

INTEGRATING CONTENT AREAS

Share the following article from Britannica about Blaise Pascal. What was his invention, the Pascaline? *(It was a calculator that could perform addition and subtraction by manipulating dials).*

URL

https://www.britannica.com/biography/Blaise-Pascal/Les-Provinciales

Example 8-5

An automobile tire is pressurized to 40 pound-force per square inch [psi or lb_f/in^2]. State this pressure in units of atmospheres [atm].

PRESSURE:

1 atm = 1.01325 bar
= 33.9 ft H_2O
= 29.92 in Hg
= 760 mm Hg
= 101,325 Pa
= 14.7 psi

Convert to Desired Units	Steps
(1) Term to be converted	40 psi
(2) Conversion formula	1 atm = 14.7 psi
(3) Make a fraction (4) Multiply	$\frac{40\ \text{psi}}{} \Big\vert \frac{1\ \text{atm}}{14.7\ \text{psi}}$
(5) Cancel, calculate, be reasonable	2.7 atm

Comprehension Check 8-5

If the pressure is 250 feet of water [ft H_2O], what is the pressure in units of inches of mercury [in Hg]?

Example 8-6

Your textbook is resting on your desk. The book has a mass of 1.8 kilograms [kg] and has dimensions of 8 inches [in] by 10 inches [in] by 1.5 inches [in]. How much pressure is the book exerting on the surface of your desk, in units of pascals [Pa]?

Convert to Desired Units	Steps	
(1) Term to be converted	8 inches	10 inches
(2) Conversion formula (3) Make a fraction (4) Multiply	$\frac{8\ \text{in}}{} \Big\vert \frac{2.54\ \text{cm}}{1\ \text{in}} \Big\vert \frac{1\ \text{m}}{100\ \text{cm}}$	$\frac{10\ \text{in}}{} \Big\vert \frac{2.54\ \text{cm}}{1\ \text{in}} \Big\vert \frac{1\ \text{m}}{100\ \text{cm}}$
(5) Cancel, calculate, be reasonable	0.2032 m	0.254 m

Calculate Required Parameters	Steps
(1) Determine appropriate equation	$P = \frac{F}{A}$
(1) Determine appropriate equation	$F = w = m\,g$
(2) Insert known quantities	$w = \frac{1.8\ \text{kg}}{} \Big\vert \frac{9.8\ \text{m}}{s^2} \Big\vert \frac{1\ \text{kg m}}{\text{N}\ s^2}$
(3) Calculate	$w = 17.64\ \text{N}$
(1) Determine appropriate equation	$A = L\ W$
(2) Insert known quantities	$A = (0.2032\ \text{m})(0.254\ \text{m})$
(3) Calculate	$A = 0.0516 m^2$
(2) Insert known quantities	$P = \frac{17.64\ \text{N}}{0.0516\ m^2} \Big\vert \frac{1\ \text{Pa}\ m^2}{1\ \text{N}}$
(3) Calculate	$P = 341.78\ \text{Pa}$

Comprehension Check 8-6

A wire made from Alloy A has a 3 millimeter [mm] diameter and is 1.25 feet [ft] long. A pressure, called "stress", is created by applying a force over the circular cross-sectional area of the wire, where the circular area is given by: $A = \pi r^2$. If the stress on the wire is 337 megapascals [MPa], what is the force on the wire in units of newtons [N]?

fluid

The general term **fluid** applies to a gas, such as helium or air, or a liquid, such as water or honey. In this chapter, we consider four forms of pressure, all involving fluids.

surface pressure
hydrostatic pressure
total pressure
gas pressure

- **Surface pressure**—the pressure created by the weight of a gas above an object.
- **Hydrostatic pressure**—the pressure exerted on a submerged object by the weight of a liquid above it.
- **Total pressure**—the summation of surface and hydrostatic pressure.
- **Gas pressure**—the pressure created by the movement of gas molecules inside a closed container. Gas pressure will be discussed in section 8.4.2.

8.4.1 Surface Pressure

atmospheric pressure

Surface pressure results from the weight of a gas above an object. When the gas is the air in the Earth's atmosphere, we call the surface pressure **atmospheric pressure**. Atmospheric pressure varies with both altitude and weather patterns. Standard atmospheric pressure is an average air pressure at sea level, defined as one atmosphere [atm], approximately equal to 14.7 pound-force per square inch [psi].

8.4.2 Pressure Measurement

When measuring pressure, two types of reference points are commonly used.

absolute pressure

Absolute pressure uses a perfect vacuum as a reference point. Most meteorological readings are given as absolute pressure, using units of atmospheres or bars. When solving problems dealing with hydrostatic or gas pressure, the pressure given should be considered as absolute.

gauge pressure

Gauge pressure uses the local atmospheric pressure as a reference point. Measurements such as tire pressure and blood pressure are given as gauge pressure.

NOTE:

Car tires are inflated with between 30 and 40 psi (gauge pressure).

"Normal" blood pressure is 120 mm Hg/80 mm Hg (gauge pressure).

Absolute pressures are distinguished by an "a" after the pressure unit, such as "psia" to signify "pound-force per square inch absolute." Gauge pressure readings are distinguished by a "g" after the pressure unit, such as "psig" to signify "pound-force per square inch gauge." When using instrumentation to determine the pressure, be sure to note whether the device reads absolute or gauge pressure.

Gauge pressure, absolute pressure, and atmospheric pressure are related by

$$P_{absolute} = P_{gauge} + P_{atmospheric}$$

For example, if we have a reading of 35 psig, this would be 49.7 psia assuming an atmospheric pressure of 14.7 psi.

$$13.5 \text{ psig} + 14.7 \text{ psi} = 49.7 \text{ psia}$$

vacuum pressure

If a gauge pressure being measured is less than the local atmospheric pressure, this is usually referred to as **vacuum pressure**. A perfect vacuum is defined as 0 psia. Thus, a perfect vacuum created at sea level on the Earth would read -14.7 psig, or 14.7 psig vacuum pressure.

As another example, if we have a reading of 10 psig vacuum pressure, this would be 4.7 psia assuming an atmospheric pressure of 14.7 psi.

$$-10 \text{ psig} + 14.7 \text{ psi} = 4.7 \text{ psia}$$

To illustrate the effect of local atmospheric pressure, consider the following scenario. You fill your automobile's tires to 35 psig on the shore of the Pacific Ocean in Peru, and then drive to Lake Titicaca on the Bolivian border at about 12,500 feet above sea level. The absolute pressure in the tires must remain the same in both locations, so your tire pressure now reads about 40 psig due to the decreased atmospheric pressure, which drops from 14.7 to 9.5 psi due to the altitude.

At the shore: $35 \text{ psig} + 14.7 \text{ psi} = 49.7 \text{ psia}$

At the lake: $49.7 \text{ psia} - 9.5 \text{ psi} = 40.2 \text{ psig}$

differential pressure

Occasionally in industry, it may be helpful to use a point of reference other than atmospheric pressure. For these specific applications, pressure may be discussed in terms of **differential pressure**, distinguished by a "d" after the pressure unit, such as "psid."

Comprehension Check 8-7

A car tire is inflated to a gauge pressure of 35 pound-force per square inch [psig]. What is the absolute pressure rating of the tire in units of atmospheres [atm]?

8.4.3 Hydrostatic Pressure

IMPORTANT CONCEPT:

Pascal's Law

$P_{hydro} = \rho g H$

Hydrostatic pressure (P_{hydro}) results from the weight of a liquid pushing on an object. *Remember, weight is a force!* A simple way to determine this is to consider a cylinder with a cross-sectional area (A) filled with a liquid of density ρ.

Pascal's law

The pressure (P) at the bottom of the container can be found by **Pascal's law**, named after (once again) Blaise Pascal. Pascal's law states the hydrostatic pressure of a fluid is equal to the force of the fluid acting over an area.

IMPORTANT CONCEPT:

How does the term $\rho g H$ equal a pressure?

$$P_{hydro} = \frac{F}{A}$$

Weight is a force $\quad = \dfrac{w}{A}$

$w = m g \quad = \dfrac{m g}{A}$

$\rho = m/V \quad = \dfrac{\rho V g}{A}$

$V = A H \quad = \dfrac{\rho (A H) g}{A}$

$$P_{hydro} = \rho g H$$

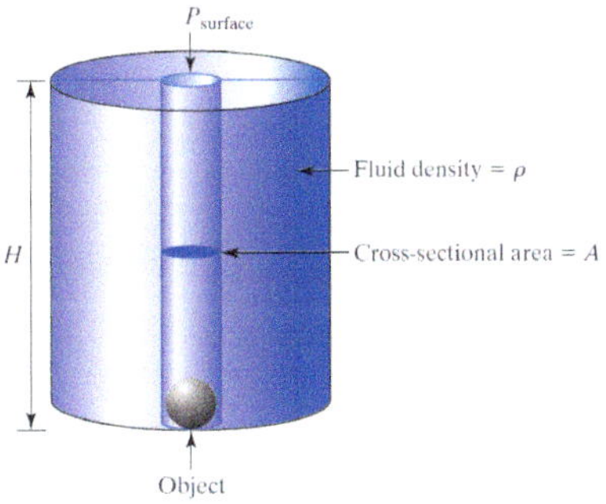

ENGAGE

To connect the formula for pressure with hydrostatic pressure, consider walking through the derivation for hydrostatic pressure from the generic pressure formula.

Recreational scuba diving takes place at depths between 0 and 20 meters. At deeper depths, additional training is usually required because of the increased risk of narcosis, a state similar to alcohol intoxication. The relationship between the depth and the level of narcosis is called the "Martini Effect," as it is said divers feel like they have drunk one martini for every 20 meters they descend.

Photo courtesy of E. Stephan

Example 8-7

We want to know the hydrostatic pressure of an object submerged in a lake at a depth of 20 feet [ft] in units of pascals [Pa].

NOTE:

Density of water = 1000 kg/m³

For hydrostatic pressure, we need to know the density of the fluid in the lake. Since a density is not specified, we assume the density to be 1000 kilograms per cubic meter [kg/m³] for water.

Convert to Base SI Units	Steps
(1) Term to be converted	20 ft
(2) Conversion formula (3) Make a fraction (4) Multiply	$\frac{20\ \text{ft}}{} \Big| \frac{1\ \text{m}}{3.28\ \text{ft}}$
(5) Cancel, calculate	6.1 m

Calculate Required Parameters	Steps
(1) Determine appropriate equation	$P_{hydro} = \rho\, g\, H$
(2) Insert known quantities	$P_{hydro} = \frac{1000\ \text{kg}}{\text{m}^3} \Big| \frac{9.8\ \text{m}}{\text{s}^2} \Big| \frac{6.1\ \text{m}}{}$
(3) Calculate, be reasonable	$P_{hydro} = 59{,}760\ \frac{\text{kg}}{\text{m s}^2}$

This is apparently our final answer, but the units are puzzling. If the units of pressure are pascals and if this is a valid equation, then our final result for pressure should be pascals. If we consider the dimensions of pressure:

		Exponents		
Quantity	**Common Units**	**M**	**L**	**T**
Pressure	Pa	1	−1	−2

A unit of pressure has dimensions P {=} M/(LT²), which in terms of base SI units would be P [=] kg/(m s²). As this term occurs frequently, it is given the special name "pascal." When we see this term, we know we are dealing with a pressure equal to a pascal.

(3) Calculate, be reasonable	$P_{hydro} = 59{,}760\ \frac{\text{kg}}{\text{m s}^2} \Big| \frac{1\ \text{Pa}}{1\ \frac{\text{kg}}{\text{m s}^2}} = 59{,}760\ \text{Pa}$

8.4.4 Total Pressure

We need to realize that Pascal's law is only a part of the story. Suppose we dive to a depth of 5 feet in a swimming pool and measure the pressure. Now we construct an enclosure over the pool and increase the pressure of the air above the water surface from 1 atmosphere [atm] to 3 atmospheres [atm]. When we dive back to the 5-foot [ft] depth, the total pressure will have increased by 2 atmospheres [atm].

IMPORTANT CONCEPT:

$$P_{total} = P_{surface} + P_{hydro}$$

$$P_{total} = P_{surface} + \rho g H$$

Consequently, we conclude that total pressure at any depth in a fluid is the sum of hydrostatic pressure and surface pressure.

Example 8-8

When you dive to the bottom of a pool, at 12 feet [ft] under water, how much total pressure do you feel in units of atmospheres [atm]?

Convert to Base SI Units	Steps
(1) Term to be converted	12 ft
(2) Conversion formula (3) Make a fraction (4) Multiply	$\frac{12\text{ ft}}{} \Big\vert \frac{1\text{ m}}{3.28\text{ ft}}$
(5) Cancel, calculate	3.66 m

NOTE:

Density of water = 1000 kg/m³

Surface pressure = 1 atm

For hydrostatic pressure, we need to know the density of the fluid in the pool. Since a density is not specified, we assume the density to be 1000 kilograms per cubic meter [kg/m³] for water.

For total pressure, we need to know the surface pressure on top of the pool. Since a surface pressure is not specified, we assume the pressure to be 1 atmosphere [atm] or 101,325 pascals [Pa].

Calculate Required Parameters	Steps
(1) Determine appropriate equation	$P_{total} = P_{surface} + \rho g H$
(2) Insert known quantities	$P_{total} = \frac{101{,}325\text{ kg}}{\text{m s}^2} + \frac{1000\text{ kg}}{\text{m}^3} \Big\vert \frac{9.8\text{ m}}{\text{s}^2} \Big\vert \frac{3.66\text{ m}}{}$
(3) Calculate	$P_{total} = 137{,}193 \frac{\text{kg}}{\text{m s}^2} \Big\vert \frac{1\text{ Pa}}{\frac{1\text{ kg}}{\text{m s}^2}} = 137{,}193\text{ Pa}$

Convert to Desired Units	Steps
(1) Term to be converted	137,193 Pa
(2) Conversion formula (3) Make a fraction (4) Multiply	$\frac{137{,}193\text{ Pa}}{} \Big\vert \frac{1\text{ atm}}{101{,}325\text{ Pa}}$
(5) Cancel, calculate, be reasonable	1.35 atm

Comprehension Check 8-8

An object is completely submerged in a tank that is open to the atmosphere, in a liquid with a specific gravity of 0.75 at a depth of 3 meters [m]. What is the total pressure on the object? State your answer in atmospheres [atm].

TEACHING TIP

Although this is mostly a repeat of an earlier section, it is potentially helpful to remind students of where pressure comes from—especially Pascal's principle. These ideas culminate in Bernoulli's Equation.

8.5 Properties of Fluids

conservation of mass

The first building block for understanding fluids is the continuity equation, which works under the idea that the flow of a fluid in should equal the flow of a fluid out. The premise of the continuity equation should sound familiar from Physics; this is simply **conservation of mass**. Although the continuity equation for a fluid is simple, it makes a powerful statement. Consider a single entry, single exit pipe that's carrying a constant flow of non-viscous fluid whose volume and density don't change with time (called incompressible). The size of the pipe, meaning its cross-sectional area, can change. If you know the cross-sectional area of the pipe A_1 and the velocity at any point v_1, you can quickly calculate the velocity at any other point v_2 by just finding the cross-sectional area at that point A_2.

$$A_1 v_1 = A_2 v_2$$

flow rate

The product of cross-sectional area and the fluid speed is called the **flow rate** (Q), which is the volume of fluid that passes through an area per unit of time.

Figure 8-6 When the end of a hose is restricted, the pressure increases.

You have probably conducted an experiment to show the effects of the continuity equation without realizing it. If you attach a hose to an outdoor faucet and turn it on, water flows out the end of the hose. You probably figured out that, if you put your thumb over the end, you can make the water shoot farther (Figure 8-6). By pressing your thumb over the hose opening, you reduce the cross-sectional area of the opening of the hose, which increases the velocity of the water as it leaves the hose.

Whether a fluid is moving or standing still, it is exerting pressure on its surroundings and objects within it. You might be more familiar with static pressure. Static pressure (P_s) is related to the magnitude of a force (F) by the area of the surface on which the force is pushing (A). In other words, it can be calculated as follows:

$$P_S = F/A$$

The classic science experiment illustrating the effect of static pressure looks like something out of a performance. In the experiment, your teacher bravely lies on a bed of nails and has a student place heavier and heavier objects on top of them. Despite the increase in weight, the teacher can get up from the bed unscathed and lecture another day. This is the principle of pressure at play! The force on the nails is distributed over a large enough area that they cannot pierce the teacher's skin.

In fluids, the formulation is a bit different; we would want to measure hydrostatic pressure. Hydrostatic pressure relies on the density of the fluid ρ (rho, pronounced "row"), the acceleration due to gravity g, and the depth of the fluid h. The hydrostatic pressure is independent of the container's shape, the amount of liquid in the container, or the surface area that the liquid touches (you'll notice no A appears like it does for the static pressure formula).

$$P = \rho g h$$

conservation of energy

Now, Bernoulli's Equation looks much more involved, but fundamentally is another statement of conservation. In this case, it is the **conservation of energy** for fluids. Here, both sides are mirror images of each other. The difference is that each side denotes two points in a fluid.

On the left side, P represents the pressure. Next, v represents the speed of the fluid. As before, ρ is the density of the fluid, g is the acceleration due to gravity, and h is the height.

$$P_1 + \frac{1}{2}\rho v_1^2 + \rho g h_1 = P_2 + \frac{1}{2}\rho v_2^2 + \rho g h_2$$

Dissecting the left side a bit further, we can give names to the different groups of variables. If we break it down, the equation is a combination of pressure energy, kinetic energy, and potential energy. Because both sides of the equation are mirrored versions of each other, it suggests that the sum of all the energy in the system needs to stay constant.

$$P_1 + \frac{1}{2}\rho v_1^2 + \rho g h_1 = \text{Pressure Energy} + \text{Kinetic Energy} + \text{Potential Energy}$$

driving force

By using the equation, we can compare the flow of a fluid at two different points. For example, we can use Bernoulli's Equation to describe how water tower height affects the pressure **driving force**, which is simply the difference in pressure. A water tower is a simple structure that serves an extremely fundamental role in a community, supplying water. A water tower is a tank of water that is elevated a certain height off the ground to provide enough pressure to distribute the water to all who need it (Figure 8-7). Municipal water supplies typically need a pressure of 50 to 100 pounds per square inch (psi).

Let's consider a simple problem where we want to see the relationship between tower height and the speed of the water at an outlet at the bottom of the tank. Assume the tank is open to the air on top. Whenever the input (or inlet) or output (outlet) is open to the air, it is said to be "open atmosphere" with 1 atm of pressure (which is 101,325 Pascals). Because both the inlet and outlet are open to the atmosphere, P_1 and P_2 are the same. That means they will cancel out in our calculation, leaving:

Figure 8-7 A typical municipality water tower.

$$\frac{1}{2}\rho v_1^2 + \rho g h_1 = \frac{1}{2}\rho v_2^2 + \rho g h_2$$

Because the tank is so large, we can consider the speed at the top of the tank to be negligible, so we'll set $v_1 = 0\ m/s$. Also, we'll assume the outlet is at ground level, which lets us conveniently choose h_2 to be zero as well, so $h_2 = 0\ m$. This means two more terms drop out of the calculation.

$$\rho g h_1 = \frac{1}{2}\rho v_2^2$$

Next, notice that ρ appears on both sides of the equation, which means we can divide it out of the calculation.

$$g h_1 = \frac{1}{2} v_2^2$$

Finally, we'll solve for v_2, leaving us with the simple formula:

$$v_2 = \sqrt{2 g h_1}$$

You'll notice here that the speed of the water at the outlet in the configuration we specified only depends on the height of the water tower! In practice, however, flow will depend on quite a few different quantities.

INTEGRATING CONTENT AREAS

To explore the mathematical relationship between the height and velocity, ask students to do the following: Write an expression for the velocity in terms of v_2 if the height is doubled.

If we double the height, we have the following result:

*v_2' = sqrt(2*g*2*h_1)*

To simplify, we can split the square root like so:

*v_2' = sqrt(2)*sqrt(2*g*h_1)*

*v_2' = sqrt(2)*v_2*

So, the velocity is scaled by a factor of sqrt(2); the velocity isn't doubled.

TEACHING TIP

You may want to emphasize that although the equation looks messy, terms are likely to drop out in certain situations. We'll see this in the example for a water tower.

ENGAGE

To help students visualize Bernoulli's Equation (or have a simulation to check calculations), you can use this simulation at the link below. Try adjusting the parameters and record how the pressure and velocity change.

Start with h_1=h_2=0, and r_1=r_2=0.5, v_1=1 and P_1=500,000. Adjust the values of r_1 and r_2, and observe how the pressure and velocity change. Then return r_1 and r_2 to 0.5 and adjust h_1 and h_2. How do the pressure and velocity change?

URL

https://ophysics.com/f12.html

TEACHING TIP

You may use the Bernoulli principle calculator to model how to properly take measurements that we discuss in section 11.2 of Chapter 11 to emphasize the need to use multiple, purposeful measurements within testing plans.

You can also use a calculator to generate practice problems for class.

TEACHING TIP

Provide students with a few example values by changing the height of the water tower.

INTEGRATING CONTENT AREAS

Although we have some insight into the behavior of fluids with the formulas presented here, there is still much work to be done. In fact, one of the Millennium Problems put forth by the Clay Mathematics Institute is about modeling the movement of viscous fluids.

URL

https://www.claymath.org/millennium-problems/navier-stokes-equation

DIFFERENTIATED INSTRUCTION

For more curious students, you may consider having them engage with a video describing the Navier-Stokes Equation.

URL

https://www.youtube.com/watch?v=XoefjJdFq6k

Poiseuille's Law

Poiseuille's Law helps us describe how fluid flows through a long cylindrical pipe in terms of the pressure drop ΔP between the two ends of the pipe, the viscosity μ (a measure of how "sticky" the fluid is), the length of the pipe L, flow rate Q, and cross-sectional area A.

$$\Delta P = \frac{8\pi\mu LQ}{A^2}$$

Therefore, the fluid flow will depend on the density, speed, and viscosity of the fluid. The characteristics of the pipe will also affect the flow, which includes the length, cross-sectional area, internal roughness, and number of bends and fittings. We'll experiment a bit with these ideas in the context of a water distribution network.

8.6 Water Distribution Systems

Broadly speaking, the purpose of a water distribution system is to move water from a water source and distribute that water to a community of people. The result is clean, drinkable water, delivered at the right flow and pressure for consumer purposes. At the same time, water from this source is also used for the needs of commercial concerns at sufficient flow and pressure for their needs (manufacturing, restaurants, agriculture, and so forth). One additional service of a water distribution system is to provide sufficient flow and pressure so that water is available to fight fires.

8.6.1 Types of Water Distribution Systems

Designing a water distribution system depends primarily on what water sources are available and the local geography. Source water can be from a man-made or natural reservoir (lake or pond), a river, or an underground aquifer. Elevation has a big role to play as well. In an ideal situation if the source is high enough above the delivery location, then gravity can serve to move the water through the system and the need for pumping stations is reduced or eliminated.

There are many ways to create a primary source reservoir, depending on the size needed. If nearby hills are suitable near the town, a surface reservoir is usually the best selection. If the particulars of the site are marginal, then additional distribution (or secondary) reservoirs can be built.

Small reservoirs may be simple excavations lined with suitable waterproof material. Larger reservoirs require a concrete lining, with retaining walls to resist soil loads while empty. Reservoirs are commonly covered with roofs to help reduce potential for contamination.

In the more generalized case, pumping stations are a requirement of the design, to maintain pressure and flow. In these cases, water can be pumped directly into closed distribution lines. More commonly, it is pumped into secondary distribution reservoirs (standpipes or water towers) which serve to equalize pumping rates throughout the day and provide extra capacity for peak use.

Pipes used to carry water into local neighborhoods are usually referred to as the main pipes or just simply water mains. Water mains generally follow the layout of the streets, with individual branch lines that supply each residence. The main pipes can be anywhere from 6" in diameter up to 16", depending on the flow that they are designed to carry. Mains are buried underneath the roadways and/or sidewalks. They need to be buried 4" to 6" below the local frost line, so they don't freeze and burst in the winter. It is good practice to have the mains loop through the neighborhood, avoiding "dead stops" since water in these areas can become stagnant. Service pipes to the customer, which branch off the main line to the individual houses or apartments are in the size range of 3/4 inches up to 6 inches.

single-main system

double-main system

There are two primary design approaches to the local neighborhood distribution strategy. First is the **single-main system**, where a single large pipe serves both sides of a street. In the other approach, a **double-main system**, there is a main on each side of the street. One main pipe supplies fire hydrants and domestic service on its side of the street, while the other (and smaller) pipe serves only domestic needs on the other side. The chief advantage of the two-main system is that repairs can be done with a minimum of interference with traffic or damage to the roadway.

An important consideration is for times when there is a local heavy demand, or during a fire. If an excessive head loss occurs, the pressure can drop below a safe level and flow will become inadequate. It is important to consider the reservoir level, the volume of secondary reservoirs, the size of the piping and the capacity of the pumps supplying the secondary reservoirs.

8.6.2 Pressure Requirements in Water Distribution Systems

When designing water distribution systems, pressure requirements for both ordinary use and firefighting must be considered. The *American Water Works Association* recommends a normal static pressure of 60 to 75 psi throughout a system.

Figure 8-8 Fire hydrant in a residential neighborhood.

residential districts

commercial districts

In **residential districts**, a minimum pressure of 60 psi at the hydrant is recommended (Figure 8-8). In **commercial districts**, a minimum pressure of 75 psi is acceptable, with the caveat that even higher pressures must be available in districts with tall buildings. Fire hydrants normally have a 2-1/2" outlet and flow at 610 GPM at a pressure of 15 psi. This is based on the minimum available pressure at the hydrant.

One option found in many cities is to use fire department motor pumper vehicles. In this way, they can develop the necessary pressure at the fire truck, so that normal operating pressure can be maintained throughout the pressure mains. This saves excessive maintenance on high pressure mains that have higher maintenance costs due to increased pumping costs and usually increased leakage.

For some large cities with lots of tall buildings, it is economical to have dual systems in business districts, a low-pressure system for ordinary use, and a high-pressure system (150 to 300 psi) for firefighting only. Finally, another approach is to use standby pumps to raise the pressure in the entire system to handle the added pressure needed whenever a fire occurs. Of course, this system must be designed to meet the high-pressure requirements, even though it will only see that pressure during an emergency.

Faucet pressures of at least 5 psi are considered satisfactory for most domestic needs. If we back-calculate from this value, we can determine the pressure required in the mains to make this work. Assuming a maximum pressure loss of 5 psi in the meter, about 20 psi in the house service pipe and plumbing, and with the main about 5 ft below ground level, a total pressure of about 35 psi in the main should be adequate for residential districts with one and two-story houses. Allowing about 5 psi for each additional story, a pressure of 75 psi would be typical for buildings up to 10 stories in height. For most domestic uses, a common household flow rate is around 8 GPM though a 1" diameter pipe with an average pressure of 50 psi.

Many cities require owners of tall buildings to install booster pumps in order to avoid the need for very high pressures in the mains. Unnecessarily high pressures should be avoided since leakage loss in the mains and from leaky plumbing fixtures will be increased.

8.6.3 Distribution Reservoirs

Distribution reservoirs are used to provide storage to meet fluctuations in use, to provide extra storage for firefighting, and to stabilize pressures in the water distribution system. In general, a secondary, or distribution reservoir should be located as close to the center of use as possible.

ENGAGE

Pressure requirements vary by appliances in the household. You may consider sharing the pressure requirements for appliances like refrigerators, dishwashers, and water softeners. For example, General Electric lists recommended pressure ranges on their website.

URL

https://products.geappliances.com/appliance/gea-support-search-content?contentId=18292

TEACHING TIP

You can supplement your discussion on water towers with the short video on how water towers work.

URL

https://www.youtube.com/watch?v=zSsxXhhGocg

ENGAGE

To provide a different perspective on water towers, you might want to share a video about New York's water towers. After watching the video, consider having a discussion about the need for maintenance of systems, especially public infrastructure.

URL

https://www.youtube.com/watch?v=P_9RLbRZJr0

You may also supplement the discussion with the American Society of Civil Engineers report on drinking water. What is the society's recommendation for increasing the efficiency of public water systems? *Implementing machine learning software and real time sensors to monitor the public water systems.*

URL

https://infrastructurereportcard.org/cat-item/drinking-water-infrastructure/

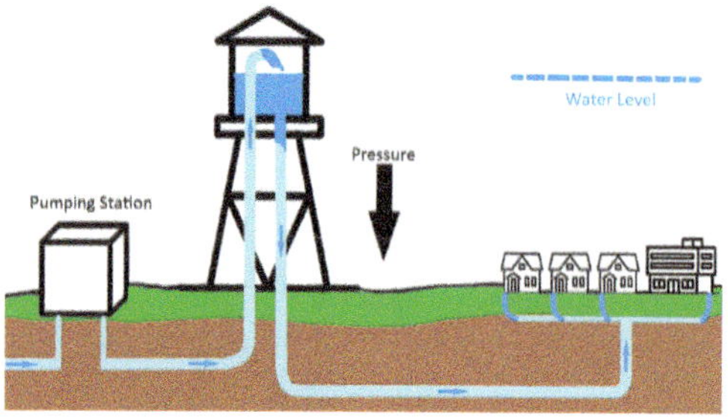

Figure 8-9 Water distribution diagram.

Figure 8-9 shows an example of a distribution reservoir. A significant consideration is the design of the water mains coming from the secondary distribution reservoirs. There should be at least two paths from the secondary distribution reservoirs, and they should be fitted with isolation valves. This will make sure you are covered if repairs must be done. This allows the repair team to isolate the secondary reservoirs to make repairs without interrupting supplies in an isolated district.

For secondary reservoirs, the water level must be high enough to permit gravity flow at satisfactory pressures to the system which it serves. Naturally for larger cities, several distribution reservoirs may be located at strategic points throughout the city, or even local to specific buildings.

Water is usually pumped into a distribution reservoir when the demand is low and withdrawn by gravity flow during periods of high demand. This type of elevated storage helps with pressure stabilization in high demand situations.

standpipe

If the topography does not permit sufficient head from a surface reservoir, then a **standpipe** or elevated tank may be used to gain the necessary hydraulic head. Steel, reinforced concrete, and timber are used for the construction of standpipes. Since tanks and standpipes are intended to reduce pressure variations in a distribution system, fluctuation of the water level in a standpipe is usually limited to 30 feet or less. The overall height of a standpipe is usually limited to 50 feet. Any higher than that increases cost without appreciably improving the function.

There are many different types of elevated tanks available. However, the selection of type, number, and location of elevated tanks as distribution reservoirs is an economic problem. The goal in designing these systems is to minimize the annual cost of reservoirs, pipe, and pumping.

8.6.4 Construction and Maintenance of Water Distribution Systems

The basic requirements of pipes for a water distribution system are adequate strength and maximum corrosion resistance. Cast iron, cement-lined steel, plastic, and asbestos-cement compete in the small sizes, while steel and reinforced concrete are competitive in the larger sizes.

Service connections to cast iron are made by tapping the distribution main with a special tapping machine which provides a threaded hole 1/2 to 2 inches in diameter. A shut off valve is then installed with a flexible gooseneck pipe leading to the service pipe. The gooseneck pipe prevents damage if there is an unequal settlement between the main and the service pipe.

Service pipes leading from the main to the consumer are usually copper tubing or galvanized steel. For single-family dwellings, 3/4 to 1-1/4-inch pipe is common for service pipes, but larger sizes may be needed for apartment houses or business establishments.

When a new pipe is first filled, all hydrants and valves are opened so that air can escape. The pipe is slowly filled (which may take several days for large systems) until a steady, uninterrupted stream issues from a hydrant, then it is closed. The procedure continues until all valves and hydrants are closed and the system is full of water.

Leakage from distribution systems will vary with the care exercised in construction and the age and condition of the system. Construction contracts usually specify an allowable leakage in the range from 50 to 250 gpd (gallons per day) per inch of pipe diameter per mile of pipe.

Leakage is tested by closing off a length of pipe between valves, and all service connections to the pipe. Then water is introduced, and normal working pressure is maintained for at least 12 hours while leakage is measured.

There are several possible methods of locating a specific leak. Leak detectors use either audio signals to "hear" the sound of escaping water or changes in the electrical field locally caused by saturated ground near the leak. Sometimes there may even be an obvious wet area where water is escaping. Once it is located, it must be repaired, and the location of all pipes, valves, and appurtenances should be entered on maps. This information is essential in case further repairs are required at a future date.

8.6.5 Disinfection of Distribution Systems

While handling the pipe during installation and stabilization in place, there are many opportunities for contamination. To avoid this, after repairs or additions the pipe is disinfected. Chlorine, calcium hypochlorite, or chlorinated lime in amounts sufficient to give an immediate chlorine residue of 50 mg/l is injected into the pipe.

One other bit of maintenance that is highly recommended when effecting repairs is to run a scraper through the pipe to reduce sediment deposits and foreign matter, and to improve the hydraulic efficiency of the pipe. Since the effects of scraping will be reversed over time, the pipes are often lined with cement mortar after cleaning to obtain longer lasting results.

TEACHING TIP

To elaborate on waste water treatment plants, the following video from National Geographic provides some additional context of how water is treated to become drinkable.

URL

https://www.youtube.com/watch?v=YW6GBciRHLg

DIFFERENTIATED INSTRUCTION

You can tweak the parameters of this map in whatever way you'd like. You can specify whatever population distribution you'd like. There's no "wrong" way to do it provided the populations in each zone add up to the total population of whatever you specified. Alternatively, you could pick the community your school is located in. You can also assign students to specific cities to introduce some variety into the mix.

DESIGN ACTIVITY

Designing a Water Distribution System

Problem Statement

The town of Sumewearin, Kansas, population of 20,468 according to the last census, recently acquired enough funding from the state and municipal bond sales to rethink how to best distribute water to different key pieces of infrastructure and houses. After years of complaints from local businesses and residents about weak pressure leading to water barely reaching a constant flow from their taps, the local government has released a request for proposals; your firm is eager to participate. In the proposals request, a map of Sumewearin is provided that outlines two residential districts, the commercial zone where shops and restaurants are located, and a few key pieces of infrastructure targeted in this water distribution project (Figure 8-10). Your goal is to design a water distribution network that would service the areas outlined in the local government's request for proposals. Here, our focus will be on the needs of the community and some initial concepts for how the network will be implemented.

Your instructor can divide up the population into the two residential districts as they see fit and vary the number of businesses (and what kinds of businesses) are in the commercial district.

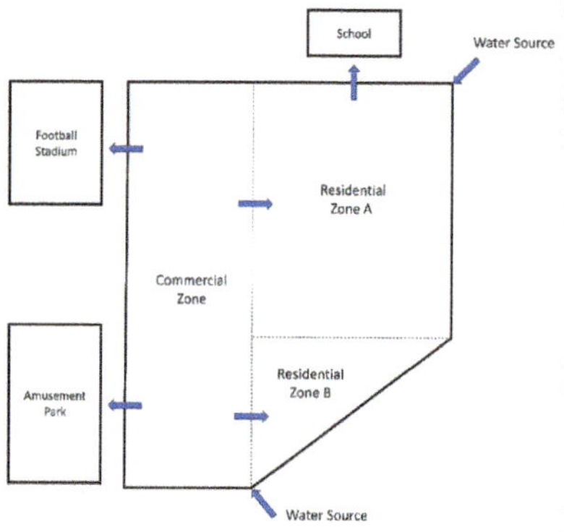

Figure 8-10 Generalization of small town into zones and major structures targeted for a water distribution network. Two suggested water sources are noted, and arrows denote which way the water is proposed to flow..

Materials (per team)

- Device connected to the Internet
- Writing utensils (different colors would be helpful)
- Paper (can be 8½ × 11 or larger, such as poster size)

Understanding Who Your Design Is For

To ensure your solution is appropriate, you need to understand the needs of the community. This involves gathering all the necessary information to complete the project appropriately. Record your findings in your engineering notebook and consider the following questions:

1. Who are the stakeholders in this design problem? List them and describe their relationship to the problem.
2. Develop a list of at least 10 questions that your team needs to consider such that the design meets the water needs of the community.
3. Research the water usage of the average household, restaurant, school, football stadium, and amusement park. How do these estimates vary by population size? What are the necessary pressure requirements for different appliances at these locations?
4. After you complete these steps, pair up with another team and compare your questions and findings. What similarities and differences did you see?

Develop Criteria, Constraints, and Design Specifications

Now that you've taken time to understand the needs of the community, next you must translate those needs into criteria, constraints, and design specifications. These will be used to guide the development of your solution and ensure it is successful. In your design notebook, address the following prompts:

1. What are the criteria you identified for this design?
2. What are the constraints limiting the possibilities of this design?
3. How will you know if your design is successful? Translate your criteria and constraints into measurable design specifications.
4. Consult with another team of students in your class and compare your criteria and constraints. What differences and similarities do you notice?
5. Develop a test plan to measure whether your design achieves the criteria you specified at the beginning of the challenge.
6. Did you revise or add design specifications during your test plan development?

Assess the Risks and Benefits of Your Design

Examine your criteria and constraints. What are the risks if your design fails to meet these? What are the benefits of implementing a correct, efficient, and functional solution? Let's address these questions:

1. What are the risks of failing to adequately meet the criteria?
2. What other risks exist?
3. What are potential benefits to an excellent design?

TEACHING TIP

We envision this project places emphasis on the first few stages of the design process. Therefore, the initial research phase is critical to emphasize. Place most of your effort in getting students to thoroughly address the "understanding who your design is for" phase.

TEACHING TIP

Ensure the students are connecting their research to the criteria, constraints, and design specifications. The design specifications must be assigned a number from their research at this point in the course.

DIFFERENTIATED INSTRUCTION

When students are coming up with potential risks, you can have students look up cases of failure in water distribution systems. The Flint Water Crisis is likely the most famous recent case of failure.

URL

https://www.cdc.gov/nceh/casper/pdf-html/flint_water_crisis_pdf.html

TEACHING TIP

The core of their analysis should be in comparing the different types of water distribution networks. You may suggest they create a table to make the benefits and drawbacks clearly distinguished.

DIFFERENTIATED INSTRUCTION

To incorporate more mathematics into the project when determining the fluid flow principles, you may elect to have students estimate the water demand for each zone. Have students research how water demand changes over the day and year for different cities to inform their estimate. This can be a way to reinforce the estimation techniques in section 3.3 of Chapter 3.

DIFFERENTIATED INSTRUCTION

To ensure a variety of assessment evidence, you may choose to let students pick their mode of communication. Or, you may elect to only check their design notebook, have them write a report, or make a presentation. It's not recommended that students be assessed using all modes for each project.

Generating Potential Concepts for a Prototype

Now it is time to generate some potential concepts for your water distribution system. We'll also perform some calculations to ensure the design is appropriate. In your design notebook, address the following items.

1. Create a sketch for three concepts and a description that captures how the concepts would work.
2. Search online for the following water distribution network types: Dead-end or Tree Distribution System, Gridiron Distribution System, Circular or Ring Distribution System, and Radial Distribution System. Describe the benefits and drawbacks of each one. How do they relate to your initial concepts?
3. How can you verify if your design will meet the needs of the community? What fluid flow principles would help you and why?

Selecting a Concept

After generating a series of design alternatives, we now must pick a concept to move forward with. In your design notebook, address the following:

1. Using your criteria and design specifications, set up a decision matrix to begin weighing the viability of the three concepts.
2. Determine weights to place on the different criteria.
3. Calculate the "winning" design using your weights.
4. Summarize your findings.

Report Your Solution

Finally, it is time to report out to the class. Your teacher may elect to have you present a PowerPoint summarizing your processes from each stage of the design process outlined here. Your presentation should focus on the preparation of the project, which would involve showing your *decision matrix* and *responsibly matrix* (included plans for conflict resolution and potential setbacks if the project was deployed in real life) created as part of your processes. You may also write a short report covering the same material. In your design notebook, create a bulleted list of the main decisions you made across the different stages of this project. If you are making a PowerPoint, create an outline for how you will present the information. Next, reflect on the following questions:

1. What was the objective of this project?
2. What worked?
3. What didn't work and why didn't it work?
4. How did the design compare with the best and worst performers?
5. What would you do differently?
6. Was the objective accomplished?

CHAPTER 8 ASSESSMENT QUESTIONS

1. Conduct research online to define the following terms: (a) *compressive strength*, (b) *resistivity*, and (c) *radioactivity*. Identify each as a physical, chemical, or mechanical property, then explain why you classified it as such. Describe the testing process for each property in two to three sentences.

2. Consider the following tests for a material's hardness: Mohr, Vickers and Knoop, Rockwell, and Brinell. We've briefly discussed Rockwell and Brinell. Conduct additional research online to define the process involved in Mohr and the Vickers and Knoop tests. Compare and contrast the processes involved in each test.

3. Explain the relationship between stress and strain in a tensile test. How can the tensile test be used to measure mechanical properties of materials?

4. Describe the following terms: voltage, current, resistance, and power. What are typical units for each of these?

5. Describe how pressure driving force affects flow rate from a pipe.

6. Describe how water tower height affects pressure driving force.

7. Describe how fittings, bends, pipe length, and pipe diameter affect flow rate in a pipe.

8. Consider a hose that lies flat on the ground. The pressure within the hose is approximately 350 kPa and the water is flowing at a rate of 2 m/s. The nozzle of the hose is open to the atmosphere. What is the velocity of the water at the nozzle? The density of water is 997 kg/m^3.

9. A 45-meter-tall water tower is open to the atmosphere and has an outlet 5 meters above the ground that is also open to the atmosphere. What is the velocity of the water at the outlet? The density of water is 997 kg/m^3.

10. An elevator has a height of 8 feet [ft] and a square floor area of 6 feet [ft] wide by 6 feet [ft] long. The pressure exerted by the passengers on the floor of the elevator cannot exceed 0.025 atmospheres [atm]. Assume each passenger weighs an average of 180 pound-force [lbf]. How many people can safely ride in the elevator per trip?

11. The voltage across a resistor is 2.3V, and the current through the resistor is 4.89mA. What is the resistance of the resistor? How much power is being dissipated?

12. A constant current of 1.8mA flows through a discharged 20.0pF capacitor. How long will it take to charge the capacitor to 3.3V?

13. A current of 54mA is flowing through a resistor. The voltage across the resistor is

14. 8V. What is the minimum required power rating (in Watts) required for this resistor? Round your answer DOWN to the nearest whole number.

15. A box containing basketball equipment is sitting on a shipping pallet. The box plus the contents weighs 25 newtons [N], and the pallet weighs 155 newtons [N]. The box can be modeled as a rectangular parallelepiped and has the following dimensions: 3 inches [in] by 15 inches [in] by 20 inches [in]. Assume the pallet is also a rectangle and has the following dimensions: 4 feet [ft] by 3.5 feet [ft] by 0.25 feet [ft]. The box is sitting on the pallet in the configuration shown. What is the pressure exerted by the box onto the pallet in units of pound-force per square inch [psi].

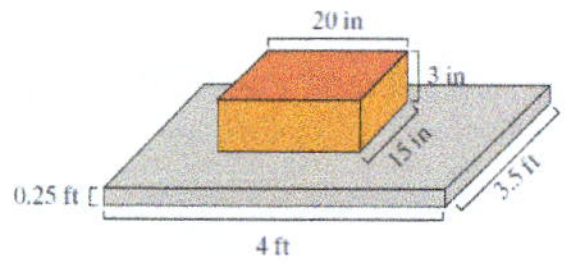

16. Assume the box described in Problem 11 is moved to the configuration shown in the picture below. What happens to the pressure exerted by the box onto the pallet?
    (a) The pressure is increased.
    (b) The pressure remains the same.
    (c) The pressure is decreased.
    (d) The pressure cannot be determined from the information given.

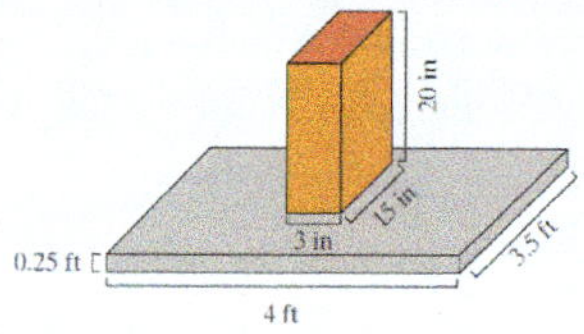

17. A pinball game is built with a plunger spring with a spring constant $k = 400\ N/m$ and a ball with a mass of 80 grams. Find the initial velocity of the ball upon launch if the player pulls the plunger:
    (a) 2 cm
    (b) 10 cm
    (c) 20 cm

18. A pinball machine is built with a 50 g ball and a plunger with $k = 200\ N/m$. Would the initial launch velocity be higher, lower, or the same if the plunger were pulled 8 cm and:
    (a) the ball were replaced by a ball with mass 60 g?
    (b) the spring were replaced by a spring with a spring constant of 175 N/m?
    (c) the spring were replaced by a spring with a spring constant of 175 N/m and the ball were replaced by a ball with mass 60 g?

OVERVIEW

Writing software (computer programs) involves describing processes and procedures; it involves the authoring of algorithms. Computer programming involves developing lists of instructions—the source code representation of software. These instructions manipulate different types of objects, e.g., numbers, words, images, sounds, etc. Creating a computer program can be like composing music, like designing a house, or like creating a variety of items.

CHAPTER 9

Computer Programming

TOPICS

In order to program a computer to do a meaningful task, we must first speak one of its languages—programming languages, that is. Communication with computers can be easily compared to everyday verbal interactions. There are thousands of spoken languages in the world; while the number of programming languages is not nearly as massive, the sheer number can be intimidating to new coders. A sizable number of the languages can be considered to have dialects, a form of a language only spoken in a specific region. The same effect can happen with programming languages, except the type of language will depend on the application. Here, we will develop the most fundamental skill to start speaking with a computer, algorithmic thinking.

algorithm

Learning to create effective algorithms is a crucial skill for any aspiring engineer. In general, an **algorithm** is a well-defined sequence of instructions that describe a process. Algorithms can be observed in everyday life through oral directions ("Simon says: raise your right hand"), written recipes ("Bake for 15 minutes at 350 degrees Fahrenheit"), graphical assembly instructions, or other graphical cues. As an engineer, writing any algorithm requires a complete understanding of all the necessary actions and decisions that must occur to complete a task.

9.1 Algorithms

When writing an algorithm, you must answer a few questions before attempting to design the process. To even begin thinking of a strategy to describe a process, you must have carefully defined the scope of the problem. The **scope** of an algorithm is the overall perspective and result that the algorithm must include in its design.

scope

For example, if we are required to "sum all numbers between 1 and 5," before thinking about an approach to solve the problem, we must first determine if the scope is properly defined. Does the word "between" imply that 1 and 5 are included in the sum? Do "numbers" include only the integer values? What about the irrational numerical values? Clearly, we observe that we cannot properly define the scope of the charge to add all numbers between 1 and 5 since there are an infinite number of such values.

Likewise, imagine you were charged to design a device that transports people from Atlanta, Georgia, to Los Angeles, California. How many people must the device transport? Does the device need to travel on land? Should it travel by air? Should it travel by water? Does the device require any human interaction?

263

OBJECTIVES

- Identify the main internal and external components of a computer
- Explain the basic components of a computer's programming design
- Understand various elements of coding structure
- Explain the difference between a variable's name and its value
- Work in teams to design and build a project related to computer engineering
- Write a reflection of the project
- Use problem-solving techniques to develop technological solutions such as product, process, or system
- Apply the design process, including decision matrices, as a team participant
- Perform different roles within the project as a team member
- Formulate decisions using collaborative strategies such as decision and design matrices and conflict resolution
- Maintain an engineering notebook for the project
- Develop and test the model for the project
- Demonstrate communication skills by preparing and presenting the project, including building consensus setback resolution and decision matrices
- Apply structured techniques such as a decision tree, design matrix, or cost-benefit analysis to select and justify a preferred solution to a problem
- Test and evaluate proposed solutions using engineering methods such as creating models, prototypes, mock-ups, or simulations, or performing critical design review, statistical analysis, or experiments
- Predict performance, failure modes, and reliability of a design solution
- Prepare a project report that clearly documents the designs, decisions, and activities during each phase of the engineering design process

OBJECTIVES

- Translate design requirements into a design solution
- Build and test designs against design specifications, evaluate the results of those tests, and present their analyses.

ENGAGE

Use the article by Ronij Pandey about Programming Languages and Their Types. He describes two categories of languages, high-level and low-level, which are distinguished by how much control the user has over the computer, readability, and efficiency.

URL
https://medium.com/@ronij.2508/programming-languages-and-their-types-d6aa9f7a361b

Explain to your students that a single computer programming language cannot do everything. There are limitations; different languages are suited for specific tasks. Students may recognize several of the languages; you can provide samples or discuss when and where they are used today.

ENGAGE

Visit the Scriptol.com programming website with your students. Navigate to the programming pages and ask your students to read the article entitled *History of Algorithms and Algorithmics.*

URL
https://www.scriptol.com/programming/algorithm-history.php

Ask your students who Abu Abdullah Muhammad ibn Musa Al-Khwarizmi was. *(A ninth-century Persian Muslim mathematician often considered to be the "father of algebra")* Ask your students what algorithm Archimedes is famous for. *(An algorithm that gives an approximation of the Pi number)*

written algorithms
graphical algorithms

This section covers two methods of defining a process: with **written algorithms** and with **graphical algorithms**. Both methods require properly identifying the scope of the problem and all the necessary input and output of the process.

9.1.1 Scope

One of the most difficult steps in designing an algorithm is properly identifying the entire scope of the solution. Like solving a problem on paper involving unit conversions and equations, it is often necessary to state all the known and unknown variables in order to determine a smart solution to the problem. If information is left out of the problem, it might be necessary to state an assumption in order to proceed with a solution. After all variables and assumptions about the problem have been identified, it is then possible to create a sequence of actions and decisions to solve the problem.

To clearly understand the scope of the problem, we often find it helpful to formally write out the known and unknown information, as well as state any assumptions necessary to solve the problem. In the following examples, notice that as the problem statements become more and more refined, the number of necessary assumptions decreases and eventually disappears.

Example 9-1

For the problem statement, list all knowns, unknowns, and assumptions.
Problem: Sum all numbers between 1 and 10.

Known:

- The minimum value in the sum will be 1.
- The maximum value in the sum will be 10.

Unknown:

- The sum of the sequence of numbers.

Assumptions:

- We will only include the whole number values (e.g., 1, 2, 3, . . .) in the sum.
- The sum will include the starting value of 1 and the ending value of 10.

Example 9-2

For the problem statement, list all knowns, unknowns, and assumptions.
Problem: Sum all numbers between (and including) 1 and 10.

Known:

- The minimum value in the sum will be 1.
- The maximum value in the sum will be 10.

Unknown:

- The sum of the sequence of numbers.

Assumptions:

- We will only include the whole number values (e.g., 1, 2, 3, . . .) in the sum.

TEACHING TIP

It is tempting to introduce programming in an introductory engineering course, but it can become a slippery slope where programming becomes one of the driving topics. We use MATLAB in this chapter because it is a common computational platform in engineering. However, it is possible to communicate the concepts in this chapter using flowcharts/flow diagrams exclusively. It's much more important to focus on the concepts than the minutiae of specific programming languages. If you do choose to introduce a programming language that's free, you may elect to teach Python.

DIFFERENTIATED INSTRUCTION

Because learning programming is so accessible through online resources, many students may come into class leaps and bounds beyond the expectations for the course.

It's crucial that you remind students that not everyone is starting from zero programming knowledge. It's okay if they are struggling and others are completing tasks in less than five minutes. The distribution of coding knowledge among students is likely very skewed. Therefore, ability grouping is appropriate here.

Tell your coding savvy students to avoid recommending more elegant or streamlined solutions that use commands or functions that are beyond the scope of the class. This can become too much for those with little programming knowledge to process and will cause more confusion than intended (usually because the suggested method is tailored to the problem).

Example 9-3

For the problem statement, list all knowns, unknowns, and assumptions.
Problem: Sum all whole numbers between (and including) 1 and 10.

Known:

- The minimum value in the sum will be 1.
- The maximum value in the sum will be 10.

Unknown:

- The sum of the sequence of numbers.

Assumptions:

- [None]

Comprehension Check 9-1

For the problem statement, list all knowns, unknowns, and assumptions. Problem: Sum all even numbers between (and including) 2 and 20.

Comprehension Check 9-2

For the problem statement, list all knowns, unknowns, and assumptions. Problem: Multiply all powers of 5 between (and including) 5 and 50.

9.1.2 Written Algorithms

A written algorithm is a narrative set of instructions required to solve a problem. In everyday life, we encounter written algorithms in the form of oral instructions or written recipes. However, it is extremely common for humans to "fill in the blanks" on a poorly written algorithm. Imagine you are handed a strongly guarded family recipe for tacos. One of the steps in the archaic recipe is to "cook beef on low heat until done." To the veteran cook, it is apparent that this step requires cooking the prepared ground beef on a stovetop in a saucepan for approximately 10 minutes on a burner setting of 2 to 3. To a first-time cook, the step is poorly defined and could result in potentially inedible taco meat.

Engineers and Written Algorithms

As an engineer, to write effective algorithms you must ensure that every step you include in a written algorithm must not be subject to misinterpretation. It is helpful to write an algorithm as if it were to be read by someone completely unfamiliar with the topic. Each step in the written algorithm should be written such that the stepwise scope is properly defined. The stepwise scope is all the known and unknown information at that point in the procedure. If a step in an algorithm contains an assumption, you must formally declare it before proceeding with the next step. By ensuring that the stepwise scope is well defined, you ensure that your algorithm will not be subject to misinterpretation.

All written algorithms should be expressed sequentially. The most effective algorithms are written with many ordered steps, wherein each step contains one piece of information or procedure. While the author of an algorithm may consider each step in an algorithm to be "simple," it might not be trivial to an external interpreter. When writing an algorithm, it is helpful to assume that the reader of your algorithm can only perform small, simple tasks. Assume that your algorithm can be interpreted by a computer. A computer can execute small tasks efficiently and quickly, but unlike a human, a computer cannot fill in the blanks with information you intended the reader to assume.

ENGAGE

Visit the TED-Ed YouTube website with your students. Ask your students to watch the video entitled *What's an algorithm?*

URL

https://www.youtube.com/watch?v=6hfOvs8pY1k

Explain to your students that an algorithm is a mathematical method of solving problems both big and small. Though computers run algorithms constantly, humans can also solve problems with algorithms. This video explains how algorithms can be used in seemingly simple situations and also complex ones.

ENGAGE

Visit the British Broadcasting Corporation website with your students. Navigate to the education pages and ask your students to read through the six-page article entitled *Introducing Algorithms.*

URL
http://www.bbc.co.uk/education/guides/z22wwmn/revision

The students can gain a better understanding of the algorithm by reading about current examples used today.

Except for the simplest algorithms, most will involve either asking questions and taking different actions based on the answers or repeating the same series of steps two or more times.

Format of Written Algorithms

The first step in writing any algorithm is defining the scope of the problem. After you define the scope of the problem, create an *ordered* or *bulleted list* of actions and decisions. Imagine taking an English class and writing a research report on the influence of 19th century writers on modern-day fiction authors. Before writing the paper, you would create an outline to ensure that your topics have connectivity and flow. Just like the outline of an English paper, an algorithm is best expressed as a sequential list rather than as complete paragraphs of information.

Example 9-4

Determine the pressure [atm] in a sealed container with a volume of 5 gallons containing 25 moles of methane at a temperature of 155°F.

Known:

- Volume $V = 5$ gallons
- Temperature $T = 155$°F
- Amount of methane $n = 25$ moles

Unknown:

- Pressure [atm]

Assumptions:

- Ideal gas constant $R = 8.314$ L kPa $K^{-1} mol^{-1}$

Algorithm:

1. Convert volume to liters
2. Convert temperature to kelvin
3. Calculate pressure [kPa] using $P = nRT/V$
4. Convert pressure to atmospheres

Comprehension Check 9-3

The total distance traveled from position zero by an object undergoing constant acceleration can be determined by

$$d = v_0 t + \frac{1}{2} a t^2$$

where v_0 is the initial velocity, t is time, and a is acceleration. Develop a written algorithm to calculate the total distance (in kilometers) traveled at time t in hours, assuming initial velocity is given in miles per hour and acceleration is given in meters per second squared.

9.1.3 Graphical Algorithms (Flowcharts)

flowchart

To visualize a process, a graphical representation of algorithms is used instead of a written algorithm. A **flowchart** is a graphical representation of a written algorithm that describes the sequence of actions of a process. Designing a flowchart forces the author of the algorithm to create small steps that can be quickly evaluated by the interpreter and enforces a sequence of all actions. Flowcharts are used by many different disciplines of engineering to describe different types of processes, so learning to create and interpret flowcharts is a critical skill for a young engineer. In fact, in the United States, any engineer who discovers a new, innovative algorithm can submit the concept to a patent office by representing the process in terms of a flowchart.

Three different shapes are used in the creation of flowcharts in this book; a number of other widely used operators are encountered across the world. In this book, we describe all actions with rectangles, all decisions with diamonds, and all connections between shapes with directional arrows.

Rules for Creating a Proper Linear Flowchart

- The flowchart must contain a START rectangle to designate the beginning of a process.
- All actions must be contained within rectangles.
- All shapes must be connected by a one-way directional arrow.
- The flowchart must contain an END rectangle to designate the end of a process.

Actions

actions

Actions are any executable steps in an algorithm that do not require a decision to be made. Based on this definition, any defined variables, calculations, and input or output commands (getting information from or sending information to the external world) would all be contained within action rectangles.

All simple actions are contained within a single rectangle on the flowchart. For each rectangle, two arrows are always associated with the shape, with two exceptions. The inward arrow to the rectangle represents the input to the action. It is assumed that any variables defined in the stepwise scope of an action rectangle are accessible and can be used in the action. The outward arrow from the rectangle represents the output of the action. If any new variables or calculations are performed within the rectangle, those values are passed along to the next shape's stepwise scope.

- Exception One: The START rectangle represents the beginning of the flowchart and does not contain an inward arrow. An oval shape is also commonly used to represent the start of an algorithm.

- Exception Two: The END rectangle represents the end of the flowchart and does not contain an outward output arrow. An oval shape is also commonly used to represent the end of an algorithm.

Example 9-5

Create a flowchart for the problem of Example 9-4.

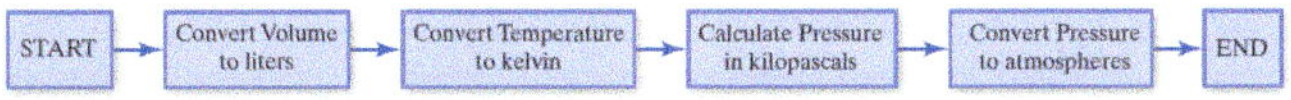

Comprehension Check 9-4

The total distance traveled from position zero by an object undergoing constant acceleration can be determined by

$$d = v_0 t + \frac{1}{2}at^2$$

where v_0 is the initial velocity, t is time, and a is acceleration. Create a flowchart to calculate the total distance (in kilometers) traveled at time t in hours, assuming initial velocity is given in miles per hour and acceleration is given in meters per second squared.

ENGAGE

Visit the BreezeTree Software website with your students. Navigate to the article page and ask your students to read the article *How to Create a Flowchart in Microsoft Word 2007-2019.*

URL
https://www.breezetree.com/articles/how-to-flowchart-in-word

Note: The instructor can research methods for developing flowcharts in any software package of his or her choice.

ENGAGE

Visit the Draw.io website with your students. Draw.io is free online diagram software for making flowcharts, process diagrams, org charts, UML, ER, and network diagrams.

URL
https://www.draw.io/

ENGAGE

There are many free apps available for the students to use on their devices. Ask the students to research a few of the available products for download. (iDiagrams, Digrammix Lite, and MetaFlow)

ENGAGE

Visit the Ruwix.com website with your students. Ruwix provides a large collection of free online Rubik's Cube and other puzzle-related programs, tutorials, and articles. Navigate to and read the article entitled *How to solve the Rubik's Cube.*

URL
http://ruwix.com/the-rubiks-cube/how-to-solve-the-rubiks-cube-beginners-method/

"There are two symmetric algorithms we use. They're called the F2L Right and F2L Left algorithms. These algorithms insert the Up-Front edge piece from the top layer in the middle layer, while not messing up the already solved white face."

9.1.4 Algorithm Best Practices

The remaining part of this section details specifics on how to begin planning and writing algorithms from scratch. It is not intended to be a definitive resource on algorithm development, but it may provide guidance if you are struggling to break down a process into small, achievable steps.

Actions

In every action within an algorithm, there must be a key verb that defines the purpose of that step within an algorithm. The remaining subsections discuss different types of actions and list some of the common verbs associated with that category of action.

Establishing Variables and Constants

After defining the scope of a problem, it might become obvious that there are intermediate calculations or assumed constants that must be contained throughout the process. Along with the explicitly defined known values, these intermediate and constant values are referred to as variables (even if they are constant!). Algorithmic variables are different from the mathematic definition of a variable because algorithmic variables are treated more like containers to store known values and results of calculations rather than being some unknown entity in a mathematical expression. They are called variables because the stored value can be written, overwritten, and used by other actions or decisions in the algorithm.

Example	Action
We assume the acceleration due to gravity is 9.8 meters per second squared.	Set variable *g* to be 9.8.

Other Verbs					
Set	Define	Assign	Write	Store	Designate
Label	Name	Cast	Insert	Save	Initialize

User Interaction

It is often necessary to write algorithms that can be executed with prompts for input from the person using the algorithm, provide feedback on results, or display any error messages generated in the algorithm.

Example	Action
We want the user of the algorithm to provide the amount of water in gallons.	**Input** the amount of water in gallons, save in variable *W*.

Other Verbs					
Input	Ask	Load	Request	Query	Prompt

User Output:

Example	Action
We want the algorithm to inform the user that the amount of water can't be negative.	**Display** error message to user "Warning: amount of water can't be negative!"

Other Verbs				
Output	Display	Reveal	Write	Warn

Calculations and Conversions

When algorithms involve calculating a value using an equation, it is helpful to write out the full equation and identify which variables in the algorithm correspond to the variables in the expression. For unit conversions, it is not necessary to write out the conversion factors since those are published standards that are readily available to anyone executing your algorithm.

When using conversions, it is best to list them individually, so they are easily recognizable to the user. For example, when converting from feet to centimeters, the expression $L = L/3.28 * 100$ is not difficult to recognize as the conversion from feet to meters, and then from meters to centimeters. It is harder to recognize the conversion of $L = L * 30.48$. Furthermore, it is easy to make a calculation error when lumping conversion factors together; it is easier to allow the program to calculate for you. When dealing with unit conversions, it is often best to save the converted value back into the original variable to reduce the number of variables you need to keep track of in your algorithm.

Calculations:

Example	Action
We want to calculate the thermal energy of a substance using the expression $Q = m\,C_p\,\Delta T$, where m is the mass, C_p is the specific heat, and ΔT is the change in temperature.	**Compute** the thermal energy: $Q = m\,C_p\,\Delta T$ All variables should appear in the variable list.

Other Verbs

Calculate	Adjust	Count	Measure	Add	Multiply
Subtract	Divide	Compute	Increment	Decrement	

Conversions:

Example	Action
We want to convert a variable t from minutes to seconds and save the result back in the variable t.	Convert t from minutes to seconds, save in t.

Other Verbs

Convert	Change	Alter	Revise	Switch

Referencing Other Algorithms

When developing a large program, it is sometimes helpful to break that program into several smaller programs, and then reference the smaller programs within the large program. In MATLAB, these smaller programs are called *functions*. As a rule of thumb, each custom function you create should have its own separate algorithm. If you have separate algorithms for a program and the different functions referenced in the code, it makes the algorithms simpler to understand and easier to debug. When calling a function within an algorithm, it is critical to list the variables passed to the function and variables returned by the function. We often refer to this process of returning variables as "capturing" the variables. In general, the most common verb used with functions is "call." If you know the name you plan to use for your function, list it; otherwise, this can be set later.

TEACHING TIP

It would be useful here to emphasize the concept of *modularity*, which involves the separation of programs/algorithms into smaller, independent algorithms/programs. Reference the idea of functional decomposition (block diagrams) and how we break down bigger problems into smaller ones that we can solve (to some extent, at least).

DISCUSSION

Consider reading the following article about the social media platform TikTok, and it's content recommendation algorithm. This can be done as pre-work before coming to class as a way to introduce the concept of algorithms.

URL
https://www.theguardian.com/technology/2022/oct/23/tiktok-rise-algorithm-popularity

The following video from the Wall Street Journal about TikTok's algorithm would also be a good supplement.

URL
https://www.youtube.com/watch?v=nfczi2cI6Cs&t=32s

How is TikTok's algorithm differ from others in the social media industry? *TikTok actively tests its predictions as you use the app. It bases its recommendations on the amount of time you linger on a given piece of content.*

Have students describe how the test account Kentucky96 was learned by TikTok. What content was the account interested in? *The test account was interested in sadness and depression. TikTok served a video about depression, which it intermittently recommended. Later, the user watches a video about heartbreak and then one about emotional pain. TikTok begins serving videos about relationships, but the user isn't interested. Kentucky96 pauses on one about mental health and swipes on videos about relationships. After lingering on videos about anxiety, 36 minutes since beginning the session, Kentucky96's feed is 93% videos about depression.*

INTEGRATE CONTENT AREAS

You may elect to have a discussion about the ethics of TikTok's algorithm, especially in the case of Kentucky96.

Example	Action
We want to use a function named *Poltocar* that converts coordinates from polar to Cartesian. We will pass in the variable *Z* as the radius and the variable *T* as the angle. We will capture the *x*-coordinate in the variable *X* and the *y*-coordinate in the variable *Y*.	**Call** *Poltocar* In: *Z*, *T* Out: *X*, *Y*

Testing Your Algorithm

The last step in writing an algorithm is developing test cases that will reveal whether or not your algorithm behaves as expected. The key to writing test cases is figuring out how many test cases are necessary to confirm whether or not your code works. In general, there should be at least one or two test cases that will demonstrate the proper behavior of the algorithm given good input values. Not only should a test case include a list of all the inputs used to generate the output, but you should also compute the expected output of the algorithm by hand in order to verify that the algorithm works.

Example 9-6

Create an algorithm to determine the volume and surface area of a cylinder, given the radius and height.

Known / Input:	Unknown / Output:	Assumptions:
Radius = r Height = H	Volume = V Surface area = SA	• All units will be in centimeters • Object is a cylinder. • Surface area does not include ends.

Algorithm: Written

1. Enter values of r and H
2. Calculate volume of a cylinder using $V = \pi r^2 H$
3. Calculate the surface area of a cylinder using $SA = 2\pi r H$

Algorithm: Flowchart

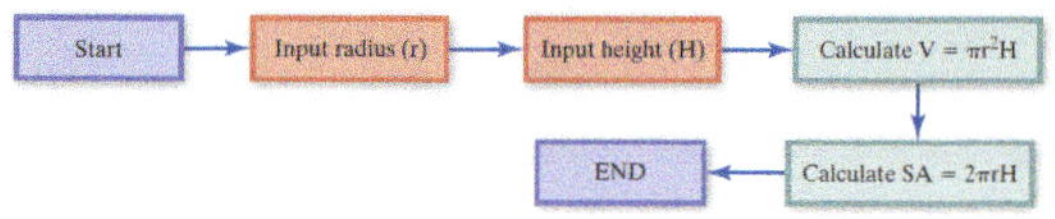

Test Cases:

Input:	Output:	
r = 3 cm; h = 4 cm	$V = 113.1\ cm^3$	$SA = 75.4\ cm^2$

Once you have developed an algorithm for a problem, it is time to convert the algorithm into programming code. To some extent, each item in your algorithm will be translated into one programming statement, but this is not always a one-to-one correspondence.

Before we actually write our first programs, we need to address a few concepts involving programming in general and MATLAB in particular.

9.1.5 Programs and Functions

programming

Programming is the process of expressing an algorithm in a language that a computer can interpret. To correctly automate a process on a computer, a programmer must be able to correctly speak the language both grammatically and semantically. In this text, we solve problems with a computer by using the MATLAB programming language. In the first part of this chapter, we introduced the idea of creating algorithms to represent a process. In the next section, we look at the process of creating variables containing different types of information in MATLAB. Later in this chapter, we will combine these two ideas to create programs and functions.

ENGAGE

Define the *Luhn algorithm. (The Luhn algorithm or Luhn formula, also known as the "modulus 10" or "mod 10" algorithm, is a simple checksum formula used to validate a variety of identification numbers, such as credit card numbers, IMEI numbers, National Provider Identifier numbers in the US, and Canadian Social Insurance Numbers. It was created by IBM scientist Hans Peter Luhn in 1960.)*

URL
https://en.wikipedia.org/wiki/Luhn_algorithm

The algorithm is in the public domain and is in wide use today.

9.2 Introduction to MATLAB

MATLAB

At the most simplistic level, the **MATLAB** programming language can be used like a powerful calculator. Not only can it do everything a fancy calculator can, it can also do far more with a single simple command than a calculator can as we will see in this section. First, we need to learn about two things:

- The MATLAB interface—what you see on the screen when you open MATLAB.
- The basic TYPES of data that MATLAB can deal with, since this encompasses considerably more than simple numeric values.

9.2.1 The Basic MATLAB Interface

When you launch MATLAB, you will see a window similar to the one that follows unless the default arrangement has been modified. You can return to the default layout by opening the Layout menu in the tool ribbon and selecting Default if your window looks a lot different.

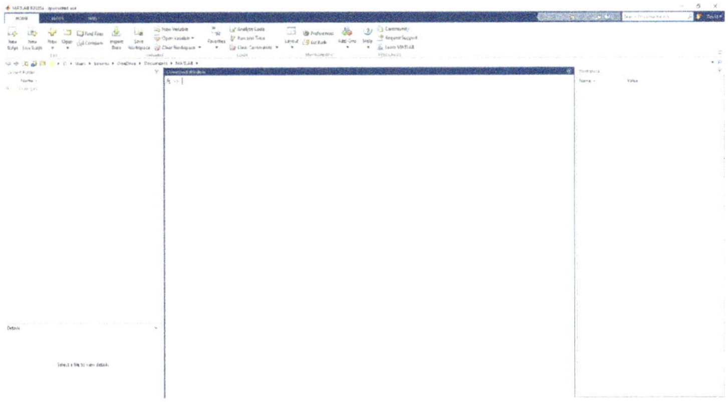

The MATLAB interface window is divided into several panes (or subwindows). These can be moved around, removed, others added, etc. We will discuss the use of the interface in more detail as we progress, but for now, we are going to focus on two of the panes.

Command Prompt

The large pane in the middle is called the Command Window. This is where you can type instructions to tell MATLAB what you want it to do, and is where you will type instructions when you simply want to use MATLAB as a calculator, although you can do many other things in the Command Window. After MATLAB completes whatever task you told it to do, it displays the Command Prompt, two greater-than signs (>>). The **Command Prompt** is MATLAB's way of telling you, "OK, I am ready to accept a command." *If you do not see the prompt AFTER the last text in the Command Window, then either it is off the screen (often to the right) or MATLAB is busy doing something else such as running a really long program and cannot accept another command at that time.*

workspace

The smaller pane on the right is called the **workspace**. The workspace shows all the variables (defined later in this chapter) in use at the current time.

TEACHING TIP

If you do not intend to teach programming in MATLAB, you may skip this section.

TEACHING TIP

As we mentioned at the beginning of the chapter, you should avoid having students memorize certain commands—especially for assessments. Emphasize that the important thing to learn is how to communicate what you want the programming language to help you accomplish. This allows you to find the appropriate functions.

We will discuss both of these windows in more detail shortly, but first we will discuss a feature of considerable significance.

Using MATLAB To Learn MATLAB

NOTE:

In general, `doc` and `help` should only be used as single-line commands in the Command Window, not included in an M-file. M-files will be discussed later in this chapter.

The two functions `help` and `doc` can be invaluable in learning what various MATLAB functions do and how they are used (their syntax). They can also help you explore the often labyrinthine sets of built-in functions to discover new and useful features of the language.

To illustrate the use of the `help` function, at the prompt (>> in the Command Window) type `help sqrt`. A short explanation of `sqrt` appears, along with the syntax used. Near the end of the information is a section that says "See also" followed by a list of hyperlinks to related functions.

To illustrate the use of the `doc` function, at the prompt in the Command Window type `doc sqrt`. This launches the documentation browser and displays information on the specified function. Not only does the documentation browser give you much more information about a function than using `help` does, usually including examples, it also makes it relatively easy to explore MATLAB. In the following figure, note the following features:

- The **Main Information** window on the right. This area typically lists the function and its basic syntax, a description of what the function does, examples of using the function, and a few other items. Of particular note is See Also near the very bottom, which lists closely related functions.
- The **Contents** outline on the left. This shows an outline of where within the structure of the MATLAB documentation the particular item being viewed is found. If the Contents outline is not visible, click the little outline icon (three horizontal lines) near the top left to expand it. In the example shown for the square root function, it resides within the group for Exponents and Logarithms, which is within the Elementary Math group, which is in turn in the Mathematics group. By clicking on one of these more-inclusive topics, you can explore related functions. At the bottom of the Contents outline are hyperlinks to each area within the selected function displayed in the Main Information window. Of particular note are Examples, from which you can learn a lot about using the function, and See Also.
- The **Search Box** near the top right of the Documentation window. Here, you can type function names or keywords to search for functions that do particular types of tasks.

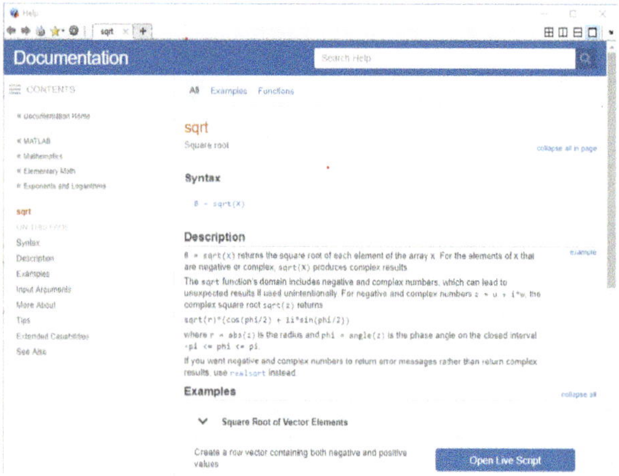

Table 9-1 lists a few functions that are useful in dealing with the MATLAB interface. It is strongly recommended that you spend some time early in your study of MATLAB using the documentation browser to familiarize yourself with its organization. It will be helpful in learning the language, but it takes practice to navigate.

Table 9-1 Useful MATLAB Interface Functions

Function	Description	Function	Description
`clc`	Clear the Command Window	`help`	Display basic information about a function in the Command Window
`clear`	Clear variables from the workspace	`doc`	Open the Documentation Browser
`close`	Close figure windows	`save`	Store a copy of workspace variables on the hard drive or other media
`;`	Do not display result of operation in Command Window	`load`	Get stored variables from the hard drive or other media and place in the workspace

9.2.2 MATLAB Variables and Data Types

A variable in MATLAB is not like a variable in mathematics. In math, a variable is typically an unknown for which you wish to determine a value. In MATLAB, on the other hand, a variable is simply a label for a container (or location) in which information is stored. Variables can be classified according to the type of information they hold. In the workspace, the data type of each variable is represented by an icon shown in the leftmost column of the workspace beside the variable name. Table 9-2 illustrates the different variable classifications in MATLAB.

Table 9-2 Variable Classifications in MATLAB

ICON	Data Type	Description	Example

	Numeric array	Scalars, vectors, and matrices containing numbers	`A = 5;`
			`B = [1, 2, 3];`
			`C = [1, 2; 3, 4];`

c h	Character array	Sequence of characters. Each character occupies one element and is individually accessible by numeric **indexing** (discussed later in this chapter).	`D = 'I Love Coding';`
			`Dprime = D(1)`
			`Dprime = 'I'`

str	String array	Text representation data type. String scalar (entire text) represents one element of a string array. Individual characters not directly accessible.	`E = "I Love Coding";`
			`Eprime = E(1)`
			`Eprime = "I Love Coding"`

	Logical array	0 means false; 1 means true Assigned using `true` or `false`	`F = true` `F = logical 1`

{ }	Cell array	Contains mixed data types. Indexed numerically.	`G = {'Age', 5}`
			`G = 1×2 cell array`
			`{'Age'} {[5]}`

	Structure array	Contains mixed data types. Data organized into named fields.	`H.age = 18;`
			`H.name = 'Juan'`
			`H = struct with fields:`
			`age: 18`
			`name: 'Juan'`

	Sparse array	Saves memory by storing only nonzero elements	*Not covered in this text*
	Symbolic array	Allows symbolic mathematics within MATLAB	*Not covered in this text*

ENGAGE

Visit the MathWorks.com website with your students. Navigate to the MATLAB section and consult the page on Variable Names.

URL
http://www.mathworks.com/help/matlab/matlab_prog/variable-names.html

"A valid variable name starts with a letter, followed by letters, digits, or underscores. MATLAB® is case sensitive, so A and a are *not* the same variable. The maximum length of a variable name is the value that the namelengthmax command returns."

DISCUSSION

Emphasize the necessity of informative variable names. Why might it be more helpful to use descriptive names when creating variables than shorthand to reduce keystrokes? *It enhances readability, especially for those who did not code the program. It also helps with debugging. If you do not have clear variables names, knowing the intention of why a variable is being defined or used can uncover issues that otherwise might have gone unnoticed.*

ENGAGE

Visit the TutorialsPoint.com website with your students. Navigate to the MATLAB section and consult the index page.

URL
http://www.tutorialspoint.com/matlab/index.htm

"This tutorial has been prepared for the beginners to help them understand basic to advanced functionality of MATLAB."

9.2.3 Variable Basics

Every variable in MATLAB must have a name. This allows us to keep track of large amounts of information and gives a means whereby our code can use the stored information relatively easily. MATLAB names must follow certain rules to avoid unintended side effects. These rules apply to *any* named item in MATLAB, not just variables.

NAMING RULES

- All names must consist only of letters, numbers, and the underscore character (displayed by holding Shift and typing a hyphen).
- Names can *only* begin with alphabetic letters (no numbers or special characters).
- Names cannot be longer than 63 characters. MATLAB will automatically truncate the variable name and display a warning message, but it will not generate an error message. As a result, the variable name will be different from what you expected. Be careful!
- Names in MATLAB are case sensitive: `Frog` and `frog` are different.
- Names cannot have the same name as any other identifier (program name, function name, built-in function, built-in constant, reserved word, etc.), because MATLAB's order of execution may prevent the program from finding the correct item.

NOTE:

A common error in naming variables (or other things) in MATLAB is to include a space character. MATLAB names cannot have a space in them—use an underscore instead.

Always choose names that have meaning within the context of the program, not just random letters. We cannot overemphasize the importance of well-chosen variable names in creating self-documenting code—code that explains itself. For example, if a variable contained the diameter of a bolt, the name `BoltDiam` is immediately recognizable for what it contains; a randomly chosen variable such as `X1` or `Q` has no meaning in the context of the program.

Comprehension Check 9-5

Which of the following are valid MATLAB variable names? For those that are not valid, explain why they are invalid.

(a) `m`
(b) `mass6`
(c) `4mass`
(d) `m&m©`
(e) `Dog Breath`
(f) `MyFile-docx`
(g) `clear`
(h) `ReactorYield`

The Assignment Operator

assignment operator

Before beginning our discussion of data types in MATLAB, we need to introduce the concept of the **assignment operator**. The purpose of an assignment operator is to tell the computer to put a value into a variable. Perhaps unfortunately, the most commonly used assignment operator in modern programming languages is =, or the equal sign. We say unfortunately because this notation often confuses novices. In mathematics, a statement such as `A=A+1` is an equation and has exactly two solutions for A: infinity and negative infinity. However, in programming this means "add one to whatever is in variable `A`, then place the result back in `A`."

NOTE:

To avoid the dual problem of mistaking = for "equals" and to explicitly indicate the direction of the movement of information, some languages use a left-pointing arrow (or other symbols) for the assignment operator.

It is important to note in MATLAB the = symbol should be read as "is assigned the value of." Thus, the programming statement `V2=V2+4` is read " `V2` is assigned the value of (the original value of) `V2` plus four." Note that the original value of `V2` is lost in this process.

It is also very important to note that the transfer of data in an assignment statement is always from right to left.

Consider the programming statement `Ex1=V3-V4^P2`.

When the computer executes this statement, it recognizes that it is an assignment statement. Therefore, it first looks at the expression to the right of the equal sign and determines its value: Raise the value in `V4` to the value in `P2` and subtract that from the value in `V3`. Next, it looks to the left of the equal sign to find the destination, thus placing the calculated result into the variable `Ex1`.

It is also important to understand that whatever appears to the left of the assignment operator MUST be a variable or a group of variables. For example, you will get an error if you type `A+1=A` since you cannot have a computation on the left of the equal sign. Also, any variables that appear on the right must already be defined before the statement is executed.

Comprehension Check 9-6

Which of the following assignment statements are valid? For those that are not valid, explain why. For those that are valid, show what value is placed into which variable. Assume the following variables (and no others) have already been defined: `A=2 B=-1 C=3 D=2.5`

Note that the priority of operators is the same as in Microsoft Excel. (See the Note below.)

(a) `X = A`	**(e)** `V = D - Q^2`	**(i)** `D = D^B`
(b) `B = Y`	**(f)** `U + 13 = 65`	**(j)** `A = 9999`
(c) `Z = A + 55`	**(g)** `T^2 = 4`	**(k)** `A + B = C - 5`
(d) `W = A^B + C`	**(h)** `Part#3 = C`	**(l)** `QPDoll = C^C`

NOTE:

See the note in Comprehension Check 9-6 above. There is a minor variation in the order of operations between MATLAB and Excel, which is the "unary minus sign" case. In Excel, = -4^2 results in a value of 16, whereas in MATLAB, -4^2 results in a value of -16. Excel considers the negative sign to be a property of the number 4, whereas MATLAB gives a higher priority to the exponent and applies the negative sign afterwards. MATLAB will match Excel if parentheses are used.

Saving and Restoring Variables

For a variety of reasons, we sometimes would like to be able to save all or some of the variables in the workspace and then restore them later. For example, we might wish to archive results, we might want to use the same data in another program, or we might need to share a set of variables with colleagues.

Variables can also be saved as .txt files in ASCII format, but this is beyond the scope of this text. See the MATLAB documentation files for more information.

To save all variables in the workspace, type

```
save('filename')
```

In this case, the filename must have an extension of .mat on the end. If you do not include .mat, MATLAB will add it for you.

As an example, the command

```
save('TestVar.mat')
```

will save all the variables currently in the workspace in a file named `TestVar.mat` placed in the current directory.

TEACHING TIP

At the beginning of the programming unit, you may choose to use "<-" as your assignment operator to distinguish it from the conventional equals sign. In fact, the notation "<-" is used in languages like R (although = will still work).

WARNING:

If your workspace already has variables with the same names as any of the variables being loaded, they will be replaced, thus losing the contents in those variables prior to the load. No warning will be generated.

To place a set of stored variables into the workspace, type

```
load('filename')
```

Again, the filename must have an extension of .mat on the end. For example,

```
load('TestVar.mat')
```

will place all the variables stored in `TestVar.mat` into the current directory. As with saving variables from the workspace, if you do not type the .mat suffix, it will be assumed.

To save only selected variables in a .mat file, use the format

```
save('filename', 'Var1', 'Var2', . . .)
```

As an example:

```
save('TV2.mat','Dens','Temp','Vol')
```

will store the three workspace variables `Dens`, `Temp`, and `Vol` in the file `TV2.mat`.

You can also load only selected variables from a .mat file. The format is similar to the save format. For example, assume TV2.mat was created as in the previous example.

```
load('TV2.mat','Dens','Vol')
```

will place the two variables `Dens` and `Vol` from the file TV2.mat into the workspace.

The commands `save` and `load` also support wildcard variable names using the asterisk (*) symbol.

For example,

```
save('WC1.mat','Q*')
```

will save all workspace variables beginning with `Q` in the file `WC1.mat`.

```
load('WC2.mat','*3')
```

will load all variables stored in the file WC2.mat that end with 3.

```
save('WC3.mat','*_*')
```

will save all workspace variables containing an underscore character in the file `WC3.mat`.

Comprehension Check 9-7

(a) Store all workspace variables in the file `TempData.mat`.
(b) Load all variables stored in `PressData.mat` into the workspace.
(c) Save the workspace variables `Pr1`, `Pr2`, and `Tmp3` in the file `PTData.mat`.
(d) Store all variables whose names end with "`al`" in the file `alVar.mat`.
(e) Load all variables stored in `VelData` whose names contain "`XS`" into the workspace.

9.3 Programs

Certain principles apply to programming, regardless of what language is being used. Entire textbooks have been written on this subject, but a few important concepts are addressed here.

9.3.1 Notes About Programming in General

- **Programming style**: Just like technical reports should follow a particular format to ensure they can be understood by someone reading them, all programs should have a few common elements of programming style, particularly the inclusion of comments that help identify the source of the program and what it does. For any program, a proper header should be written describing the scope of the problem, including a problem statement and definition of all input and output variables used in the program. Properly commenting the source code is also critical for ensuring that someone else can follow your work on the program.

  As a rule of thumb, for each rectangle and diamond in a flowchart, there should be a few comments in the source code where that action occurs.

- **Program testing**: When you write a program, testing it is a critical step in making sure that it does what you expect. Lots of things can go wrong.
  - *Syntax errors*: These violate the spelling and grammar rules of the programming language. Compilers (programs that interpret your computer program) generally identify their location and nature before the code is actually executed.
  - *Runtime errors*: These occur when an inappropriate expression is evaluated during program execution. Such errors may occur only under certain circumstances. Examples of runtime errors are a variable not being defined before a line of code tries to use it; incorrect matrix dimensions, such as trying to add two matrices of different sizes; or the wrong kind of input, such as a user entering letters when numbers are expected. Many languages will report an error when an infinite result occurs in a calculation, such as division by zero or taking the logarithm of zero, but MATLAB will simply calculate this as inf (infinity) and continue without error. Computer programs should anticipate runtime errors and alert users to the error, rather than having the program print angry red letters on the screen and unexpectedly terminate.
  - *Formula coding errors*: It is not uncommon to enter a formula incorrectly in a form that does not create either a syntax or a runtime error. For example, if you wrote `Q=X+1^2` with the intention of adding one to X, then squaring the result, no error would be generated, but the answer would not be correct due to priority of operators. Evaluating even a few simple calculations by hand and comparing them to the values calculated by the program will usually reveal this kind of error.
  - *Formula derivation errors*: This difficult error to discover in programming occurs when the computer is programmed correctly but the programming is based on a faulty conceptualization or erroneous formulae. Programmers generally assume that a program that executes without generating any errors is correctly written, and if the hand-verified sample calculations are based on the same misconceptions, this type of error can be very difficult to detect. This situation underscores the importance of checking a solution strategy before starting to write a computer program. For particularly complicated situations, it is helpful to have someone else check your approach to the solution.
- **Keeping track of units**: Some computer programs can be designed to have the user keep track of units as long as they use consistent units. Other computer programs require input data to be in a particular set of units. Regardless of which approach you use, the way in which units are managed must be clear to the user of the program and in the program's comments. In MATLAB, there are also specialized data types that carry units directly on the variable, such as the duration and datetime variable types.

ENGAGE

Define the *static testing* and *dynamic testing* approaches to software testing. *(Reviews, walkthroughs, or inspections are referred to as static testing.)*

Static testing *is often implicit, like proofreading. It also takes place when programming tools/text editors check source code structure or compilers (pre-compilers) check syntax and data flow as static program analysis.*

Dynamic testing *takes place when the program itself is run. Dynamic testing may begin before the program is 100% complete in order to test particular sections of code and is applied to discrete functions or modules.)*

TEACHING TIP

If user input is required in the program, it's useful to specify the expected units in whatever message is displayed to the user. Otherwise, it's possible the user could get a number that is off by a particular conversion factor and they may not catch the error. This becomes even trickier when the program is used in other larger programs, which could propagate errors further.

ENGAGE

Define the *comments* in computer programming. *(A comment is a programmer-readable annotation in the source code of a computer program. Comments are added with the purpose of making the source code easier to understand, and are generally ignored by compilers and interpreters. The syntax of comments in various programming languages varies considerably.)*

ENGAGE

Visit the MATLAB YouTube website with your students. Bookmark and utilize the videos developed by MathWorks.com.

URL
https://www.youtube.com/@MATLAB

- **Documentation**: In any programming language, there is usually a special character or pair of characters that tell the program interpreter to ignore everything to the right of the special character to allow the programmer to write human-readable notes within the code of the program. These notes are commonly referred to as "comments" within the code. A great deal of controversy surrounds how much documentation is necessary in any program, but this text proposes the following guideline to properly document a program:

  For every numbered item in a written algorithm, you should include that item as a comment in your code.

In addition, at the top of each program, it is extremely helpful to at least write a problem statement and document any major variables used within the program. It is helpful to cluster your description of variables into three groups: input variables, output variables, and function variables.

9.3.2 Notes Particular to MATLAB

Before proceeding with the material on MATLAB, it would be wise to review matrices. Since variables in MATLAB are inherently matrices, it is critical that you understand not only some basic matrix concepts, but also the specific notation used in this text (as well as MATLAB) when referring to matrices. Certain details specific to MATLAB are important to note early on.

- **The MATLAB interface**: The MATLAB program window has a number of subwindows. The specific subwindows displayed and their organization can be customized using the Layout menu in the Home tab of the tool ribbon.

NOTE:

`clc` = clear the Command Window

To clear all text from the Command Window, enter the special reserved MATLAB expression `clc` in the Command Window.

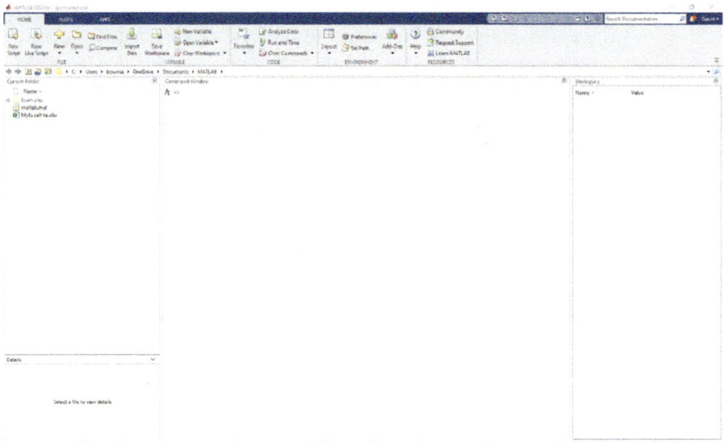

Individual subwindows can be undocked from the main window so that they appear in their own separate windows. The small arrow in a circle near the top right of each subwindow can be clicked and various actions involving that subwindow, including undock, may be selected.

– Previously in this chapter, we typed individual commands in the Command Window. We could type several commands in a sequence to achieve a larger purpose than was possible with a single command. If we wanted to repeat that same sequence of commands with perhaps one minor change in one of the commands, we could use the up arrow to select the previous commands in sequence to edit and re-execute. For long sequences of commands, this method would become extremely cumbersome, however. MATLAB provides a special interface, called the Editor/Debugger, which is where you will normally write your programming code. Then, if a minor change is necessary, you only need to edit that one line of code and tell MATLAB to re-execute the entire sequence.

FILE TYPE:

Word = .docx
Excel = .xlsx
PowerPoint = .pptx
MATLAB program files = .m
MATLAB saved variables = .mat

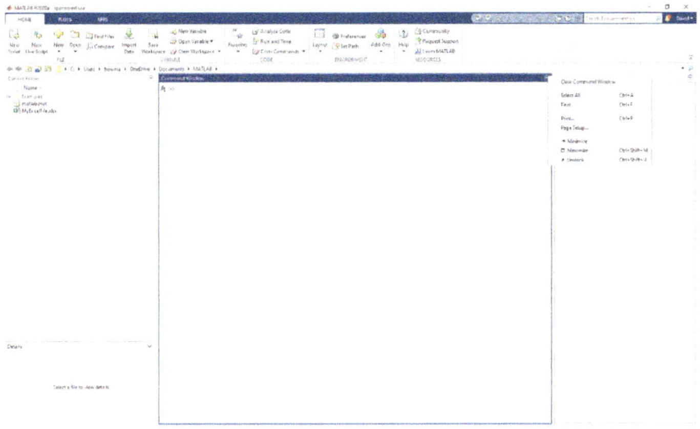

NOTE:

Order of search:
- Variable
- Built-in function
- Program or function in current directory
- Program or function in current path

– The *Editor/Debugger window* is the interface of choice for developing MATLAB programs. The Editor stores your list of commands (your program) in what are called M-files, since they have a file extension of .m. This permanent record of your commands allows you to make small changes to either the program or the inputs and to re-execute a complicated set of commands very easily. The Editor also assists in the formatting of MATLAB programs, using both indentation and color to distinguish program elements. To launch the Editor window, click the **New Script** button on the **Home** ribbon or press CTRL+N. The Editor interface will launch in a new window.

- **Naming programs, functions, and variables in MATLAB**: In MATLAB, program names and function names must follow the same rules for naming variables that we introduced previously in this chapter. Note that they cannot include space characters.
- **Order of search**: Order of search is the sequence MATLAB goes through when you type a name in either the Command Window or in a program or function. If you type `Frog` in the Command Window:
  1. MATLAB first checks if `Frog` is a variable and displays its value if it is.
  2. If it was not a variable, MATLAB checks if `Frog` is one of MATLAB's built-in functions and executes that function if it is.
  3. If it was neither a variable nor a function, MATLAB checks if `Frog` is a program or function in its current directory and executes the program or function if it is.

TEACHING TIP

When students ask if something is possible or would work as part of their program, encourage them to use the command window to test their idea!

TEACHING TIP

If a student wants to have a longer name for a variable or program, the underscore (_) is a popular way to introduce a space without violating the naming convention rules.

ENGAGE

Another popular method of avoiding spaces is the use of *Camel Case,* which is accomplished by switching between upper and lower case to distinguish words from one another. The following article can be shared to show how Camel Case is used outside of programming context.

URL

https://www.techtarget.com/whatis/definition/CamelCase

ENGAGE

Emphasize the need for comments in programs. Like informative variable names, comments greatly increase the readability of the code. Without descriptive comments, the code itself may not be very enlightening. Combined with poor variable name choices, the code may become too dense to decipher in a reasonable amount of time.

4. If it still has not found Frog, MATLAB checks if Frog is a program anywhere in its path and executes the program or function if it is. The location where you store your MATLAB programs or functions must be included in MATLAB's path. You can check where MATLAB is looking for programs and functions by typing path in the Command Window. Any M-files in the directories listed will be visible to MATLAB. If you try to execute an M-file outside any of the directories listed, MATLAB will ask if you want to switch current directories or add that directory to the list of paths MATLAB can see.
5. If it cannot find Frog anywhere, it will send angry red letters to the screen stating that the name was an undefined function or variable.

NOTE:

In programming, the term Easter Egg refers to an unexpected "feature" that the programmers included just for fun. One Easter Egg in MATLAB is the why function. Type **why** in the Command Window and see what appears. Type **why** again, and again, and again...

- **what and who**: To display a list of all MATLAB programs in the current directory, type what into the Command Window. To display all variables currently in MATLAB's workspace, type who in the Command Window. Remember that all variables automatically clear after you exit MATLAB.
- **Comments**: For documenting or commenting any code in MATLAB, a special character is used to "comment" out text: the percent symbol (%). The comment character can appear anywhere in a MATLAB program or function, but preferably either at the beginning of a line or at the middle of a line of MATLAB code. Everything after the percent sign to the end of the line is ignored by MATLAB, so you as a programmer can "comment out" any broken lines of code during code testing in addition to providing comments. Program or function file headers will usually have 6–10 lines of comments at the top of every program and more throughout the rest of the program or function.

```
% This is an example of a comment line.
% Comment lines are shown in the Editor window in green.
```

block comment

A **block comment** is a block of nonexecuted text between the special block characters %{ and %} that allow the user to type full-length paragraphs as comments without needing to type the % symbol before each sentence. For example:

```
%{
This is an example of using
a block of comments in a code
%}
```

9.3.3 Program Structure and Use

program
M-files

A **program** is a set of instructions given to a computer to perform a specific task. In this section, we create MATLAB program files, or **M-files**, to automate our processes. To create a program in MATLAB, launch the MATLAB Editor. By default, the Editor will create a new, blank document named Untitled, or Untitled2, or Untitled3, and so on, depending on how many unsaved programs are open in the Editor at once. The following example provides a walkthrough for creating a basic program, as well as screenshots of the actions described in the text to help you to follow along while re-creating these steps on your own computer running MATLAB.

Example 9-7

Video Solutions

We want to create a MATLAB program to convert Cartesian coordinates (x, y) to polar coordinates (r, θ). Before we can even try to write the program, we need to recall some information about the coordinate systems.

We know the following:

- In the Cartesian coordinate system, the coordinates (x, y) represent where a particular data point is in relation to the origin $(0, 0)$ by specifying the horizontal and vertical distance with respect to the origin. The x-value represents the horizontal distance and the y-value represents the vertical distance.
- In the polar coordinate system, the coordinates (r, θ) represent the exact distance (r) from the point to the origin $(0, 0)$ as well as the angle of elevation (θ) with respect to the positive horizontal axis, typically in units of radians.
- The conversion from polar coordinates to Cartesian coordinates is

$$x = r \cos \theta$$
$$y = r \sin \theta$$

- The conversion from Cartesian coordinates to polar coordinates is

$$r = \sqrt{x^2 + y^2} \qquad \theta = \tan^{-1}\left(\frac{y}{x}\right)$$

We now have the equations necessary to write a program to convert Cartesian coordinates to polar coordinates, and vice versa.

NOTE:

CTRL + C = force quit a MATLAB program.

First, we need to learn how to ask MATLAB for more information on doing things like calculating a square root or calculating an inverse tangent of a value if we do not already know how to do these operations.

In MATLAB's Command Window, we can search for built-in functions to calculate a square root and an inverse tangent. The `lookfor` *function allows you to search MATLAB help documentation for certain key words. Note that if you type* `lookfor square root` *into the Command Window, MATLAB displays an error message. The* `lookfor` *command is expecting only one keyword following the command "lookfor," so we decide to try our search again, this time including only the word "square."*

Because MATLAB is searching through every built-in function, we need a way to tell MATLAB to stop, because when the function `sqrt` *appears, it seems to be the exact function we need. To terminate a MATLAB command early, we press* **CTRL + C** *(i.e., the CTRL key and the "C" key at the same time) on our keyboard, and the prompt will return to the Command Window.* `CTRL + C` *can also be used to stop a program caught in an infinite loop or that otherwise wandered away and got lost.*

To ask for more information on the function `sqrt`*, we can type* `help sqrt` *or* `doc sqrt` *to double-check that the function is going to perform our intended calculation as well as details of how to use it.*

Repeating this process to find the inverse tangent function, we determine that `atan` *is the correct function necessary to calculate the inverse tangent. We note, from the help documentation, that the* `atan` *function will return the inverse tangent in radians.*

TEACHING TIP

Writing out a flowchart before attempting the program tends to be a useful way to structure a coding example if you will be coding live in your class.

Example 9-7 *(continued)*

To create a program, we click the ***New Script*** *button on MATLAB's Home tab. After creating a new blank M-file, we need to write out our problem statement as code comments. Comments are any non-code text that you would like to include in your code file. All comments start with the percent symbol (`%`), and everything to the right of the symbol will be ignored on a per-line basis. Comments can follow any executable statements in our code, but for now, we will add the first couple of lines in our MATLAB program by writing out a short problem statement with the `%` symbol before any text we write. We can also use block comments as mentioned previously.*

Next, we start writing our code. In a MATLAB program, it is good practice to start the program by clearing all of the variables from the workspace (`clear`) and clearing the Command Window (`clc`). That way, we guarantee that if we run our program on a different computer or run it after restarting MATLAB, our program will create all the variables we need to solve the problem, and we need not rely on the workspace having variables predefined before code execution. Also, this will prevent values left over from a previous run of this or even a different program from affecting the results. Commands like this are often referred to as "housekeeping" commands.

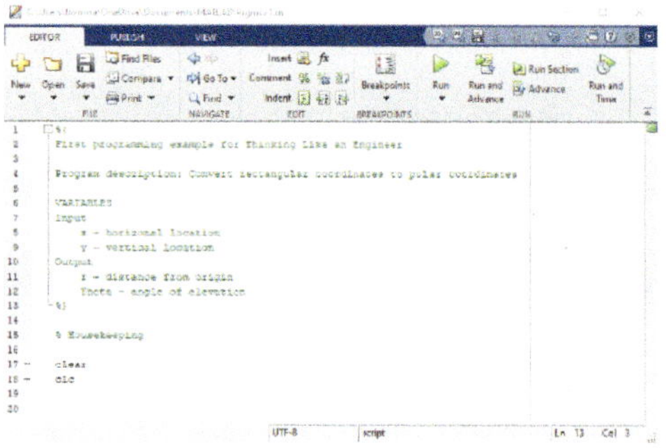

Note that all comments are shown in green, whereas our two program commands (`clear` and `clc`) are shown in black. As we will see later, MATLAB uses other colors for other specific types of things in your program. It is also helpful to use blank lines to visually separate different sections of your code.

Now we are ready to start our program. Let us look at the step-by-step written algorithm necessary to solve this problem:

1. *Create a variable to contain the x-coordinate as a variable named `x`.*
2. *Create a variable to contain the y-coordinate as a variable named `y`.*
3. *Calculate the radius by taking the square root of the sum of the squares of x and y and store the result in a new variable called r.*
4. *Calculate the angle by taking the inverse tangent of the result of dividing y by x and store the result in a new variable called `Theta`.*

In the written algorithm, the first thing we have to do is to manually assign the values of x and y. For this example, we assign x to contain the value of 5 and y to contain the value of 10, but in later chapters, we show how the user of a program can enter values. Note that in addition to creating the variables x and y, we have also added comments in the header of our program documenting what these variables represent.

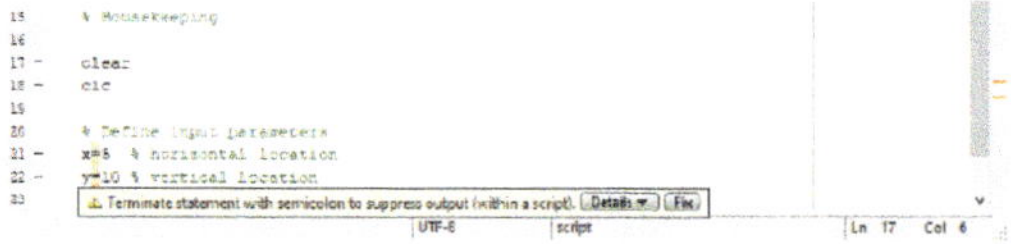

Example 9-7 *(continued)*

Note that MATLAB highlights the = symbol in the Editor. When we move the cursor over the assignment operator (=), a pop-up message tells us that we can suppress the output of the assignment expression by using a semicolon. For now, we will omit the semicolons and observe MATLAB's behavior when we leave the code unsuppressed.

Next, we type the equations to calculate the radius and the angle, using the built-in MATLAB functions we discovered with `lookup` *and* `help`*. In addition, we've added extra comments about the new variables representing the radius and angle of the polar coordinate.*

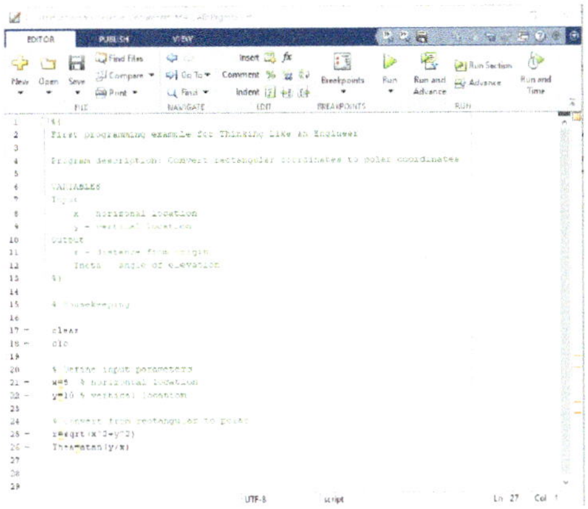

NOTE:

Remember, MATLAB programs and functions *cannot* have any spaces in their file names, so be sure that a file is named without spaces, for example, CartesianToPolar.m —*not* "Cartesian To Polar.m."

Next, we save the program in MATLAB's path with a file name that follows the program naming convention. We choose the name "CartesianToPolar" since that name nicely describes the purpose of the program.

To save a program, we click the **Save** *icon in the Editor window.*

⌘ ***Mac OS:*** *MATLAB is one of the few programs that does not place the main menu items (such as File and Edit) at the top of the screen. Instead, its main menu is at the top of the MATLAB window, just as in the Windows implementation. This is because MATLAB is actually running under an X Windowing System, similar to Linux or UNIX®.*

Running the Program

There are a number of different ways to run a MATLAB program located within MATLAB's path.

Method 1: *In the main MATLAB window (not the Editor), type the name of the program in the Command Window and press* ***Enter.***

Method 2: *In the main MATLAB window (not the Editor), right-click the program you want to run while in the Current Directory box and select* ***Run***.

Example 9-7 *(continued)*

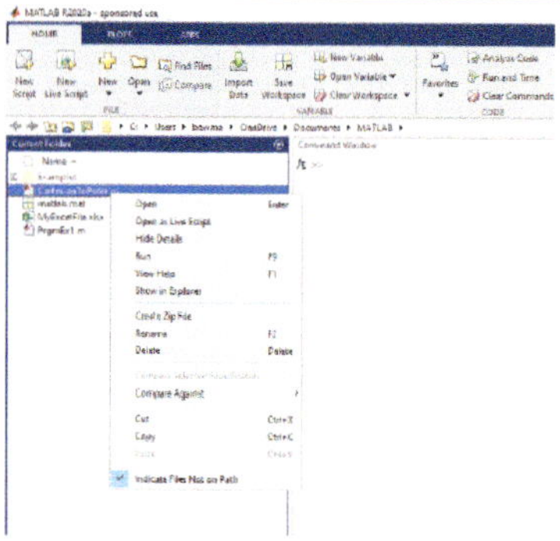

NOTE:

To suppress output, end the code line with a semicolon (;).

Method 3: *With the program loaded in the Editor window, click the* ***Run*** *button in the toolbar.*

Looking in the Command Window, we see that MATLAB spits out the results of every assignment statement in the program.

Let us assume we want to show only the output from the calculations of `r` *and* `Theta`*, and not the output of the specification of x and y. To do this, we add a semicolon (;) at the end of each line of code we do not want showing up in the output of the program.*

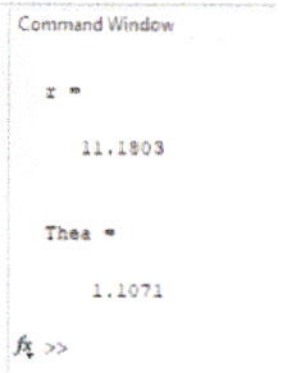

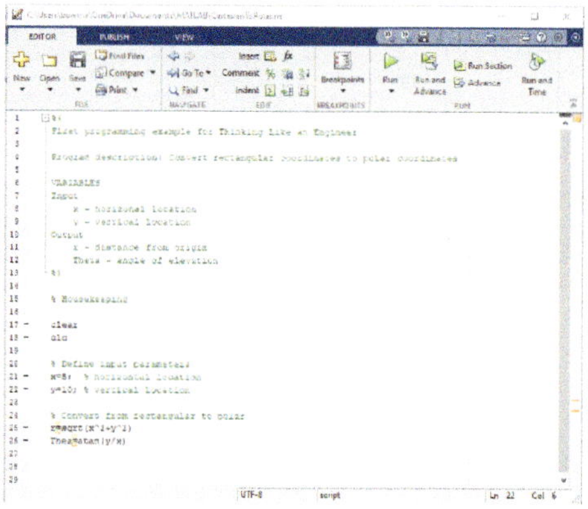

Now when we save and run our modified program, we will see only the output from the calculations of `r` *and* `Theta`*.*

Comprehension Check 9-8

For the problem given in Comprehension Check 9-4, write a complete program to implement the flowchart you created. Be sure to include appropriate comments in your code. Suppress all assignment statements other than the final answer, which should appear on the screen. Use the following values to test your code: $V_0 = 10$ mph, $t = 0.02$ hr, $a = 3$ m/s².

9.3.4 Programming Features in Our Coverage of MATLAB

MATLAB has many advanced programming features that make sophisticated numerical analysis, including simulation, possible. In this text, you will learn basic programming structures to solve engineering analysis problems.

- **Data types**: There are different data types that can be stored and manipulated by MATLAB. Some of the more common data types used by programs include numeric and character or string array variables, but many applications also use vector and matrix variables for advanced calculations, as well as special types like cell arrays for storing a combination of different types within a single structure.
- **Input and output**: By gathering input from the external world, whether from a human user or some sort of sensor, the program can solve a different problem each time it is executed. Output statements allow you to produce formatted text to communicate with the user, as well as to plot graphs to clearly communicate data relationships. Likewise, being able to read and write Microsoft Excel files using MATLAB code allows your program to perform calculations on existing data without needing to enter it manually or to permanently save results to special MATLAB data files (.mat files).
- **Conditional statements**: These allow a program to take different paths for various user inputs, intermediate calculations, or final results. Conditional statements allow the programmer to ask questions and have the code take different actions based on the answers.
- **Looping**: Repetition is critical in breaking down complicated problems into many simple calculations.

9.4 Functions

function

A **function** is a special type of program that is not designed to be used by itself but to be used as an element in a larger program. The primary reason for using functions is to simplify and modularize programming by allowing commonly used code to be expressed as a single word instead of having to paste all the code into the larger program. For example, when we need to perform a square root calculation, we do not have to write code to accomplish this, nor even paste the code that does the calculation into our program; we simply use the name `sqrt` followed by the values for which we want the square roots. When we write something like `Q=sqrt(Vals)`, the program knows to send the data in the variable `Vals` to a piece of code named `sqrt` that has already been written and debugged to do the calculation. This reduces the amount of redundant code and makes our programs cleaner and easier to read.

In general, functions require the user to pass input information to them—the data on which the function will perform its calculations and obtain the final results created by these calculations. However, some functions do not require any input to do their job. If you type `RV=rand`, for example, a single random number between 0 and 1 will be placed in RV. Some functions do not generate any data as output. An example would be `pause(60)`, which simply stops program execution for one minute (60 seconds), after which it resumes. Some functions require neither input nor do they generate data as output. Make sure the volume on your computer is not muted and type `beep` in the Command Window, for example.

TEACHING TIP

Emphasize that creating functions will help students generalize their code and ensure that debugging is more straightforward. Encourage students to create functions that handle a specific task. In other words, if the student describes a function that does something *and* something else, it's possible that the function can be split into two or more other functions. This, of course, does not apply to outputs. A function can have several outputs (the same applies with multiple inputs).

TEACHING TIP

You may choose to simply say "input" instead of "argument." Or you can alternate between the two, emphasizing that you are referring to the same concept.

TEACHING TIP

It's typically helpful to distinguish variable names in the function from those used in other programs. This extends beyond the recommendation here, which states that new functions should not have the same name as existing MATLAB functions.

TEACHING TIP

This is a common error students make. Make sure to introduce the idea of matching the name of the function with the filename as a convention in class.

Sometimes our application requires a certain type of calculation that is not common enough to have a built-in function, but we need to use it in multiple places in our code. Rather than place the same code multiple times in our program, we can write our own function to accomplish the task, and once it is properly tested and debugged, we can simply use the name we give to our new function in the larger program. User-defined functions are similar to programs in that they contain a series of executable MATLAB statements in M-files that exist independently of any program we create in MATLAB. To allow any MATLAB program to call (use) a user-defined function, however, we must follow a few guidelines to ensure that our function will run properly.

NAMING RULES

In computer programming, and often in math as well, the word "argument" does *not* refer to an altercation or vehement disagreement. In this context, it means the information that is given to a function as input to be processed. If you use the function `sqrt` to find the square root of 49, you would write `sqrt(49)`. In this case 49 is the argument of `sqrt`— the value upon which the `sqrt` function performs its calculations. Functions may have zero arguments (e.g., `Z=zeros` will place a single zero into Z—no input is required), one argument (e.g., `sqrt(4)`), or two or more arguments (e.g., `power(a,b)` raises a to the power b). Some functions may have different numbers of arguments depending on what you wish them to do. For example, `Ox=ones` will place a single 1 into `Ox`, `Ox=ones(5)` will place a 5 × 5 matrix filled with ones into `Ox`, and `Ox=ones(3,6)` will place a 3 × 6 matrix filled with ones into `Ox`.

9.4.1 Function Creation Guidelines

- The first line in a function other than comments **must** be a function definition line:

```
function[output_variables]=function_name(input_variables)
```

  - ***output_variables***: A comma-separated list of output variables in square brackets.
  - ***input_variables***: A comma-separated list of input variables in parentheses. The list of input variables to a function are commonly referred to as **arguments**.

arguments

  - ***function_name***: Function names should not copy the names of built-in functions, user-created variables, or any reserved word in MATLAB. Avoid creating variables with the same name as a user-defined function because you will not be able to call your function.
- **The name of the function must be the same as the file name**. For example, if the function name is `sphereVolume`, the function must be saved in an M-file named `sphereVolume.m`, where `.m` is just the extension of the M-file. In addition to the aforementioned function-naming rules, the name of the function must also follow all program-naming rules, including not using spaces in the file name.
- To test a function, you should call it from MATLAB's Command Window or from a different program or function by passing in input variables and saving the resulting output into different variables. If you click the Run button in the Editor, a dialog box will appear asking you to enter necessary arguments, but then it will use those values as the default when testing in the future. This may or may not be something you wish to do. We will discuss this in more detail shortly.
- Do *not* use the built-in function `clc` within one of your own functions. The functions you are writing are intended to be general purpose and can be called from other programs (or functions), so if text is written to the screen prior to calling a function that contains the `clc` command, it will wipe the Command Window clean—losing all the previous output on the screen.

- For a different reason, the `clear` command should not be used within a function since functions usually contain input arguments, so calling the `clear` command would delete any input variables and their values. In general, the `clear` and `clc` commands should only appear at the top of programs, not functions!
- In general, all assignment statements in a function should have the screen echo suppressed by adding a semicolon at the end of each assignment statement. Otherwise, you are likely to fill the screen with unnecessary output. The main exception would be during debugging, where having key values echo to the screen may assist in finding errors. However, once the function is working properly, the screen echos should be suppressed before the final version is saved.

9.4.2 Function Structure and Use

The following examples demonstrate the structure of different functions and how MATLAB programs can use the functions to reduce code redundancy.

Example 9-8

Assume we are required to create a function, `areaCircle`, to calculate the area of a circle given the radius. The function should accept one input (radius) and return one output (area).

Source Code

```
% Problem Statement: This function calculates the area of a
% circle
% Input:
% r - Radius [any length unit]
% Output:
% A - Area [units of input^2]
function [A]=areaCircle(r)
% Calculation of the area of a circle
A=pi*r^2;
```

Typical Usage (typed in Command Window or a Separate Program):

You can call the function by typing the name of the function and passing in a value for the radius. Note that the result is stored in a variable called `ans` *when you do not specify a variable to capture the output.*

```
>> areaCircle(3)
ans=
     28.2743
```

You can call the function by typing the name of the function and passing in a variable.

```
>> Rad=4;
>> CircleA=areaCircle(Rad)
CircleA=
   50.2655
```

Comprehension Check 9-9

What is the output when you "pass in" the value 10 to the function in Example 9-8?

Assume that you want to store the result of the previous function call in a variable named `Dogs`. What is the command you would type in MATLAB?

Comprehension Check 9-10

Write a function named `RAC` that will accept two parameters, `N` and `A`. The function should return a single value `R` calculated as $R = 3 \sin(N^A)$. For this example, you do not need to include any comments with this code.

TEACHING TIP

Including comments at the beginning of a function to explain the function's purpose, inputs, and outputs is a good practice to adopt.

ENGAGE

MATLAB's function inputs are positional, meaning that the order of inputs matters. Switching in1 and in2 will produce a different answer unless the function is commutative. Typically though, a program will not have commutative inputs.

9.4.3 Executing Functions

NOTE:

Just a reminder—you should be following along on your own computer as we work through these exercises.

MATLAB's Editor contains functionality to allow you to efficiently test functions that you write by allowing you to create different test scenarios and providing a shortcut to run those test cases you can create directly in the Editor window. Assume we have created and saved the following function in MATLAB:

```
function [out1,out2,out3]=myFunction(in1,in2)
out1 = in1 * 2;
out2 = out1 + in2;
out3 = [out1, out2]*3;
```

This function named `myFunction` accepts two input arguments `(in1, in2)` and returns three output variables: `out1` and `out2` are both scalars and `out3` is a vector. Assume we want to test to see what the resulting output variables will contain when the value 4 is used for `in1` and 7 is used for `in2`. To do this, we can use the same Run button we used for programs. The first time you click the Run button, an error message will appear in the screen indicating that the input variables are missing:

```
>> myFunction
Error using myFunction (line 2)
Not enough input arguments.
```

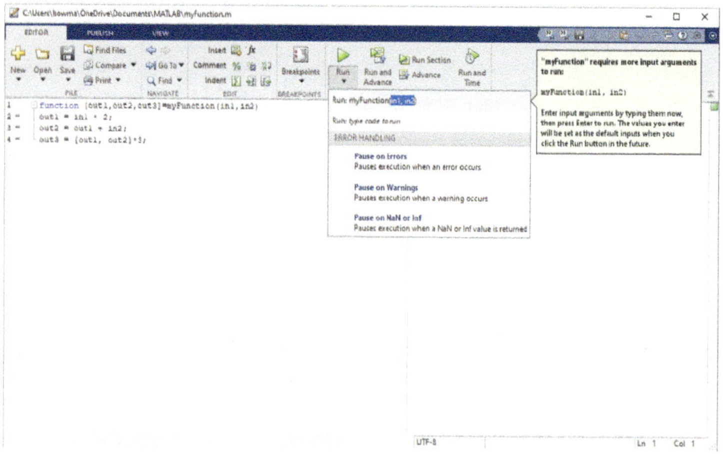

However, a little pop-up dialog appears in MATLAB's Editor window that allows you to type in values for the test case. Type in 4, 7 within the parentheses on the dialog, representing the values of `in1` and `in2` respectively, and press the Enter key. You'll notice that output has appeared in the Command Window:

```
>> myFunction(4,7)
ans =
    8
```

NOTE:

Functions can be defined to have optional input values, for example `rand`, `rand(n)`, and `rand(n,m)`. We do not cover writing functions with optional inputs in this text.

This output shows that the function was successfully executed with the value of 4 for `in1` and 7 for `in2`, but since this function didn't specify any output variables, the value 8 is stored in the default variable `ans`, which is actually the value of the `out1` variable calculated in the function.

By default, MATLAB requires you to provide values for all the input variables, but it does not require you to capture all the output variables with each call. If you do not specify any variable or variables in which to place the results, the first returned value in the list (`out1` in this case) is placed in `ans`. However, the Editor is smart enough to be able to set up test cases where all the output variables are assigned to variables when the function is called. To save all three output variables (for testing, we will save `out1` into variable a, `out2` into variable `b`, and `out3` into variable `c`), select the Run drop-down menu in the Editor tool ribbon (NOT the Run icon), right-click the line that reads `myFunction(4,7)`, which was stored by the Editor when you entered those values earlier, select Edit, and edit it to read:

```
[a,b,c]=myFunction(4,7)
```

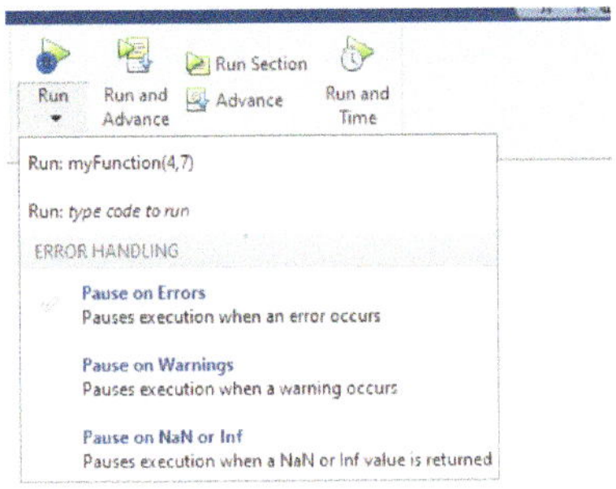

After we execute the function with our new test case, all three output variables are displayed in the Command Window:

```
>> [a,b,c]=myFunction(4,7)
a =
  8
b =
  15
c =
  24 45
```

You may have noticed that after you first entered the input values 4 and 7 the Run icon in the Editor tool ribbon changed appearance—it now has a blue circle containing ellipses (three dots…). This indicates that there is at least one edited version of the function call stored. If you modify the function code, you may retest it with the same values by opening the Run menu and selecting the version you wish. To add another version with different input values, click "type code to run" in the Run menu, and that will be added to the selections in the future.

ENGAGE

Describe the difference between a local variable and a global variable. *A local variable is defined in the context of a function and is destroyed once the function completes its tasks. A global variable is one that is accessible by any program, which can be defined using the word "global."*

TEACHING TIP

It is critical to mention that the functions have their own private workspace. Variables that are created in the program will not make it to the main workspace (i.e., at the global level). Therefore, if a variable "x" is assigned in the context of the function and is not an output, there's no way to retrieve that value.

Example 9-9

Assume we are required to create a function, `volumeCylinder`, to calculate the volume of a cylinder given the radius. The function should accept two inputs (radius, height) and return one output (volume). For this example, you may assume that the user will input the radius and height in the same units. As usual, any calculations in the functions must be suppressed.

Source Code

```
% Problem Statement: Calculate the volume of a cylinder.
% Input:
%  r – Radius [any length unit]
%  H – Height [same length unit as radius]
% Output:
%  V – Volume [units of input^3]
function [V]= volumeCylinder(r, H)
% Calculation of the volume of a cylinder
V=pi*H*r^2;
```

Typical Usage

```
>> Vol1 = volumeCylinder(1,2)
Vol1=
  6.2832
>> Vol2= volumeCylinder(2,1)
Vol2=
  12.5664
```

Note that the order of arguments passed in to the function matters. The order of variables expected by the function is based on the order they are listed in the function header. Therefore, the first input argument is the radius, followed by the height.

Comprehension Check 9-11

For the problem given in Comprehension Check 9-4, write a function named `dist` that accepts the initial velocity, time, and acceleration values as input variables and returns the calculated distance in units of meters. Be sure to include appropriate comments in your code. Remember to suppress all assignment statements. Test your code with at least three different sets of input values.

9.4.4 Local and Global Variables

NOTE:

Local variables are only visible to the program or function using them.

Global variables are visible within any program or function.

One important fact to note about variables is that the variables in the workspace are only available within MATLAB's Command Window and to any running program. Functions create a private workspace and cannot access any of the variables available in MATLAB's main workspace. The only way to access any variables in MATLAB's main workspace is to "pass" the variable to the program as function input.

The reason for having "main workspace" variables inaccessible by functions is because functions exist within a different scope from the programs. This way, the programmer does not need to know the names of all variables used in all functions to avoid conflicts. For example, if the `sqrt` function defined a variable named `x1`, and your program also used `x1`, whenever the `sqrt` function was used, it would destroy the value of `x1` being used in your program.

local variables

The variables created within the "private workspace" of a function are commonly referred to as **local variables** since they are destroyed after the function executes and do not appear in the "main workspace" unless they are passed back as function output.

In some situations, this localization of variables could be very inconvenient if there were not a mechanism to circumvent it. For example, it would be a nuisance if every time you needed to use the predefined variable `pi` in a function, you had to include it in the input parameter list. Another case would be when both the main program and some of the functions it uses share a large number of variables, making the input parameter list excessively long.

global variables

global

In the first case, special constants like `pi` are predefined to exist within the scope of all programs and functions. Variables that are visible within any scope are commonly referred to as **global variables**. Global variables are created in functions using the reserved word **global** followed by a comma-separated list of the variables that should be considered global variables.

```
function [x] = functionName(y)
global z
z=30;
```

In addition, the same variables must be declared global in the main program and in any other functions that will rely on them.

In this example, after the function `functionName` is called, the value of `z` is accessible by other programs or functions that also include the statement `global z`.

On the left side of the following diagram, the inner box is a program (`Prgrm1`) and the outer box represents all of the variables in MATLAB's workspace. Since we do not include any commands to clear the variables from the workspace at the top of the program `Prgrm1`, the program is able to "see" and use all of the variables within MATLAB's workspace. In addition, it will create a new variable in MATLAB's workspace (`c`) when we execute program `Prgrm1`. The workspace then contains the three variables `a`, `b`, and `c`.

In the diagram on the right, again the outer box represents the workspace, which this time contains the variables `e` and `g`. The large box inside the outer box represents a program (`Prgrm2`) that uses a function (innermost box, `Func1`) to calculate a value for a new variable `h`. The function cannot "see" or use any of the variables in MATLAB's workspace and creates its own separate workspace to execute the code inside the function. In order to run the function, the programmer must provide the values of variables `x` and `y` by "passing" the variables `e` and `g` into the function. The function code then defines a variable `k`, calculates the value of `d`, and returns `d` as function output. When the value of `d` is returned to `Prgrm2`, it is stored in the variable `h`. After the function has executed, the separate function workspace is destroyed, along with variables `d`, `k`, `x`, and `y`, which were never stored in MATLAB's main workspace.

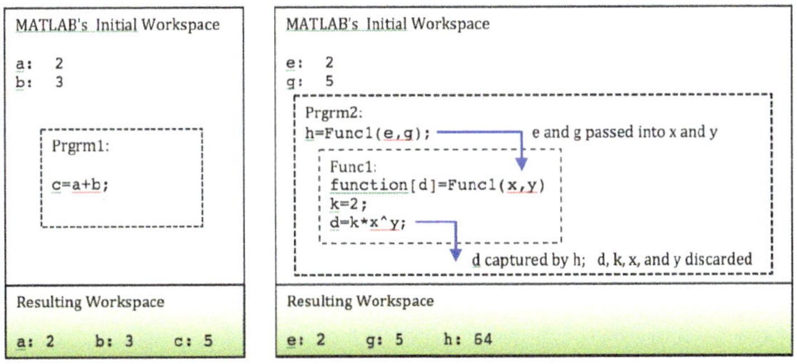

Example 9-10

Assume we are required to create a function, `sphereCalculations`, to calculate the volume and surface area of a sphere given the radius. The function should accept one input (radius) and return two outputs (volume, surface area). As usual, any calculations in the functions must be suppressed.

Source Code

```
% Problem Statement: This function calculates the volume and
% surface area of a sphere.
% Input:
%  r-Radius [any length unit]
% Output:
%  V-Volume [units of input^3]
%  SA-Surface Area [units of input^2]
function [V,SA]=sphereCalculations(r)
% Calculation of the volume of a sphere
V=4/3*pi*r^3;
% Calculation of the surface area of a sphere
SA=4*pi*r^2;
```

Typical Usage

```
>> [Vol,SurfArea]= sphereCalculations(7)
Vol=
  1.4368e+03
SurfArea=
  615.7522
```

Example 9-11

Assume we are required to create a function, `cylinderCalculations`, to calculate the volume and lateral surface area of a cylinder given the radius and height. The function should accept two inputs (radius, height) and return two outputs (volume, surface area). For this example, you may assume that the user will input the radius and height in the same units. Any calculations in the functions must be suppressed.

Source Code

```
% Problem: Function calculates volume and surface area of
% cylinder.
% Input:
%  r-Radius [any length unit]
%  H-Height [same length unit as radius]
% Output:
%  V-Volume [units of input^3]
%  SA-Surface area [units of input^2]
function [V,SA]=cylinderCalculations(r,H)
% Calculation of the volume of cylinder
V=pi*H*r^2;
% Calculation of the surface area of a cylinder
SA=2*pi*r*H;
```

Typical Usage

```
>> [Volume,Surface]=cylinderCalculations(3,2)
Volume=
  56.5487
Surface=
  37.6991
```

Example 9-12

Write a program that uses the two functions developed in Exercises 9-10 and 9-11 that will define a radius and a height, then place the volume and surface area of a sphere with the specified radius in `SphVol` and `SphArea` respectively, and the volume and surface area of a cylinder with the specified radius and height in `CylVol` and `CylArea` respectively.

```
% Problem: Calculate the surface area and volume of
% both a sphere and a cylinder.
% Input:
%  radius-Radius [any length unit]
%  ht-Height [same length unit as radius]
% Output:
%  SphVol-Volume of sphere [units of input^3]
%  SphArea-Surface area of sphere [units of input^2]
%  CylVol-Volume of cylinder [units of input^3]
%  CylArea-Surface area of cylinder [units of input^2]
clear
clc
radius=5; % define radius
ht=7; % define height
[SphVol,SphArea]=sphereCalculations(radius);
[CylVol,CylArea]=cylinderCalculations(radius,ht);
```

This yields the following values:

```
SphVol=
  523.5988
SphArea=
  314.1592
CylVol=
  549.7787
CylArea=
  219.9115
```

9.5 Conditionals

Outside the realm of computing, logic exists as a driving force for decision making. Logic transforms a list of arguments into outcomes based on a decision.

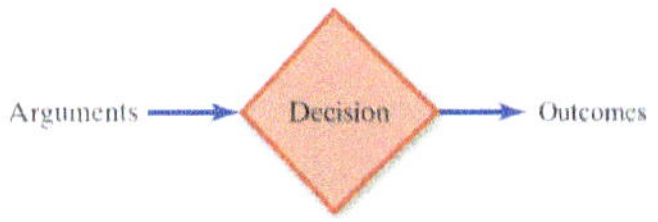

Some examples of everyday decision making follow:

- If the milk has passed the expiration date, throw it out; otherwise, keep the milk.

  Argument: expiration date

  Decision: before or after?

  Outcomes: garbage, keep

- If the traffic light is red, stop. If the traffic light is green, go. If the traffic light is yellow, decide if you can safely stop before reaching the intersection.

  Argument: three traffic light colors

  Decision: which bulb lit?

  Outcomes: stop, go, ask can I stop safely and act accordingly

ENGAGE

Provide the class with a list of potential conditional statements and ask whether the question is an appropriate conditional.

Is the temperature above 100 degrees F? *Yes*

What is the value of temperature? *No, the answer will be a number, not a yes/no.*

Is the configuration of resistors parallel or series? *Yes, we can have compound logic using AND and OR connectives. As long as the final answer is a yes/no.*

Is the configuration of resistors parallel? *Yes*

What are the units of distance? *No, the answer to this would be a word (or string, in programming).*

Consider coming up with additional examples or having students develop their own and quiz others.

Boolean logic
binary behavior

To bring decision making into our perspective on problem solving, we need to first understand how computers make decisions. **Boolean logic** exists to assist in the decision-making process, where each argument has a binary result, and the overall outcome exhibits binary behavior. **Binary behavior**, depending on the application, is any sort of behavior that results in two and only two possible outcomes. Boolean logic, Boolean algebra, and related terms are named for George Boole (1815–1864), the English mathematician whose writings formed the basis of modern computer science.

Algorithms Revisited—Representing Decisions

decisions

Decision making can be expressed in a written algorithm. Assume you are designing a process to determine if the value read from a temperature sensor in a vehicle indicates it is unsafe for operation. To express the decision in a written algorithm, you should phrase your **decisions** in terms of questions that have a Yes or No response.

When an algorithm requires that different actions be taken based on the answer to a question, it is helpful to indent the actions that might be taken relative to the question. This visually indicates that these actions are only associated with a particular answer to the question.

Example 9-13

Create a written algorithm to express a temperature given in relative units [°F or °C] in the corresponding absolute units [K or °R].

Known:

- *Temperature in relative units (degrees Celsius or degrees Fahrenheit).*

Unknown:

- *Temperature in absolute units (kelvins or degrees Rankine).*

Assumptions:

- *Since the problem does not explicitly state the temperature of interest, assume that the interpreter of the algorithm will input the temperature and units.*

Algorithm:

1. *Input the numeric value of the temperature.*
2. *Input the units of the numeric value of the temperature.*
3. *Ask if the input unit is degrees Fahrenheit.*
   - **(a)** *If yes, calculate the value in degrees Rankine.*
   - **(b)** *If no, calculate the value in kelvins.*
4. *Display the new value and absolute unit.*
5. *End the process.*

Note that only one of the actions (a or b) under the question in item 3 will be completed depending on the answer.

Sometimes an action that is to be taken only if a question has a specific answer requires that another question be asked.

Example 9-14

Create a written algorithm to express the decision process a person goes through when approaching a stoplight.

Known:
- *Status of stoplight, distance from light, speed of vehicle*

Unknown:
- *Vehicle motion: Stop or continue*

Assumptions:
- *This is not an emergency situation requiring violation of traffic laws*

Algorithm:
1. *Determine color of light.*
2. *Is the light red?*
   - **(A)** *If yes, stop before reaching intersection.*
   - **(B)** *If no, is the light yellow?*
     - (i) *If yes, can I safely stop before the intersection?*
       - **(a)** *If yes, stop.*
       - **(b)** *If no, continue through intersection with caution.*
     - (ii) *If no, continue through intersection.*
3. *End the process.*

Note that instead of asking a single question, "What color is the light?" this was broken down into questions that could be answered with Yes or No. The reasons for this will become clearer as we proceed.

Also note that each time a question was asked, the actions to be taken based on the two possible answers were indented. We cannot overemphasize the importance of maintaining proper indentation, not only in algorithms, but also in the code that implements those algorithms.

Comprehension Check 9-12

Create a written algorithm that determines the dose of a medicine to be given based on whether the person is a child (≤ 14 years old) or an adult. Children get one pill, adults get two.

Comprehension Check 9-13

Create a written algorithm that determines the state of water given the temperature in °C.

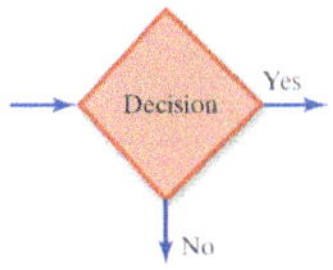

To include decision making in flowcharts, we will introduce another symbol, the decision diamond. All decisions in a flowchart must be represented within a diamond shape.

There must be at least one arrow entering the decision diamond, although there may be more than one if different paths in the algorithm lead to the same question. The question itself is written inside the diamond. The decision diamond represents a binary question and thus has two arrows that leave it, one for each possible answer, typically labeled either Yes and No or True and False. If the answer to the question is Yes, the algorithm follows the arrow labeled Yes, and similarly for No. By convention, arrows enter and leave decision diamonds at the vertices as shown. It is assumed that any variables defined in the stepwise scope of a decision diamond are accessible and can be used in the decision. Since no new variables are created in a decision diamond, the stepwise scope that enters the decision diamond is passed on to the next shape of each conditional branch.

TEACHING TIP

It may be helpful to go through this example with students by breaking down the question of "what color is the light?" using the algorithm described in Example 9-14. Creating a flowchart to illustrate the process would be useful.

Moreover, it's essential to communicate that once a condition is met, then the commands associated with that case will run.

ENGAGE

Consider a change sorting algorithm that takes in a dollar amount and returns the minimum number of coins (quarters, dimes, nickels, and pennies) to make up the inputted dollar amount. Part of this program will require four cases using a condition that checks if the total amount is bigger than the denomination of each coin. What are the conditional statements if the user input is assigned to the variable *amount*?

Is amount >= 0.25?
    Add one to quarter count
Is amount >= 0.10?
    Add one to dime count
Is amount >= 0.05?
    Add one to nickel count
Is amount >= 0.01?
    Add one to penny count

Do you need the last statement to specify the final case? *No, we don't need it. If the nickel case is false, then we could immediately include that case. Consider making a flow diagram to illustrate this point.*

How would the computer interpret our algorithm?

The change counter would first check if the amount is bigger than or equal to 25 cents (if so, we add one to our quarter count). For our dimes, we would also have to check if the amount is bigger than or equal to 10 cents. If our amount is 30 cents, then both of these conditions are true. This may seem like a problem, but the computer will check cases until one is true. Once it's checked, it will exit the decision, even if a later condition is true. This makes ordering extremely important.

296 Chapter 9 Computer Programming

Example 9-15

Create a flowchart to express a temperature given in relative units [°F or °C] in the corresponding absolute units [K or °R].

Known:

- *Temperature in relative units (degrees Celsius or degrees Fahrenheit).*

Unknown:

- *Temperature in absolute units (kelvins or degrees Rankine).*

Assumptions:

- *Since the problem does not explicitly state the temperature to be determined, the interpreter of the algorithm will ask for the temperature and units.*

Flowchart:

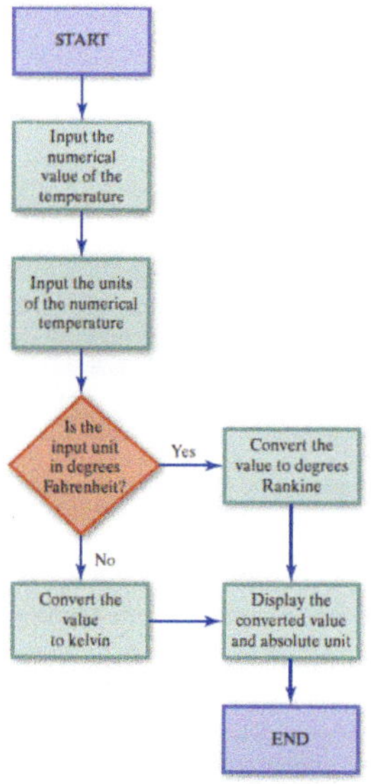

Example 9-16

Create a flowchart to express the decision process a person goes through approaching a stoplight.

Known:

- *Status of stoplight, distance from light, speed of vehicle*

Unknown:

- *Vehicle motion: Stop or continue*

Assumptions:

- *This is not an emergency situation requiring violation of traffic laws*

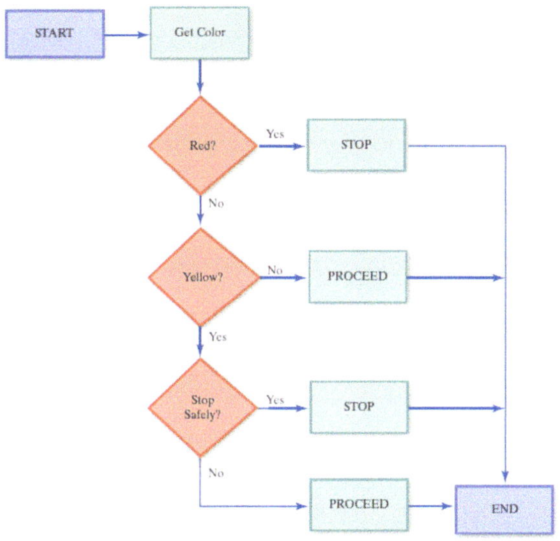

Comprehension Check 9-14

Create a flowchart that determines the dose of a medicine to be given based on whether the person is a child (≤ 14 years old) or an adult. Children get one pill, adults get two.

Comprehension Check 9-15

Create a flowchart that determines the state of water given the temperature in °C.

TEACHING TIP

Use this diagram to showcase how a decision structure terminates. Because the computer is checking line by line, the flow will be determined by the order of the conditions. How would the logic be impacted if we switched "yellow?" and "red?" cases while keeping everything else the same? Assume the color is "red." *The car would proceed even though the color of the light should indicate that the driver should stop. Obviously not a good outcome!*

TEACHING TIP

Sometimes students struggle when conditional statements are included, especially with multiple steps. It may help to have students pretend they are the computer and read the commands line by line. This forces them to follow the arrows in the diagram and not look at the process globally.

Also, if students are not following the appropriate conventions for flowcharts, then having the students pretend they're the computer is also effective. Students often add more arrows than they should, which introduces ambiguity. By having them act like the computer, they'll run into a situation where they have two or more potential paths to take but potentially no decision block. This is a good time to point out that the computer wouldn't know what to do, because it's lacking explicit instructions.

TEACHING TIP

We'll introduce the "while" and "for" loops, but not by name. We distinguish them using iteration to define the "for" loop. If you'd like to call them by name, especially if you're planning to teach a programming language in this unit, then please do so.

9.6 Loops

loops

It is extremely common to want a group of lines in a program to be executed multiple times instead of just once, and the structures that allow us to do this are called **loops**. Typically, each time the code inside a loop is executed, it uses different values, generating different results. This ability to execute the same set of instructions many times extremely rapidly is perhaps the most important factor in making computers the powerful tools that they are.

To incorporate looping in a written algorithm, we simply add an instruction that specifies to which step in the algorithm we wish to return and under what conditions.

Example 9-17

Ask the user to make a selection from a list dialog. If the user did not make a correct choice but closed the list dialog instead, issue a warning message and ask the user to try again. This process should repeat until the user makes a correct selection, after which the selection is printed on the screen.

Known:

- *User's Choice*

Unknown:

- *None*

Assumptions:

- *None*

Algorithm:

1. *Display list dialog.*
2. *If the user closed the list dialog.*
   (a) *Warn users that they must make a valid selection.*
   (b) *Return to Step 1.*
3. *Display user's response on the screen.*
4. *End the process.*

Example 9-18

Create a written algorithm to calculate the sum of all integers between a lower bound and an upper bound.

Known:

- *Upper bound of whole number sequence.*
- *Lower bound of whole number sequence.*

Unknown:

- *Sum of all whole numbers between the upper and lower bound.*

Assumptions:

- *Since the problem does not explicitly state the upper and lower bounds, assume the program will ask for the values.*
- *Include the boundary values in the summation.*

Algorithm:

1. *Input the lower bound of the sequence.*
2. *Input the upper bound of the sequence.*
3. *If the lower bound is larger than the upper bound,*
   - *(a) Warn the user that the input is invalid.*
   - *(b) End the process.*
4. *If the upper bound is larger than the lower bound,*
   - *(a) Create a variable to keep track of the sum (S).*
   - *(b) Create a variable to keep track of the location in the sequence (L).*
   - *(c) Set the initial value of S to be zero.*
   - *(d) Set the initial value of L to be the lower bound.*
   - *(e) If the value of L is less than or equal to the upper bound,*
     - *(i) Add L to the current value of S.*
     - *(ii) Add one to the current value of L.*
     - *(iii) Return to step 4.e. and ask the question again.*
   - *(f) Display the sum of the sequence (S).*
5. *End the process.*

In step 4.e.iii, we required that the interpreter return to an earlier step in the algorithm after changing the values of our variables. This allows us to create a feedback loop necessary to calculate the sequence of values. A feedback loop is a return to an earlier location in an algorithm with updated values of variables. It is important to note that if we failed to update the variables, the feedback loop will never terminate. A nonterminating feedback loop is also known as an **infinite feedback loop**.

infinite feedback loop

Comprehension Check 9-16

Create a written algorithm to multiply all integer powers of 5, 5^x, for x between (and including) 5 and 50.

To implement looping in a flowchart, we simply add a directional arrow that eventually returns the flow of the chart back to a previous step. This is almost always associated with a decision of some form that answers the question, "Should this block of code repeat again?" The one notable exception is the rare case where you actually want an infinite loop that will never terminate.

ENGAGE

Whenever we're implementing a loop, we'd like to keep track of one or more variables. In this case, we need 4(e)(i) and 4(e)(ii) because we want to track the value of the sum and the current integer value we're adding to the total sum.

What happens if we eliminate 4(e)(i)? *The sum will be zero.*

What happens if we eliminate 4(e)(ii)? *The loop will never stop.*

Example 9-19

Create a flowchart to ask the user to make a selection from a list dialog. If the user did not make a correct choice but closed the list dialog instead, issue a warning message and ask the user to try again. This process should repeat until the user makes a correct selection, after which the selection is printed on the screen.

Known:

- *User's Choice*

Unknown:

- *None*

Assumptions:

- *None*

Flowchart:

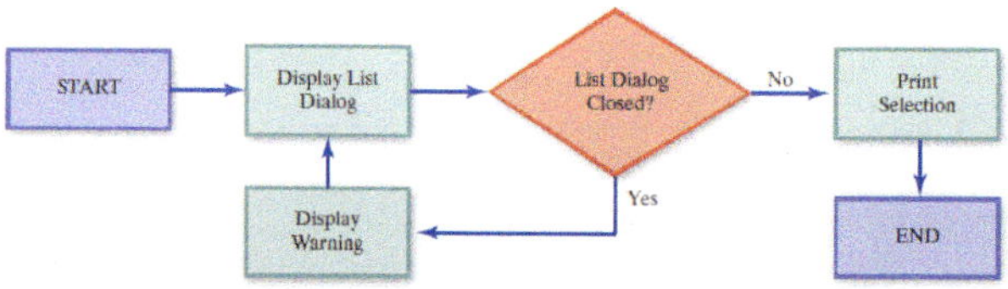

Example 9-20

Create a flowchart to calculate the sum of a sequence of whole numbers, given the upper and lower bounds of the sequence.

Known:

- *Upper bound of whole number sequence.*
- *Lower bound of whole number sequence.*

Unknown:

- *Sum of all whole numbers between the upper and lower bound.*

Assumptions:

- *Since the problem does not explicitly state the upper and lower bounds, the interpreter will ask for the values. Include the boundary values in the summation.*

Flowchart:

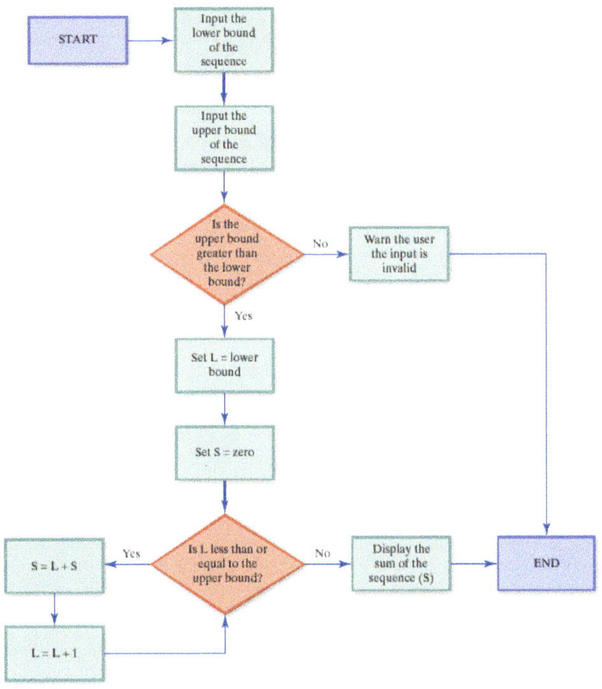

ENGAGE

Walk students through test cases for this program. Have them predict the output given the lower and upper bound.

Lower = 5 and Upper = 10:
Output = 45

Lower = 1 and Upper = 100:
Output = 5050

Lower = -5 and Upper = 5:
Output = 0

Lower = 23 and Upper = 12:
Output = invalid inputs

ENGAGE

Checking for valid inputs and values throughout a function or program is essential for code that is used in professional settings. If there is no way for the code to check for incorrect values, then the user will receive an incorrect or implausible result. Consider the following article about the scenario of software inside a medical device that administers a dosage of medication. This is an example of where the stakes for checking valid and invalid inputs are important:

URL
https://blog.bytellect.com/software-development/software-testing/why-is-invalid-input-testing-important-isnt-valid-input-testing-sufficient/

ENGAGE

When you include a counter, you must update the value of that counter in the loop, otherwise the loop will not end. Consider asking students for examples of loops that wouldn't end if run in MATLAB (or whichever language you're using).

Comprehension Check 9-17

Create a graphical algorithm to multiply all integer powers of 5, 5^x, for x between (and including) 5 and 50.

Example 9-21

Ask the user to enter a strictly positive value for time in seconds. If the user enters a nonpositive value, warn them and ask them to try again. This should be repeated until the user enters a valid value. Create a section of a flowchart to implement the logic necessary for this problem.

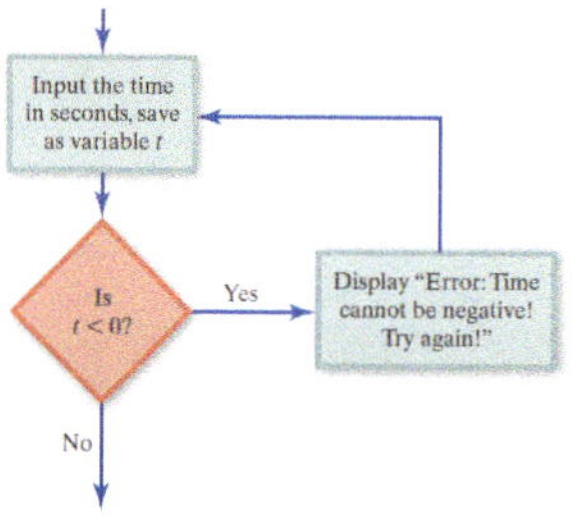

This is an example of a loop in which there is no way to predict how many times the loop code will be executed, since one cannot predict how many times a user will enter invalid information. The usual type of loop to use in cases like this is the `while` loop.

Comprehension Check 9-18

Modify the flowchart in Example 9-20 so that if the user enters an upper bound that is less than the lower bound, they are warned and asked repetitively to try again until a correct pair of values is entered.

Iteration

Some algorithms require repeated calculations that typically involve the use of a sequence of values or some operation on the individual values stored in a vector or matrix. Such algorithms are considered to be iterative because they require an index variable to keep track of when to terminate. For example, assume we have a vector, V, which contains positive and negative values in random order. If we want to create two new vectors, VN and VP, that contain the negative and positive values of V respectively, we will need to iterate through each element of the vector V, make a decision about each value, and store it in the corresponding vector. To do this, we would need to create a **counter variable** often called an **index variable**, that will keep track of the number of times we have repeated a calculation or decision. If we create a counter variable, X, and initialize it to be the number 1, X will actually serve two purposes. In addition to keeping track of the number of times we have repeatedly made decisions and stored new values into VN and VP, it will also serve as the index variable into the V vector so that we can access element V(1), V(2), and so on until we reach the last element in V. Example 9-22 demonstrates this scenario as a flowchart, including the iterated counter variable X.

counter variable
index variable

Algorithms that will require iteration typically have a scenario where you must repeat some decision or calculation "for each" or "for every" element or value within a sequence or vector. Since there is no "for each" or "for every" building block within an algorithm, this type of structure must be constructed out of the following steps:

- Initialization of a counter variable (e.g., Set X to be 1)
- A decision that involves the value of a counter variable (e.g., if X is less than or equal to the number of elements in V)
- Some action block that increments the counter variable (e.g., Set X equal to the current value of X plus 1)

Example 9-22

Create a flowchart that implements the sorting of positive and negative values in a vector as previously described.

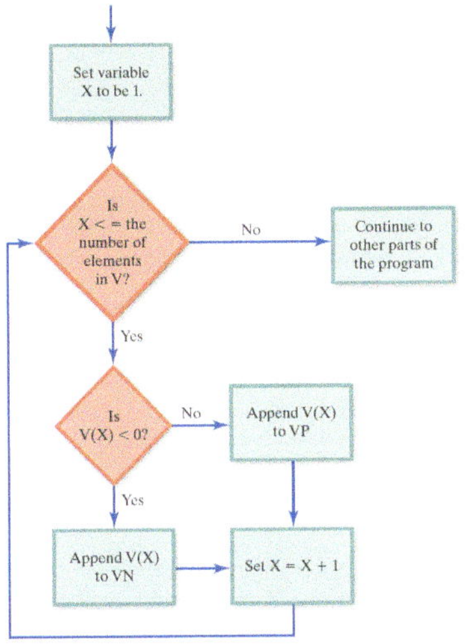

This is an example of a loop in which we can predict how many times the loop code will be executed, in this case the number of elements in the vector V.

9.7 Debugging Tips

debug

If a program has syntax errors, MATLAB (and other compilers) will give you useful feedback regarding the type and location of the error to help you fix, or **debug**, your program. If this information is not sufficient for you to understand the problem, check with others who might have seen this error previously—some errors are particularly common. If you still cannot identify the error, you might need to use a more formal process to study the error.

DIFFERENTIATED INSTRUCTION

If you're using "for" and "while" loops explicitly, a common question students have is whether a "for" or "while" loop should be used in a given situation. You might pose this question to the class, especially if students are anxious about choosing the *right* kind of loop.

The general answer is that a for loop could be rewritten as a while loop, so it doesn't matter in most cases. If you know how many iterations you need to do or have a list of numbers/strings/groups, then a for loop is more appropriate. If you don't know how many iterations you need to do or your loop is based on a condition that can't be represented as a list (e.g., is temperature bigger than 100 degrees?), then a while loop is more appropriate.

DIFFERENTIATED INSTRUCTION

Some loops may take a long time to complete. Consider the Fibonacci sequence, which is a *recursive* sequence (meaning it is defined in terms of itself) that involves starting at two values (usually f(0)=1 and f(1)=1) and defining the next number in the sequence, f(n), to be f(n)=f(n-1)+f(n-2). This can be programmed using loops. To show students how long computers might spend on a loop, considering having your students watch the video on the Ackermann function, produced by Computerphile.

URL
https://www.youtube.com/watch?v=i7sm9dzFtEI

TEACHING TIP

When discussing debugging, remind students of the best practices for variable names and commenting. The more descriptive comments they have in their code, the easier it will be to parse out the purpose of each line or section of code.

ENGAGE

The MATLAB YouTube channel has a video on common debugging strategies if you'd like to include it as supplemental material for this unit.

URL
https://www.youtube.com/watch?v=VEqevFAZIQo

NOTE:

Why do we say that computers (and programs) have bugs? Grace Murray Hopper is quoted in the April 16, 1984, issue of Time Magazine as saying, "From then on, when anything went wrong with a computer, we said it had bugs in it"—referring to when a 2-inch-long moth was removed from an experimental computer at Harvard in 1947.

For other kinds of errors, this formal process is essential for diagnosing the problems. When you first write a program, you must test the output of the program with a set of inputs for which you know the result. Such test cases would include:

- Simple cases for which you can quickly compute the expected output;
- Cases provided by an instructor or textbook with a published solution;
- Cases for which results have already been produced by a previously tested program that does the same thing;
- Test cases that are customarily used to test programs.

When the output of a program is different from what you expect, the debugging process begins. It is common to debug shorter programs simply by reading them and writing a few notes. Longer programs may require the use of the MATLAB Debugger, which is available as part of the MATLAB Editor. Whether you are using the MATLAB Debugger or are debugging "by hand" or in the Command Window, the same techniques apply.

Preparing For Debugging

When using the Editor/Debugger, open the program for debugging. If it is open, make sure changes are saved—otherwise MATLAB will run the most recently saved version without including changes made since the last time the program was saved. If you are debugging from a printed program and output, make sure that the printed output came from the version you are reading. In preparing to debug a program, it is critical to be able to reproduce the conditions that caused the bug in the first place. If the bug occurs in processing data from a large data set, you may have to split the data set to find the specific data that triggered the bug.

Setting Breakpoints

breakpoints

When using the Editor/Debugger, establish **breakpoints** that allow you to check your agreement with the program at various stages. A breakpoint stops the program at the specified location to allow you to examine the contents of variables and other elements before the program continues. This will help you find the specific location when the program does something you were not expecting. When debugging shorter programs, write down everything the program does or make all program results display in the Command Window by removing semicolons that were used to suppress output to the Command Window.

NOTE:

A breakpoint stops the program at the specified location to allow you to examine the contents of variables before the program continues.

MATLAB has two main types of breakpoints: standard (set by location in the program) and conditional (triggered at a specified location by specified conditions). We focus on standard breakpoints in this section. Standard breakpoints are normally shown as little red dots next to the line of code where the executing should pause, and conditional breakpoints will appear as little yellow dots. If the breakpoint dot displays in gray, either the file has not been saved since changes were made to it or there is a syntax error on the line or somewhere in the file. All breakpoints remain in a file until you clear (remove) them or until they are automatically cleared.

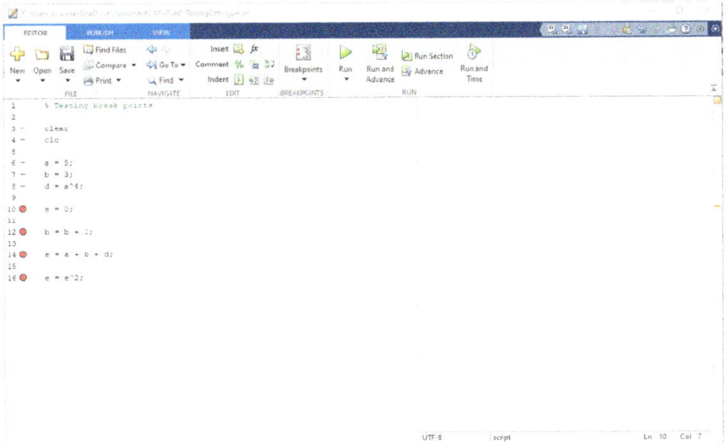

Stepping Through a Program

When using the Editor/Debugger, you must start by running the program or function. You can choose to step through the program one line at a time or have it continue until a breakpoint is encountered. Typically, you will set a breakpoint and execute the program and the Editor will automatically stop before the line containing the breakpoint is executed. A green arrow will point to the program line when execution is paused. The controls in the Editor window will allow you to "step in" to functions you've written, or "step out" of functions you're currently in. Likewise, pressing the Continue button will resume execution—to completion or until the debugger encounters the next breakpoint.

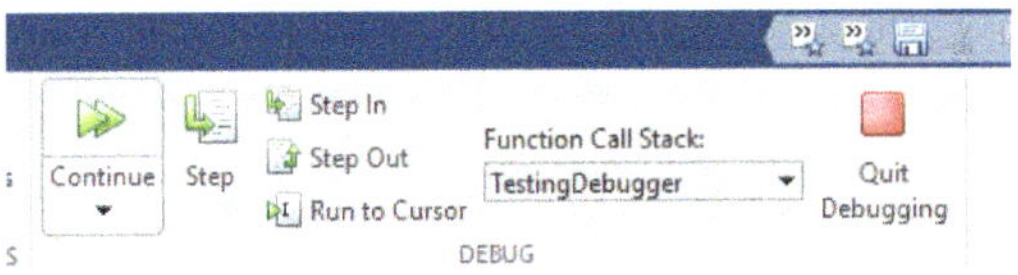

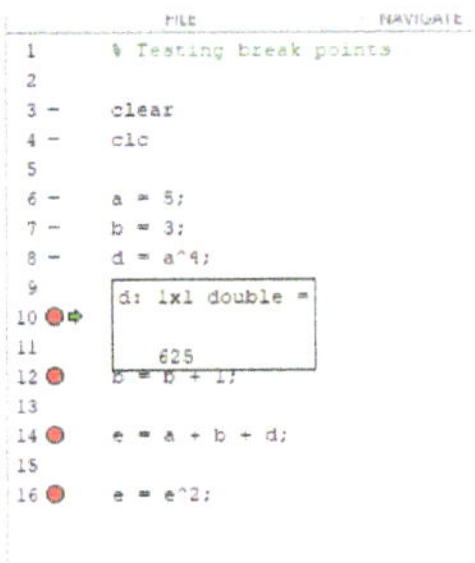

Examining Values

When the program is paused at a breakpoint or when you are stepping through the program one line at a time, you can view the value of any variable by hovering your mouse over the variable to see whether a line of code has produced the expected result. If the result is as expected, continue running or step to the next line. If the result is not as expected, then that line, or a previous line, contains an error. You can also examine the value of a variable in the Workspace window in the main MATLAB program window.

ENGAGE

If you'd like to discuss some of the limitations of computer-aided computation, it might be worth talking about the issues with floating point numbers. Because of the way computers store information, we can't get absolute precision on our numbers. Thus, it's possible that floating point numbers could cause havoc. Consider sharing this video from Computerphile as a supplement, especially if you'd like to dig deeper in computer science topics.

URL
https://www.youtube.com/watch?v=PZRI1IfStY0

TEACHING TIP

Sometimes you may believe your code is working because you've been running it to check for correct outputs. However, if you make changes later and try to rerun your code, you may find that things are no longer working! Clear your workspace when testing your code to ensure no values from a previous run are still there. It's also helpful to keep different versions of your code in case you make significant changes that end up being an unproductive effort.

ENGAGE

Many students have developed a strong interest in hardware by buying or building gaming systems. Consider asking if any students have a gaming system and would be willing to share the specs (specifications) of their systems, and why some characteristics are important.

Correcting Problems and Ending Debugging

If a problem is difficult to diagnose, one method to help define or correct problems is to change the value of a variable in a paused program to see if continuing with the new value produces expected results. A new value can be assigned through the Command Window, Workspace browser, or Array Editor. Do not make changes to an M-file while MATLAB is in debug mode—it is best to exit debug mode before editing an M-file. If you edit an M-file while in debug mode, you can get unexpected results when you run the file.

Some time spent now familiarizing yourself with the debugger on simple programs will be well worth the effort when the programs become a lot more complicated.

9.8 Hardware

The programs we write are designed to run on hardware. This can mean a microcontroller (like a Raspberry Pi or Arduino) or a computer. Many types and brands of computers are available, sometimes with a seemingly endless array of options. Do you need more RAM? More hard drive space? Dedicated graphics? Whether the system is a desktop or a laptop, the components found inside are the same. Let's look inside a computer to see what its fundamental components are.

Figure 9-1 A computer motherboard.

motherboard

The first thing we notice inside the computer is the large circuit board that probably covers most of the area inside—this is the **motherboard** (Figure 9-1). Everything in the computer connects to the motherboard, passing all information from one component to another.

9.8.1 Processor

Central Processing Unit (CPU)

The computer's brain is the processor, or the **Central Processing Unit (CPU)**, diagrammed in Figure 9-2. It's also most likely the largest item on the motherboard. It will have a fan or heat-sink attached since it can get very hot, and it needs to dissipate heat. The processor not only drives the programs you run or the applications you use, but it also accepts and processes the input from the mouse and keyboard, tells time, and sometimes figures out what to put on the display given what the user is working on. Suppose you are working on a shared document on the Internet and need to type the letter k. In that case, the processor accepts the *k* from the keyboard, works with the Ethernet or wireless port of the computer to interact with the Internet Service Provider, and 'sends' the letter through the Internet to the document. At the same time, the processor is dealing with other programs and applications, and keeping track of hardware—monitoring the keyboard and mouse, possibly the temperatures throughout the computer, timers for dimming the monitor, detecting if some other peripheral is plugged in, communicating with Internet traffic, etc.

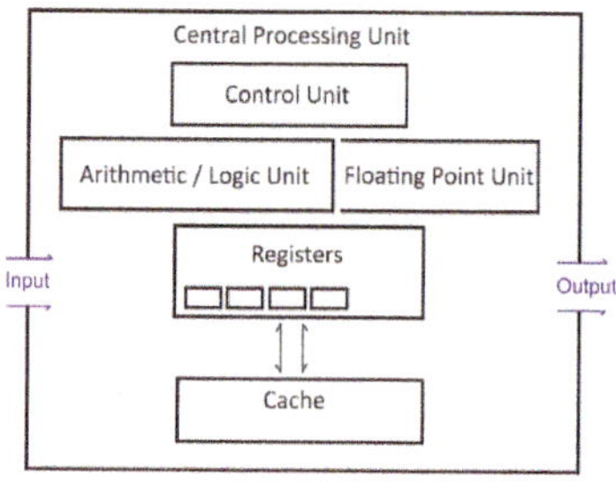

Figure 9-2 Architecture of a Central Processing Unit (CPU).

The processor is comprised of billions of electronic transistors. However, when we analyze processors, we look at their architecture rather than trying to break them down electrically. Processors contain a number of cores, which can be thought of as individual processors. Each core can tackle multiple processes or can perform multiple tasks; the more cores a processor has, the more tasks it can handle at the same time. Browsing the Web and typing in a document can be done well with a few cores (2 or so), while engineering or serious gaming may be best with 10 or more cores.

Arithmetic Logic Unit (ALU)

The processor also has an **Arithmetic Logic Unit (ALU)**, which takes care of math and operations involving integers. Integer operations happen far more often than you may expect—as data is manipulated within a computer, each **byte** of data (chunk of 8-bits of data), or larger unit of data (64-bits of data, for example) has a value and a location. Even a simple operation may involve grabbing some data from memory, grabbing more data from memory, adding the data which is at those memory locations, and putting them into another specific location in memory. Working with the addresses (locations) of the data requires a lot of integer math, and manipulating the data itself requires more integer math. The processor also has a **Floating Point Unit (FPU)** which tackles floating point math (using real numbers with decimals).

byte

Floating Point Unit (FPU)

registers

cache

control unit

The processor also contains **registers**, which are temporary holding positions for data being manipulated, which the ALU or FPU are working with. There will also be a small amount of memory which can be accessed very fast, designed for the registers to pull data from. This is called cache memory, or just **cache**. Finally, the CPU will have some space dedicated to a **control unit**, which controls what is happening inside the CPU itself.

integrated graphics

The processor may also have a section dedicated to graphics, figuring out what needs to be displayed. If the processor takes care of this task, the computer uses **integrated graphics**.

9.8.2 Graphics Card

A computer may have a graphics card if the graphics needs are intense, as found in gaming, running multiple monitors, etc. (Figure 9-3). The graphics card will have some memory and processor(s), all of which are dedicated to figuring out what belongs on the monitor(s). Powerful graphics cards may have fans or heat sinks attached if they generate significant heat. A computer with a graphics card frees up capability in the main processor to do more work, since it doesn't need to run the graphics.

Figure 9-3 A graphics card.

9.8.3 Memory

We saw that the processor itself has some extremely fast memory (cache). This memory is a relatively small amount of memory. It pulls data from the **Random Access Memory (RAM)**. The RAM is usually easily seen on the motherboard. It is mounted on smaller circuit boards, sometimes called 'sticks' of RAM (Figure 9-4). The more RAM a computer has, the faster it operates.

Random Access Memory

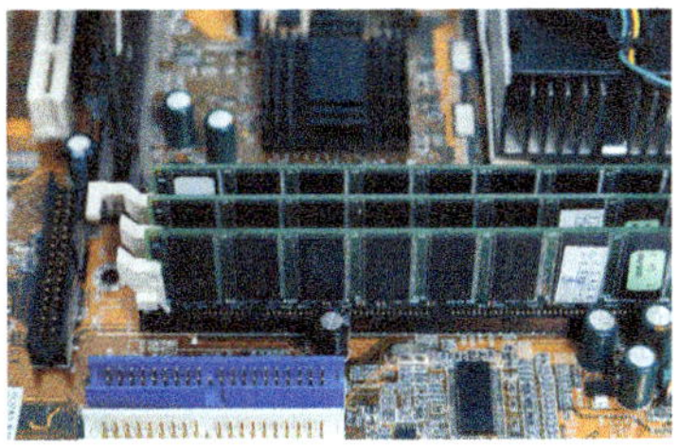

Figure 9-4 Memory, or RAM, on a motherboard.

When the processor is working with some data, it will likely need the next piece of data, then the next, and so on. When the processor calls for a byte (or a few bytes) of data, the data nearby that will be loaded into RAM, and the data immediately near the data being used is sent to the cache. As the cache data is used, more is pulled from RAM. As the contents of RAM are used, more data is pulled from the hard drive.

ENGAGE

Graphics cards somewhat suddenly skyrocketed in price as cryptocurrency became popular. These cards were used in mining crypto. As the price of crypto went up, demand for graphics cards went up so quickly that demand outpaced supply. Consider a project to understand how graphics cards were used in cryptomining. Sites like this may be helpful:

URL

https://www.adamfowlerit.com/2021/09/expensive-graphics-cards-and-mining-cryptocurrency/

ENGAGE

The internal operation of non-SSD hard drives is interesting to see. This video provides a a good explanation, with a lot of detail of the internal operation:

URL
https://www.youtube.com/watch?v=NtPc0jI21i0

Here's an interesting video of a hard drive operating without a cover:

URL
https://www.youtube.com/watch?v=9eMWG3fwiEU

9.8.4 Hard Drive

hard drive

The **hard drive** is where data are stored (Figure 9-5). When you install a program or download or store data like a game or song, a document, or images, they are stored on the hard drive or saved to the cloud, which can be pictured as thousands of giant hard drives located elsewhere. Your programs and your data, pictures, music, and documents which are used and manipulated by those programs, are stored on the hard drive. The **operating system** (usually Windows, Linux, or a Mac operating system) is also stored on the hard drive since the hard drive will hold its contents even if power is lost.

operating system

form factor

The hard drive is mounted somewhere in the computer and attached to the motherboard with a cable. The physical size of the hard drive is called the **form factor**.

Data is stored on the hard drive in one of two ways. A spinning hard drive has multiple magnetic discs with magnetic read/write heads. The disks spin very fast—typically around 7,200 revolutions per minute (or rpm).

Solid State hard drives use integrated circuits to store the data, so there is no spinning disk and no moving parts.

Figure 9-5 Hard drives.

The size of the hard drive in a system decides how much data or how many programs can be stored. If you run low on storage, most computers can easily add another hard drive (Figure 9-6), or storage can be found in the cloud.

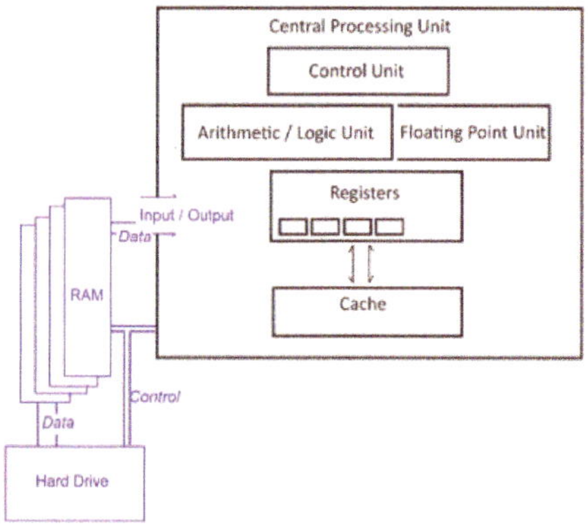

Figure 9-6 Memory, from the hard drive, to RAM, to cache and registers inside the Central Processing Unit.

9.8.5 Power Supply

power supply

The large unit inside the computer where the power cord is attached is the **power supply** (Figure 9-7). This unit takes in power from the outlet and allows it to be distributed throughout the computer.

Figure 9-7 Computer power supply.

9.8.6 Input/Output and Peripherals

There are many things that plug into the computer from outside of the case. Some input/output devices, like a mouse and keyboard, plug in through a USB port, a port for connecting peripherals. Besides mice and keyboard, you may plug in flash memory (also called a thumb drive), a webcam (camera), a microphone, or a gaming controller. The computer should recognize what type of device is plugged in; if not, you may need to find a driver (a program telling the computer how to operate the device that was plugged into the USB port).

The keyboard and mouse are considered input devices because these allow a user to give input to a program or game. Your speakers (possibly connected via a specific jack in the back), printer, etc. are output devices—a program can make them do something or send data out to them.

9.8.7 Software

software
BIOS

There are multiple types of **software** in a computer. We describe software from low-level to high-level. Low-level software, or **BIOS**, runs with the hardware of your system. As a program, it will check to see that the memory (RAM) exists and is installed properly as the computer starts. It keeps track of specific things the operating system needs to know. You can modify some parameters within the BIOS by interrupting the computer as it powers on. Typically, we leave this alone because it talks specifically to the hardware in your system.

The next higher-level software is the operating system (like Windows, Mac, or Linux). The OS communicates with hardware and with the user. It allows you to start a program, print from that program, use the mouse to move around the monitor, and chat using a webcam and microphone.

Programs, applications, games, and programs you write (like MATLAB programs) are all at the highest level. These programs run within the operating system, and are usually stored on the hard drive, ready to run.

DESIGN ACTIVITY

Designing a Rehabilitation Game

When we think of rehabilitation, the first image that may come to mind is physical therapy (more formally called physiotherapy): someone flexing their muscles repeatedly, cycling through a set of exercises prescribed by a physician. The purpose of rehabilitation is to regain motor skills, which could involve several weeks, months, and maybe even years of work to reach the same level of performance that the individual had before the need for rehabilitation. Whether or not you have personally experienced physical therapy before, it is easy to see how boring and tedious the process can become. Enter **console-based therapy,** which is an area of research about leveraging technology to enhance physical therapy interventions. Games can be an extremely powerful way to help patients who have experienced both physical and cognitive issues, such as muscular dystrophy or a traumatic brain injury, to regain functions they might have lost through the adverse health event. Motion controlled interactive devices such as the Wii from 2006 to the headset apparatus from Meta in 2020 with eye and face-tracking have provided new media for individuals to play and socialize in the comfort of their homes. Because games often require repetitive motions to accomplish actions in the virtual environment, console-based therapy provides a link from these motions to necessary exercises the patient must perform. It should be mentioned that console-based therapy is not supposed to replace the physical therapist; instead, it supplements the overall rehabilitation plan for the patient.

Different kinds of devices and games have advantages and disadvantages to their use in the context of console-based therapy. The game does not need to be complicated to be effective, for instance. Simple games like Scrabble can be beneficial to someone who is in rehabilitation for a stroke, for example. For patients needing more whole-body or limb-specific exercises, therapists have leveraged Microsoft's Kinect and Nintendo Wii. In a paper published in 2006, Jonathan Halton described activities a patient could conduct with the Nintendo Wii to accomplish several different rehabilitation goals. For example, to increase a patient's range of motion in their arms, it is recommended that the associated games have large but natural movements like tennis. Or distal motor control improvements should be aligned with games that require steady and controlled movements. As these platforms improve, the devices can capture more information about the player's movement. Such improvements allow for therapists to more accurately measure if the patient's exerted effort is applied to correct body parts, which allows for specific games to be recommended or developed by third parties. For example, the Microsoft Kinect did not require any auxiliary devices to detect the player's three-dimensional motions and found acceptance due to its low cost and accessibility for developers to create their own games to align with rehabilitation outcomes. The Nintendo Wii Fit Plus balance board, an optional accessory for the Nintendo Wii that was released in 2009 for North America, has delivered positive results for patients with Multiple Sclerosis (MS) in a study by Antonio Celesti and colleagues.

Problem Statement

As we've seen, the use of console-based therapy has the potential to be a useful supplement to rehabilitation. Your objective in this project is to design a rehabilitation game for patients who have suffered a stroke. According to the American Stroke Association, stroke is the fifth leading cause of death and is the leading cause of disability in the United States—yet *80 percent* of strokes are preventable. A stroke is caused by a clot in or burst of a blood vessel that delivers necessary nutrients and oxygen to the brain, leading cells in the brain to die. Because the brain controls so much of our body's essential functions, such cell death can have a significant impact on an individual's quality of life.

There are some initial design considerations that can be leveraged in developing a rehabilitation game for the population who have suffered strokes. For instance, Ya-Zuan and colleagues conducted a study in 2016 to understand what stroke patients seek in console-based rehabilitation. Of the 30 patients and 19 occupational therapists they surveyed, they found that patients desire diverse and engaging games, systems that are cost-effective, and opportunities to socialize. Twelve patients wanted games that are intuitive, ten wanted games that leverage their prior game experience, and seven wanted the game to be challenging.

In this challenge, you will conduct initial research to understand the broader landscape of console-based therapy, consider the needs of the relevant stakeholders, and develop a working flowchart to outline the main functionality of the game. Depending on the programming expectations in the course, you will develop a minimum viable product for some of your game's functions.

Materials

- Device connected to the Internet with MATLAB installed
- Writing utensils (different colors would be helpful)
- Paper (can be 8½ × 11 or larger, such as poster size)

Understanding Who Your Design Is For

To ensure your solution is appropriate, you need to understand the needs of the user. Conduct initial research on common issues experienced by patients who suffered a stroke. Consider the following questions in your design notebook:

1. What are the common outcomes of a stoke that individuals experience?
2. What challenges do stroke patients face on a daily basis?
3. What are the current rehabilitation options for those who have suffered from a stroke?
4. How would the proposed design complement current methods?
5. Who are the stakeholders for your design?

DIFFERENTIATED INSTRUCTION

You may elect to have students complete a research report on the intended audience of their design, stroke patients. This report would include the current rehabilitation options and an argument for why their rehabilitation therapy game would fill a need.

DIFFERENTIATED INSTRUCTION

For this project, we see two main paths forward. It's possible to complete this project using flowcharts alone, especially to walk through the basic functionality of their code. If you have a class that will be engaging in programming (MATLAB or otherwise), the essential criteria will be implementing the appropriate logical structures to conduct the basic functions of their game.

If you have students who are savvy coders, consider putting them in an ability group and ask them to create a full prototype of the game, such as adding a graphical user interface beyond text prompts. Placing one advanced student per team may run the risk of that student completing the entire project for the team. If you do decide to distribute these students to different teams, ensure they have a code of cooperation statement to have everyone contribute to the code in a meaningful way.

DIFFERENTIATED INSTRUCTION

To increase the diversity of possible games, you can replace the medical condition with a variety of other issues that can be treated using some kind of physical therapy: lymphedema, headaches, Parkinson's disease, joint replacement, tendinitis, Cerebral Palsy, concussion, or Alzheimer's disease. You may either give them the list here or have them research and justify a medical condition that could be reasonably treated using a rehabilitation therapy game.

TEACHING TIP

In this context, the functional decomposition and block diagram will eventually become a set of flowcharts. Have your students consider the various elements of the game they plan to design. What are the smaller functions they'd need to create in MATLAB to accomplish them?

TEACHING TIP

Hold your students accountable for doing the brainstorming activities. At this stage, they should have advanced to become more systematic in their ideation processes.

DIFFERENTIATED INSTRUCTION

If you'd like to give more advanced students a challenge, you might elect to have them create flowcharts of different functions present in the games they research. Then, they can judge the extent to which those functions can be translated to their game. This is another application of modularity.

Develop Criteria, Constraints, and Design Specifications

Now that you've taken time to understand your user's needs, next you must translate those needs into criteria, constraints, and design specifications. These will be used to guide the development of your solution and ensure it is successful. In your design notebook, address the following prompts:

1. What are the criteria you identified for this design?
2. What are the constraints limiting the possibilities of this design?
3. How will you know if your design is successful? Translate your criteria and constraints into measurable design specifications.
4. Consult with another team of students in your class and compare your criteria and constraints. What differences and similarities do you notice?
5. Develop a test plan to measure whether your design achieves the criteria you specified at the beginning of the challenge.
6. Did you revise or add design specifications during your test plan development?

Functional Decomposition

With our criteria, constraints, and design specifications outlined, now we can decompose our potential solution into functions that it must accomplish to be successful. Using your answers in the previous section, record the following in your design notebook.

1. What are the inputs, outputs, and functions involved in the potential design solution? Be sure to include the implications for programming while thinking through the general functional requirements.
2. What subsystems are necessary to generate the desired output?
3. Develop a functional block diagram relating the input, outputs, and functions.
4. Compare your functional block diagram to a different team of students than you consulted in the "Develop Criteria, Constraints, and Design Specifications" step. What differences and similarities do you notice?

Generating Potential Concepts for a Prototype

Now that you have considered the different functions and subsystems needed for your design to meet its criteria and constraints, we can turn to generating potential concepts. In your design notebook, address the following prompts.

1. Use a systematic procedure, such as 6-3-5 Brainwriting or Wrong Theory Protocol, to generate ideas for each of the functions and subsystems identified in the functional decomposition. Be sure to ideate what the user interface would look like for the user; that is, what would the user be looking at when they are interacting with the game?
2. What kind of games would best suit the needs identified through your initial research? Conduct additional research on games that are used in rehabilitation contexts and create a chart that organizes the types of games by the impacts of strokes on patients and their needs. Append your brainstorming ideas to the chart in the appropriate places.
3. Organize the results from your ideating session into a morphological chart using the functions and subsystems identified in your functional decomposition. Pick at least three methods per function or subsystem to include from your brainstorming.

4. Once your morphological chart is complete, identify five potential concepts using the methods you specified.
5. From the five concepts, create a sketch and a description that capture how the concept would work. Next, search online to see if that concept has been implemented in an existing rehabilitation therapy game. If there is an existing product, paste the image into your design notebook and explain how your concept is similar and different. If you cannot locate an existing product, justify why you believe your concept is unique.

Selecting a Concept

After generating a series of design alternatives, the concepts from the morphological chart, we now must pick a concept to move forward with. One tool for making a viable prototype selection is a decision matrix. In your design notebook, answer the following:

1. Using your criteria and design specifications, set up a decision matrix to begin weighing the viability of the five concepts.
2. Determine weights to place on the different criteria.
3. Calculate the "winning" design using your weights.
4. Summarize your findings.
5. Compare your selected concept to a different team of students than you consulted in the "Develop Criteria, Constraints, and Design Specifications" step. What differences and similarities do you notice? Discuss the process by which the other team arrived at the solution.

Developing the Prototype

Now that your team has chosen a design to move forward with, it is time to prototype the solution. Because this design is for a piece of software, it would be beneficial to outline how your team believes the logic of the game would work. Address the following prompts in your design notebook.

1. Develop an outline in the form of a flowchart that details the main functions of the game. For example, if there is a score, how is the score incremented as the game progresses? What are the conditions that affect the score, like multipliers that double points for a period of time?
2. Consider how the different flowcharts you developed fit together. Are there any functions that you found as you were developing the logic that are dependent upon each other that weren't before?
3. Use a platform like MATLAB to implement at least three of the functions you identified in your functional decomposition.

DIFFERENTIATED INSTRUCTION

Students often do not link explicitly back to their criteria/constraints when deciding on attributes for a decision matrix. For students who need additional help here, consider having them create a table of the criteria/constraints/specifications and simplify them into attributes that can be assigned weights in the decision matrix. At this stage, you should expect them to be doing this anyway.

TEACHING TIP

Most of your attention should be placed in supporting students in developing their prototypes, considering this unit is more focused on programming than design. Coding/flowcharting processes should be the focus of your assessment.

TEACHING TIP

Check in with student teams when they begin their prototyping to ensure the functions they're planning to develop flowcharts for are sufficiently complex to warrant full credit for the project. You may adjust this level depending on the composition of your class and ability group organization.

DIFFERENTIATED INSTRUCTION

We use three functions as an absolute minimum here. You can expand the number of necessary functions if you'd like to challenge the whole class or ability groups in the class.

If you're not emphasizing programming in your class, you can skip the coding in MATLAB or a similar platform prompt.

TEACHING TIP

If students are programming their games using MATLAB or some other platform, have them revisit the strategies for debugging. Chances are, there will be issues in their code, so include debugging throughout the duration of the project.

Encourage students to record bugs in their design notebook.

DIFFERENTIATED INSTRUCTION

We suggest having students reflect considerably on how they would extend their functionality to complete the game. You may elect to have students research how their game can incorporate physical movements using sensors like the Microsoft Kinect and Nintendo Wii, and make that market research a deliverable for their project and part of their final reporting assessment.

DIFFERENTIATED INSTRUCTION

To ensure a variety of assessment evidence, you may choose to let students pick their mode of communication. Or, you may elect to only check their design notebook, have them write a report, or make a presentation. It's not recommended that students be assessed using all modes for each project.

Testing the Prototype

Once you have a workable prototype, it's time to test whether the design meets the specifications outlined earlier in your process.

1. Implement the test plan with your prototype and record your results in your design notebook.
2. Interpret the results: what worked and what didn't? What do you need to do to revise your prototype to meet the specifications?
3. Your prototype currently performs some essential functionality that your game needs to function properly. If you were to develop the game further, what other functionality is needed to reach a product that is ready for broader distribution?

Report Your Solution

Finally, it is time to report out to the class. Your teacher may elect to have you present a PowerPoint summarizing your processes from each stage of the design process outlined here. You may also write a short report covering the same material. Your teacher may choose to use online tools such as Google Docs or Google Slides to develop your report or presentation. Be sure to focus your written, verbal, and visual communication to be clear, concise, and effective. In your design notebook, create a bulleted list of the main decisions you made across the different stages of this project. If you are making a PowerPoint, create an outline for how you will present the information. Next, reflect on the following questions:

1. What was the objective of this project?
2. What worked?
3. What didn't work and why didn't it work?
4. How did the design compare with the best and worst performers?
5. What would you do differently?
6. Was the objective accomplished?
7. Was your communication clear (visual, verbal, and written)?
8. Was your communication concise (visual, verbal, and written)?
9. Was your communication effective (visual, verbal, and written)?

IN-CLASS ACTIVITIES

Chapter 9

ICA 9-1

Describe the steps necessary to create a paper airplane. You may assume that you are starting with a single sheet of 8½ × 11-inch paper. When you are finished, hand your algorithm to the instructor and wait for further instruction.

ICA 9-2

Describe the steps necessary to create a jelly sandwich. You may assume that you are starting with a loaf of bread, jar of jelly, a knife, and a plate on the table in front of you. When you are finished, hand your algorithm to the instructor and wait for further instruction.

ICA 9-3

Determine the length of one side of a cube of solid gold, in units of inches. You may assume that the specific gravity of gold is 19.3 and that the user will provide the mass of the cube in units of kilograms. For example, you might assume that the user has a 0.4-kilogram cube.

ICA 9-4

Which of the following are not valid program/function file names? Circle all that apply.

(a) `2b_solved.m`

(b) `calc_circum.m`

(c) `graph-data.m`

(d) `help4me.m`

(e) `MATLAB is fun.m`

(f) `matrix*matrix.m`

(g) `Mult2#s.m`

(h) `pi.m`

(i) `ReadFile.m`

(j) `SuperCaliFragiListicExpiAliDocious.m`

ICA 9-5

Without running these code segments in MATLAB, what will be the value stored in the specified variable after executing the program?

What is stored in Dogs?

```
clear
clc
Greyhounds = 30;
G_Cost = 250;
Dalmatians = 10;
D_Cost = 150;
Dogs = Greyhounds * G_Cost + Dalmatians * D_Cost;
```

ICA 9-6

Write a MATLAB program to evaluate the following mathematical expression. The equation should utilize a variable for x. For example, you might run the program with $x = 30$.

$$A = \frac{x^2 \cos(2x + 1)}{(6x)\log(x)}$$

ICA 9-7

Write a MATLAB program to evaluate the following mathematical expression. The equation should utilize variables for x, μ, and σ. For example, you might run the program with $x = 0$, $\mu = 0$, and $\sigma = 1$.

$$P = \frac{1}{\sigma\sqrt{2\pi}} e^{-(x-\mu)^2/(2\sigma^2)}$$

ICA 9-8

Create a written algorithm and flowchart to determine if a material is in the solid, liquid, or gaseous state given the temperature. The algorithm should acquire the name of the material, its freezing and boiling temperatures, and the actual temperature of the material from the user.

ICA 9-9

The Occupational Safety & Health Administration (OSHA) defines safety regulations on working environments to protect workers from unsafe conditions. The following flowchart shows how OSHA categorizes the safety level of the working temperature given the environmental temperature in degrees Fahrenheit. Given the flowchart, for what range of heat index (in degrees Fahrenheit) will the risk level be Lower (Caution), Moderate, High, or Very High to Extreme?

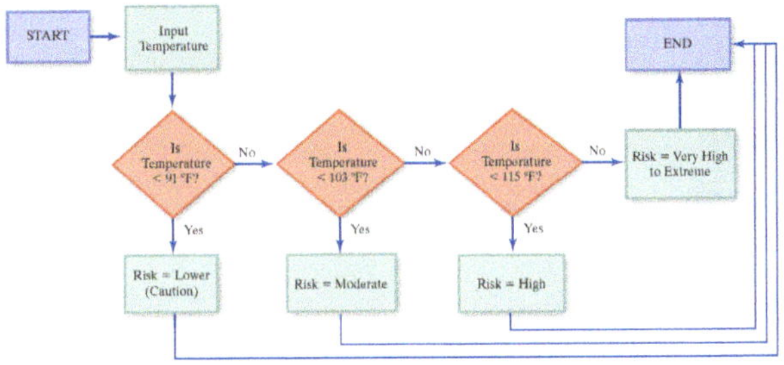

ICA 9-10

Answer the following questions.

(a) For what integer values of G will the code between `if` and `end` execute?

```
if G < 2 && -5 <= G
   code
end
```

(b) For what values of R will the code between `if` and `end` execute?

```
if R < 75 || R >= 95
   code
end
```

(c) What combinations of values for W and M will cause a warning to be displayed?

```
if W<50 || (W>120 && M~=0)
   fprintf('WARNING! Parameters out of bounds.\n')
end
```

ICA 9-11

Create a written algorithm or flowchart as assigned to ask the user to draw a card from a deck of cards until the user pulls out the queen of hearts.

ICA 9-12

Create a written algorithm or flowchart as assigned to ask the user to enter the number of Internet-enabled devices they own. If the user enters a negative number or a number greater than 500, the message "This is not reasonable. Try again." should appear to the user and they should be asked to enter another value. The warning message should continue to appear each time they enter an invalid value. When the user provides an acceptable value, the message "Thank you." should be displayed and the algorithm should terminate.

ICA 9-13

Create a written algorithm or flowchart as assigned to first ask the user to enter a positive number. After checking that the value entered is positive, determine all the even numbers between 0 and that value that are also evenly divisible by both 3 and 5. If the user enters a nonpositive value, issue a warning and ask the user to enter a new value repetitively until a positive value is entered. At the end of your algorithm, your algorithm should display the total number of values that meet the specified criteria.

CHAPTER 9 ASSESSMENT QUESTIONS

1. Create an algorithm to cook your favorite meal.

2. Create an algorithm to convert a temperature provided by the user in units of Fahrenheit to units of kelvin. For example, you might assume the user provides the temperature of –129 degrees Fahrenheit, which is the world's lowest recorded temperature.

3. Create an algorithm to determine how long, in units of seconds, it will take a motor to raise a load into the air. Assume the user will specify the power of the motor in units of watts, the rated efficiency as a percentage (in whole number form—for example, 50 for 50%), the mass of the load in kilograms, and the height the load is raised in the air in units of meters. As a test case, you may assume the user provides 100 watts for the power of the motor, 60% for the efficiency of the motor, 100 kilograms for the mass of the load, and 5 meters for the height the load is raised.

4. Write a program to convert a temperature provided by the user in units of Fahrenheit to units of kelvin. For example, you might assume the user provides the temperature of –129 degrees Fahrenheit, which is the world's lowest recorded temperature.

5. Write a MATLAB program that implements the quadratic equation. Recall the quadratic equation:

$$r = \frac{-b \pm \sqrt{b^2 - 4ac}}{2a}$$

6. Write a MATLAB program that implements the Pythagorean theorem. Recall that the theorem states that the square of the hypotenuse (z) can be calculated by the sum of the squares of the adjacent sides (x and y), or

$$z = \sqrt{x^2 + y^2}$$

7. Without running these code segments in MATLAB, what will be the value stored in the specified variable after executing the program?

   What is stored in `W`?

```
clear
clc
X = 2;
Y = 4;
Z = 2;
W = (Y + Z/X + Z^1/2)
```

8. Without running these code segments in MATLAB, what will be the value stored in the specified variable after executing the program?

   What is stored in X?

```
clear
clc
A = 0;
B = 5;
X = A^B + B^A;
```

9. Mach number is defined as the speed of an object divided by the speed of sound. Assuming the speed of sound is 343 meters per second, write a MATLAB function that will calculate the speed in units of miles per hour given the Mach number. For example, an unmanned X-43A scramjet test vehicle has achieved a maximum speed of Mach number 9.68 in a test flight over the Pacific Ocean, so you might test this function with a Mach number of 9.68.

10. Write a function to calculate the volume of a cone which accepts the diameter of the cone's end and the height as inputs.

11. Create an algorithm to classify a given altitude [miles above sea level] entered by the user as in the troposphere, stratosphere, mesosphere, thermosphere, or exosphere. The algorithm should include a check to ensure the user entered a positive value. If a nonpositive value is entered, the program should warn the user and terminate. If the altitude entered is beyond the exosphere, the algorithm should report "Not in atmosphere." The atmospheric regions are defined by the following criteria:

    Troposphere < 12 km
    12 km ≤ stratosphere < 50 km
    50 km ≤ mesosphere < 80 km
    80 km ≤ thermosphere < 700 km
    700 km ≤ exosphere < 10,000 km

12. What is stored in variable A after each of the following code segments is executed? If the program returns an error, indicate why.

(a)
```
A=5;
if A/2>2
  A = A*-1;
else
  A = A/2;
end
```

(b)
```
x=2; y=4; A=0;
if x+y<4
  A = A+2;
elseif x+y<6
  A = A+4;
end
```

(c)
```
T=6; A=-1;
if T/3>2
  A = A*-1;
elseif T/3<=2
  A = A+2;
end
```

(d)
```
x=2;y=0;A=10;
if x<=0&&y<=0
  A = A+2;
elseif x>=0&&y<0
  A = A/2;
elseif x>=2&&y<=0
  A = A*2;
end
```

13. Answer the following questions.

 (a) For what integer values of D will the code between `if` and `end` execute?

```
if D >= -1 && 3 > D
code
end
```

 (b) For what values of F will the code between `if` and `end` execute?

```
if F >= 100 || F <= 5
code
end
```

 (c) What combinations of values for B and Q will cause a warning to be displayed?

```
if B>100 || (B<20 && Q~=0)
fprintf('WARNING! Parameters out of bounds.\n')
end
```

14. Create a written algorithm or flowchart as assigned to ask the user to roll 5 dice until all 5 dice are all 6's on a single roll.

15. Create an algorithm to allow the user to enter a list of courses they have taken along with the grades received and the credit hours for each course, and then calculate the user's GPA. The program should first ask the user to type either Yes or No in response to the question, "Do you want to enter a grade?" If the answer is Yes, the user should be asked to enter the name of the course (text), the grade in the course (A, B, C, D, or F), and the number of credit hours for the course (a positive integer). If the user enters an invalid grade or an invalid number for the credit hours, he or she should be warned and asked to try again a maximum of three times, after which an appropriate error message should be generated, and the program terminated. If they successfully enter the information for a course, the course name, the letter grade, and the credit hours should be displayed in a single row on the screen in an appropriate format, and then they should be asked again if they wish to enter a grade, and the entire process repeats if they answer Yes. If they answer No, the GPA should be calculated and displayed in an appropriate format, and then the program terminated unless no grades were entered (they typed No the first time) in which case an appropriate message should be generated, and the program terminated.

16. Your team has been tasked to determine the number of employees needed for a manufacturing line. For every 100 products per hour produced, one employee is needed. For every 10 employees, an additional employee (a supervisor) is needed. If the plant expands and adds another device, that line will need 1.2 times the number of employees as the previous line. Write an algorithm to determine how many employees are needed. Remember that this must be an integer (after all, we can't hire 8.2 employees!).

322 Chapter 9 Computer Programming

17. Find a computer for sale online. Find the specifications, including:
    (a) Does it have a graphics card or integrated graphics?
    (b) How much memory does it have?
    (c) What technology is the hard drive, and how large is it?
    (d) What is the speed of the processor?
    (e) What is the cost?

    Include a link to the system.

18. If cache memory is much faster than memory from a hard drive, why not just make all memory cache memory?

19. List and briefly describe five different input/output peripherals. What does each one do? Is it strictly input, strictly output, or I/O?

20. Find an image of a motherboard online, or if your school has some older computers you can disassemble, look at a real motherboard. Identify each component on the motherboard.

21. If you are using a Windows computer, look at the "devices" inside. What are some of the important characteristics inside the computer you are currently using?

22. Explain the basic components of a computer's programming design.
    (a) What is the purpose of software?
    (b) Describe the difference between software and data.
    (c) What is the purpose of computer programming languages?

OVERVIEW

Since the Industrial Revolution, people all over the globe have been pushing more and more towards having machines simplify repetitive tasks. This trend has only increased in modern times. Truly, the evolution and growth of technology has expedited this process. Through the use of technology, industry has been able to explore and implement automation for many of its tools, thus simplifying repetitive tasks even more.

CHAPTER 10

Manufacturing and Automation

TOPICS

manufacturing

People have been **manufacturing** for thousands of years, which is the process of creating a product or good using a machine or by hand. Before the development of the advanced equipment we have today, clay was formed and heated to create pottery, and homes and weapons were created from natural materials such as stone and wood. People learned effective manufacturing techniques from others who had discovered an innovative technique. Certainly, this is still true today, but what is different about manufacturing methods of the past? For thousands of years, items were manufactured *one at a time* and as needed. This process is called **craft manufacturing**.

craft manufacturing

durable goods

Eventually, factories emerged as a means to manufacture goods much more efficiently. **Durable goods** (such as cars, smartphones, and tables) are now manufactured almost exclusively in factories. There are still cases where people create items by hand, but these are typically sold as hand-crafted specialty items, usually at a higher price. **Nondurable goods** (such as paper products, food, and toiletries) are goods that have a short life; they are designed to be used over a short period of time and usually discarded. They are also manufactured in factories.

nondurable goods

mass production

Producing goods one at a time is not particularly effective, especially when several copies of the goods are needed. **Mass production** is defined as the systematic production of a large quantity of goods using a combination of workers and machinery. The Industrial Revolution, starting in the 1700s, ushered in the capability for factories to decrease the amount of time and effort needed to produce a good. The key to mass production is the standardization of processes. Unlike the craft manufacturing model where goods are effectively one of a kind and cannot be replaced or fixed without specialized assistance, mass production utilizes interchangeable parts.

interchangeable part

An **interchangeable part** is a component of a product that is designed so that it can be replaced with another version of itself. While not invented by Eli Whitney in in the 19th century, he popularized the concept of interchangeable parts through his assembly of 10,000 muskets for the U.S. Army—which was intended to be completed in two years but arrived several years late. In fact, when Congress questioned him after failing to deliver their order, he put on a fake demonstration of reassembling ten different rifles to convince them of his approach. It turned out that he had marked the parts before the demonstration, so he knew which component belonged to each rifle. The idea of interchangeable parts had been used several years prior—also in the context of weaponry—by Jean-Baptiste Vaquette de Gribeauval, a French general, and Honoré Blanc, a gunsmith. What follows describes the tools engineers use in this process, providing detail about each of these processes.

OBJECTIVES

- Describe applications of robotics, process control, and automation systems
- Apply design concepts to problems in robotics, process control, and automation systems
- Identify fields and career opportunities related to robotics, process control, and automation systems
- Identify emerging trends in robotics, process control, and automation systems
- Determine the design parameters such as materials, personnel, resources, funding, manufacturability, feasibility, and time associated with an engineering problem
- Identify the requirements for and role of intellectual property in design
- Understand the role of specialists in solving bioengineering problems
- Work in teams to design and build a project related to bioengineering
- Write a reflection of the project

ENGAGE

Visit the History.com website. Read about the Industrial Revolution in the 19th century and how interchangeable parts affected industry at the time.

URL
http://www.history.com/topics/inventions/interchangeable-parts

Ask your students to explain the impact of interchangeable parts. *(Higher productivity for factories)*

ENGAGE

Discuss when and why a craft approach can still be marketable. *(High-quality products are typically handcrafted. Handcrafting allows the skilled artisan to become close with the product he or she is creating. The product shows the quality, care, and dedication the crafter must maintain to produce the most ideal product. Some of the most expensive cars can even be handcrafted!)*

ENGAGE

What is the main advantage of design for manufacturability? *By focusing on how the product will be manufactured, it is possible to reduce the time and cost to create the product at scale.*

TEACHING TIP

In this chapter, we use a biomedical engineering example to illustrate design for manufacturability. This would be a useful chapter to highlight biomedical engineering career opportunities.

10.1 Designing for Manufacturability

design for manufacturability

An effective engineering design should take into account how the item which was designed will be manufactured. This is called **design for manufacturability**. We can look at the steps needed to take an idea from initial concept to a product ready for production. In general, the steps are ideation, concept development, preliminary design, prototype and test, acceptance test, assembly, and initial production. Each of these steps has a specific outcome that advances the product idea until it moves to its initial production stage.

Ideation

This first step begins with examining a particular problem that needs a new or better solution than the existing one. Let's suppose you have been working with individuals that have single arm amputations below the elbow. One of the known conditions of amputees is something called Phantom Limb Pain (PLP), which can be present in as many as 60% of amputees. It has been known for some time that if a mirror is oriented so that the intact arm reflects in the mirror and it appears to be in its rightful place on the other side of the body, then the phantom pain is reduced or eliminated in some patients. This gives your team the idea that if you could produce a pair of Augmented Reality glasses that are programmable so that the phantom limb always appears when wearing those glasses, it might alleviate the PLP problem at least for some. You decide to move forward with this design. At this stage, you have a problem to solve and the start of an idea in where to begin.

Stakeholders and Specialists

Successful designs rely on seeking the expertise of both experts and stakeholders. Failing to consider all of the stakeholders in a design can lead to failure, or even dangerous situations. For example, car safety features such as seatbelts and airbags have historically been designed by teams consisting of men. Safety tests have featured crash test dummies since the 1950s. Prior to 1971, all crash test dummies were modeled after the 50th percentile male. The average dummy was 5' 9" tall and weighed about 171 pounds, far larger than the 50th percentile woman. Researchers in the 1980s argued for the inclusion of a crash test dummy representing the average woman, but this didn't happen until 2003, when the National Highway Traffic Safety Administration (NHTSA) used a scaled-down crash test dummy to represent a woman. It was not until 2011 that testing for both men and women was required. Further, women tend to sit closer to the front of the car than men, which affects the success rate of air bags. Failing to consider all stakeholders in this case resulted in women being 71% more likely to be moderately injured and 47% more likely to be seriously injured in an automobile accident.

If our team is developing a video game to help with PLP, we should make sure all necessary specialists and appropriate engineering disciplines are represented on the design team. A project like this may be considered a mix of bioengineering and software development. Experts should be sought in biomedical engineering and computer science, and perhaps in software engineering, the medical field, and mechanical engineering. In a pure bioengineering design, where existing tissue may be used to interface with an artificial limb, for example, these experts plus medical doctors and bioengineers would need to be on the team. Doctors would bring the understanding of how to operate on the existing tissue to preserve it, and bioengineers would bring the knowledge of how to interface to the new limb.

Concept Development

To take this idea to reality will require some research into a few areas, and require definition of the goals, objectives, and requirements to solve this problem. Your team will need to investigate AR glasses which are available on the market and how to program them. It is possible that the hardest part will be to get the phantom limb to appear in the right location and orientation. While that function can be developed and tested on someone who is not an amputee, at least initially, it will eventually require testing on someone who is an amputee. At this point, you may need to bring someone from the medical community to your team to advise you.

At this stage, it is also important to develop some idea of how the physical parts might look, what they cost, and how they work together. Having a first conceptual representation is critical to being able to explain your idea to others and have it become the baseline design. It is usually at this stage that in order to move ahead, you will need to get funding to research potential solutions through grants or investment. This is a difficult stage and whole books have been written about this subject. For our example, let's assume that from this point forward, at every stage of the development, you are able to receive a grant or investment to cover your costs.

Your engineering team will spend significant time in this stage as different ideas and subsystems are developed. The team will go through most of the engineering design process in this stage.

Preliminary Design

It is very important to reduce your concept down to a working model. You learn a lot at this stage—mostly you learn about areas where you assumed the solution would be straightforward, but it turns out that some aspects are more difficult than you imagined.

Assume you have identified the following issues:

1. The amputated arm must have some sort of arm cuff on the stump with reference marks on it, so the AR glasses have a reference from which to reconstruct the phantom arm.
2. Special cases (arms crossed for example) are hard to reliably reproduce and can be confusing.
3. The AR glasses must have a wider field of view to be effective.
4. The AR glasses must be light in weight as they quickly get uncomfortable.
5. The AR glasses have a small camera that needs mounting in a different location so that it can pick up the reference marks on the arm cuff.

You could expect a much longer list of issues in the real world, but for our example, we will end our list here as this tells you where the team will need to put the most effort in the next stage.

Prototype and Test

This is where you address all the known issues with the AR PLP reduction device. The arm cuff needs to be manufactured (perhaps a 3-D printer would be a good choice). A custom mount for the small camera that is adjustable could be manufactured—again with a 3-D printer. Changes to the glasses themselves may need to be addressed by the manufacturer. As you make these changes, you also want to keep detailed records, comparing between the original concept and the post-prototype version, keeping track of changes to cost, materials, design specifications, and performance. This will help in moving the product to production. Actual users should also test this design to develop a sense of what kinds of variability the final version needs. This is the stage at which you start thinking about manufacturing, product assembly, packaging, potential programming instructions for the user, or some way to "teach" the software by doing a series of structured movements while wearing the arm cuff.

Acceptance Testing

In this stage, you decide what level of performance is acceptable and what the design team and the manufacturer are promising that this product will do. Part of this testing involves a prediction of the performance and reliability of the initial design, as well as the potential failure modes. Typically, acceptance testing is done in batches, featuring multiples of the finished product which have been assembled according

TEACHING TIP

Remind students that although some of these terms may appear new, such as acceptance testing, this is just the design process we've been reiterating throughout the text. Now we've put a spotlight on what needs to happen for a design to make it to market.

TEACHING TIP

If you have already covered the chapter on statistics, you can highlight the need for rigorous testing plans when developing products.

ENGAGE

To have students generate the list of technological resources, have groups brainstorm what they would need to start a lemonade stand business.

326 Chapter 10 Manufacturing and Automation

to the manufacturing plan. The purpose of acceptance testing is to demonstrate that any variability in the product is within an acceptable range. It is also to demonstrate the specified operating environment (high and low temperature, for example). Certain devices may have to comply with specific industry standards (software, electronics, etc.) and this would also be a chance to verify that those standards apply, in the final assembly and in individual suppliers' parts.

Assembly and Initial Production

Often, the batch of product that is used for acceptance testing is also used during the initial assembly and production phase. They were tested to the limits of their guaranteed specifications, so they are not considered to be saleable hardware. It is time to design and test any fixtures needed for assembly. If there are any specialized tools that need to be acquired, processes that can be automated, manufacturing instructions that can be finalized, and any inspection and packaging steps that are required, these should be completed. Once all of these steps have been identified and finished, you can now produce your first batch of saleable hardware. If all goes as expected, it is time to manufacture.

10.1.1 Technological Resources

technological resources

To make the design into a final product, manufacturing systems make use of several physical and intangible resources, called **technological resources**, that enable the creation of a physical product.

capital

Capital is the primary technological resource that makes manufacturing possible, which is typically thought of as cash or debt that is taken on by the company from a lender like a bank to finance their operations. However, capital encompasses much more than just cash and credit; **capital** is defined as anything that increases an individual, group, or company ability to generate value. Therefore, capital includes the land the company occupies, any equipment, and even things like intellectual property or patents.

angel investors

Often, companies will sell shares of their company to people outside of the company to raise money for its operations. These individuals then become financially intertwined with the company, complete with voting rights for company decisions. If the company performs well, then those individuals will see the value of the stock they bought increase. Other types of investors, called **angel investors**, are individuals who invest at the very beginning of a company, before stock is even available to the public, which is inherently riskier than the investment approach general shareholders take. In many cases, angel investors do not require that their investment be paid back if the company fails.

intellectual property (IP)

Because of the broad definition of capital, other technological resources are nested underneath it, such as *materials* and *tools/equipment*. *Information* is another crucial technological resource. **Intellectual property (IP)** describes anything created, whether that is a design, image, symbol, or an invention. Since the creation of designs almost always means creation of intellectual property, it is good to know who owns the IP. Typically, if you work for an engineering firm, the firm itself owns the IP. If you create something of real value, it's a good idea to investigate who owns the final design. Without knowing the market for the product, reasonable cost of materials, *energy* types that are best suited for manufacturing (e.g., natural gas, fossil fuels, electricity, heat), or best practices for design or trade secrets, the manufacturing process would be extremely inefficient or potentially impossible to be successful. Of course, materials and equipment or tools are needed for manufacturing, but who conceptualizes the product, gathers information, and is involved in building the product? *People*. As we'll soon discuss, including people as a technological resource during the actual assembly of the product is rapidly decreasing due to the advancement in other technological resources, specifically the automation of equipment and information gathering. However, people will continue to remain integral to the generation of new product ideas and elements of the design process that do not require repetitive actions.

productivity

During all these processes, it is important to remember that a key element of manufacturing is the *time* it takes to produce the output and the cost of the production process. The idea tying how quickly and cost-effectively a product is produced is **productivity**. Therefore, time is the last technological resource critical in manufacturing systems.

In summary, the seven essential technological resources that enable effective manufacturing systems are capital, materials, tools/equipment, information, energy, people, and time.

10.1.2 Rapid Prototyping

rapid prototyping

Rapid prototyping involves the use of computer-aided design data, like we saw in Chapter 6, to quickly create a scale model of parts or assemblies before committing resources to manufacturing the product at scale. At this stage, engineers are focused on testing the fundamentals of the design to make sure the prototype addresses the criteria. User feedback can also be solicited to ensure the prototype is a promising lead or if it is unlikely that the product would ultimately be marketable. Therefore, the engineers can get a wealth of valuable data about their design, iterate upon it, and create another prototype without significant financial or material investments. There are a couple of different techniques for rapid prototyping, such as stereolithography and fused deposition modeling. One of the main methods of rapid prototyping today is **3-D printing**, which is the construction of three-dimensional objects from a digital 3-D model. Note that rapid prototyping is *not the same as 3-D printing*; 3-D printing is one method of rapid prototyping. Generally, 3-D printers use a process called **additive manufacturing**, which involves building a product by adding layer upon layer of material, such as a thermoplastic, representing cross-sections of the product until the product is fully assembled (Figure 10-1). Back in the 1980s, 3-D printing used to involve a massive machine, costing tens of thousands of dollars to purchase and maintain, and only had limited use. Today, a decent 3-D printer can fit right on a desk for only a couple hundred dollars. The lower cost to entry has made 3-D printing a much more accessible way to mock up preliminary concepts than other methods.

3-D printing

additive manufacturing

Figure 10-1 A 3-D printer.

10.1.3 Computer-Aided Manufacturing and Computer-Integrated Manufacturing

computer-aided manufacturing (CAM)

computer-integrated manufacturing (CIM)

Computer-Aided Manufacturing (CAM) describes software that allows computers to control machines in factories. **Computer-Integrated Manufacturing (CIM)** describes a complete system incorporating systems beyond manufacturing, including billing, inventory control, purchasing, shipping, etc.

10.1.4 Just-in-Time Manufacturing

just-in-time manufacturing

Just-in-time manufacturing requires sufficient planning to keep materials and products in storage for as little time as possible (Figure 10-2). Typically, materials are delivered in bulk (which is usually much less expensive). In just-in-time manufacturing, the necessary quantity of material is ordered and delivered only as it is needed. The necessary machinery and factory processes, and the necessary assemblers, technicians, and engineers must be on hand and ready when needed. While materials may cost more since quantities ordered are usually much smaller, the manufacturer should also pay less for storage of items, which should be completed and immediately shipped. This allows manufacturers to be very responsive to customers' needs.

Figure 10-2 Just-in-time manufacturing keeps storage to a minimum. Materials arrive at the factory only when they are needed.

TEACHING TIP

To define productivity using an equation, you may share that economists measure labor productivity by taking the output of goods and services and dividing it by the hours of labor taken to create the goods and services.

ENGAGE

3-D printing has advanced to the point of printing meals. Consider sharing the following video from Origins of Food about advances in the production of printable food.

URL

https://www.youtube.com/watch?v=2kCjSq_l-0s

ENGAGE

Have the students compare and contrast the different types of automated manufacturing. What are the advantages? The disadvantages? When would you use the different types of manufacturing? Have students visit the following website after they have thought about the questions, to gain more information on the different types of automated manufacturing.

URL

https://www.ibm.com/topics/automation

ENGAGE

Define *automation. (A system that is able to work independently from major human interaction and input.)*

ENGAGE

Visit the PBS website. Watch the video about the assembly line.

URL
https://www.pbs.org/video/american-experience-ford-assembly-line/

Ask the students to explain how Henry Ford's ideas revolutionized current industry. *(Higher productivity)* Also, ask students why many people did not like his ideas. *(Loss of jobs for skilled workers, less personalized products)*

QUESTION

Ask students what happens when the automation in a factory fails. How does this affect its productivity? *Many times, when automation fails in a factory, production is forced to stop. If the automation in the factory is minimal, production may still be able to proceed, but at a slower pace. Manufacturers must factor automation fails into their total budget to gain an accurate picture of their fiscal year.*

INTEGRATING CONTENT AREAS

To integrate with the history of computer science, share the following article summarizing Alan Turing's contributions to society. You may also highlight the personal challenges he faced to humanize him as a significant figure in computer science.

URL
https://sitn.hms.harvard.edu/flash/2012/turing-biography/

What was Turing's contribution during WWII? He was a codebreaker, creating machines called bombes to translate messages encoded by the German's Enigma Machine.

10.2 Automation in the Manufacturing Space

assembly line

In mass production, an **assembly line** is used to divide up manufacturing tasks into repetitive, standardized processes. Each worker is responsible for performing a specific operation, where it is then passed to the next worker who performs their task. As technology has progressed, the need for workers to stand on the assembly line all day has decreased. In most contexts, manufacturing robots are responsible for performing an action that contributes to the construction of some product (Figure 10-3).

Figure 10-3 A robot says "hello!".

The loss of these assembly line jobs has created a need for more specialized workers, such as those who maintain and design the robots. For example, robotic arms have found several applications on the factory floor on the assembly line (Figure 10-4). These robotic arms can perform many of the rote tasks previously performed by a human, including lifting, twisting, compressing, or placing objects in a specific location. In a general sense, robots have ushered in an age of automation, the use of systems to operate equipment with limited to no human interaction—the robot crowded assembly line being the most concrete example. Compared to workers, the robotic arms can be more predictable, accurate, and less expensive in the long run.

Figure 10-4 Robotic arms are used to assemble most automobiles.

The concepts of robots and artificial intelligence permeated our culture long before the advent of the first real android. Science fiction novelists and screenwriters have toyed with the ideas of the modern-day robot—mainly with what could go wrong. For instance, the science fiction writer Isaac Asimov created the "three rules of robotics" which have been modified throughout the years since their original formulation.

- A robot may not injure a human being or, through inaction, allow a human being to come to harm.
- A robot must obey the orders given it by human beings except where such orders would conflict with the First Law.
- A robot must protect its own existence as long as such protection does not conflict with the First or Second Laws.

Robots are engrained in popular culture but may be further integrated into our lives in the future. In fact, it is becoming harder to distinguish machine intelligence from human intelligence. The first major step occurred in 2017, where Sophia, a robot created by David Hanson and Hanson Robotics, was awarded citizenship in Saudi Arabia. In 2022, a software engineer at Google claimed that LaMDA (Language Model for Dialogue Applications), an artificial intelligence algorithm, had achieved sentience. Both events have been critiqued, creating debate as to where we draw the line between human and machine. One of the simplest tests to determine if a machine or artificial intelligence can demonstrate human intelligence is called a **Turing test**. In the Turing test, there are three people: a human called the interrogator, another human called player A, and a computer called player B. The interrogator is tasked with determining which of the two players is a computer and which is a human based on written responses to the interrogator's questions. Whether a machine can think is a deeply philosophical question that is still debated.

Turing test

Do not fall into the trap of imagining an electrically powered human-like machine (these are often called *androids*); plenty of household objects can fall under the robot umbrella. Domestic vacuum cleaners that independently clean the floor are indeed robots but are only programmed to do a set of simple tasks. The continual improvement of technology could bring about new luxuries, perhaps a robot able to clean our entire house automatically.

The displacement of workers is an often-cited economic consequence of robots. When companies can replace an unskilled worker with a machine able to perform the same task without needing to pay it a salary, it is no wonder why the working world demands a different type of employee. The threat of displacement in the workforce has created anxiety about "future-proofing" one's career. Although it is scary that a robot could eventually take away jobs we accept as part of our everyday lives, automation has created more jobs than it has destroyed—and these new jobs are more desirable than those we lost. In fact, the World Economic Forum estimated that technology will generate at least 12 million new jobs more than it will eliminate from the economy in a few short years. For example, instead of doing the robot's mind-numbingly repetitive job, robotics engineers will be designing it and robotics technicians will be performing the necessary maintenance. These jobs will require the integration of several disciplines of engineering, including mechanical engineering, computer science, software engineering, and electrical engineering.

We should mention that while the robots are not paid a salary, they still require regular maintenance. Robots do have lifespans, even though they are not biologically "living" beings. Without proper maintenance, the expected lifespan of the robot can decrease in addition to reduced efficiency—meaning it performs its job worse than before. Moreover, if the robot is improperly installed, it will undergo similar losses. Also, plenty of robots still require a human operator. Mechanized arms have found their way into medicine as electronic surgeons. Advances in robot-aided surgery allows for numerous patient benefits including shorter hospital stays, decreased post-surgery pain, and less scarring. Automation often does not mean replacing a human with a robot; instead, it means a robot helping a human do a job much more efficiently.

The New Frontiers of Automation

With reduced computing costs and increased computing power, a new field called data science emerged in the 1990s and more recently accelerated in prominence to reimagine how we handle all the data collected every day. **Data science** is a field that is concerned with extracting usable information from structured or unstructured data using mathematical, algorithmic, and statistical methods. At the time of this writing, the job market for roles in this field is red hot. In 2022, the American Statistical Association reported that the average salary for these roles ranges from \$92,000 to \$98,000 with an unemployment rate of 2%. The job outlook for 2020 to 2030 in the field of mathematicians and statisticians, which a data science job would fall under, is 33% for the percent change in employment. But wait, this course is about engineering, not mathematics and statistics…so why do we care about data science?

data science

First, some disciplines of engineering, such as electrical, computer, and mechanical engineering, will interface with many of the tools in data science. It could be a tangential interaction; for example, if you work as a data engineer, you oversee developing, maintaining, and testing the architecture used to collect the information that the data scientist will use for analysis. More involved applications of data science methods will take the form of machine learning. **Machine learning** is a field concerned with developing models that use data to improve their performance. If you've interacted with email filters, recommendations popping up in your Netflix or Hulu queue, an Amazon Alexa to set a timer or turn on your bedroom lights, or autopilot in a Tesla vehicle, you've seen machine learning. In manufacturing systems, these data-driven models allow for technological resources—especially time, capital, and people—to be used more efficiently.

machine learning

Machine learning uses mathematical models, most heavily drawn from statistics, to allow the machine to "learn" the best way to perform a task. These models allow a machine to perform tasks at a level likely impossible for a human to do. For example, a model called MarI/O created by Pastebin (a website used to share text and code between programmers) user SethBling, uses techniques called neural networks and

ENGAGE

If you'd like to dive more into the Turing Test, you may consider showing the following video from Computerphile that explains the test fully.

URL

https://www.youtube.com/watch?v=Qbp3LJvcX38

ENGAGE

To underscore the point about automation and the labor market, you may consider having students watch the following video from Vox about the future of work.

URL

https://www.youtube.com/watch?v=TUmyygCMMGA&t=270s

The video provides two points of view, from economics and futurists. Which arguments resonate more with your class?

ENGAGE

It might be helpful to share the occupational outlook for data scientists from the Bureau of Labor Statistics to introduce students to this growing profession.

URL

https://www.bls.gov/ooh/math/data-scientists.htm

What is the projected growth from 2021 to 2031? *36%*

Visit the following article from Coursera about the seemingly similar position, data engineer. What's the difference between a data scientist and a data engineer? *A data engineer builds and maintains the systems that data scientists leverage in their analyses.*

URL

https://www.coursera.org/articles/what-does-a-data-engineer-do-and-how-do-i-become-one

ENGAGE

For a brief overview of machine learning, consider showing the following video, which describes the general process of developing a machine learning model.

URL
https://www.youtube.com/watch?v=HcqpanDadyQ

What are some situations in which a machine learning model would improve our personal or professional lives?

ENGAGE

The video link for MarI/O is here if you'd like to share it with the class.

URL
https://www.youtube.com/watch?v=qv6UVOQ0F44

What machine learning models did the author use to create the artificial Mario player? *He combined neural networks and genetic algorithms. Neural networks are mathematical models that are based on an analogy of how the human brain learns (i.e., neural pathways). The genetic algorithm is based on the premise of evolution, where models are created in an initial generation and are chosen to survive and breed for the next generation, passing on traits that made those models successful.*

genetic algorithms to play Super Mario World. Most surprisingly, MarI/O was not "told" anything about the game it was playing; it was only given an objective of moving the player from the beginning of the stage to the end. It learned how to play completely from scratch; that's how powerful these algorithms can be. Take a minute to think about how these principles could be applied to a manufacturing scenario.

There are several aspects of manufacturing that benefit from machine learning. For example, recall that equipment, like robotic arms, will require maintenance. Machine learning algorithms can be used to predict when a robotic arm will experience a malfunction long before it occurs. Logistics and inventory can also be managed using these algorithms. In fact, Google uses DeepMindAI to monitor the energy use at their data centers. These data enabled Google to save up to 40% on energy used to cool the machines. Defects can also be caught more efficiently using machine learning algorithms than human quality control employees. By giving the algorithm examples of acceptable and unacceptable product qualities, the algorithm can identify common defects automatically and with considerable accuracy. The act of monitoring quality or performance and adjusting it based on the observed data is called **process control**.

process control

As we reimagine what's possible with machine learning, products that originally did not have these capabilities are being augmented with models to help them perform their functions more effectively. For example, consider the iRobot Roomba, a robotic vacuum that used a process called visual simultaneous localization and mapping (vSLAM) to take information from its various sensors while cleaning to create a map of its surroundings. In 2020, iRobot announced a substantial upgrade in software that the CEO Colin Angle called a "lobotomy and replacement of the intelligence systems" in the next generation of Roombas. Instead of bumping into walls and furniture aimlessly, the new software recognizes specific types of furniture and identifies clean zones so the Roomba can more intelligently navigate the home. With Amazon's $1.7 billion acquisition of iRobot in 2022, it's possible these robotic vacuum cleaners will continue to evolve and use more machine learning methods to keep our homes clean (Figure 10-5).

Figure 10-5 A robotic vacuum cleaner in action.

Augmenting products with more and more technologically complex features can create issues for the manufacturer, however. These products require specific material to enable the more complex behaviors, not just those enabled by advances in machine learning. Because of the COVID-19 pandemic, global supply chains were severely interrupted in 2020. This led to a chip shortage, decreasing the availability of several consumer products like vehicles, laptops, and smartphones. Given the ubiquity of machine learning, it's becoming more apparent that competency in programming will be necessary to engage with these models.

10.3 Using an Arduino to Automate Processes

programming

Programming, or coding, is the process in which a problem is formulated in a way that a computer can understand and solve. Programming is crucial to the use of computers or automation in manufacturing. In the same way mathematicians view math as an art form, computer scientists find beauty in the ways of programming. The problem could be a simple mathematical exercise like finding the closest value of $n!$ (pronounced n factorial) for any input, or more applied, like creating a video game. The introduction of machine learning across the engineering disciplines has enabled us to use the beauty of mathematics and programming to automate day-to-day processes most efficiently in manufacturing.

In order to program a computer to do a meaningful task, we must first speak one of its languages—programming languages, that is. Communication with computers can be easily compared to everyday verbal

ENGAGE

Though the premise of Roomba being smarter may sound appealing, the acquisition of the company that designed the Roomba, iRobot by Amazon, has drawn substantial critique. The following article from Business Insider can provide an opportunity to talk about ethics and machine learning.

URL
https://www.businessinsider.com/amazon-roomba-vacuums-most-dangerous-threatening-acquisition-in-company-history-2022-8

What are the issues at the core of the concerns about Amazon's acquisition? *Data privacy and imbalanced market competition. Machine learning using personal data is not a trivial task. When using human subjects data for research (not even just machine learning), scientists must seek approval from a review board to ensure the subjects of the research are protected. Thus, people are concerned about unfettered use of personal data in predictive models, especially in companies that already know a lot about us.*

interactions. There are thousands of spoken languages in the world; while the number of programming languages is not nearly as massive, the sheer number can be intimidating to new coders. A sizable number of the spoken languages can have *dialects*, a form of a language only spoken in a specific region. The same concept happens within the world of programming languages, except the type of language will depend on the application.

open source

For the project in this chapter, we will be discussing a particular programming platform called Arduino. Arduino is a popular **open source** electronic prototyping platform that is based on a microcontroller board and is useful for a variety of applications. Since the platform is designated as open source, the licensing to the software and internal workings of the platform are free. In addition, since the Arduino environment is open to the public, anyone can make improvements to the overall design and functionality and then share it to other users. This means that it is widely used. You might even use it at school and home!

sketch

Physically, the Arduino is a programmable device. This is an integrated circuit or a "chip"—physical hardware usually mounted on a circuit board (Figure 10-6). The platform works by taking inputs from a series of sensors and switches and regulating various physical outputs like lights and motors. The software associated with Arduino has a text editor for the code that is saved as a **sketch**—Arduino's name for files containing code—with a file extension of .ino. The individual files are called sketches (Figure 10-7) since Arduino uses the concept of a "sketchbook." The sketchbook is like the folder where the files are stored.

Figure 10-6 An empty Arduino board.

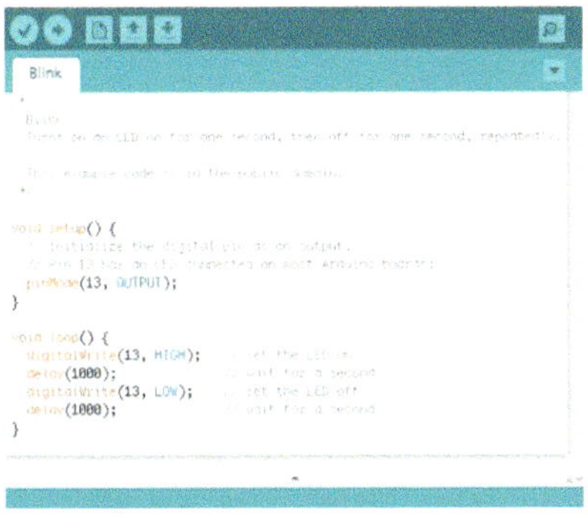

Figure 10-7 Arduino environment.

In the project for this chapter, we will explore the basic functionalities of Arduino by designing a system that could be considered a part of a manufacturing process. For example, we may need physical hardware that indicates each time a completed part flows by a point in our manufacturing line. We will build and program a device that mimics this by testing an LED and automating the manner in which it turns on and off.

QUESTION

Ask your students how computer programming can be useful outside of factory automation. *(Answer: Programming deals with automation but not necessarily in a factory setting. Calculators use automation to calculate simple and complex mathematical formulas. Programming is also behind the graphics and functionality of both console and PC video games!)*

ENGAGE

Have the students search online where Arduinos can be used. Students may be surprised by how versatile Arduinos can be!

TEACHING TIP

If you do not have access to Arduino kits, there are several Arduino simulators available online. One simulator is Tinkercad, which is provided by Autodesk.

URL
https://www.tinkercad.com/

TEACHING TIP

Have the students work individually or with their peers. Either way, encourage them to talk with their peers about questions or findings they have while working through the project.

PROJECT
Using an Arduino to Automate Processes

Problem Situation

In order to better understand how programming can be used to automate processes, we will learn to use an Arduino board to control the behavior of a circuit. Arduinos can be used to enable functionality in basic robotics, such as simple biped robots, physically manipulated clocks, and laser engravers. This section will familiarize you with the Arduino environment and some basic commands.

Your Challenge

Using the Arduino board, your team will be responsible for programming the board to turn a light on for 5 seconds and off for 3 seconds repeatedly.

Materials You Will Need

- Computer or laptop connected to the Internet

PROGRAMMING THE ARDUINO REQUIRES THE ARDUINO DRIVERS TO BE ON YOUR COMPUTER.

The software is free of cost and may already be installed on the computer you will use.
Please check with your teacher if working on a school computer.
To download, visit Arduino.cc/en/Main/Software
This site also has a lot of sample code you may want to try.

- Arduino board kit
- Arduino Uno, USB to mini-USB cable, breadboard, resistors, LED lights

1. Clarify the Design Specifications and Constraints

In this challenge, we need to find a way to have a circuit interact with the Arduino board. Our plan will be to use the Arduino development environment to write code to "talk" to the board. We need to ensure that the blinking criterion is met; that is, the LED lights up for 5 seconds and turns off for 3 seconds.

In order to meet the specified challenge, we will need to learn how to connect circuit components to the Arduino and decide whether to build the circuit on the board itself using the provided pins or by other means. Our next step will involve basic programming within the Arduino environment. The code will be written in a file called a "sketch," which you may recall is part of a larger "sketchbook" that contains all of your other relevant sketches. Actions performed by the Arduino will be facilitated through built-in functions, but it is up to us to figure out how the functions interact with one another.

2. Research and Investigate

Knowledge and Skill Builder 1: **Building the LED Circuit**

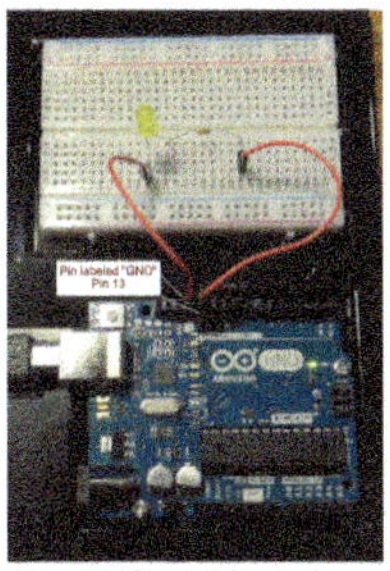
Figure 10-8 First Arduino circuit.

There are two methods of creating the LED circuit, one with the Arduino board itself and one using a breadboard. Arduino boards often have an LED already integrated into the board on pin 13, so if you try to run the experiment with no additional hardware, that light will blink. Nevertheless, the circuit can be constructed using only two circuit elements, a 250Ω (or 330Ω) resistor and an LED (Figure 10-8).

1. Connect one leg of the resistor in pin 13 on the Arduino board. Note that it may be helpful to connect a wire from pin 13 to the same row as the resistor leg.
2. Attach the longest leg of your LED to the other leg of the resistor; again, this is easiest to do in a breadboard.
3. Take the other leg of the LED and connect to the ground pin (GND).
4. Done!

Knowledge and Skill Builder 2: **Programming in Arduino**

If you want to add comments in the sketch to explain certain pieces, then you can do so by pressing the forward slash twice, //, and then typing your comment. For example,

```
// This is a comment.
```

NOTE

Whatever comes after // is ignored as part of the program, even if it 'makes sense' to you:

```
// This is a comment, and it is ignored
// delay(7000)-this is also ignored because of the // at the start
of the line
```

In order to speak Arduino, we need to know a few basic commands. Before beginning, we need to include two specific functions to ensure that our sketch functions properly. We add these lines to the *setup* area of the code:

```
void setup() {
Serial.begin (9600);
pinMode(13, OUTPUT);
//whichever pins you'd like to initialize, but in this case, let's
use 13…
}
```

TEACHING TIP

Check how students are connecting the components into the breadboard. Breadboards are configured in a very specific way. The upper two rows and bottom two rows in Figure 10-8 are connected horizontally and the rows between them are connected vertically. The gap in the middle creates two sets of vertical connections. This means that the legs of circuit components that follow one another in series should be in the same column. You should announce this disclaimer to the class if this is the first time they're using breadboards.

TEACHING TIP

Make sure students understand that they are using a programming language to code their Arduinos. There are many different programming languages that digital devices use. These languages can be used to create many different programs on many different devices.

TEACHING TIP

Make sure students understand that they are using pre-programmed functions that their programming language already knows. Without these functions, the students would have to create code for every function that they would want their program to complete.

The "setup" function is used to initialize any of our variables or pins. Within these beginning lines is the "Serial.begin" function as well, which defines the rate at which data is transferred to the board. For this application, a rate of 9600 bits per second is sufficient.

For the rest of the code, we need to know how to *initialize* the pin containing the LED, essentially telling it to get ready for commands. We can prepare a pin to receive commands using one function, "pinMode." For example, to *initialize* pin 13, we write,

```
pinMode(13, OUTPUT);
```

Without this line of code, the Arduino will not know how to treat its various pins. If we write a command and forget to specify whether the Arduino is supposed to treat the pin as an input or output, digital headaches—errors—are bound to occur.

Since we need to initialize a pin for our project, the pinMode function should go in the "setup" function.

We need one more concept, the idea of a loop. A loop is simply a function in Arduino that allows us to repeat a certain bit of code repeatedly. To introduce this into our program, we write,

```
void loop() {
… code / commands you want to loop…
}
```

We know from our knowledge of circuit that we need to provide a voltage drop over the LED to produce a current. Without a current, we have no hope of lighting up the LED. Therefore, we will need to use the Arduino as a power source. The Arduino can provide the necessary voltage for us by using a simple line of code. The function will supply 5 volts to the pin of our choosing. If we wanted to supply a 5 volt drop for pin 10, we would write,

```
digitalWrite(10, HIGH);
```

Why HIGH? Since we want to supply a voltage at a *higher* potential than what is currently in the circuit, we need to tell the Arduino to do so. On the other hand, if we want to cut the supply, we can replace 'HIGH' with 'LOW.' The meaning of LOW is precisely the opposite; this command takes the pin back down to the ground state of zero volts.

To tell the Arduino to "wait" before going to the next command in programming terms, we need to introduce a delay. The delay command uses milliseconds as units, so be sure to keep that fact in mind when declaring the amount of time Arduino needs to hold. For example, if we want Arduino to wait for 7 seconds, we write,

```
delay(7000);
```

In more complex sketches and programs in general, the use of comments can help immensely when organizing these functions. Due to the multiple ways in which problems can be solved in computing, you need to be able to communicate your solution to others. For example, suppose you are working for a large software company and are working on a team of programmers to complete a certain project. If you do an initial pass and write a horrendously complicated script without comments, then the next programmer who lays his or her eyes on your work could become confused, frustrated, or both.

Writing in comments can also be a useful method of reinforcing concepts while learning to program. By forcing yourself to explain the line of code, you are engaging in a form of reflective practice, which further cements the concepts in your memory. If you do happen to become "rusty" with programming, returning to an old script with your comments can help refresh your memory.

Practice

1. Open the Arduino development environment and enter the code as shown in Figure 10-9.
2. Describe what happens and why (based on the code).

```
void setup() {
  // put your setup code here, to run once:
Serial.begin (9600);
pinMode(13, OUTPUT);

}

void loop() {
  // put your main code here, to run repeatedly:
digitalWrite(13, HIGH);
delay(1000);
digitalWrite(13, LOW);
delay (100);

}
```

Figure 10-9 Practice code.

Project

Perform the following steps and respond to the prompts in your engineering notebook:

1. Write the script to meet the criteria specified in the problem statement.
2. Why are comments useful when programming? Are there other benefits beyond what was mentioned in the passage?
3. Can you modify your code to make a different pattern for the light?

TEACHING TIP

Have the students sketch out some of their different alternative designs. If they are not working in groups, have the students talk to their peers to generate additional designs and pick the optimal solution.

TEACHING TIP

Have the students sketch out some of their prototypes. You can talk to the students about the different levels of prototyping that industry uses (low, medium, and high).

3. Generate Alternative Designs

Another parallel with mathematics is the flexibility in which problems can be solved. The solution your team used to make the LED blink is not the only method of achieving the same result.

In your engineering notebook, respond to the following prompts:

1. Describe two other ways we could have made the LED blink using Arduino. Include any appropriate diagrams, figures, or sketches. The approaches could be similar but be sure to describe how the methods differ. If possible, use the Internet to find alternate solutions.
2. Computer scientists often describe pieces of code as elegant. Elegant code uses the simplest possible algorithm and design needed to solve the problem at hand efficiently. Is your code elegant? How can you make it elegant?

4. Choose and Justify the Optimal Solution

Pick your preferred solution. In your engineering notebook, respond to the following prompts:

1. Explain how your solution meets the requirements and constraints.
2. What trade-offs, if any, did you make in selecting your solution?

5. Develop a Prototype

Our prototype in this activity is the draft script you will be writing. Note, however, that in any technological activity, you will use seven resources: people, capital, time, information, energy, materials, and tools and machines.

In your engineering notebook, indicate which resources were most important in this activity and how you made trade-offs among them.

6. Test and Evaluate

Testing the Arduino is as simple as uploading the sketch you created onto the board. To do so, click the "Upload to I/O Board" button (Figure 10-10). If you cannot locate the button, it looks like an arrow pointing to the right. The Arduino environment will alert you if the upload was successful by displaying, "Done uploading." on the status bar.

Figure 10-10 Upload and verify.

Answer the following questions in your engineering notebook:

1. Our goal was to illuminate the pin 13 LED for an 8 second periodic interval where it is on for 5 seconds and off for 3 seconds. What happened?
2. If your LED did not follow the specified pattern, why do you think an incorrect cycle or pattern was displayed?

Expand on Your Design

We aren't limited to using pin 13 for our Arduino environment (assuming we are using a breadboard to build our design).

1. Change the code to that shown in Figure 10-11. Record the new code in your engineering notebook, along with your observation (what happened?) and explanation (why?).
2. Add more LEDs. Given the sample code and picture provided, and referencing Figures 10-12 through 10-14, do the following:
   (a) Create the sketch and circuit.
   (b) Test the sketch and circuit.
   (c) Record the new code and describe the new circuit in your engineering notebook. What did you observe? How can it be explained?

```
void setup() {
  // put your setup code here, to run once:
Serial.begin (9600);
pinMode(13, OUTPUT);
}

void loop() {
  // put your main code here, to run repeatedly:
digitalWrite(13, HIGH);
delay (1000);
digitalWrite(13, LOW);
delay (1000);
digitalWrite(13, HIGH);
delay (100);
digitalWrite(13, LOW);
delay (100);
digitalWrite(13, HIGH);
delay (100);
digitalWrite(13, LOW);
delay (100);
}
```

Figure 10-11 Modified code.

DIFFERENTIATED INSTRUCTION

Encourage the students to be creative with their design expansions! Let them know that it is okay to fail when trying to get to a correct answer. You can have them create all sorts of different light patterns with multiple LEDs. In fact, you may consider having groups try distinct patterns, and then have them present their code and discussion similarities and differences.

One application that is used in digital design courses is the design simulating traffic lights at an intersection. Have teams design traffic lights. Be sure to consider realistic timing; for example, yellow is displayed for a comparatively short time.

TEACHING TIP

Encourage the students to think about different ways that they can communicate their achievements. Students could model their prototypes, complete an oral presentation, or even think of a different method to communicate their results!

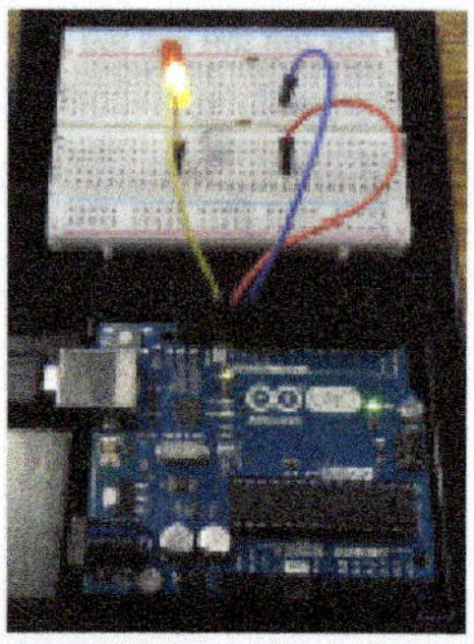

Figure 10-12 Image of Arduino using pins 12 and 13.

Figure 10-13 Closeup of breadboard.

```
void setup() {
  // put your setup code here, to run once:
  Serial.begin (9600);
  pinMode(13, OUTPUT);
  pinMode(12, OUTPUT);

}

void loop() {
  // put your main code here, to run repeatedly:
  digitalWrite(13, HIGH);

  for (int i=0; i <= 10; i++){
    delay(100);
    digitalWrite(12, HIGH);
    delay(100);
    digitalWrite(12,LOW);
  }

  digitalWrite(13, LOW);
  delay (1000);

}
```

Figure 10-14 Code for dual-LED circuit.

7. Redesign the Solution

More often than not, the first draft of code will not be correct. While we often allow grammatical errors when speaking without much recoil, making a mistake in the context of talking to a computer could be disastrous. Unlike verbal communication in various languages where one error makes a sentence sound a tad awkward, computers will act like the strictest writing or speech teacher you have ever had. Imagine if your teacher stopped reading your term paper completely upon finding the slightest error. In the same manner, computers will halt whatever process they are doing upon encountering an error, so we undergo a stage of debugging.

Through debugging, we can identify and correct any errors in the code. Note that we could still be receiving an output, but it is different from what we intended. This issue can also be corrected during the debugging stage.

In your engineering notebook, answer the following questions:

1. Is there a way your code can be improved? Explain.
2. Implement the necessary changes in your code and retest. Did your results improve?
3. How can this device be used in a manufacturing environment? Be creative in your response.

8. Communicate Your Achievements

Write a three-page report using the data collected during this investigation and responses in your engineering notebook. Your report must have the following sections:

Introduction: What is this report about?

Purpose: Why did you do this project?

Procedure: How did you go about completing the objective?

Results: What happened? What did you discover? Include relevant tables, charts, and equations.

Discussion: Comment on your results. What were the limitations of this project?

Conclusion: Summarize the report and provide the reader with a sense of finality.

ENGAGE

Sustainability in manufacturing intersects with the idea of the *circular economy*. Have students watch the following video from Sustainability Illustrated.

URL
https://www.youtube.com/watch?v=X6HDcubgxRk

What is meant by substitution and dematerialization? *Substitution is using a different material in a product to achieve the same outcome. Dematerialization is using less of a given material to achieve the same outcome.*

ENGAGE

Have students visit the EPA's Waste Management Hierarchy website. What are the four categories in order of preference from least preferred to most preferred? *Treatment and disposal, energy recovery, recycling/composting, source reduction and reuse.*

What is meant by energy recovery? *Converting non-recyclable materials using a process like combustion to generate energy.*

URL
https://www.epa.gov/smm/sustainable-materials-management-non-hazardous-materials-and-waste-management-hierarchy

10.4 Sustainability in Manufacturing

The most common definition of sustainability is "meeting the needs of the present without compromising the ability of future generations to do the same." Notice that this definition is fundamentally about people. Notice also that the definition includes future generations as well as the present one. Sustainability is not just an issue for our children and grandchildren, it is an issue that is affecting all of us right now.

The terms "environmental" and "sustainability" are often used interchangeably. However, sustainability also has social and economic dimensions. All three of these dimensions must be balanced for truly sustainable engineering solutions. Our society would not exist if our environment did not support human life. Our economy would not exist if we did not have a stable society (most people do not want to start businesses in failed states). These relationships seem quite obvious but can be overlooked if we just focus on one dimension of sustainability. As business leader Peter Senge points out: "the economy is the wholly owned subsidiary of nature, not the other way around."

You can apply a basic understanding of sustainability to your own engineering solutions. Sustainability is not a stand-alone topic. It cannot be bolted onto an engineering design at the end of the project. For example, in a new building project, one of the first sustainability considerations should be whether this project is even necessary. Perhaps similar goals could be achieved by more efficient use of existing facilities. This is quite an ethical dilemma! Imagine telling a potential client they do not need to hire and pay your engineering firm to design a new building. Assuming the building project is necessary, some of the best sustainability opportunities occur early on in the project, during project planning and design. This is where engineers play a key role. Teamwork and communication in the process are vital because sustainable solutions require consideration of multiple issues. We must be able to work with engineers from different disciplines and with non-engineers, such as architects, contractors, and lawmakers. We must be able to communicate with the end users who will occupy and operate the building. After all, the end user is the recipient of your design. These basic ideas apply across disciplines, whether you are designing a building, an engine, or a new material.

Even if we decide the building is necessary, we must understand different constraints. There are many types of potential constraints to be considered, and some may be more relevant than others with any given project. When designing a new building, we may need to consider, for example:

- **Health**: Materials in the new building must not pose any health risk, and the building cannot adversely affect the health of the community.
- **Safety**: Considerations on potential earthquakes or hurricanes in parts of the country have an effect on our design, and appropriate safety exits and fire extinguishing capability need to be included.
- **Social**: The social fabric of the location of the building needs to be considered. If the building is to be in an area known for its natural beauty, we would want to avoid a design that would be unpopular with residents.
- **Environmental**: The effect of water runoff or wildlife displacement must be considered.
- **Political, regulatory, and legal**: Any building site must be properly zoned, and local, state, and federal regulations must be followed. It may be a stretch to see any political considerations for a building in the United States, but there have been instances of buildings painted the wrong color in different countries, causing political strife.

Some of these constraints may prove to be more important than others, but all should be considered before moving forward.

You may be wondering how much of humanity currently considers sustainability. Are we already on a sustainable path? Unfortunately, this is not the case. Our use of critical resources, such as energy and water, and less critical resources, such as tequila and chocolate, cannot be sustained at current rates. Increasing population and affluence will stress these resources even more. Allocation of resources for the present and future is a huge ethical question engineers must consider. Should you build a reservoir that will provide water for an impoverished area, but restrict availability downstream? Do the risks associated with nuclear energy outweigh the fact that it is a carbon-free source of energy?

Creating solutions for sustainability issues requires expanding the boundaries of single discipline thinking, being able to recognize relationships between systems and the associated problems and opportunities. For example, our fossil-fuel-based energy system has increased standards of living all over the world, but this same system also contributes to climate change, which is already having significant negative impacts, with more predicted for the future. In addition, the system contributes to inequalities between those who have energy and those who do not, which is a major source of poverty and conflict. These complex relationships can make problems seem overwhelming; however, these relationships also offer opportunities. Engineers creating sustainable energy solutions will have positive impacts in multiple areas, such as helping to curb climate changing emissions, while reducing energy poverty and resource conflict. You can make a conscious effort to build your skills in the broad systems-thinking needed to identify these opportunities.

Sustainability affects manufacturing as well as design. Minimizing resources used in the manufacturing process is an important step in sustainable manufacturing. For example, electronic manufacturing generates significant waste. Circuit boards are often made with copper-coated glass-epoxy boards (basically, a mix of fiberglass and glue with a copper coating on one or both sides). The electronic design that will be on the surface of the board requires the unwanted copper to be stripped away from the board. This is typically done chemically, generating a solution of copper mixed with the material used to dissolve the unwanted copper. In the United States, regulations prohibit dumping this toxic waste directly into a waterway; however, in other countries, this may be legal, or officials may simply look the other way. In the U.S., this toxic residue must be stored and taken away. Designing an assembly process to minimize this waste is a positive step in sustainability. Circuit board geometry may be changed slightly so more boards can be cut from a single piece of substrate. Selecting materials that make a product recyclable in its end-of-life is a positive step in sustainability.

10.5 Designing a Prosthetic Arm for Developing Countries

The technology associated with artificial limbs has progressed over the past few years. A Roman general received a prosthetic arm in 77 AD, allowing him to return to battle. An iron prosthetic with articulating fingers was used by German knight Götz von Berlichingen in the early 1500s. It was so heavy that it had to be affixed to his armor, and the other, intact hand was used to control its grip. As early as 1916, a prosthetic hand which was operated by muscles in the arm had been developed. By the 1980s, designs featured hands made from silicon (much more lifelike than previous designs) and operation based on sensors on the outside of the arm, measuring signals from muscle movements, making movements far more natural. These myoelectric devices need to be powered, requiring batteries. Significant training was required to isolate muscle movement due to lack of sensory feedback.

Today, it is possible to use faint signals from arm nerves to control movement and enable real-time finger-level control. This involves surgery in which individual nerves are isolated from nerve bundles and microscopic muscle grafts. Moving individual fingers is intuitive for users—controlling finger movement from brain activity (the typical way fingers are controlled) is possible, due to work from medical researchers, electrical and computer engineers (with significant Artificial Intelligence programming), biomedical engineers, and mechanical engineers.

ENGAGE

You can supplement this case study with the following video from Wired about the development of prosthetic limbs over the years.

URL
https://www.youtube.com/watch?v=UOJ0lkr2SFc

What are the different classifications of prosthetic arms? *Passive, body powered, robotic/myoelectric.*

These techniques are new and very expensive. A functional arm can cost $10,000, while a myoelectric arm can be between $20,000 and $100,000. For people living in developing countries, this price can mean a suitable prosthetic is nearly impossible.

The design of a prosthetic limb, especially one which interfaces with the body, requires very delicate and precise instrumentation. Nerves in the arm may be 0.3 mm thick, while blood vessels may be 1 mm in diameter. Working with structures where tremendous precision is necessary requires extremely precise measurements. If you have cooked, you've seen measuring cups which may be marked every ¼ cup. If your measurement is slightly off, there is probably no real issue. If you are dealing with a blood vessel, you must be far more accurate. How accurate can we be?

You have probably also used a graduated cylinder or similar instrument as shown here. The precision of this instrument depends on how finely graduated it is. In this case, we can read this accurately as 85 units. We may be able to tell that it is just above the 85 line, and could perhaps specify this as 85.5 units, but we can't reliably read it more precisely. If we needed greater precision, we would need a new cylinder with a finer scale.

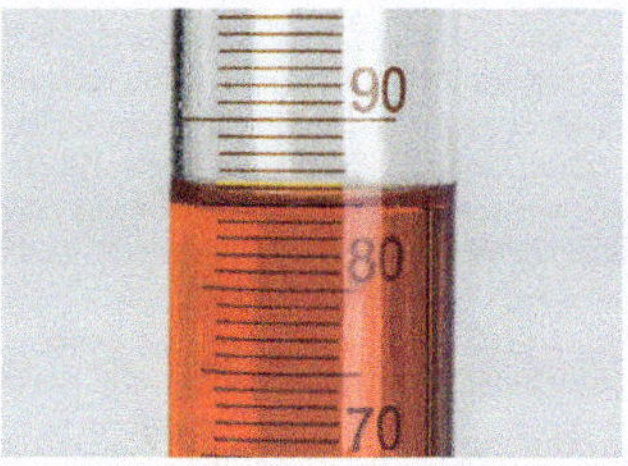

Case Study of a Prosthetic Limb

What happens if no medical services are available to someone who loses a limb? In one example, a person needed a prosthetic leg after a motorcycle accident. This amputation was at the knee, so the homemade limb didn't offer the typical use of the leg (it did not bend at the knee). The person still relied on crutches, but could get around and work.

He made a homemade prosthetic leg from wood, scrap plastic, and a filled shoe (Figure 10-15). Through a donation, he was able to be fitted for a basic prosthetic leg which would be far more comfortable and less intrusive (Figure 10-16).

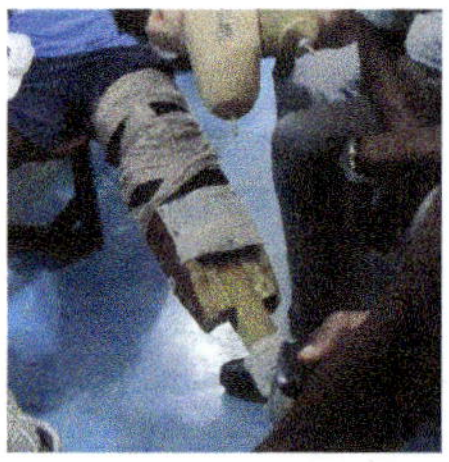

Figure 10-15 A homemade prosthetic lower leg.

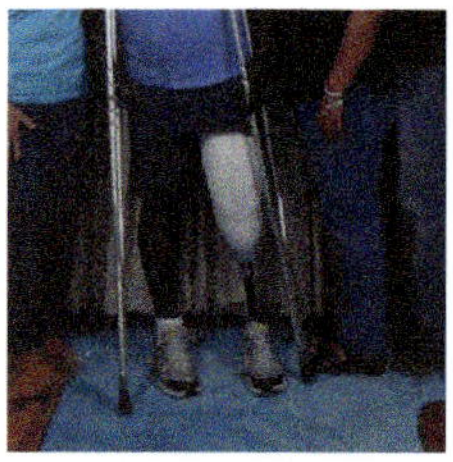

Figure 10-16 A new, fitted prosthetic leg.

DESIGN ACTIVITY

Design a Prosthetic Arm for a Person in a Developing Country

Problem Statement

Let's examine a scenario in which we need to design a prosthetic arm for someone in a developing country. In this scenario, consider someone who lost part of an arm below the elbow (so the elbow is intact). Your team is to explore a design for a prosthetic limb that is low-cost and easy to manufacture for a patient in a developing country.

Understanding Who the Design Is For

The annual salary for someone working in a developing country, or a middle-income country, is typically under or around $1,000/year (which works out to be about $19 per week). This will affect the type and functionality of prosthetics that is possible, but it shouldn't mean, ethically, that someone in one of these countries should not be helped.

Your team should spend some time online and identify some questions you will need to guide the design, including:

1. What are some of the activities that someone receiving one of these prosthetics needs to do from day-to-day?
2. What designs already exist? Are there existing solutions to this problem? (Note: If we do find an existing solution, we can use this data, but we want to design improvements to that prosthetic for this exercise.)
3. How will the prosthetic be attached and secured?
4. What technological resources will you need to develop and manufacture the device?
5. If your team selected a particular country, are there any societal, political, or cultural characteristics which must be considered in your design?

Your team should answer these questions about the entire prosthetic arm but concentrate on the part that will attach the prosthetic to the existing arm.

Develop Criteria, Constraints, and Design Specifications

After investigating your user's needs, your team will need to specify the criteria used to determine an effective design, constraints to which your design must adhere, and specifications you will need to meet. Use your design notebook to discuss and develop these.

1. What are the criteria you identified for your design? Be sure to consider health, safety, social, environmental, ethical, political, regulatory, and legal criteria.
2. What are the constraints limiting the possibilities of this design? Remember to be specific: "low cost" doesn't give us a suitable constraint, but "must cost less than $42.50 to build" is meaningful. (Note: Don't use $42.50—it's just an example.)
3. How will you know if your design is successful? Translate your criteria and constraints into measurable design specifications.

TEACHING TIP

Because this chapter is about manufacturing and automation, particular attention should be paid to how students conceptualize the manufacturability of their prosthetic arm.

DIFFERENTIATED INSTRUCTION

To emphasize the concept of design for manufacturability, you may challenge students to adopt the design for manufacturability framework throughout their project.

You may also choose to emphasize sustainability as a criterion. This would elevate the challenge of the project to have students employing design for manufacturability with a specific focus on sustainable materials and processes.

DIFFERENTIATED INSTRUCTION

You may elect to challenge students by having them fully explore the cost of production associated with their prosthetic arm, which would also include labor costs. If they were to scale this to a full business, what resources would they need?

TEACHING TIP

In step 6 of the "Generating Potential Concepts for a Prototype" section, emphasize that this is one example of applying design concepts in robotics. You can call attention to the Arduino exercise to get them thinking about how to accomplish the opening and closing action.

344 Chapter 10 Manufacturing and Automation

4. Consult with another team of students in your class and compare your criteria and constraints. What differences and similarities do you notice?
5. Develop a test plan to measure whether your design achieves the criteria you specified at the beginning of the challenge.
6. Did you revise or add design specifications during your test plan development?

Again, develop the criteria, constraints, and specifications for the entire device, but be sure you consider the mechanism to attach the device to the arm.

Generating Potential Concepts for a Prototype

Now it is time to generate some potential concepts for your prosthetic. Depending on your available materials, criteria, constraints, and specifications, you may be able to design and prototype a complete device or a subsystem of the device.

1. Perform a functional decomposition and construct a function block diagram for your prosthetic.
2. Create a sketch for three concepts and a detailed description that captures how the concepts would work. Be sure sketches are neat and legible, and drawn to correct proportions. Also, be sure to show accurate dimensions. Sketches may be orthogonal or isometric; this may be a good time to try both types of sketches to see which conveys your idea more clearly.
3. For each of the three concepts, discuss how they meet the criteria and are within the constraints.
4. How can you verify if one of your devices will be acceptable? How can you determine if it would be the best possible design among those in your class?
5. Focusing on the component of the design that attaches the prosthetic to the arm,
   (a) What materials might be suitable for this component? Consider materials you may be able to use to build a prototype and whether that material would work for production.
6. Suppose a robotic prosthetic arm incorporated a small Arduino to help manage the operations of opening and closing a fist. What design considerations need to be made for such a prototype to be feasible?

Selecting a Concept

After generating a series of design alternatives, we now must pick a concept to move forward with. In your design notebook, answer the following:

1. Using your criteria and design specifications, set up a decision matrix to begin weighing the viability of the three concepts.
2. Determine weights to place on the different criteria.
3. Calculate the "winning" design using your weights.
4. Summarize your findings.
5. Select a material to use for a prototype of the component of the design that attaches the prosthetic to the arm.
   (a) What materials will you use? If you have a 3-D printer available, is this a good choice? If you don't, might a low-fidelity prototype made from cardboard and duct tape work? Something else?

Building a Prototype

Your team should build a functional prototype of, at least, the component of the arm that attaches the prosthetic to the arm. Can your team build a prototype to show the effectiveness of your complete final design? Your design may have required an injection molding process which may be difficult to implement in a standard classroom; if so, can you substitute 3-D printing or something for flexible plastic (if your design requires this)?

Testing the Prototype

Because we are sticking to a conceptual design, we will assume you developed a full or partial prototype based on the concept selected from your decision matrix. How do we know if it works? In your design notebook, address the following prompts for each component:

1. How did the prototype perform?
2. Interpret the results: What worked and what didn't? What do you need to do to revise your prototype to meet the specifications?
3. Pair up with another team in the class. Compare your prototypes to one another, noting the similarities and differences in each prototype's performance.
4. Can different components of both teams be used together to improve the design?
5. If your prototype meets all specifications, can it be manufactured? Are there any concerns in manufacturing the final product?

Report Your Solution

Finally, it is time to report out to the class. Your teacher may elect to have you present a PowerPoint, create a short video, or develop a poster summarizing your processes from each stage of the design process outlined here. You may also write a short report covering the same material. In your design notebook, create a bulleted list of the main decisions you made across the different stages of this project. If you are making a PowerPoint, create an outline for how you will present the information. Next, reflect on the following questions:

1. What was the objective of this project?
2. What worked?
3. What didn't work and why didn't it work?
4. How did the design compare with the best and worst performers?
5. What would you do differently?
6. What are the ethical implications of your design being manufactured and sold in a developing country? How will your team address these ethical issues?
7. What considerations for sustainability did your team incorporate into your design?
8. Was the objective accomplished?

ENGAGE

What kind of prototyping does the suggestion of 3-D printing encompass? *Rapid prototyping.*

DIFFERENTIATED INSTRUCTION

You may ask students to research materials that are used in the manufacture of prosthetics, especially high-performing prosthetics, such as those which would be used in athletics or those with a great deal of control for the user.

TEACHING TIP

To bring back the idea of ethics from earlier in the course, make sure to emphasize step 6 in "Report Your Solution."

CHAPTER 10 ASSESSMENT QUESTIONS

1. Write a short block of code (or an Arduino patch) to light three LEDs in a sequence such that only one is on at any given time, and each different state lasts ½ second. If possible, program and test your design.
2. Research: Why is the fact that an Arduino is "open-source" important?
3. Describe the differences between agile, just-in-time, and flexible manufacturing.
4. Draw a Gantt Chart to plan moving everything in your bedroom to a different bedroom. Include each large item that needs to be moved, clothing packing/carrying/moving, set up, etc.
5. List five examples of how manufacturing technology positively affects lives on a daily basis.
6. Describe how you might create a new video game controller, and list any challenges you might find producing it in a manufacturing plant.
7. Select a device you use every day and explain how it has progressed through history. How is it manufactured today?
8. Write a paragraph on how your life would be different if goods were produced one-by-one instead of being manufactured.
9. Give two advantages and two disadvantages of using robotics in manufacturing processes.
10. Write a paragraph on this topic: Assuming manufacturing capabilities advance, what does the manufacturing process for a smartphone look like in ten years?

    a) What emerging trends in automation systems do you see influencing future processes?

    b) What emerging trends in robotics do you see influencing future processes?

    c) What emerging trends in process control do you see influencing future processes?
11. List five durable goods and five nondurable goods that are manufactured using automation in a factory.
12. Describe some of the necessary acceptance testing your team would need to complete as part of the project to develop Augmented Reality glasses to overcome Phantom Leg Pain. How can you automate the acceptance testing when production of the Augmented Reality glass scales?
13. Describe how some of the concepts for the design of a very low-cost artificial arm would meet the necessary constraints and criteria for a successful design. How could you predict performance and reliability of your design, and what are some possible failure modes?
14. A team working on attaching a prosthetic leg develops an innovative attachment mechanism and builds a prototype from Styrofoam. Write a paragraph on whether this is effective, is not effective, or if it has positives and negatives.

15. Your team has been tasked to redesign to-go containers for local restaurants to be more sustainable.
    (a) List some possible constraints for this problem.
    (b) List some possible criteria defining a good design.
    (c) Brainstorm some ideas you may want to pursue.

16. Identify the technological resources impacted by COVID-19 and the global chip shortage in 2020. Research the effect of these limitations on the automotive industry and describe solutions that different manufacturers took to stay competitive in terms of technological resources.

17. A big-box store has decided to sell products that are designed with sustainability in mind. Your team has been assigned to the small appliance section, where the focus will be on toasters, panini makers, and waffle irons.
    (a) How can these devices be manufactured to be more sustainable?
    (b) How can the manufacturing process be changed to be more sustainable?

    (Note that you may need to make some assumptions in how these are manufactured. If so, be sure to state your assumptions.)

18. Write one paragraph explaining why angel investing in a new company would be riskier than investing through stocks.

19. Consider a device you use daily. In what way could adding artificial intelligence to this device improve its performance?

20. Your team has been tasked to work on a bioengineering project to develop an artificial limb which uses existing tissue, muscle, and nerves to interface to the body. Describe the role that each of these experts might play in the design:
    (a) Medical doctor
    (b) Materials engineer
    (c) Bioengineer

    Then, in one paragraph, describe the goals, objectives, and requirements of this problem.

21. Your team designed a virtual reality chess game using mythical creatures that you drew. What intellectual property was created?

22. Your team has been tasked with the development of an artificial hand that will be able to sense heat and cold. What potential constraints do you face of each type?
    (a) health
    (b) safety
    (c) social
    (d) environmental
    (e) ethical
    (f) political
    (g) regulatory
    (h) legal

23. There have been considerable advancements in fast food automation, and a major company in the industry is attempting to come up with a new way to ensure that their french fries maintain the right shape and consistency - an issue of process control. Design a system to automatically check for a french fry's shape and consistency and report the results to the line cooks in the restaurant.

    a) What are the criteria, constraints, and design specifications for this product?

    b) Develop a functional block diagram for your product.

    c) Develop a flowchart to outline how the process control will be handled.

    d) Describe how this system can be automated and be integrated into an automated fry slicer.

24. Examine engineering notebook entries for any projects you've worked on thus far. Assess the effectiveness of your notebook. Describe whether your notebook meets each of these criteria. If it could be improved, note how it could be improved.

    (a) Entries are sequential and chronological.

    (b) Entries are an accurate and complete reflection of the progress being recorded.

    (c) Sketches or pictures are included where appropriate.

    (d) There are no loose entries or pages.

    (e) Each page is dated and witnessed.

    (f) Unused spaces are identified and lined out.

    (g) Errors are not erased or obliterated.

    (h) Test data and calculations are included.

25. How much volume is shown in each of these images?

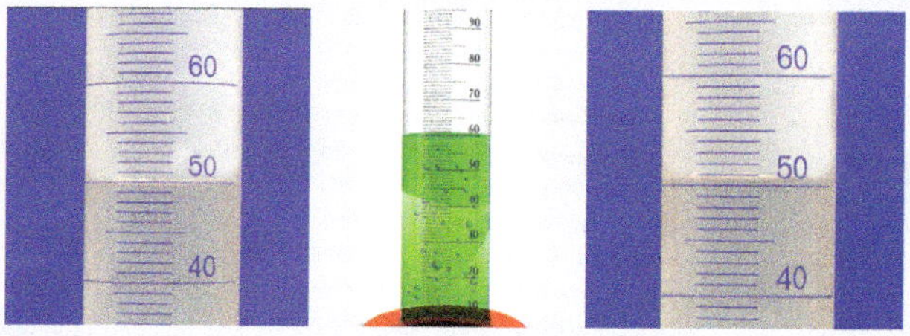

(a) (b) (c)

OVERVIEW

Statistics are important to engineering and engineering design. Through the exploration of number relationships, students gain a general idea of the connections within the physical world. This chapter discusses different tests and key information within statistics, allowing students to grasp the basics of the subject.

CHAPTER 11

Basic Statistical Analyses and Experiments

TOPICS

probability

Probability is associated with assessing the likelihood that an event will or will not occur. For example:

Airplane crash	River breaching a levee	Nuclear reactor accident
Tornado	Failure of equipment	Terminal cancer
Earthquake	Microprocessor failure	Space probe data reception

statistics

Statistics are used for design-concept evaluation because they provide quantitative measures to "things" that behave in a random manner. This evaluation helps us make rational decisions about everything from natural events to manufactured products. Statistics, as well as probability, use numerical evidence to aid decision making in the face of uncertainty. Roles of statistics in engineering include the following:

- Evaluation of new or alternative designs, concepts, and procedures
- Estimation of amount to bid on projects
- Management (human uncertainty, economic uncertainty, and others)
- Determination of degree of acceptable item-to-item variation (quality control)

Often, the best way to analyze an engineering problem is to conduct an experiment. In this chapter, we will explore basic statistical principles that can be used to predict the performance of your design and its reliability. Through these techniques, you will be able to pinpoint the potential of a design to succumb to one of its failure modes. The main questions we'll be asking here include:

- How many tests do we need?
- How confident are we in the results?
- Can we extrapolate the results to other conditions?
- Can we estimate how often the result will lie within a specified range?

There are many other related issues but addressing all of them requires a separate book. Many readers will take or have taken an entire course in probability or statistics, so for now we just touch on some of the important fundamentals.

OBJECTIVES

- Perform a design-of-experiments
- Define probability
- Define statistics
- Create and define histograms
- Define and use basic statistical behavior (mean, median, mode, variance, etc.)
- Identify and explain different types of distributions
- Identify and explain different skewed distributions

DIFFERENTIATED INSTRUCTION

We view this chapter as a supplement to instruction related to constructing viable and appropriate test plans. We recommend incorporating this chapter whenever you believe it is critical for students to construct more advanced test plans for their designs.

TEACHING TIP

If you are pressed for time in the curriculum, you may focus on sections 11.1 and 11.2. The rest can be supplemented as needed. Sections 11.1 and 11.2 will provide the necessary concepts to conduct a proper test plan for a design project.

ENGAGE

Define and discuss statistics within everyday life. *(Statistics is the numeric description of relationships. These relationships can happen within everyday life and are used to describe groups of people. For example, statistics can be used to determine the average height of a class of students. You can use statistics to see if there were anyone in the population that was statistically different in height, either taller or shorter than the average. Encourage the students to think of examples of statistics in their own life.)*

TEACHING TIP

One type of engineering job called a *quality engineer* deals with the quality of manufactured products. You may refer to Chapter 10, which introduced this profession. These engineers specifically oversee processes and quality standards, and perform experiments to test the quality of manufactured products.

DISCUSSION

You may choose to show an example of someone in a quality engineer job, such as the following YouTube video from BAE Systems.

URL

https://www.youtube.com/watch?v=cSFiruvdLg4

What surprises you about the role of a quality engineer?

ENGAGE

Define the terms *repeatability*, *precision*, and *accuracy*.

Repeatability *measures how close multiple measurements are to one another under the same conditions.*

Accuracy *is how close a measured value is to the actual (true) value.*

Precision *is how close the measured values are to each other, including how close it is to the true value (often encapsulated in the number of significant digits).*

- **Repeated tests**: When a test is conducted multiple times, we will not get the same (exact) result each time. For example, use a ruler to measure the length of a particular brand of shoe manufactured by Company X. You can produce a table of values, all of which will be nearly, but not exactly, the same. This is because you make slight errors in measurement, so even if every shoe is identical, there will always be some errors in your measurement. Moreover, every shoe is not identical.
- **Differences in a population**: What is the heart rate of all students in a class? Obviously, not everyone will have the same heart rate. We do expect, however, that everyone's rate will lie between, say, 40 and 140 beats per minute. Through measurement, we can determine this variation. In fact, we may find that the average rate for females and males differs. Statistical procedures help us analyze situations such as this.
- **Manufacturing errors**: Suppose you are manufacturing a run of widgets and a buyer wants all of them to be exactly alike. Obviously, this is impossible, but you can make them almost alike and then tell the buyer how much variation to expect. If you measure each widget as it comes off the assembly line, there will certainly be some variation.
- **Design criteria for products**: When you build a house, you would like it to stand safely for some period of time. For example, you might specify that the house be designed to withstand a windstorm that would occur, on average, every 50 years. For that case, you must be able to calculate the wind speeds associated with such a storm. Statistical methods allow you to do this.

11.1 Accuracy and Precision

Although laypersons tend to use the words *accurate* and *precise* interchangeably, an aspiring engineer should understand the difference in meaning as applied to measured or calculated values.

accuracy

Accuracy is a measure of how close a calculation or measurement (or an average of a group of measurements) is to the actual value. For measured data, if the average of all measurements of a specific parameter is close to the actual value, then the measurement is accurate, whether or not the individual measurements are close to each other. The difference between the measured value and the actual value is the error in the measurement. Errors come about due to lack of accuracy of measuring equipment, poor measurement techniques, misuse of equipment, and factors in the environment (e.g., temperature or vibration).

repeatability

Repeatability is a measure of how close together multiple measurements of the same parameter are, whether or not they are close to the actual value.

precision

Precision is a combination of accuracy and repeatability and is reflected in the number of significant figures used to report a value. The more significant figures, the more precise the value is, assuming it is also accurate.

NOTE:

Most common measuring devices can reliably measure the parameter in question to only three or four significant figures at best and sometimes less than that. To obtain a really repeatable measurement (e.g., to seven significant figures) in general requires sophisticated (and expensive) equipment.

To illustrate these concepts, consider the distribution of hits on a standard bullseye target. The figure shows all four combinations of accuracy and repeatability.

- Neither repeatable nor accurate
- Repeatable, but not accurate
- Accurate, but not repeatable
- Both repeatable and accurate. This is called precise.

TEACHING TIP

How to Remember?

aCcurate is Correct (a bullseye). pRecise is Repeating (hitting the same spot, but maybe not the correct spot).

TEACHING TIP

Sometimes accuracy and precision are juxtaposed without singling out repeatability, which is often coupled with precision. Thus, you would replace "repeatability" with "precision" in the bullseye figures, then change the lower right quadrant to "accurate and precise" instead of precision.

When considering if the precision of a numeric value is reasonable, consider the following:

- How many significant figures do I need in my design parameters? The more precision you specify in a design, the more it will cost and the less competitive it will be unless the extra precision is really needed.
- You can buy a really nice 16-ounce hammer for about \$20. If you wanted a 16 ± 0.0001-ounce hammer, it would probably cost well over a hundred dollars, possibly thousands.
- What are the inherent limitations of my measuring equipment? How much is the measurement affected by environmental factors, user error, etc.?
- The plastic ruler you buy at the discount store for considerably less than a dollar will measure lengths up to 12 inches with a precision of better than 0.1 inch, but not as good as 0.01 inch. On the other hand, you can spend a few hundred dollars for a high-quality micrometer and measure lengths up to perhaps 6 inches to a precision of 0.0001 inch.
- Pumps at gas stations all over the United States often display their gas price and the amount of gas pumped to three decimal places. When gas prices are high, it is extremely important to consumers that the pumps are correctly calibrated. The National Institute of Standards and Technology (NIST) requires that for every 5 gallons pumped, the amount must not be off by more than 6 cubic inches. To determine if a gas pump is calibrated correctly, you need to be able to see to three decimal places the amount pumped, since 6 cubic inches is approximately 0.026 gallons.

You should report values in engineering calculations in a way that does not imply a higher level of accuracy than is known. Use the fewest number of decimal places without reducing the usefulness of the answer. Several examples that illustrate this concept follow.

- We want to compute the area (A) of a circle. We measure the diameter (D) as 2.63 centimeters. We calculate

$$A = \frac{1}{4}\pi D^2 = \frac{1}{4}\pi(2.63)^2 = 5.432521 \text{ cm}^2$$

  The value of π is known to as many places as we desire, and ¼ is an exact number. It seems reasonable to give our answer as 5.4 or 5.43 square centimeters since the original data of diameter is given to two decimal places. Most of the time, reporting answers with two to four significant digits is acceptable and reasonable.

- We want to compute the area (A) of a circle. We measure the diameter (D) as 0.0024 centimeters. We calculate

$$A = \frac{1}{4}\pi D^2 = \frac{1}{4}\pi(0.0024)^2 = 0.0004446 \text{ cm}^2$$

  If we keep only two decimal places, our answer would be 0.00 square centimeters, which has no meaning. Consequently, when reporting numerical results, particularly those with a magnitude much smaller than 1, we use significant figures, not decimal places. It would be reasonable to report our answer as 0.00044 or 4.4×10^{-4} square centimeters.

- We want to determine a linear relationship for a set of data, using a standard software package such as Excel. The program will automatically generate a linear relationship based on the data set. Suppose that the result of this exercise is

$$y = 0.50236x + 2.0378$$

ENGAGE

Define the terms *micrometer* and *ruler*.

A ***micrometer****, sometimes known as a micrometer screw gauge, is a device incorporating a calibrated screw that is widely used for precise measurement of components.*

A ***ruler*** *is an instrument used in geometry, technical drawing, printing, engineering, and building to measure distances or to rule straight lines.*

QUESTION

Which of the two is designed to be more precise?

The accuracy of a ruler is limited by the eye's ability to see the spacing of the measurement marks and to line up the item to be measured with the marks on the ruler.

A micrometer uses a very accurate screw thread to get the measurement.

While we do not necessarily have proof that the coefficients in this equation are nice simple numbers or even integers, a look at the equation above suggests that the linear relationship should probably be taken as

$$y = 0.5x + 2$$

NOTE:

In general, it is a good idea to set your calculator to show answers in engineering format (or generally less desirable, scientific format) with two to four decimal places.

- If calculations and design procedures require a high level of precision, pay close attention to the established rules regarding significant digits. If the values you generate are small, you may need more significant digits. For example, if all the values are between 0.02 and 0.04 and you select one significant figure, *all* your values will read 0.02, 0.03, or 0.04; going to two significant figures gives values such as 0.026 or 0.021 or 0.034.
- Calculators are often set to show eight or more decimal places.
  - If you measure the size of a rectangle as 2 1/16 inches by 5 1/8 inches, then the area is calculated to be 6.4453125 square inches since the calculator does not care about how many significant digits result. It is unreasonable that we can determine the area of a rectangle to seven decimal places when we made two measurements, the most accurate of which was 0.0625 inches, or four decimal places.
  - If a car has a mass of 1.5 tons, should we say it has a mass of 3010.29 pound-mass?
- Worksheets in Excel often have a default of six to eight decimal places. Two important reasons to use fewer are that (1) long decimal places are often unreasonable, and (2) columns of numbers to this many decimal places make a worksheet difficult to read and unnecessarily cluttered.

Comprehension Check 11-1

In each of the following cases, a value of the desired quantity has been determined in some way, resulting in a number displayed on a calculator or computer screen. Your task is to round each number to a *reasonable* number of significant digits—*up* if a higher value is conservative, *down* if a lower value is conservative, and to the *nearest* value if it does not make a difference. Specify why your assumption is conservative.

(a)	The mass of an adult human riding on an elevator	178.8 pounds
(b)	The amount of milk needed to fill a cereal bowl	1.25 cups
(c)	The time it takes to sing *Happy Birthday*	32.67 seconds

Increasingly, engineers are working at smaller and smaller scales. On the left, a vascular clamp is compared to the tip of a match. The clamp is made from a bio-absorbable plastic through the process of injection molding.

Photo courtesy of E. Stephan

11.2 Experimental Design

Experiments enable engineers to develop a creative solution to a problem and test the validity of the proposed idea. An experiment is a test of a proposed explanation of a problem. A good design of an experiment is a critical part of the scientific method.

11.2.1 What Constitutes the Scientific Method?

1. Observation: Observe the problem and note items of interest.
2. Hypothesis: Search for a known explanation of the phenomenon or attempt to formulate a new explanation.
3. Prediction: Create a model or prediction of behavior based on that hypothesis.
4. Experiment: Test your predictions. If necessary, modify your hypothesis and retest.

11.2.2 Why is Experimental Design Important?

As you move through your college career, you will be inundated with many equations and theories. These are useful in solving a wide variety of problems. However, as you will see, often the equations are only really useful in solving the most basic type of problems.

As an example, suppose you are interested in the speed of a ball as it rolls across the floor after rolling down a ramp. In physics, you will learn the equations of motion for bodies moving under the influence of gravity. If you are good, you can use these to examine rolling balls. What you will quickly find, however, is that numerous complicating factors make it difficult to apply the basic equations to obtain an adequate answer. Let us suppose you are interested in smooth balls (such as racquetballs), rough balls (tennis balls), heavy balls (bowling balls), and lightweight balls (ping pong balls). The simplified equations of motion predict that all these will behave in essentially the same way. You will discover, however, that the drag of the air affects the ping pong ball, the fuzz affects the tennis ball, and the flexible nature of the racquetball will allow it to bounce at steep ramp angles. It is difficult to predict the behavior analytically. Often, one of the quickest ways to learn about the performance of such complex situations is to conduct experiments.

11.2.3 What are Experimental Measurements?

independent variable

dependent variable

control variable

Most scientific experiments involve measuring the effect of variability of an attribute of an object. In an experiment, the **independent variable** is the variable that is controlled. The **dependent variable** is a variable that reacts to a change in the independent variable. A **control variable** is part of the experiment that can vary but is held constant to let the experimenter observe the influence of the independent variable on the dependent variable. Keeping control variables constant throughout an experiment eliminates any confounding effects resulting from excess variability.

NOTE:

Do you know what these unique measurment instruments do?

- Durometer
- Dynamometer
- Euldometer
- Galvanometer
- Gyroscope
- Manometer
- Oplsometer
- Pycnometer
- Tachymeter
- Thiele tube

Any measurement acquired in an experiment contains two important pieces of information. First, the measurement contains the actual value measured from the instrument. In general, a measurement is some physical dimension that is acquired with some man-made data-collection instrument. As with any man-made device, there may be some imperfection that can cause adverse effects during data collection. Thus, the second piece of information that goes along with any measurement is the level of uncertainty.

Any uncertainty in measurement is not strictly by instrumentation error. Systematic error is any error resulting from human or instrumentation malfunction. Random error is caused by the limits of the precision of the data-collection device. It is possible to minimize the systematic error in an experiment, but random error cannot be completely eliminated.

TEACHING TIP

Use this chapter to emphasize the need for systematic testing during the design process

INTEGRATING CONTENT AREAS

In essence, constructing an experiential design is simply the scientific method, but with a focus on the details of what exactly you are trying to measure and what statistical tests you need to employ to justify your conclusions. You can communicate this relationship to the class to tie into their knowledge of the scientific method from their previous science courses.

ENGAGE

What is the difference between an independent, dependent, and control variable?

An independent variable is a variable that does not depend on other variables in the experiment/analysis.

A dependent variable is a variable that is impacted by one or more independent variables.

A control variable is a variable that is under the researcher's control, which can be manipulated for different experimental conditions.

11.2.4 What Measurements Do You Need to Make?

You need to develop a coherent experimental program. You should make enough measurements to answer any anticipated questions, but you do not usually have the time or money to test every possible condition. Points to consider:

- What are the parameters of interest?
- What is the range of these parameters—minimum values, maximum values?
- What increments are reasonable for testing (every 10 degrees, every 30 seconds, etc.)?
- What order is best to vary the parameters? Which should be tested first, next, etc.?

The acronym PERIOD can help you to remember these important steps. As an example, it is applied to the problem of the ramp and rolling balls previously described.

P—Parameters of interest determined

- Parameter 1 is the ramp angle.
- Parameter 2 is the distance up the ramp that we release the ball.
- Parameter 3 is the type of ball.

E—Establish the range of parameters

- Ramp angle can vary between 0 and 90 degrees in theory, but in reality can only vary between 10 degrees (if too shallow, the ball would not move) and 45 degrees (if too steep, the ball will bounce).
- The distance we release the ball up the ramp can vary between 0 and 3 feet in theory, assuming that the ramp is 3 feet long. We cannot release the ball too close to the bottom of the ramp or it would not move. In reality, we can only vary between 0.5 and 3 feet.
- We will test as many types of balls as we have interest in.

R—Repetition of each test specified

- The ramp angle will be set according to the height of the ramp from the floor, so there is not much room for error in this measurement; only one measurement is needed for such geometry.
- Each placement of the ball before release will vary slightly and may cause the ball to roll slightly differently down the ramp; this is probably the most important factor in determining the speed, so three measurements at each location are needed.
- We will assume that every ball is the same, and the actual ball used will not change the outcome of the experiment; only one ball of each type is needed.

I—Increments of each parameter specified

- We will test every 10 degrees of ramp angle, starting at 10 degrees and ending at 40 degrees.
- We will release the balls at a height of 0.5, 1, 1.5, 2, 2.5, and 3 feet up the ramp.
- We will test five types of balls: racquetball, baseball, tennis ball, ping pong ball, and bowling ball.

ENGAGE

Ask students to name the different considerations for an experimental design: *parameters, range of parameters, increments of parameters, order of variations, number of measurements.*

TEACHING TIP

Another nontrivial detail that students must understand is that they should test their prototypes in the expected conditions of the final design. For example, if they are designing a line following robot that needs to navigate over uneven terrain, only testing on flat terrain with no obstacles does not align with the criteria. While it is valuable to test designs in ideal conditions, the testing should not end there!

ENGAGE

Some potential test plans you can propose to the class include:

How would you evaluate the effectiveness of a(n)...
- coffee maker?
- ATM?
- water bottle?
- knee pad?

The students should begin by asking about the criteria/constraints for these different products. From there, they must come up with some way to make those criteria/constraints measurable using design specifications. For example, knee pads should cushion the wearer's knees from the impact of hitting the ground. Therefore, a design specification should quantify the magnitude of impact from which it should reliably insulate the wearer.

TEACHING TIP

It's very important that students understand that one measurement per condition is not sufficient. Have them recall the ideas of repeatability, accuracy, and precision. Is it possible to know if a measurement is accurate with only one data point? Of course not.

TEACHING TIP

These answers will vary depending on how the students establish their design specifications, but should demonstrate that they understand the parameters are independent variables. For example, the magnitude of the impact can be a parameter of the experiment.

TEACHING TIP

To demonstrate how the experimental conditions can be accounted for, you may choose to represent the potential combinations of variables in a tabular form. In the case of the ramp and ball example, the rows can be the ramp angles (10, 20, 30, 40) and columns can be the heights (0.5, 1, 1.5, 2, 2.5, 3), then have the type of ball recorded on top of the table. This should impress upon the students how many different combinations must be tested to complete the experiment.

O—Order to vary the parameters determined

- We will set the ramp angle and then test one ball type by releasing it at each of the six different distances up the ramp.
- We will repeat this process three times for each ball.
- We will then repeat this process for each type of ball.
- We will then change the ramp angle by 10 degrees and repeat the process.
- This process is repeated until all conditions have been tested.

D—Determine number of measurements needed and do the experiment

It is always important to determine before you start how many measurements you need to make. Sometimes you can be too ambitious and end up developing an experimental program that will take too much effort or cost too much money. If this is the case, then you need to decide which increments can be relaxed, to reduce the number of overall measurements.

The number of measurements (N) you will need to make can be easily calculated by the following equation for a total of n parameters:

N = (# increments parameter 1 * number of repetitions for parameter 1)

(# increments parameter 2 * number of repetitions for parameter 2) * . . .

(# increments parameter n * number of repetitions for parameter n)

Continuing the preceding examples, the number of actual measurements that we need to make is calculated as

N = (4 angles) * (6 distances * 3 repetitions) * (5 types of balls) = 360 measurements

In this example, 360 measurements may be extreme. If we examine our plan, we can probably make the following changes without losing experimental information:

- We decide to test every 10 degrees of ramp angle, starting at 20 degrees and ending at 40 degrees. This will lower the angle testing from four to three angles.
- We will release the balls at a height of 1, 2, 2.5, and 3 feet up the ramp. This will lower the distances from six to four.
- We will test three types of balls: racquetball, ping pong ball, and bowling ball. This will lower the type of balls from five to three.

The number of actual measurements that we now need to make is calculated as

N = (3 angles) * (4 distances * 3 repetitions) * (3 types of balls) = 108 measurements

This result seems much more manageable to complete than 360!

INTEGRATING CONTENT AREAS

How would you describe the rate at which the number of measurements increases as we increase the number of parameters? Suppose there are "i" increments for each parameter, "r" repetitions for each parameter, and "p" parameters.

*The number of measurements would be N = ir*ir*ir*...*ir, which is repeated p times. Therefore, N=(ir)^p. This means that the growth is exponential. You can plot this graph to show how additional parameters will drastically increase the number of measurements needed.*

ENGAGE

Ask students to define a histogram. *(A histogram is a diagram that is used to display statistical information. Typically, it consists of multiple rectangles that depict the frequency of a particular variable as it relates to an equal interval.)*

11.3 Histograms

To illustrate several common statistical concepts, we use data representing the height of several freshman engineering students. Table 11-1 shows the height, to the nearest inch, of each student in a typical class. Table 11-2 shows the same data, summarized by number of students at each height.

Table 11-1 Student Height

Student ID	Height (H) [in]
A	67
B	73
C	71
D	69
E	68
F	64
G	70
H	72
I	67
J	71
K	70
L	68
M	66
N	71
O	74
P	71
Q	68
R	72
S	67
T	64
U	75
V	74
W	72

Table 11-2 Summary of Height Data

Height (H) [in]	Number of Students
62	0
63	0
64	2
65	0
66	1
67	3
68	3
69	1
70	2
71	4
72	3
73	1
74	2
75	1
76	0
77	0
Total	23

When we graph the values shown in Table 11-1, we end up with a scatter plot with data that is exactly the same: scattered, as shown in Figure 11-1.

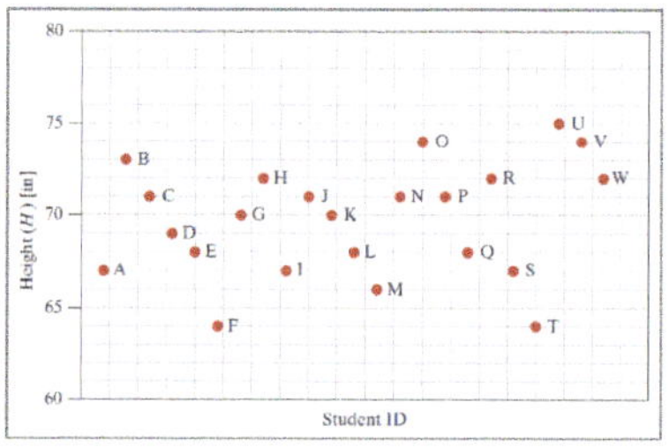

Figure 11-1 Example of student height, shown on a scatter plot.

bin

histograms

Instead of using a scatter plot, we can group the data and plot the group values in a chart similar to a column chart, shown in Figure 11-2. Using the summarized data shown in Table 11-2, we will place two height ranges into a single column or **bin**. The first bin will contain all student-height values less than 63 inches. The next bin will contain student-height values of 64 and 65 inches. The next bin will contain student-height values of 66 and 67 inches, and so on. The abscissa of the graph is the height values; the ordinate is the number of students measured at each height range. Graphs of this nature are called **histograms**. By counting the number of blocks, we find the area under the curve represents the total number of samples taken, in this case, the total number of students (23) observed.

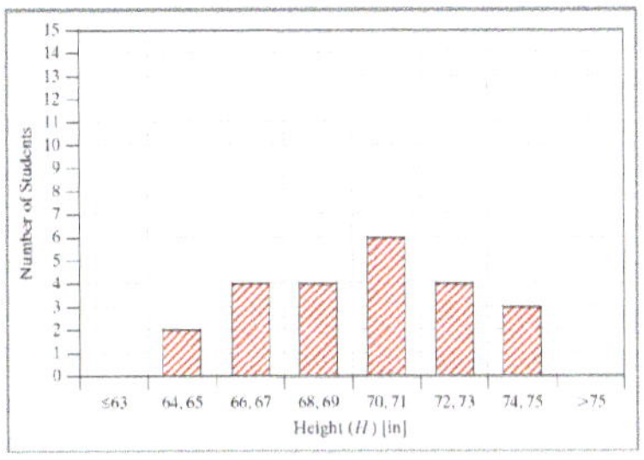

Figure 11-2 Example of student height, shown on a histogram.

Technically, before beginning this example we should have mathematically determined a bin size, rather than arbitrarily grouping the measurements in pairs (64 and 65 in one bin, 66 and 67 in the next bin, etc.). There are several ways to calculate the bin size that will best display the information; following is one method.

DETERMINATION OF BIN SIZE

Step One: Determine the number of bins needed.

Number of bins = Square root of number of data points, rounded to whole number

Step Two: Determine the range of the data.

Range = $X_{max} - X_{min}$

Step Three: Determine the number of items in each bin.

Bin size = Range divided by number of bins, rounded to whole number

Let us apply this to our example.

Step 1: As shown, we have a class of 23 students, so we would need five bins, since the square root of 23 is about 4.8, which rounds to 5. Four would probably also work fine, as would 6. Remember this is just a rule of thumb.

Step 2: The shortest person is 64 inches tall and the tallest is 75, so the range is 11 inches.

Step 3: Dividing the range determined in step 2 by the number of bins determined in step 1, we get 2.2, or about 2 inches per bin. On the other hand, we might instead decide to have four bins. If we divide the range by 4, we have 2.75 or 3 inches per bin.

ENGAGE

Ask students to survey their peers and record their heights. Have the students use this data to create a histogram. You can allow students to work in groups or individually. When everyone has completed the task, have them share their results with each other.

QUESTION

What is the difference between the average, the mean, and the median?

Answer:

- *The average is the same as the mean. To calculate it, have students add a group of numbers together and divide by the number of data points they added.*
- *The median is the number that is directly in the center when all of the data points are ordered from least to greatest. If the data set contains an even number of data points, take the middle two numbers, add them, and divide by two. This will be the median number for this data set.*
- *All of these measures can be classified as measures of central tendency.*

distribution

Depending on the number of bins, we sometimes get two different, but acceptable, bin sizes. By changing the bin size, we can change the appearance of the data spread, or the data **distribution**.

What happens to the student height data if we alter the bin size? The chart in Figure 11-2 shows a 2-inch bin interval, and Figure 11-3(a) shows a 3-inch interval.

In Figure 11-3(b), we have used a 4-inch bin interval, and while it is not what we obtained from the "rule of thumb" (2- or 3-inch intervals), it is still mathematically correct but not as informative as the other two.

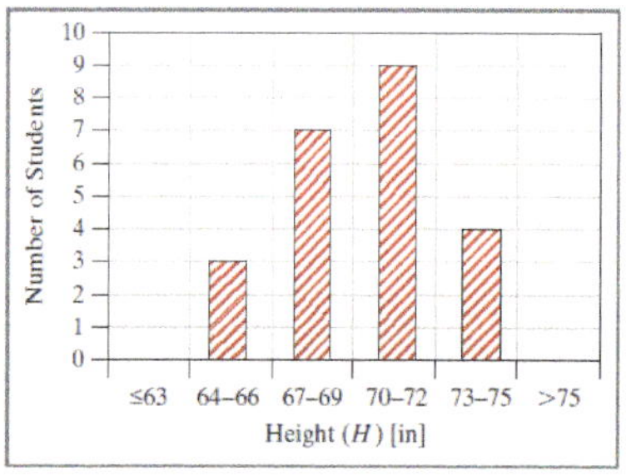

Figure 11-3(a) Bin size of 3.

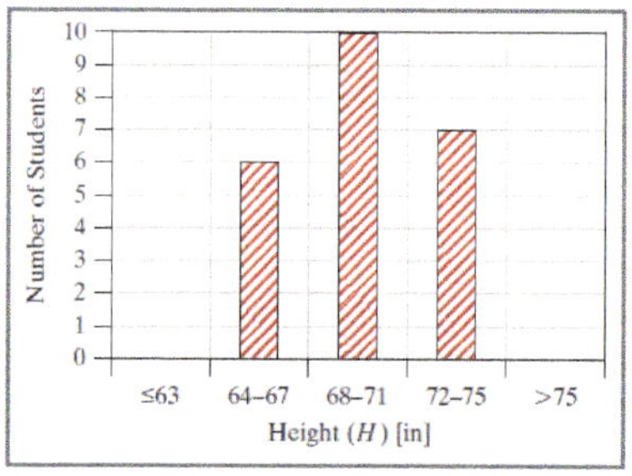

Figure 11-3(b) Bin size of 4.

Comprehension Check 11-2

The following table lists the number of computer chips rejected for defects during random testing over the course of a week on a manufacturing line. Four samples of 20 parts are pulled each day.

Use the following data to generate a histogram by hand.

1	1	8	0	2	0
0	2	10	1	3	2
0	1	12	0	2	1
1	6	15	0	0	
3	8	1	2	5	

11.4 Statistical Behavior

NOTE:

Average or Mean = typical, expected value of the data set; sensitive to outliers.

Median = value representing the exact middle value of the list; typically unaffected by outliers. Data must be in ascending or descending order to determine median!

When we have gathered the data and plotted a distribution, the next step is to explain the outcome to others. For convenience, we identify a set of parameters to describe distributions.

average

mean

central tendency

One parameter of a distribution is the average value. The **average**, or **mean**, is an estimate of the value most representative of the population. This is often called the **central tendency** of the data. The computation of the mean ($\bar{X}$) of a data set containing N values is given in the following equation.

$$\text{Mean} = \bar{X} = \frac{1}{N}(X_1 + X_2 + \cdots + X_N) = \frac{1}{N}\sum_{i=1}^{N} X_i$$

In other words, the mean is the sum of all of the values divided by the total number of values.

median

The **median**, another measure of central tendency, is the value between the lower half and the upper half of the population. In other words, if all data points are listed in numerical order, the median is the value exactly in the middle of the list. If the number of data points is odd, the median will be the middle value of the population. If the number of data points is even, however, the median will be the average of the two values at the center. A few examples should clarify this.

NOTE:

Although average and mean are synonymous in common language, their formal definitions differ. An average is calculated from a sample and approximates the entire population value, which is the mean.

Review the data shown in set 4. It would seem logical if every data point has a value of 50 except one, the average of the data should be about 50; instead, it is 208!

Set	Data	Mean	Median
1	1, 2, 3, 4, 5, 6, 7	4	4
2	1, 50, 70, 100	55	60
3	5, 10, 20, 40, 80	31	20
4	50, 50, 50, 50, 50, 1000	208	50

extreme values

outliers

This illustrates the sensitivity of the mean to **extreme values**, or **outliers**. Note that the median is unaffected or only slightly affected. It is for this reason that the mean is insufficient to describe the central tendency of all distributions.

variance

standard deviation

Two other terms useful in describing a distribution are **variance** and **standard deviation**. Both terms quantify how widely a set of values is scattered about the mean. To determine the variance (V_x^2), the difference between each point and the mean is determined, and each difference is squared to keep all terms positive. This sum is then divided by one less than the number of data points.

NOTE:

Variance = measure of data scatter; has SQUARED UNITS of the original data set. Standard deviation = square root of the variance; has units of the original data set.

$$\text{Variance} = V_X^2 = \frac{1}{N-1}((\bar{X}-X_1)^2 + (\bar{X}-X_2)^2 + \cdots + (\bar{X}-X_N)^2)$$

$$= \frac{1}{N-1}\sum_i^N(\bar{X}-X_i)^2$$

The standard deviation (SD_x) is found by taking the square root of the variance:

$$\text{Standard deviation} = SD_X = \sqrt{V_X^2}$$

If we again examine the data found in Table 11-2, we can calculate the mean, median, variance, and standard deviation for our height data.

ENGAGE

Ask students to survey their peers and record their heights. Have the students use this data to find the mean. You can allow students to work in groups or individually. When everyone has completed the task, have them share their results with each other.

ENGAGE

A common question about the formula for variance is why there is an N-1 in the denominator instead of N. This is called *Bessel's correction*. The purpose of subtracting 1 from the sample size is to reduce the bias in the estimate of the variance.

ENGAGE

Have your students use the peer height data to find the variance and standard deviation. Once again, they can work in groups or individually. When everyone has completed the task, have them share their results with each other.

Height (*H*) [in]	Number of Students
62	0
63	0
64	2
65	0
66	1
67	3
68	3
69	1
70	2
71	4
72	3
73	1
74	2
75	1
76	0
77	0
Total	23

Calculation of the Mean:

Total number of points (N)
= 23 students

The sum of all heights
= (2 students * 64 inches/student) + (1*66) + (3*67) + (3*68) + (1*69) + (2*70) + (4*71) + (3*72) + (1*73) + (2*74) + (1*75)
= 1604 inches

Mean
= 1604 inches/23 students
= 69.7 inches/student

Calculation of the Median:

Put data in order of value, listing each entry once
64, 64, 66, 67, 67, 67, 68, 68, 68, 69, 70, 70, 71, 71, 71, 71, 72, 72, 72, 73, 74, 74, 75

Find the center value since the total number of students is odd
64, 64, 66, 67, 67, 67, 68, 68, 68, 69, 70, 70, 71, 71, 71, 71, 72, 72, 72, 73, 74, 74, 75

Median = 70 inches

Calculation of Variance: *Note that the variance will have the same units as the variable in question squared, in this case, inches squared.*

$$Variance = \frac{1}{23-1}((69.7 - 64)^2 + (69.7 - 64)^2 + (69.7 - 66)^2 + \cdots + (69.7 - 75)^2) = 9.5 \text{ in}^2$$

Calculation of Standard Deviation: *The standard deviation has the same units as the variable in question, in this case, inches.*

$$Standard\ deviation = \sqrt{9.5} = 3.08 \text{ in}$$

Example 11-1

Consider the following velocity data, listed in units of feet per second. Determine the mean, median, variance, and standard deviation of the data.

1	28	14	32	35	25	14	28	5
16	42	35	26	5	33	35	16	14

Calculation of the Mean:

Total number of points (N) = 18
Sum of all data $(\sum X_i) = (1) + (2*5) + (3*14) + \cdots + (42) = 404$
Mean $= 404/18 = 22.4$ *feet per second*

Calculation of the Median:

Put data in order, listing each entry once, including all duplicate values.
1, 5, 5, 14, 14, 14, 16, 16, 25, 26, 28, 28, 32, 33, 35, 35, 35, 42
Find the center two values and average them, since total number of entries is even (18).
1, 5, 5, 14, 14, 14, 16, 16, 25, 26 , 28, 28, 32, 33, 35, 35, 35, 42
Median $= (25 + 26)/2 = 25.5$ *feet per second*

Calculation of Variance:

$$Variance = \frac{1}{18-1}((22.4 - 1)^2 + \cdots + (22.4 - 42)^2) = 147\ (ft/s)^2$$

Calculation of Standard Deviation:

Standard deviation $= \sqrt{147} = 12.1$ *feet per second*

Example 11-2

Consider the following energy data, given in units of joules. Determine the mean, median, variance, and standard deviation of the data.

159	837	618	208	971	571	379	220	31

Calculation of the Mean:

Total number of points (N) = 9
Sum of all data $(\sum X_i) = 159 + 837 + \cdots + 31 = 3994$
Mean $= 3994/9 = 443.7 = 444$ *joules*

Calculation of the Median:

Put data in order, listing each entry once.
31, 159, 208, 220, 379, 571, 618, 837, 971
Find the center value since the total number of students is odd (9; center value at entry 5).
31, 159, 208, 220, 379, 571, 618, 837, 971
Median = 379 joules

Calculation of Variance:

$$\textit{Variance} = \frac{1}{9-1}[(444-31)^2 + \cdots + (444-971)^2] = 105{,}059 \textit{ joules}^2$$

Calculation of Standard Deviation:

$$\textit{Standard deviation} = \sqrt{105{,}059} = 324 \textit{ joules}$$

Comprehension Check 11-3

For the following mass data given in units of kilograms, determine the mean, median, variance, and standard deviation.

8	7	9	11	16
12	2	9	10	9

Comprehension Check 11-4

For the following temperature data given in units of degrees Celsius [°C], determine the mean, median, variance, and standard deviation.

105	120	110	100	102
103	58	110	100	118

11.5 Distributions

From Figure 11-2 and Figure 11-3, you can see a similarity in the histogram shape of all three plots. The values start small, increase in size, and then decrease again. In the case of student height, this means that a few people are short, most people have some "average" height, and a few people are tall. This same conclusion is true in many things we can measure. For example:

- If we weigh many standard-size watermelons (neither miniature nor giant), we will find that most weigh between 20 and 30 pounds. A few weigh less than 20 pounds, and a few weigh more than 30 pounds.
- As we look through a dictionary, we find that there are many words with between four and six letters. There are a few with one, two, or three letters and a few with more than six, but clearly most have between four and six letters.
- To improve efficiency in the office, we had an expert monitor the length of phone calls made by the staff. The expert found that most of the time, phone calls lasted between 3 and 5 minutes, but a few were longer than 5 and some others lasted only a minute.

QUESTION

Why are normal distributions called "normal"?

(When large populations are surveyed, the data tends to take the shape of the normal distribution curve when plotted. This can be explained further by the Central Limit Theorem.)

11.5.1 Normal Distributions

We wanted to know how many "flexes" it takes to cause a paper clip to fail, so we asked volunteers to test the bending performance of paper clips by doing the following:

- Unfold the paper clip at the center point so that the resulting wire forms an S shape.
- Bend the clip back and forth at the center point until it breaks.
- Record the number of flexes required to break the clip.

Using these data, we created Figure 11-4. This is the same as the earlier histogram, but the boxes are replaced by a smooth curve through the values.

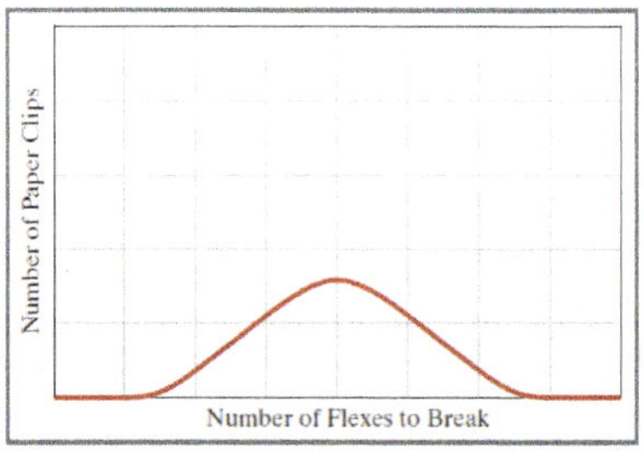

Figure 11-4 Distribution of paper clip failure.

normal distribution

68-95-99.7 rule

When you are interested in the *shape* of the curve rather than the exact data values, you can replace the bars of the histogram with a smooth curve and rename the graph a distribution. A distribution is considered *normal* if the following rules hold true. This **normal distribution** is also known as the **68-95-99.7 rule**, shown in Figure 11-5.

- 68% of values are within one standard deviation (1σ) of the mean (μ).
- 95% of values are within two standard deviations (2σ).
- 99.7% of values are within three standard deviations (3σ).

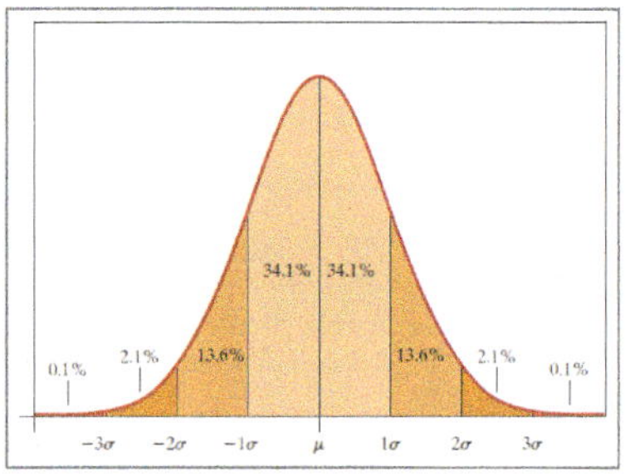

Figure 11-5 "Normal" distribution, showing the 68-95-99.7 rule.

Example 11-3

Suppose we ask a class of students how many states they have visited. The results might appear as shown below.

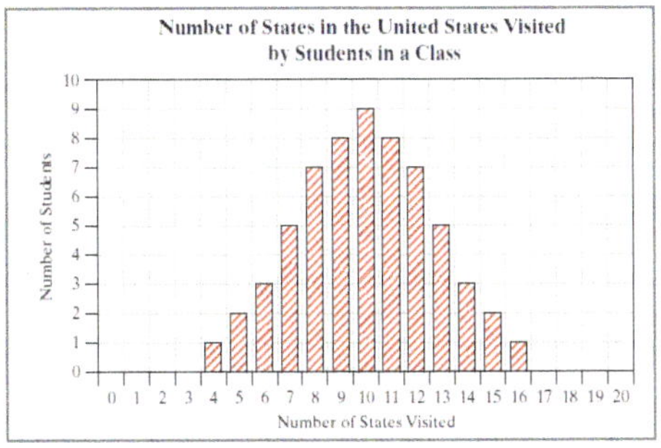

It seems that most have visited between 8 and 12, and that as many have visited more than 10 as have visited fewer than 10. A few have visited as many as 16 states, and all the students have visited at least 4. Let us calculate some values pertinent to this situation.

How many students are there in the class?

To do this we simply add the number of students represented by each bar, or

$$1 + 2 + 3 + 5 + 7 + 8 + 9 + 8 + 7 + 5 + 3 + 2 + 1 = 61 \text{ students}$$

What is the cumulative number of state visits?

*We answer this by totaling the product of the bar height with the number of states represented by the bar. For example, 5 students have visited 7 states, so those 5 students have visited a total of 5*7 = 35 states. Or, 8 students have visited 11 states, so those students have visited a total of 88 states. We calculate*

$$1*4 + 2*5 + 3*6 + \cdots + 2*15 + 1*16 = 610 \text{ states}$$

What is the average number of states visited by a student?

Once we have the values from our first two answers, this is straightforward division: the total number of visits divided by the total number of students.

$$610/61 = 10 \text{ states per student}$$

Notice that the value 10 is in the center of the distribution. For distributions that are symmetrical (such as this one), the average value is the one in the center, the one represented by the largest number of occurrences.

11.5.2 Decrease in Variance

Let us examine the shape of the distribution of paper clip failures discussed earlier. How would the distribution change if we brought in a machine that did it "exactly" the same way each time? Both the distribution from the data class and the distribution of the machine are shown in Figure 11-6. The same number of clips was tested in each case, so the areas under each curve must be the same.

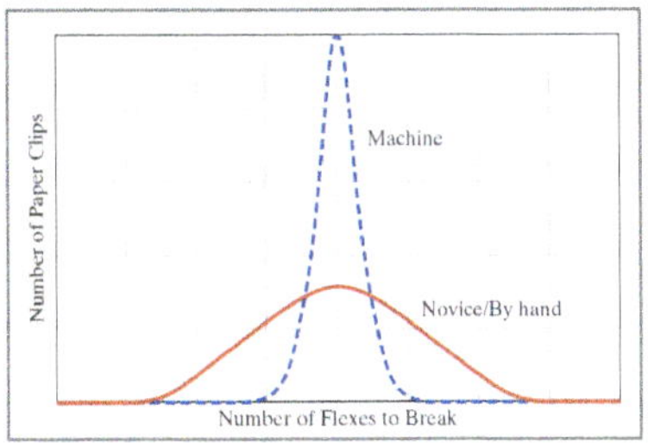

Figure 11-6 Distribution of paper clip failure after a decrease in variance.

This exercise illustrates that distributions that have the same mean (and median) can look very different. In this case, the difference between these two distributions is in

ENGAGE

You may decide to collect actual student responses for how many states they have visited using a Google Form or Microsoft Form. Depending on your school, your data may be skewed or have outliers, which you can discuss as a class.

ENGAGE

If variance increases, how would you expect the graph of the distribution to change? *The distribution will stretch and become flatter.*

QUESTION

What would cause data to be skewed?

(Data is skewed when factors pull the mean from the typical center of the normal distribution graph, to either the left or the right. If the mean is pulled to the right, larger numbers are affecting the mean. If the mean is pulled to the left, smaller numbers are affecting the mean. An example of skewed data could be age in a family with young children.)

364 Chapter 11 Basic Statistical Analyses and Experiments

their *spread*, or their variation about the mean. The effect of using a machine to break the paper clips was a decrease in the variance. Similarly, many machine manufacturers may publish expected variance in results through illustrations like distribution graphs, but in general, the lower the variance, the higher the cost of the machine, and likewise, the higher the variance, the lower the cost.

11.5.3 Shift in Mean

Redraw the paper clip distribution; then on the same plot, sketch the distribution if each volunteer tested the same number of clips that were manufactured by the same manufacturer as before with the same variance, but were stronger and typically required 10 more flexes to fail. The result is shown in Figure 11-7. The stronger material caused the distribution to shift to the right. Since the variance and number of clips remained the same, the shape and size of the curve remain the same.

11.5.4 Skewed Data

skewed

positively skewed

negatively skewed

range restriction

It is often easy to place an upper or lower limit on the value of the possible outcome. In these cases, the distribution is no longer symmetric—it is **skewed**. A population is **positively skewed** if the mean has been pulled higher than the median, and **negatively skewed** if the mean has been pulled lower than the median (see Figure 11-8 for an example of a positively skewed graph). You have probably heard news reports that use the median to describe a distribution of income in the United States. The median is used in this case because the distribution is positively skewed. This skew is caused by two factors, the presence of extreme values (millionaires) and the **range restriction**, the latter because income cannot be lower than $0. The extreme values causing the positive skew are not shown on the graph. Most of these would be far off the page to the right.

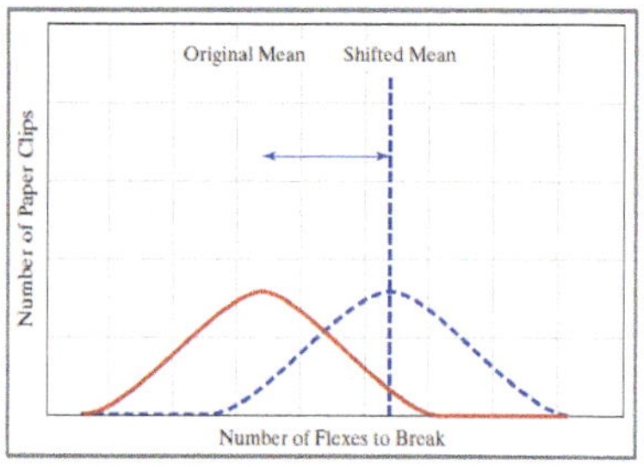

Figure 11-7 Distribution of paper clip failure with a shift.

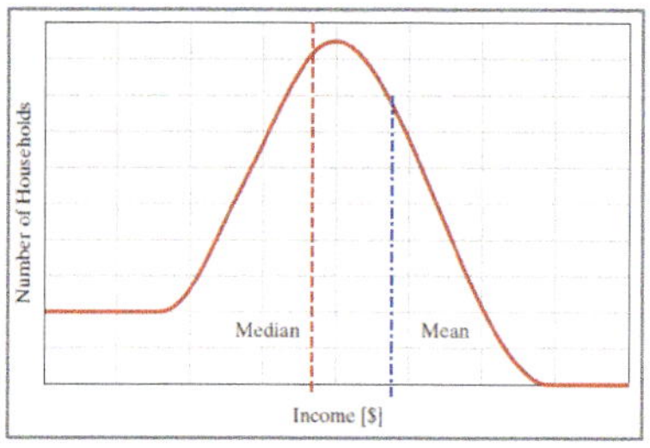

Figure 11-8 Distribution of positively skewed data.

Comprehension Check 11-5

For each of the following graphs, decide if the mean, variance, or population size has changed.

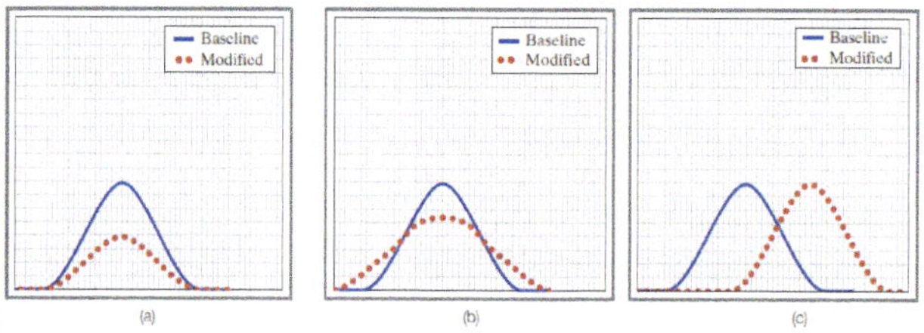

Example 11-4

For each scenario, identify one graph from the following set that best illustrates how the baseline curve would change under the conditions of that scenario. Each graph shows the usual distribution (labeled baseline) and the way the distribution would be modified from the baseline shape (labeled modified) under certain conditions.

The graphs show SAT composite (verbal + quantitative) scores, for which 400 is generally considered to be the minimum possible score and 1600 is considered to be the maximum possible score.

(a) The designers of the SAT inadvertently made the test more difficult, while the variance of the scores remains the same.

This is shown by curve F: Area and variance the same; mean shifted to left.

(b) The variability of scores is reduced by switching to true/false questions, while the average remains the same.

This is shown by curve D: Area and mean are the same; distribution is narrower.

(c) A population boom increases the number of students seeking college admission.

This is shown by curve B: Area increases, distribution stays the same.

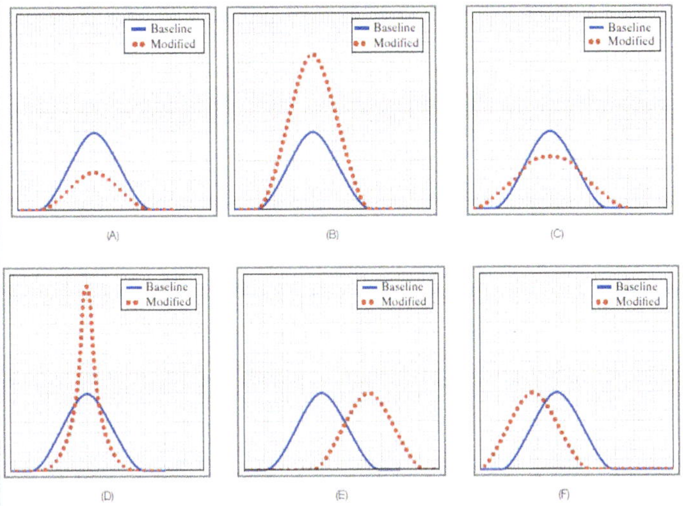

Comprehension Check 11-6

Use the scenario described in Example 11-4. For each scenario, identify one graph from the graph choices shown in Example 11-4 that best illustrates how the baseline curve would change under the conditions of that scenario.

(a) The economy declines, so more students decide to enter the workforce instead of attending college. The variance in the SAT scores remains the same.

(b) As a performance measure, all high school seniors are required to take the SAT; the variance remains the same.

(c) Due to mandatory test preparation courses, the mean of the SAT increases for the same number of students taking the exam. The variance remains the same.

QUESTION

What is a normalized plot?

Answer: Where the scale for a data set is adjusted to show a more accurate representation of the data.

ENGAGE

Have your students use the peer height data to create a histogram. Then have them create a normalized plot for this data set. You can allow students to work in groups or individually. When everyone has completed the task, have them share their results with each other.

11.6 Cumulative Distribution Functions

normalized plot

For the earlier student height plot using two heights per bin, we will graph the bin data but now show the values on the ordinate as a fraction rather than a whole number. To do this, we divide the number of students in each bin of the histogram by the total number of students. If we now add the heights of all the bars in the new plot, they should equal 1. This is called a **normalized plot**, shown in Figure 11-9. We "normalized" the values by dividing by the total number of data points. This graph holds no new information; it is simply a rescaling of the histogram we drew earlier.

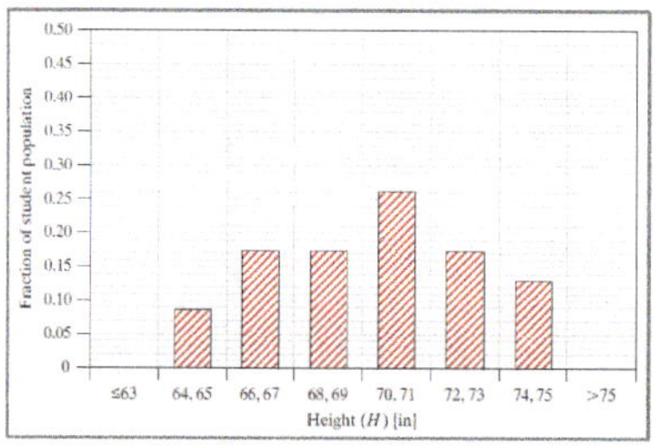

Figure 11-9 Normalized plot of student height, originally shown in Figure 11-2.

cumulative distribution function (CDF)

This plot can be used as an intermediate step to obtain a final plot called a **cumulative distribution function (CDF)**. We derive this plot by summing the values for each bin on the normalized plot from the first bin up to each individual bin. For example, suppose the values in the first three bins were 0, 0.08, and 0.18. By adding the values, we get new "cumulative" values: bin 1 = 0; bin 2 = 0 + 0.08 = 0.08; and bin 3 = 0 + 0.08 + 0.18 = 0.26. It should be obvious that we can obtain the CDF value for each bin by adding the normalized value of that bin to the CDF value of the bin before it. The CDF values are usually shown as percentages rather than fractions, for example, 50% instead of 0.5.

As we move across the plot, the values should go from 0 to 1. Using the height data in the normalized plot, we have produced a cumulative distribution shown in Figure 11-10. Sometimes, the CDF is shown as a continuous, curved line rather than a column chart. Both the original histogram and the cumulative distribution plot are useful tools in answering questions about the composition of a population.

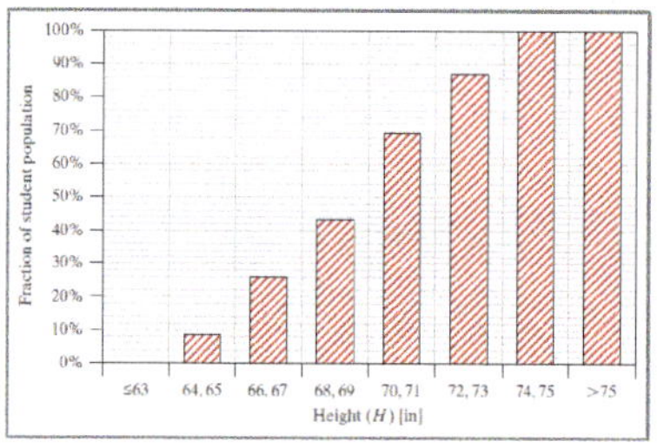

Figure 11-10 CDF of student height, originally shown in Figure 11-2.

Example 11-5

Consider the following pressure data, given in units of pascals. Draw the histogram and CDF of the data.

36	9	33	11	23	3
34	39	56	51	39	1
27	25	2	1	53	32
14	41	55	28	29	19
51	15	25	10	35	38

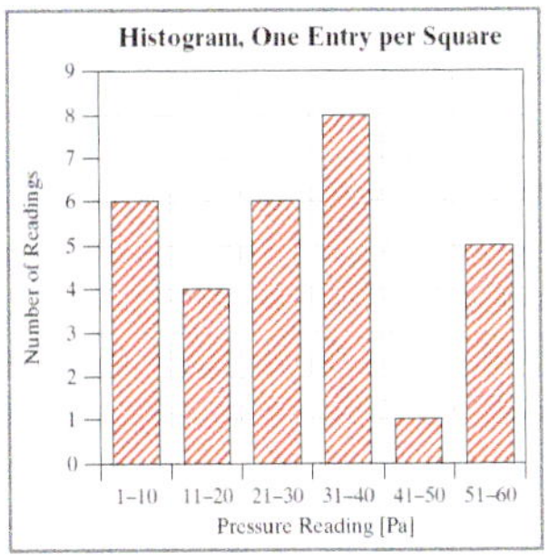

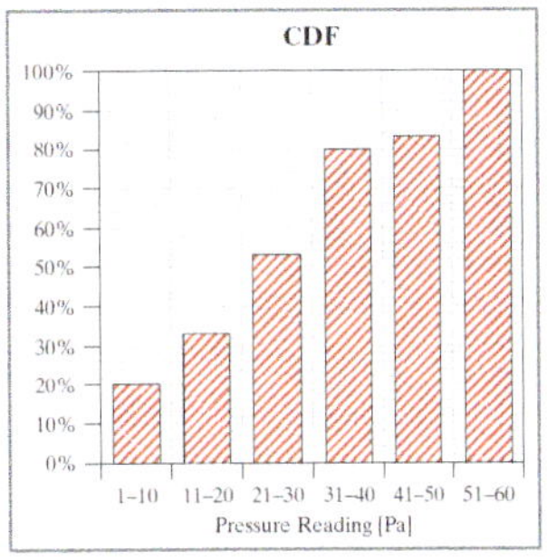

Comprehension Check 11-7

Consider the weight of shipping boxes sent down an assembly line, given in units of newtons. Draw the histogram and CDF of the data.

38	103	20	42	16
20	74	63	90	61
114	79	61	50	64

Comprehension Check 11-8

A team testing widget power usage in a trial of 25 experiments presented the results of their experiment as a CDF. Use the information from the CDF to create the histogram of their experiment.

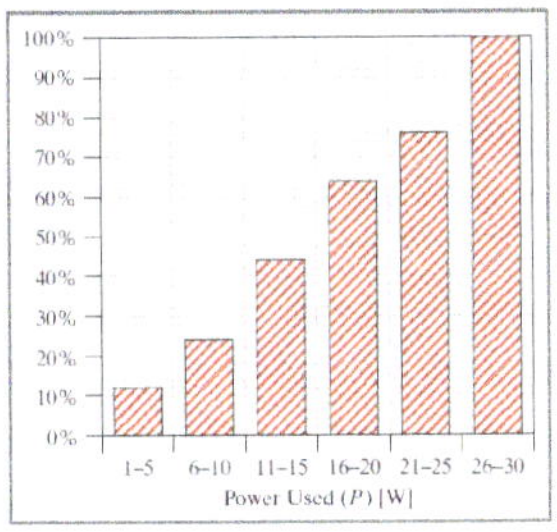

ENGAGE

Working in groups or individually, have your students use the peer height data histogram to create a cumulative distribution function for this data set. When everyone has completed the task, have them share their results with each other.

QUESTION

What is a quality control chart? Why are quality control charts important?

Answer:

- *A quality control chart is a plot that shows the differences from the mean, and if the values are outside a certain standard deviation.*
- *These charts can be useful within industry when looking at products, to make sure the company machines are creating the products within a certain standard deviation.*

ENGAGE

Working in groups or individually, have your students use the peer height data histogram to create a quality control chart for this data set. When everyone has completed the task, have them share their results with each other.

368 Chapter 11 Basic Statistical Analyses and Experiments

11.7 Statistical Process Control

We showed that a histogram such as the one shown in Figure 11-11 visually summarizes how a set of values is distributed. Sometimes, however, we are not only interested in the values themselves, but also in how the distribution changes over time.

For example, as a machine in a factory operates, it may slowly (or occasionally quickly) lose proper alignment or calibration due to wear, vibrations, and so on. If a machine was making bolts with a mean length of 1 inch and a standard deviation of 0.01 inch when it first began operating, after it had made 100,000 bolts, the alignment may have drifted so that the mean was only 0.95 inches with a standard deviation of 0.02 inch. This may be unacceptable to the customer purchasing the bolts, so the parameters of the process need to be monitored over time to make sure the machine is readjusted as necessary.

quality control chart

A graph called a **quality control chart** is often used to show how close to the mean the results of a process are when measured over time. The graph is usually a scatter plot, with the abscissa shown as time or another indicator that would change with time, such as batch number. Figure 11-12 shows a sample control chart, the mean, and the standard deviation.

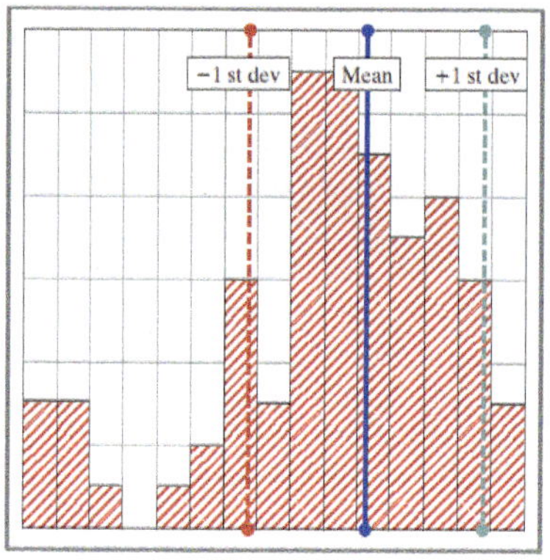

Figure 11-11 Sample histogram.

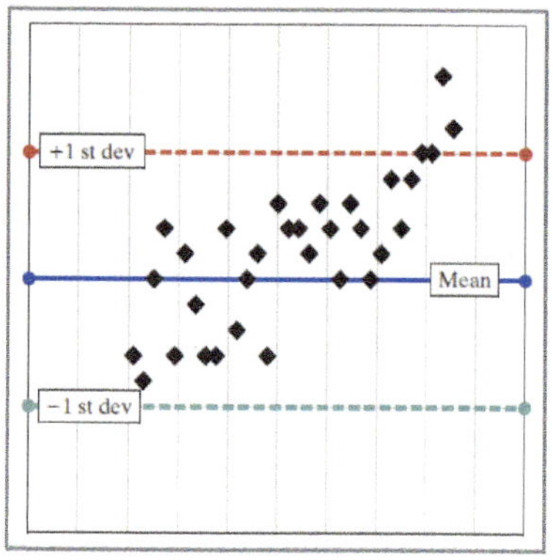

Figure 11-12 Sample quality chart (different data from 11-11).

statistical process control (SPC)

When tracking a manufacturing process, engineers are often concerned with whether the process is "in control," or behaving as expected. **Statistical process control (SPC)** is a method of monitoring, controlling, and improving a manufacturing process. In some situations, the desired mean and acceptable deviation limits may be preset for a variety of reasons (chemistry, safety, etc.). Often, the upper or lower limits of control are determined by the desired end result. For example:

- The reactor temperature must not rise above 85 degrees Celsius, or the reactant will vaporize.
- The injection pressure should be between 50 and 75 kilopascals to ensure that the part is molded properly.
- A bolt must be machined to ±0.02 inches to fit properly in a chair leg.

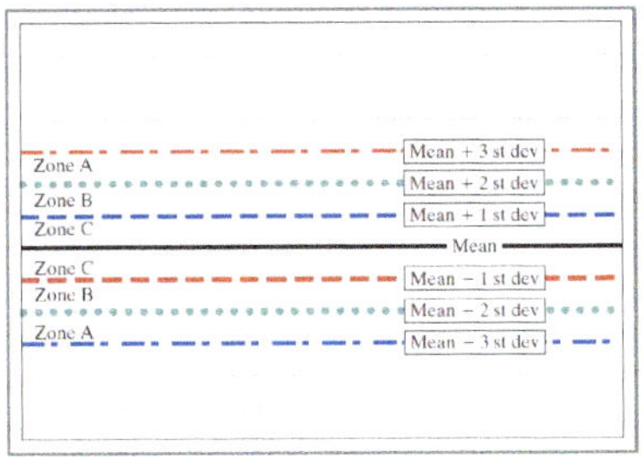

Figure 11-13 Standard deviation ranges.

An engineer will study how the process relates to the control limits and will make adjustments to the process accordingly. To discuss whether a process is in control, we can divide a chart into zones, shown in Figure 11-13, to create a control chart. The mean is determined either by the desired end result (the iron content of the product must average 84%) or by the process itself (the reactor temperature should average 70 degrees Fahrenheit for an optimum reaction to occur). The standard deviation is most often determined by experimentation. For example:

- The purity range for this product is ±0.0005%.
- The standard deviation for the reactor temperature must not exceed 5 degrees Fahrenheit or the reaction will create unwanted by-products.
- The standard deviation of the current gain of the transistors being produced must be less than 15.

Eight Ways To Be Out of Control

Nelson rules

A variety of conditions can indicate that a process is out of control. First published by Lloyd S. Nelson in the October 1984 issue of the *Journal of Quality Technology*, the **Nelson rules** follow, with examples and graphs. In the graphs, solid points indicate the rule violations. The actual conditions may vary slightly from company to company, but most take the same standard form. For example, a company may operate with rule 3 stated as seven or eight points in a row instead of six.

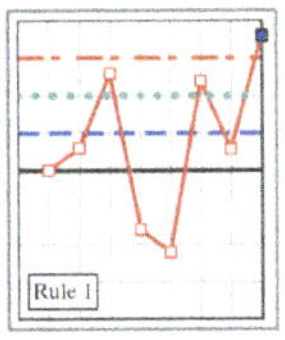

1. **A point falls anywhere beyond Zone A.** The value is more than three standard deviations away from the mean. May occur on either side of the mean.

   Example: The mean temperature of a reactor is 85 degrees Celsius with a standard deviation of 5 degrees Celsius. If the temperature exceeds 100 degrees Celsius, the reactor vessel may explode. If the temperature falls below 70 degrees Celsius, the reaction cannot proceed properly.

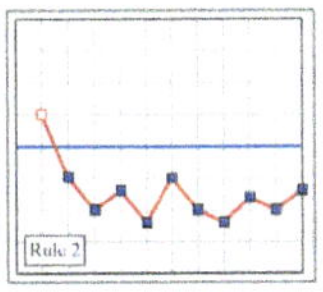

2. **Nine points in a row occur on the same side of the mean.** The actual value seems to be drifting away from the mean.

   Example: The percentage of boron in a semiconductor should be 250 parts per billion. Nine consecutive samples have boron contents less than this value. The machine incorporating the boron into the semiconductor material may need to be cleaned or recalibrated.

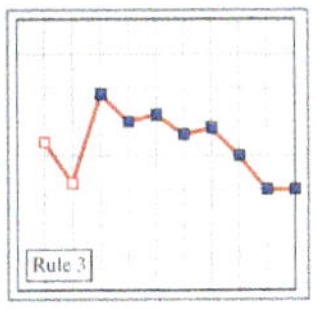

3. **Six points occur with a consistently increasing or decreasing trend.** If this pattern continues, the values will eventually become unacceptable.

   Example: The shaft length of a part is increasing with each successive sample; perhaps the grinding wheel needs to be changed

ENGAGE

Statistical software packages such as Minitab have features for implementing statistical process control using the Nelson Rules. Students can explore the following website to see how Minitab implements this kind of approach in their platform.

URL

https://blog.minitab.com/en/statistics-in-the-field/using-the-nelson-rules-for-control-charts-in-minitab

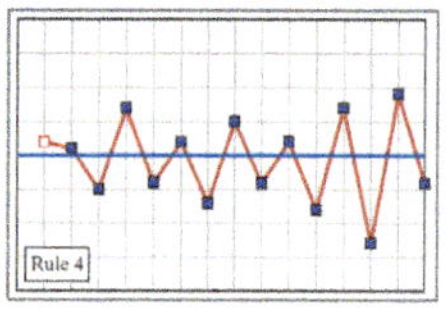

4. **Fourteen points in a row alternate from one side of the mean to the other.** The process is unstable.

Example: The control system for a crane errs from one side to the other. This may indicate a sensor failure or the need to reprogram the controller.

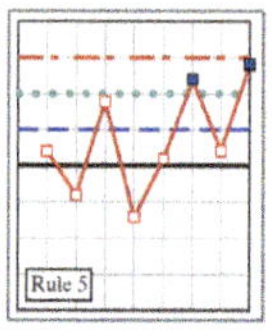

5. **Two out of three points in a row occur in Zone A.** The process is close to the upper limit; take preventive measures now.

Example: A robot that is spot-welding parts in an automobile is coming close to the edge of the material being welded. It probably needs attention.

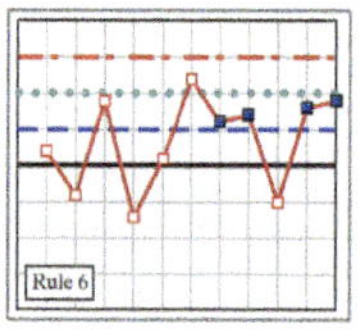

6. **Four out of five points in a row occur in Zone B.** The process is very close to the upper limit; take preventive measures now.

Example: Four out of five customers in the bank teller queue have waited more than one standard deviation to be helped. Perhaps another teller is needed.

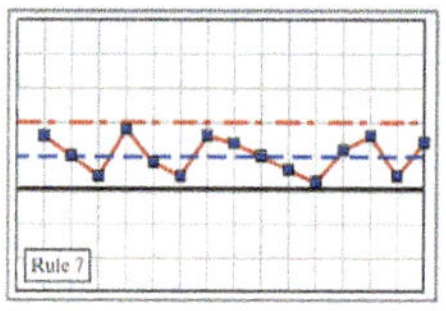

7. **Fifteen points in a row occur in Zone C; points can occur on either side of the mean.** The process is running too perfectly; in many applications the restrictions can be loosened to save time and money.

Example: The thickness of all washers being manufactured for quarter-inch bolts is within 0.0005 inch of the desired mean. Very few applications require washers with such close tolerances. Perhaps the process could be set to process the washers faster.

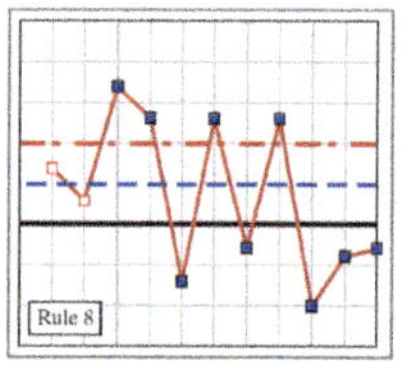

8. **Eight points in a row occur beyond Zone C; points can occur on either side of the mean.** The process does not run close enough to the mean; the parts are never quite on target; may indicate a need for a process adjustment.

Example: The postmark machine in a regional postal distribution center is stamping the envelopes too high or too low; it probably needs attention.

Example 11-6

The data shown in the table were collected from a manufacturing process that makes bolts.

Assume the process specifies an average bolt length of 10 inches, with a standard deviation of 0.25 inches. Is this process under statistical control?

Part	Length (*L*) [in]	Part	Length (*L*) [in]
1	10.00	11	10.25
2	10.25	12	10.65
3	10.65	13	9.50
4	9.50	14	9.36
5	9.36	15	9.25
6	9.00	16	10.50
7	10.50	17	10.20
8	10.20	18	9.80
9	9.80	19	10.45
10	10.00	20	10.10

The control chart for these data is shown below. Rule 1 is violated since a point falls outside of Zone A (part 6) and rule 5 is violated as parts 14 and 15 fall inside Zone A. The process is not in statistical control.

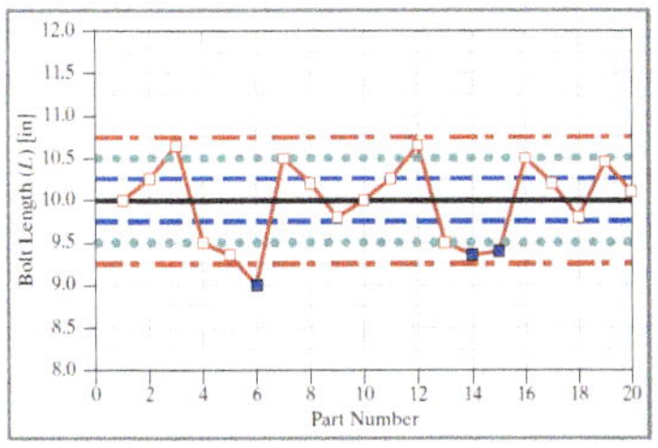

Comprehension Check 11-9

The pressure in a water filter is monitored in a chemical plant. The filter should operate at 18 pounds-force per square inch [psi], with a standard deviation of ±2 psi. Analyze the data shown to determine if the filter is behaving as expected (the process is in control; the filter does not require any attention) or if the filter needs attention (the process is out of control; the filter should be cleaned). Refer to the Nelson rules to explain your conclusion, and include the time the violations occur.

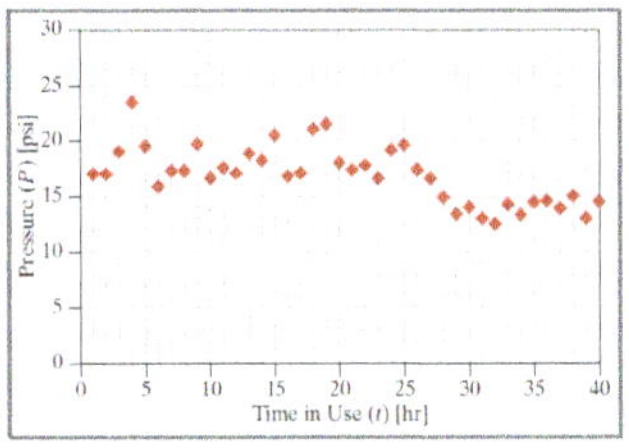

IN-CLASS ACTIVITY
Chapter 11

ICA 11-1

This exercise includes the measurement of a distributed quantity and the graphical presentation of the results. You are to determine how many flexes it takes to cause a paper clip to fail.

Test the bending performance of 20 paper clips by doing the following:

- Unfold the paper clip at the center point so that the resulting wire forms an S shape.
- Bend the clip back and forth at the center point until it breaks.
- Record the number of flexes required to break the clip.

On a copy of the following table, record the raw data for the paper clips you break. Then, summarize the data for the team by adding up how many clips broke at each number of flexes. Each team member should contribute 20 data points unless otherwise instructed by your professor. Analyze the data using one of the methods (a) through (c), as specified by your instructor. Create a histogram with an appropriate bin size and a CDF, using the data collected and creating the graphs:

(a) by hand

(b) using Excel

(c) using MATLAB

Paper Clip Flexing Data

Paper Clip	Flexes to Break
1	
2	
3	
4	
5	
6	
7	
8	
9	
10	
11	
12	
13	
14	
15	
16	
17	
18	
19	
20	

Summary of Data

No. of Flexes	No. of Clips

CHAPTER 11 ASSESSMENT QUESTIONS

1. For the following pressure data, recorded in units of pound-force per square inch, answer the following questions. Note that this data set should be treated as a sample.

3	42	6	45	30
18	9	3	54	

(a) What is the mean of the data?

(b) What is the median of the data?

(c) What is the variance of the data?

(d) What is the standard deviation of the data?

2. A technician tested two temperature probes by inserting their probes in boiling water, recording the readings, removing and drying the probes, and repeating the process. The results follow, giving temperature reading in degrees Celsius. Note that this data set should be treated as a sample.

Probe 1	90	89	90.5	92	89.5	91	91.5
Probe 2	98	102.5	104	100	93	94	106

(a) What is the mean of each probe?

(b) What is the median of each probe?

(c) What is the variance of each probe?

(d) What is the standard deviation of each probe?

3. One of the NAE Grand Challenges for Engineering is **Develop Carbon Sequestration Methods**. According to the NAE website: "In pre-industrial times, every million molecules of air contained about 280 molecules of carbon dioxide. Today that proportion exceeds 380 molecules per million, and it continues to climb. Evidence is mounting that carbon dioxide's heat-trapping power has already started to boost average global temperatures. If carbon dioxide levels continue upward, further warming could have dire consequences, resulting from rising sea levels, agriculture disruptions, and stronger storms (e.g., hurricanes) striking more often."

The Mauna Loa Carbon Dioxide Record is the longest continuous record of atmospheric concentrations of carbon dioxide (CO_2), the chief greenhouse gas responsible for global climate warming. These data are modeled as the Keeling curve, a graph showing the variation in concentration of atmospheric CO_2 based on measurements taken at the Mauna Loa Observatory in Hawaii under the supervision of Charles David Keeling. It is often called the most important geophysical record on Earth and has been instrumental in showing that mankind is changing the composition of the atmosphere through the combustion of fossil fuels.

The Keeling curve also shows a cyclic variation in each year corresponding to the seasonal change in the uptake of CO_2 by the world's land vegetation. Most of this vegetation is in the northern hemisphere, where most of the land is located. The level decreases from northern spring onward as new plant growth takes CO_2 out of the atmosphere through photosynthesis and rises again in the northern fall as plants and leaves die off and decay to release the gas back into the atmosphere.

Data and wording for this problem set were obtained from *www.esrl.noaa.gov/gmd/ccgg/trends/*. Additional information on the Mauna Loa Observatory can be found at: *http://scrippsco2.ucsd.edu/*.

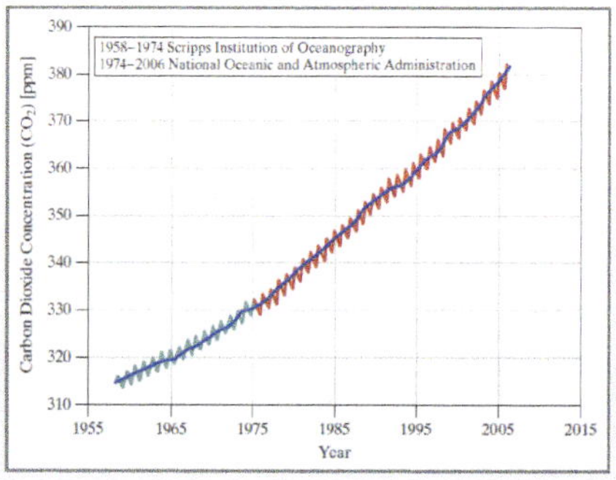

Examine the estimated increase in monthly CO_2 emissions for 2019, taken from the Mauna Loa data set. All values given are in parts per million [ppm] CO_2 as the difference between the December 2018 and the monthly 2019 reading.

2.38	3.64	-1.77	4.23	3.20	1.45
2.87	4.89	-2.16	4.92	-0.37	2.73

(a) What is the mean of these data?

(b) What is the median of these data?

(c) The variance of the data set shown here is 5.87 parts per million squared [ppm^2]. What is the standard deviation of these data?

(d) The estimated annual growth rates for Mauna Loa are close, but not identical, to the global growth rates. The standard deviation of the differences is 0.76 parts per million per year [ppm/year]. What is the variance?

Chapter 11 Assessment Questions 375

4. You use the data from the Mauna Loa observatory in the previous question to create the following histogram and CDF. These data reflect the observed yearly increase in CO_2 emissions for the past 51 years. The annual mean rate of growth of CO_2 in a given year is the difference in concentration between the end of December and the start of January of that year. If used as an average for the globe, it would represent the sum of all CO_2 added to, and removed from, the atmosphere during the year by human activities and by natural processes.

NOTE:

In the graph, A and B are *not* drawn to scale, and the locations of C, D, and E are approximate. In other words, you cannot *guess* the value based upon the graph—you must *calculate* the value.

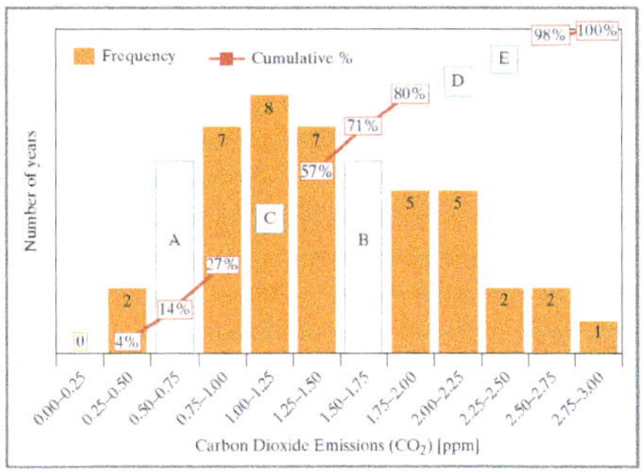

(a) What is the value of point A?

(b) What is the value of point B?

(c) What is the value of point C?

(d) What is the value of point D?

(e) What is the value of point E?

5. Polyetheretherketone (PEEK)™ polymers are resistant to both organic and aqueous environments; they are used in bearings, piston parts, and pumps. Several tests were conducted to determine the ultimate tensile strengths in units of megapascals [MPa]. The following CDF shows results from 320 points.

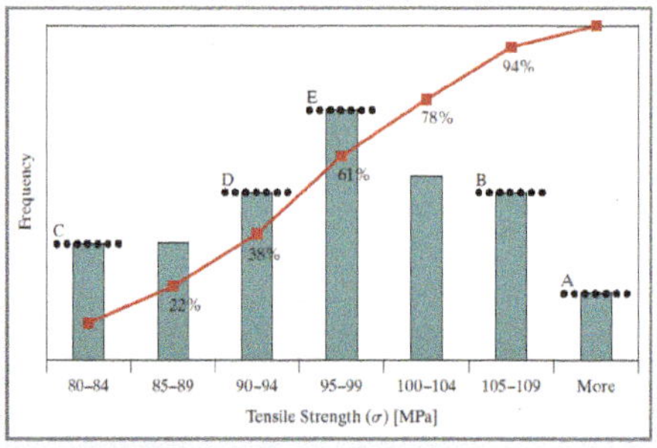

(a) What is the frequency value of A on the chart?

(b) What is the frequency value of B on the chart?

(c) What is the frequency value of C on the chart?

(d) What is the frequency value of D on the chart?

(e) What is the frequency value of E on the chart?

6. A technician tested a temperature probe by inserting it in boiling acetic acid (theoretical boiling point is 118 degrees Fahrenheit), recording the readings, removing and drying the probe, and repeating the process. The data are shown in the following table.

Temperature, Probe 1 (*T*) [°F]	122	120	107	115	107	122	122
Temperature, Probe 2 (*T*) [°F]	102	92	97	107	92	Missing Point

(a) Determine the mean of probe 1.

(b) Determine the median of probe 1.

(c) A second probe was tested, yielding a mean of 97 degrees Fahrenheit and a median of 94.5 degrees Fahrenheit. If the data from probe 2 are as shown in the table, determine the missing data point.

(d) If a probe has a standard deviation of 10 degrees Fahrenheit, what is the variance of the probe?

7. A technician tested two temperature probes by inserting them in boiling water (theoretical boiling point is 100 degrees Celsius), recording the readings, removing and drying the probe, and repeating the process. The CDF for both probes is shown in the following graph.

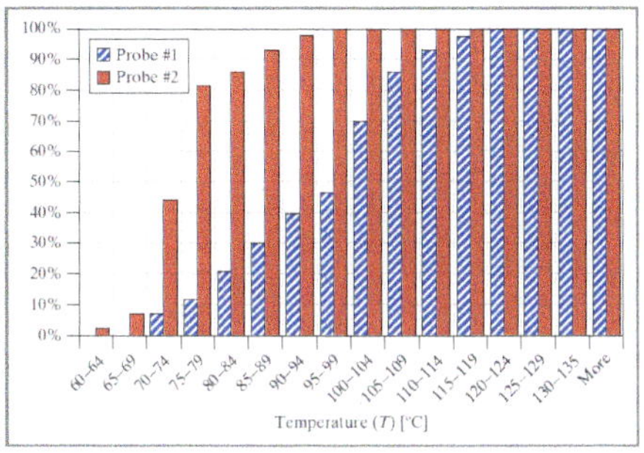

(a) Using this information, which probe would you choose to use? Explain why.

(b) How would you use the probe you chose to ensure that you found the correct boiling point?

(c) Which probe has the higher standard deviation?

8. During November, the heating system in your apartment appeared to be broken. To prove this, you record the following daily high temperatures in degrees Fahrenheit, taken every other day:

58	54	60	58	55	55	60	60
86	80	85	85	82	85	91	93

(a) Calculate the mean and median of the data.

(b) Draw the associated histogram and CDF for the data by hand; plot temperature on the abscissa and days on the ordinate.

(c) When you take your complaint to the apartment manager, he fails to see the problem; according to his heating bill, your apartment had an average temperature of 72 degrees Fahrenheit. Show both the mean and the median on the graph drawn for part (b). Which is a better presentation of the data, part (a) or part (b)? Justify.

9. You are assigned to inspect metal-composite beam trusses for a new bridge being built over a nearby lake. The manufacturer has run a prototype set of 500 beams and conducted strength tests, which you consider to be the baseline case, shown by the solid line in all graphs. Examine each graph and explain the changes to the baseline curve observed in the dashed line by choosing a cause from the following list. The strength of the beam is shown on the abscissa.

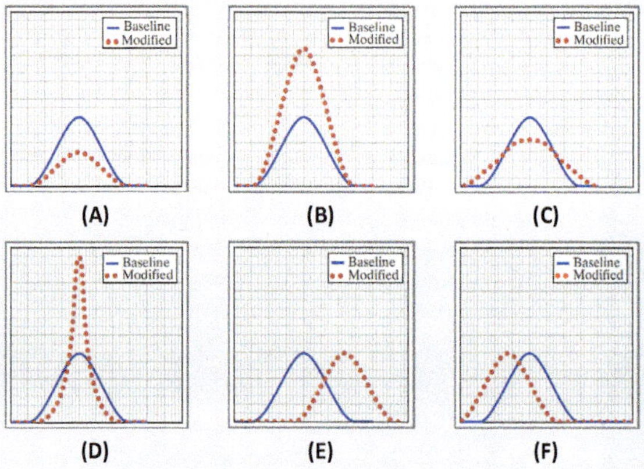

(a) The manufacturer tested 1000 beams instead of 500 beams.

(b) A reinforcing coating was used on a sample of 500 beams.

(c) The manufacturer upgraded processing equipment to lower the variability of the metal-composite strength.

10. Use the scenario described in the preceding problem. Examine the graphs shown and explain the changes to the baseline curve observed in the dashed line by choosing a cause from the following list. The strength of the beam is shown on the abscissa.

(a) The manufacturer tested 200 beams instead of 500 beams.

(b) An impurity, which caused the beams to weaken, was discovered in a sample of 500 beams.

(c) During a plant strike by union workers, substitute workers manufactured the beams.

11. You test several temperature probes by inserting them in boiling ethanol (theoretical boiling point is 78.4 degrees Celsius), recording the readings, removing and drying the probe, and repeating the process 20 times. The distribution curves for the probes are shown in question 9. The solid line "baseline" curve in every graph is the same curve as for a previous probe tested 20 times in boiling ethanol.
    (a) Which probe was tested 40 times instead of 20 times?
    (b) Which probe has the highest standard deviation?
    (c) During the testing of one probe, you suspect your assistant of using formic acid (which boils at 101 degrees Celsius) instead of ethanol. Which probe did your assistant incorrectly test?

12. Use the scenario described in the previous problem. Answer these questions based on the graphs shown in question 9.
    (a) Which probe has the lowest standard deviation?
    (b) Which probe was tested 10 times instead of 20 times?
    (c) During the testing of one probe, you suspect your assistant of using chloroform (which boils at 61 degrees Celsius) instead of ethanol. Which probe did your assistant incorrectly test?
    (d) If you could choose between probes C and F, which probe would you choose to use? In a single sentence, describe how you would use the probe to ensure you determine the correct boiling point.

13. The following data were collected from a manufacturing process for making plastic cylinders. According to the specifications, the cylinder diameter should be 100 inches (the average diameter is 100 inches) and the standard deviation is ±5 inches.

    Graph the data on a control chart. Use graph paper as directed by your instructor. Be sure to clearly indicate the zones of control.

    Using the eight SPC rules, determine whether the process is in statistical control. If it is not in statistical control, indicate which rule or rules are violated and list the part numbers that violate those rules.

Part	Diameter (*D*) [in]	Part	Diameter (*D*) [in]
1	100	10	96
2	106	11	89
3	103	12	89
4	99	13	87
5	90	14	92
6	95	15	94
7	105	16	87
8	107	17	96
9	97	18	98

14. The following table lists the number of resin patio furniture chairs rejected for defects during random testing over the course of a week on a manufacturing line. Four samples of 20 chairs are pulled each day. Use the following data to generate a histogram and CDF in Excel.

0	1	0	0	2	0
0	2	0	1	0	2
0	0	0	3	2	1
1	2	35	0	0	
0	0	1	2	1	

15. A company that fabricates small, custom machines has been asked to generate a machine that throws darts at a dart board as precisely and accurately as possible. To assess the precision and accuracy of each proposed design, the engineers build a model and record the distance from the bullseye of the dart board to the location of each dart thrown—both the straight-line distance (A) and the horizontal (B) and vertical (C) distances are recorded separately with regard to the bullseye. The engineers throw 15 darts with their prototype machine and record the three data points for each dart.

A	B	C
5.13	-5.13	-0.08
8.29	-3.08	-7.70
2.26	0.08	2.26
1.92	-1.92	-0.08
8.60	6.69	5.41
1.60	-0.70	-1.44
1.59	-0.07	1.59
2.37	-2.25	0.74
0.47	-0.28	-0.38
0.52	0.27	0.45
8.26	4.09	-7.17
4.10	4.10	0.20
2.85	-2.10	1.93
4.22	-2.03	-3.70
1.44	-0.35	-1.40

Using the data collected for a design in the starting workbook, create a histogram and a CDF in Excel for the straight-line distance (A), as well as the horizontal (B) and vertical (C) distances, and determine which graph or graphs are better for assessing the performance of the design if we were interested in (a) if the machine is throwing accurately to hit the bullseye or (b) if the machine needs to be calibrated (or adjusted) to correctly hit the bullseye. Justify your answer with a few sentences about why you selected the graph or graphs.

16. Your engineering team rework the design process and develop a different dart-throwing model. Repeat problem 15 with this data recorded with a different model of dart-throwing machines, using the data shown.

A	B	C
0.14	-0.12	0.06
0.02	-0.01	-0.02
0.02	0.02	-0.01
1.32	-0.03	1.32
0.23	0.09	0.21
0.23	-0.08	-0.21
0.15	-0.05	0.14
1.24	1.24	0.06
1.12	-1.05	-0.39
0.36	-0.34	0.09
1.74	-1.14	-1.32
1.03	-1.02	-0.19
0.25	0.15	0.20
1.31	0.71	-1.10
0.55	0.35	-0.42

17. The data provided online was collected by Ed Fuller of the NIST Ceramics Division in December 1993. The data represent the polished window strength, measured in units of kilopounds per square inch [ksi], and were used to predict the lifetime and confidence of airplane window design. Use the data set to generate a histogram and CDF in Excel *(http://atomic.phys.uni-sofia.bg/local/nist-e-handbook/e-handbook/eda/section4/eda4291.htm).*

18. Choose *one* of the following options—(a), (b), (c), or (d)—and collect the data required. For the data source you select, do the following using the analysis in Excel.
    - Construct a histogram, including justification of bin size.
    - Determine the mean, median, variance, and standard deviation values.
    - Construct a cumulative distribution function.

    (a) On a campus sidewalk, mark two locations 50 feet apart. As people walk along, count how many steps they take to go the 50 feet. Do this for 125 individuals.

    (b) Select 250 words at random from a book (fiction). Record the number of letters in each word. Alternatively, you can count and record the words in 250 sentences.

    (c) Go to one section of the library and record the number of pages in 125 books in that same section.

    (d) Interview 125 people to determine how far their home is, in miles, from the university.

382 Chapter 11 Basic Statistical Analyses and Experiments

19. Using the PERIOD method, design an experiment that compares how four different tires perform. You decide to run an experiment that tests all four tires on dry, damp, and wet roads for decelerations of 1, 3, 5, 7, and 9 m/s^2. If each experiment is repeated two times, how many measurements are needed?

    (a) 240 measurements

    (b) 120 measurements

    (c) 60 measurements

    (d) 30 measurements

20. Which of the following describes random error?

    (a) Random error is the uncertainty caused by instrumentation malfunction.

    (b) Random error is the uncertainty inherently present in the instrumentation.

    (c) Random error is the uncertainty that we try to eliminate from our experiment.

    (d) Random error is the uncertainty caused by human mistakes.

OVERVIEW

This chapter is a repeat of the exercise in Chapter 0. You may elect to use this activity as a way to reinforce the big ideas covered throughout this text. In particular, you can highlight specific steps in the design process and the fact that it is a *systematic* process.

OBJECTIVES

- Form criteria, constraints, and design specifications
- Discuss the design in terms of functions, inputs, and outputs instead of a final product
- Brainstorm ideas for how to accomplish each function
- Prototype and test against the design specifications
- Keep a record of the processes using a design notebook

CHAPTER 12

A Scenario in the Future - Revisited

TOPICS

The year is 2034. You still love your job as a computer engineer at a multinational engineering firm, although you continue to be surprised that you don't work exclusively with other computer engineers. You remember, during school, that you thought you would work with mechanical engineers in a cubicle, rarely making human contact. You are surprised by how important it is to work with a team—and that almost every project involves teamwork. You also write quite a bit and make the occasional presentation. When you were in speech class, you thought that speech you made on "How to Grill the Perfect Grilled Cheese Sandwich" was probably the last time you would have to speak to a group.

You continue to enjoy the creativity and innovation you experience regularly. Even though every design hasn't been perfect, the failures allowed you and your team to learn and improve.

You continue to hang out with some people on the team, including the people who went on that trip a few years ago. (You might recall your adventure from Chapter 0.)

In fact, you remember the radioactive water lifter you built on that trip a few years ago, and wonder—had we used some design process, could we have built a better lifter?

12.1 Happy Nature National Park

You may fondly recall the time your team visited Happy Nature National Park and discovered that the pristine water wasn't quite suitable for drinking. In fact, one lake was found to have radioactive fish and contaminated water. You successfully informed the government, and the lake has since been cleaned.

Since you and your team members all have vacation time coming up once again, you decide to revisit Happy Nature. "Remember the time we rented the cabin? Let's do it again!" Your team members agree and the plans are set.

Your team of three engineers—once again—packs up and hits the road. You make the drive to Happy Nature. The plan: Enjoy this vacation!

384 Chapter 12 A Scenario in the Future - Revisited

It's déjà vu: When you arrive at the cabin, a note on the door reassures you that this park is now SO CLEAN that there is no need for separate drinking water! The fine print on the note mentions that a few brave and inspiring engineers did discover a little contamination, but it has since been cleaned up.

A few empty bottles and a map to the local stream where you can fill them were left on the counter. Your cabin is still equipped with a beautiful, flat, study table. While it brings back memories of the trip a few years ago, you are pretty confident. They MUST have cleaned the park this time, right?

Figure 12-1 Happy Nature National Park.

The water is clear and seems perfectly safe! One member of your team likes to fish, one likes to hike, one prefers to swim—and there is time for all of it. Thirsty? Enjoy a drink of the perfectly clear and clean water.

Figure 12-2 Fresh, clean water.

12.2 Trouble in Paradise?

Not this time.

Things are going well. However, one member of the team really misses the design aspects of engineering. She has a suggestion: Let's redesign the radioactive water lifter, but this time, let's just call it a water lifter!

12.3 Engineering Design

Your task: The team must design and build a device to lift the fresh and clean water bottle into a radiation-proof container.

The team decides to build this device once again. They scour the cabin in search of supplies. They find a toolbox and assorted other materials, including tape, rubber bands, paper clips, a spool of wire, and more.

> **Important**: Switching back to our current year for a minute... your teacher may want to substitute or change quantities of the items below. Now, back to the year 2034.

They also find a toolbox with a note saying:

> **You can use these tools, but you may NOT include any of these tools in a design you build.**

12.4 Your Task

In teams of 3 or 4, gather the supplies listed in our story.

Materials list:

(Note: Your teacher may modify the first part of this list if necessary—as long as each team has the same set of supplies, the design goes on.)

5 cardboard strips: 12" long, 2.5" wide

15 rubber bands

8 clothespins

25 paper clips

20 craft sticks

6 unsharpened pencils

8 binder clips

3 feet masking tape

3 feet wire

3 feet string or twine

Also *(not to be incorporated into the water lifter)*:

Scissors

Wire clippers

Ruler

Water bottle (at least ¾ full)

Bucket (preferably small, but large enough to hold a water bottle)

This time, spend **20 minutes** designing a water lifter able to lift a full water bottle into the air and place it in a bucket. After designing, you have **10 minutes** to build your device. The operator may not get any part of his or her body within 18" of the water bottle because we want to have the same experience as we did a few years ago. You may test while you build.

After the activity, discuss as a class:

1. **(a) Was this activity "engineering"? Why or why not?**

   (b) Was it "engineering" last time?

2. **Did the teams exhibit effective teamwork this time? Did you improve over last time?**

3. **Did the design step lead to an improved device?**

DIFFERENTIATED INSTRUCTION

You may choose to increase the difficulty of the task this time, such as by introducing costs for each of the items used in the design, or increasing the distance the arm must extend. However, you may want to keep the parameters of the challenge the same to see how far the students have come since the first day.

DISCUSSION

During this part of the activity, you can use the board to map their decision-making skills to the steps in the design process. Or, you may choose to collect their responses using Google Forms or Microsoft Forms to display to the class.

If you took pictures of the student work in the first iteration of this activity, have students compare their designs to the previous attempt.

APPENDIX A Formulas, Values, Symbols, and MATLAB Functions

Conversions

Angle

1 rad	= 57.3 deg
π rad	**= 180 deg**

Area

1 acre	= 4047 m^2
	= 0.00156 mi^2

Energy

1 J	= 0.239 cal
	= 9.48×10^{-4} BTU
	= 0.7376 ft lb_f
1 kW h	**= 3,600,000 J**

Force

1 N	= 0.225 lb_f
	= 1 E 5 dyne
1 kip	**= 1,000 lb_f**

Length

1 m	= 3.28 ft
1 km	= 0.621 mi
1 in	**= 2.54 cm**
1 mi	**= 5280 ft**
1 yd	**= 3 ft**

Mass

1 kg	= 2.205 lb_m
1 slug	= 32.2 lb_m
1 ton	**= 2,000 lb_m**

Power

1 W	= 3.412 BTU/h
	= 0.00134 hp
	= 14.34 cal/min
	= 0.7376 ft lb_f/s

Pressure

1 atm	= 1.01325 bar
	= 33.9 ft H_2O
	= 29.92 in Hg
	= 760 mm Hg
	= 101,325 Pa
	= 14.7 psi

Time

1 d	**= 24 h**
1 h	**= 60 min**
1 min	**= 60 s**
1 yr	= 365 d

Temperature

1 K	**= 1 °C**
	= 1.8 °F
	= 1.8 °R

Volume

1 L	= 0.264 gal
	= 0.0353 ft^3
	= 33.8 fl oz
1 mL	**= 1 cm^3 = 1 cc**

Named Units

1 F	= 1 A s/V
1 H	= 1 V s/A
1 Hz	= 1 s^{-1}
1 J	= 1 N m

1 N	= 1 kg m/s^2
1 P	= 1 g/(cm s)
1 Pa	= 1 N/m^2
1 St	= 1 cm^2/s

1 V	= 1 W/A
1 W	= 1 J/s
1 Ω	= 1 V/A

Conversions shown in bold text above indicate exact conversions

SI Prefixes

Numbers Less Than One		
Power of 10	**Prefix**	**Prefix Abbreviation**
10^{-1}	deci-	d
10^{-2}	centi-	c
10^{-3}	milli-	m
10^{-6}	micro-	µ
10^{-9}	nano-	n
10^{-12}	pico-	p
10^{-15}	femto-	f
10^{-18}	atto-	a
10^{-21}	zepto-	z
10^{-24}	yocto-	y

Example: 1 millimeter [mm] = 1×10^{-3} meters [m]

Numbers Greater Than One		
Power of 10	**Prefix**	**Prefix Abbreviation**
10^{1}	deca-	da
10^{2}	hecto-	h
10^{3}	kilo-	k
10^{6}	Mega-	M
10^{9}	Giga-	G
10^{12}	Tera-	T
10^{15}	Peta-	P
10^{18}	Exa-	E
10^{21}	Zetta-	Z
10^{24}	Yotta-	Y

Example: 1 Megajoule [MJ] = 1×10^{6} joules [J]

Fundamental Dimensions [Base SI Units]

I	electric current [A]	M	mass [kg]	T	time [s]
J	light intensity [cd]	N	amount of substance [mol]	Θ	temperature [K]
L	length [m]				

Common Derived Units in the SI System

Dimension	SI Unit	Base SI Units = Dimensions	Derived from
Force (F)	newton [N]	$1\,N = 1\frac{kg\,m}{s^2} \{=\} \frac{M\,L}{T^2}$	F = m a Force = mass times acceleration
Energy (E)	joule [J]	$1\,J = 1\,N\,m = 1\frac{kg\,m^2}{s^2} \{=\} \frac{M\,L^2}{T^2}$	E = F d Energy = force times distance
Power (P)	watt [W]	$1\,W = 1\frac{J}{s} = 1\frac{kg\,m^2}{s^3} \{=\} \frac{M\,L^2}{T^3}$	P = E/t Power = energy per time
Pressure (P)	pascal [Pa]	$1\,Pa = 1\frac{N}{m^2} = 1\frac{kg}{m\,s^2} \{=\} \frac{M}{(L\,T^2)}$	P = F/A Pressure = force per area
Voltage (V)	volt [V]	$1\,V = 1\frac{W}{A} = 1\frac{kg\,m^2}{s^3\,A} \{=\} \frac{M\,L^2}{(T^3\,I)}$	V = P/I Voltage = power per current

Physical Constants [Value and Units]

c	speed of light in a vacuum	$= 3 \times 10^8$ m/s
e	Euler's number (base of natural logarithm)	= 2.71828 . . .
	elementary charge of an electron	$= 1.602 \times 10^{-19}$ C
φ	Golden ratio	= 1.61803 . . .
F	Faraday's constant	$= 9.65 \times 10^4$ C/mol
g	acceleration due to gravity	$= 9.8$ m/s^2 on Earth
G	gravitational constant	$= 6.67 \times 10^{-11}$ (N m^2)/kg^2
h	Planck constant	$= 6.62 \times 10^{-34}$ Js
k	Boltzmann constant	$= 1.38 \times 10^{-23}$ J/K
k_e	Coulomb's constant	$= 8.98755 \times 10^9$ N m^2/C^2
N_A	Avogadro's number	$= 6.022 \times 10^{23}$ mol^{-1}
π	ratio of circle circumference to diameter	= 3.14159 . . .
R	gas constant	= 8314 (Pa L)/(mol K) = 0.08206 (atm L)/(mol K)
σ	Stefan-Boltzmann constant	$= 5.67 \times 10^{-8}$ W/(m^2 K^4)

Greek Letters [Name, Greek Form, English Form, Common Uses]			
alpha	A, α	A, a	α angle, angular acceleration, Neper frequency, thermal diffusivity
beta	B, β	B, b	β bulk expansion coefficient, coefficient of thermal expansion, transistor current, gain
gamma	Γ, γ	G, g	γ specific weight; Euler's constant
delta	Δ, δ	D, d	Δ change; tip deflection of a beam δ small change, impulse function
epsilon	E, ε	E, e	ε dielectric constant, pipe roughness, porosity, strain
zeta	Z, ζ	Z, z	ζ dampening ratio of spring system
eta	H, η	H, h	η efficiency
theta	Θ, θ	Q, q	Θ dimensionless time θ thermal resistance, Euler angle
iota	I, ι	I, i	ι imaginary unit, consecutive integers
kappa	K, κ	K, k	κ permeability, thermal conductivity
lambda	Λ, λ	L, l	λ wavelength, arrival rate, eigenvalue
mu	M, μ	M, m	μ SI prefix micro (10^{-6}), statistical mean, dynamic viscosity, coefficient of friction
nu	N, ν	N, n	ν frequency, kinematic viscosity, Poisson's ratio, specific volume
xi	Ξ, ξ	X, x	ξ permeability, damping coefficient in circuit
omicron	O, o	O, o	
pi	Π, π	P, p	Π product π 3.141592… (circumference/diameter of circle), osmotic pressure
rho	P, ρ	R, r	ρ density, resistivity
sigma	Σ, σ	S, s	Σ summation σ conductivity, standard deviation, surface tension, stress, Stefan-Boltzmann constant
tau	T, τ	T, t	τ time constant, shear stress, torque
upsilon	Y, υ	U, u	
phi	Φ, ϕ or φ	F, f	Φ potential ϕ, φ Euler angle, Golden ratio
chi	X, χ	C, c	χ electronegativity
psi	Ψ, ψ	Y, y	ψ Euler angle
omega	Ω, ω	W, w	Ω unit of ohm ω angular frequency, angular velocity

Nomenclature [common units]

a	acceleration [m/s^2]
A	altitude [m] area [m^2]
C	capacitance [F] circumference [m] concentration [g/L]
C_d	coefficient of drag [-]
C_p	specific heat [J/(g °C)]
d	distance [m]
D	diameter [m]
E	energy [J] Young's modulus [Pa]
F	force [N]
f	frequency [Hz] friction factor [-]
G	conductance [S]
h	heat transfer coefficient [W/(m^2 °C)]
H	height [m]
I	current [A] moment of inertia [kg m^2]
k	spring constant [N/m] thermal conductivity [W/(m °C)]
K	beam stiffness [N/m]
KE	kinetic energy [J]
L	inductance [H] length [m]
m	mass [kg]
$\dot{m}$	mass flowrate [kg/s]
Ma	Mach Number [-]
n	number of moles [mol] shaft speed [Hz]
P	perimeter [m] power [W] pressure [Pa] vertical load of beam [N]
PE	potential energy [J]
Q	electric charge [C] thermal energy [cal] volumetric flowrate [m^3/s]
r	radius [m]
R	resistance [Ω]
Re	Reynolds Number [-]
S	arc length [m]
SG	specific gravity [-]
T	period [s] temperature [K]
t	time [s]
v	velocity [m/s]
V	volume [m^3] voltage [V]
w	weight [N]
W	width [m] work [J]
x	displacement [m]
α	neper frequency [Hz] thermal diffusivity [m^2/s]
β	coefficient of thermal expansion [$°C^{-1}$]
γ	specific weight [N/m^3]
Δ	delta; change in quantity tip deflection of beam [m]
ε	pipe roughness [mm] strain [-]
η	efficiency [-]
θ	angle [rad]
μ	coefficient of friction [-] dynamic viscosity [cP]
ν	kinematic viscosity [St]
ρ	mass density [kg/m^3] resistivity [Ω m]
σ	stress [Pa] surface tension [J/m^2]
τ	shear stress [Pa]
ω	angular frequency [rad/s]

Unit Abbreviations

°C	degree Celsius
°F	degree Fahrenheit
°R	degree Rankine
A	ampere
amu	atomic mass unit
atm	atmosphere
AU	astronomical unit
BTU	British thermal unit
C	coulomb
cal	calorie
cc	cubic centimeter
cd	candela
cP	centipoise
d	day
Da	dalton
deg, °	degree
eV	electron volt
F	farad
fl oz	fluid ounce
ft H_2O	feet of water
ft	foot
g	gram
gal	gallon
gpm	gallons per minute
H	henry
h	hour
hp	horsepower
Hz	hertz
in Hg	inches of mercury
in	inch
J	joule
K	kelvin
kg	kilogram
kWh	kilowatt-hour
L	liter
lb_f	pounds-force
lb_m	pounds-mass
m	meter
mi	mile
min	minute
mm Hg	millimeters mercury
mol	mole
mph	miles per hour
N	newton
Pa	pascal
psi	pound-force per square inch
rad	radian
s	second
S	seimens
St	stokes
t	ton
tbsp	tablespoon
tsp	teaspoon
V	volt
W	watt
yd	yard
yr	year
Ω	ohm

Miscellaneous Equations

Antoine equation

$$P = 10^{\left(A - \frac{B}{T+C}\right)}$$

Cantilever Beam Deflection

$$P = K\,\Delta$$

Discharging Capacitor

$$V = V_0\, e^{-t/RC}$$

Moore's Law

$$T = T_0\, 2^{\frac{t}{2}}$$

Oscillating spring

$$f = \frac{1}{2\pi}\sqrt{\frac{k}{m}}$$

Poiseuille's Equation

$$R = \frac{8\mu L}{\pi r^4}$$

Pythagoras' Theorem

$$d = \sqrt{(x_2 - x_1)^2 + (y_2 - y_1)^2}$$

Resonant frequency for one inductor and one capacitor

$$f_R = \frac{1}{2\pi\sqrt{L\,C}}$$

Shockley Equation

$$I = I_0\left(e^{\frac{V_D}{nV_T}} - 1\right)$$

Torricelli's Law

$$v = \sqrt{2\,g\,H}$$

van der Waals Equation

$$\left(P + \frac{an^2}{V^2}\right)(V - b\,n) = n\,R\,T$$

Bingham plastic

$$\tau = \mu\frac{\Delta v}{\Delta y} + \tau_0$$

Dimensionless Number Equations

Coefficient of Drag

$$C_d = \frac{F}{\frac{1}{2}\rho v^2 A_p}$$

Friction factor: Laminar Flow

$$f = \frac{64}{Re}$$

Friction factor: Blasius Approximation

$$f = 0.316\,(Re)^{-1/4}$$

Friction factor: Colebrook

$$\frac{1}{\sqrt{f}} = -2\log\left(\frac{\varepsilon/D}{3.7} + \frac{2.51}{Re\sqrt{f}}\right)$$

Mach Number

$$Ma = \frac{speed_{object}}{speed_{sound\ in\ air}}$$

Power Number

$$N_P = \frac{P}{\rho n^3 D^5}$$

Reynolds Number: Pipe Flow

$$Re = \frac{\rho D v}{\mu}$$

Reynolds Number: Use with Coefficient of Drag

$$Re = \frac{D_p \rho v}{\mu}$$

Reynolds Number: Use with Power Number

$$Re = \frac{D^2 n \rho}{\mu}$$

Statistics Equations

Mean

$$\bar{X} = \frac{1}{N}(X_1 + X_2 + \cdots + X_N) = \frac{1}{N}\sum_{j=1}^{N} X_j$$

Median

Place data in ascending order.
If N = odd, median = central value.
If N = even, median = mean of two central values.

Variance

$$V_x^2 = \frac{1}{N-1}((\bar{X} - X_1)^2 + \cdots + (\bar{X} - X_N)^2) = \frac{1}{N-1}\sum_{j=1}^{N}(\bar{X} - X_j)^2$$

Standard Deviation

$$SD_X = \sqrt{V_X^2}$$

Equations

Current
$Q = I\,t$

Conductance
$G = 1/R$

Power: Electric
$P = V\,I = I^2 R = V^2/R$

Electric charge
$Q = CV$

Inductance
$V = L\,dI/dt$

Capacitors: Parallel
$C_{eff} = C_1 + C_2 + \cdots + C_{N-1} + C_N$

Capacitors: Series
$$C_{eff} = \left(\frac{1}{C_1} + \frac{1}{C_2} + \cdots + \frac{1}{C_{N-1}} + \frac{1}{C_N}\right)^{-1}$$

Capacitance: Energy Stored
$Ec = 1/2\,C\,V^2$

Inductance: Energy Stored
$E_L = 1/2\,L\,I^2$

Inductors: Series
$L_{eff} = L_1 + L_2 + \cdots + L_{N-1} + L_N$

Inductors: Parallel
$$L_{eff} = \left(\frac{1}{L_1} + \frac{1}{L_2} + \cdots + \frac{1}{L_{N-1}} + \frac{1}{L_N}\right)^{-1}$$

Circuits: Ohm's Law
$V = I\,R$

Circuits: Parallel Resistors
$$R_{eff} = \left(\frac{1}{R_1} + \frac{1}{R_2} + \cdots + \frac{1}{R_{N-1}} + \frac{1}{R_N}\right)^{-1}$$

Circuits: Series Resistors
$R_{eff} = R_1 + R_2 + \cdots + R_{N-1} + R_N$

Density
$\rho = m/V$

Efficiency
$\eta = \text{output}/\text{input}$

Work
$W = F\,\Delta x$

Potential Energy
$PE = m\,g\,\Delta H$

Kinetic Energy, translational
$KE_T = \tfrac{1}{2}\,m\,(v_f^2 - v_i^2)$

Kinetic Energy, rotational
$KE_R = \tfrac{1}{2}\,I\,(\omega_f^2 - \omega_i^2)$

Kinetic Energy, total
$KE = KE_T + KE_R$

Thermal Energy
$Q = m\,C_p\,\Delta T$

Force: Newton's 2nd Law
$F = m\,a$

Interpolation
$$\frac{(A - A_1)}{(A_2 - A_1)} = \frac{(B - B_1)}{(B_2 - B_1)}$$

Power
$P = E/t$

Pressure
$P = F/A$

Pressure, Hydrostatic: Pascal's Law
$P_{hydro} = \rho\,g\,H$

Pressure, Total
$P_{total} = P_{surface} + \rho\,g\,H$

Pressure: Ideal Gas Law
$P\,V = n\,R\,T$

Specific Gravity
$SG = \rho_{object}/\rho_{water}$

Specific Weight
$\gamma = w/V$

Springs: Hooke's Law
$F = k\,x$

Springs: Parallel
$k_{eff} = k_1 + k_2 + \cdots + k_{N-1} + k_N$

Springs: Series
$$k_{eff} = \left(\frac{1}{k_1} + \frac{1}{k_2} + \cdots + \frac{1}{k_{N-1}} + \frac{1}{k_N}\right)^{-1}$$

Temperature
$$\frac{T\,[°F] - 32}{180} = \frac{T\,[°C] - 0}{100}$$
$T\,[K] = T\,[°C] + 273$
$T\,[°R] = T\,[°F] + 460$

Temperature: Newton's Law of Cooling
$T = T_0\,e^{-kt}$

Viscosity, Dynamic: Newton's Law of Viscosity
$\tau = \mu\,(\Delta v/\Delta y)$

Viscosity, Kinematic
$\nu = \mu/\rho$

Weight
$w = m\,g$

Young's Modulus
$\sigma = E\,\varepsilon$

Geometric Formulas

Rectangle
Area = $a\,b$
Perimeter = $2\,a + 2\,b$

Rectangular Parallelepiped
Volume = $a\,b\,c$
Surface area = $2\,(a\,b + a\,c + b\,c)$

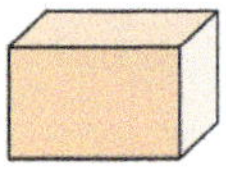

Circle
Area = $\pi\,r^2$
Perimeter = $2\pi\,r$

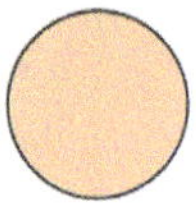

Sphere
Volume = $4/3\,\pi\,r^3$
Surface area = $4\,\pi\,r^2$

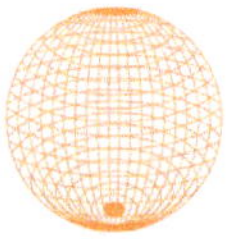

Triangle
Area = $\frac{1}{2}\,b\,H$

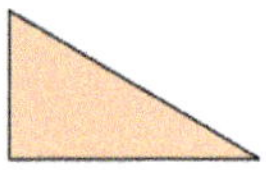

Right Circular Cone
Volume = $1/3\,\pi\,r^2\,H$

Right Circular Cylinder
Volume = $\pi\,r^2\,H$
Lateral Surface Area = $2\pi\,r\,H$

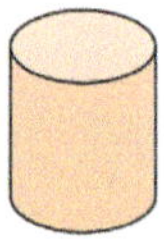

MATLAB Special Character Reference

These special characters are available for use plots (title, legend, xlabel, ylabel, etc.).

Character Sequence	Symbol	Character Sequence	Symbol	Character Sequence	Symbol
\alpha	α	\upsilon	υ	\infty	∞
\angle	∠	\phi	ϕ	\clubsuit	♣
\ast	*	\chi	χ	\diamondsuit	♦
\beta	β	\psi	ψ	\heartsuit	♥
\gamma	γ	\omega	ω	\spadesuit	♠
\delta	δ	\Gamma	Γ	\leftrightarrow	↔
\epsilon	ε	\Delta	Δ	\leftarrow	←
\zeta	ζ	\Theta	Θ	\Leftarrow	⇐
\eta	η	\Lambda	Λ	\uparrow	↑
\theta	Θ	\Xi	≡	\rightarrow	→
\vartheta	ϑ	\Pi	Π	\Rightarrow	⇒
\iota	ι	\Sigma	Σ	\downarrow	↓
\kappa	κ	\Upsilon	γ	\circ	°
\lambda	λ	\Phi	Φ	\pm	±
\mu	μ	\Psi	Ψ	\geq	≥
\nu	ν	\Omega	Ω	\propto	∝
\xi	ξ	\forall	∀	\partial	∂
\pi	π	\exists	∃	\bullet	•
\rho	ρ	\ni	∍	\div	÷
\sigma	σ	\cong	≅	\neq	≠
\varsigma	ς	\approx	≈	\aleph	ℵ
\tau	τ	\Re	ℜ	\wp	℘
\equiv	≡	\oplus	⊕	\oslash	∅
\Im	ℑ	\cup	∪	\supseteq	⊇
\otimes	⊗	\subseteq	⊆	\subset	⊂
\cap	∩	\in	∈	\o	o
\supset	⊃	\lceil		\nabla	∇
\int	∫	\cdot	·	\ldots	...
\rfloor	⌊	\neg	¬	\prime	′
\lfloor	⌋	\times	×	\0	∅
\perp	⊥	\surd	√	\mid	|
\wedge	∧	\varpi	ϖ	\copyright	©
\rceil	⌉	\rangle	〉		
\vee	∨	\sim	~		
\langle	〈	\leq	≤		

Original version: http://www.mathworks.com/help/techdoc/ref/text_props.html#String
Note: Greek letters appear in italics throughout this text but will appear as shown on this table in Matlab.

MATLAB Line Styles, Markers, and Colors

Line Style Specifiers

Specifier	Line Style
'–'	Solid line (default)
'--'	Dashed line
':'	Dotted line
'–.'	Dash-dot line
'none'	No line

Marker Specifiers

Specifier	Marker Type
'+'	Plus sign
'o'	Circle
'*'	Asterisk
'.'	Point
'X'	Cross
'square' or 's'	Square
'diamond' or 'd'	Diamond
'^'	Upward-pointing triangle
'V'	Downward-pointing triangle
'>'	Right-pointing triangle
'<'	Left-pointing triangle
'pentagram' or 'p'	Five-pointed star (pentagram)
'hexagram' or 'h'''	Six-pointed star (hexagram)
'none'	No marker (default)

Color Specifiers

Specifier	Color
r	Red
g	Green
b	Blue
c	Cyan
m	Magenta
y	Yellow
k	Black
w	White

Original source: http://www.mathworks.com/help/techdoc/ref/linespec.html

MATLAB Graphing Properties

Most graphing functions have the ability to apply special properties for additional customization of the appearance. This section contains a list of some of the different property names and property values available in MATLAB. Note the property name and property values are placed inside single quotes, with exceptions listed.

Plotting Properties:

Functions: `plot`

Example

```
plot(X,Y,'s','MarkerSize',12)
```

Property Name	Purpose	Property Values	Default
LineWidth	To change the width of lines on a graph	Numerical values, without single quotes	0.5
MarkerEdgeColor	To change the border color of all markers	Same as `BackgroundColor`	Default marker color
MarkerFaceColor	To change the fill color of all markers	Same as `BackgroundColor`	Default marker color
MarkerSize	To change the size of all markers	Numerical values, without single quotes	2.0

Text Properties:

Functions: `text`, `xlabel`, `ylabel`, `title`

Example

```
text(100,200,'Example Text','EdgeColor','red')
title('Example Title', 'FontWeight','bold','FontSize',16)
```

Property Name	Purpose	Property Values	Default
BackgroundColor	To change the background color of the textbox created by the function	`yellow` `green` `magenta` `blue` `cyan` `white` `red` `black`	transparent
Color	To change the color of the text created by the function	Same as `BackgroundColor`	`black`
EdgeColor	To change the color of the border around the textbox created by the function	Same as `BackgroundColor`	none
FontAngle	To change the font angle	`normal` `italic` `oblique`	`normal`
FontSize	To change the font size (in points)	Numerical values, without single quotes	10.0
Note: By default, all font size are given in the unit of points (e.g., 12 pt font)			

Text Properties *(continued)*

Property Name	Purpose	Property Values	Default
`FontWeight`	To change the font weight	`light` `demi` `normal` `bold`	`normal`

`HorizontalAlignment`	To change the horizontal alignment of the text within the textbox	`left`
		`center`
		`right`
			`center`

`LineStyle`	To change the style of the line bordering the textbox. Enabled when `EdgeColor` changed to a visible color	`-`
		`--`
		`:`
		`-.`
		`none`
			By default, the style is `-`, but not visible until `EdgeColor` changed

`LineWidth`	To change the width (in unit of points) of the line bordering the textbox. Enabled when `EdgeColor` changed to a visible color	Numerical values, without single quotes	0.5
`Margin`	To change the amount of space (in unit of pixels) between the text and the line bordering the textbox. Enabled when `EdgeColor` changed to a visible color	Numerical values, without single quotes	2.0

`VerticalAlignment`	To change the vertical alignment of the text within the textbox	`top`
		`cap`
		`middle`
		`baseline`
		`bottom`
			`bottom`

ASCII Code

Character	A	B	C	D	E	F	G	H	I	J	K	L	M
Value	65	66	67	68	69	70	71	72	73	74	75	76	77

Character	N	O	P	Q	R	S	T	U	V	W	X	Y	Z
Value	78	79	80	81	82	83	84	85	86	87	88	89	90

Character	a	b	c	d	e	f	g	h	i	j	k	l	m
Value	97	98	99	100	101	102	103	104	105	106	107	108	109

Character	n	o	p	q	r	s	t	u	v	w	x	y	z
Value	110	111	112	113	114	115	116	117	118	119	120	121	122

Character	0	1	2	3	4	5	6	7	8	9
Value	48	49	50	51	52	53	54	55	56	57

APPENDIX B Tolerances

Overview

Tolerances on dimensions are necessary to specify the acceptable variability in the dimension of a part. Tolerances can be indicated for all dimensions using a general note or they can be specified for each individual dimension. Tolerances are based on the function of the part or are used to ensure that mating parts fit together. However, in many cases the manufacturing process determines the tolerance. The surface finish may also be specified on a drawing to indicate the roughness of the surface. To further minimize ambiguity in engineering drawings, geometric dimensioning and tolerancing is used to specify the form of a feature, such as flatness or roundness, or to indicate the ideal position of a feature.

B.1 Why Tolerances?

In 1815, it was hoped that muskets could be produced in a number of different U.S. Government and private armories so that the parts of the musket would be interchangeable. Before that time, muskets were individually handcrafted, so parts from one musket rarely fit another musket. Because engineering drawings were not commonly used, a number of "perfect" muskets were made according to master gauges and jigs. **Gauges** are devices used to check the individual dimensions of a part. For instance, the diameter of a shaft can be checked using a gauge consisting of two holes: one of the maximum allowable diameter of the shaft and one of the minimum allowable diameter. If the shaft fits through the larger hole, but not the smaller hole, then it is within the allowable dimensions for the shaft. **Jigs** are devices used to hold a workpiece and guide the tool to ensure repeatable machining. The master gauges and jigs were sent to the armories with the instruction that "no deviations from the pattern were to be allowed." In 1824, this concept of interchangeable parts was tested by disassembling a hundred rifles from several different armories, mixing the parts, and then reassembling the parts at random. Most reassembled rifles worked as they should, proving the concept and value of interchangeable parts.

gauges

jigs

The interchangeability of parts is so common now that it is hard to imagine anything different. The concept allows parts made in various locations by different manufacturers to be successfully assembled and to function properly as an assembly. Although engineering drawings rather than "master parts" came into use in the mid-19th century, it was the two World Wars in the 20th century that brought about the development of methods of showing the acceptable variation in a dimension on a drawing, or **tolerancing**.

tolerancing

The need to control precisely the geometry of a part arises from the part's function. For instance, the cross-sectional geometry of an airfoil, such as the wing of a jet aircraft or a turbine blade, must be accurately controlled to ensure aerodynamic efficiency. The more commonplace need for controlling geometry results from the requirement for parts to fit together. But it is quite difficult to make every part the exact size that is specified in a dimension because of slight differences in tool size, machine tool wear, human operator error, and other factors. As a consequence, tolerances are used in engineering drawings to specify the limit in the variation between mating parts and to provide guidelines to control the manufacturing process. The **tolerance** is the amount that a specific dimension is permitted to vary, or the difference between the maximum and minimum limits of a dimension.

tolerance

Consider a dimension on a drawing specified as 3.750 ± .003. This dimension indicates that the part has a **nominal size**, or general size, of 3 3/4 inches (usually expressed as a common fraction). The **basic size**, or theoretical size for the application of the tolerance, is 3.750 inches (expressed as a decimal). The tolerance

nominal size

basic size

of the dimension is .006 inch. Thus, an acceptable machined part can have an actual size ranging from 3.747 to 3.753 inches, which are the minimum and maximum limits of the dimension. If the actual size is smaller than 3.747 or larger than 3.753, then the part is not acceptable.

target size

Of course, it is desirable to make the part as close as possible to the basic size, sometimes called the **target size**. But the more accuracy needed in a machined part, the higher is the manufacturing cost. Furthermore, the method of machining a part limits the tolerance that can be specified. For instance, the tolerance in drilling a .500-inch hole is between .002 and .005 depending on the quality of the drill press and drill bit. To require a tolerance smaller than .002 would require an additional machining process, such as reaming. Consequently, tolerances can play a large role in the cost of manufacturing a part. Therefore, the tolerances should not be specified tighter than necessary for the product to function properly.

An increased awareness of the importance of tolerances in manufacturing has led to an approach in which the deviation of the actual size of a part from the target size has a cost in terms of reworking of the part, scrap, customer dissatisfaction, and poor reliability. An approach to handle this is to design a product so that it is less sensitive to manufacturing tolerances. The idea is to design the part intelligently so that easy-to-manufacture tolerances, rather than exceptionally tight tolerances, are implemented. Consequently, proper fit between mating parts comes about because of good design, instead of depending on tight tolerances. This approach is known as **robust design**.

robust design

B.2 Displaying Tolerances on Drawings

Every dimension on a drawing should have a tolerance. Tolerances can be displayed on engineering drawings in several ways depending on the situation.

general tolerances

General tolerances can be given in a note on the drawing or in the title block. An example would be the note ALL DIMENSIONS TO BE HELD TO ± .003". Thus, a dimension of .375 would have a minimum limit of .372 and a maximum limit of .378. Often general tolerances are specified in terms of the number of digits following the decimal point in a dimension. In this case, a note would appear on the drawing, such as

UNLESS OTHERWISE SPECIFIED

TOL: .XX = ±.010
.XXX = ±.005
HOLES = ±.002
ANGLES = ±.5 DEG

Thus, a dimension of .50 on a drawing would indicate an actual size between .490 and .510. A dimension of .500 would indicate an actual size between .495 and .505. For metric dimension, a similar form for general tolerances would be X.X METRIC = ± .08.

limit dimensions

plus and minus dimensions

bilateral tolerance

unilateral tolerance

Two other methods of displaying tolerances on a drawing are shown in Figure B-1. The maximum and minimum sizes are specified directly when using **limit dimensions** to specify the tolerance. Usually the upper limit is placed above the lower limit. While this method clearly shows the limits of the dimension, many engineers, designers, and machinists think in terms of the nominal or basic size plus a tolerance, neither of which is immediately evident using limit dimensions. Thus, a more common approach is to use **plus and minus dimensions** where the basic size is shown followed by the tolerance values. If the plus and minus tolerance values are identical, then the basic dimension is followed by a plus-or-minus sign and half of the numerical value of the tolerance. Otherwise, the plus and minus tolerances are indicated separately with the plus above the minus. A **bilateral tolerance**, such as the tolerance for the 1.500-inch dimension in Figure B-1, varies equally in both directions from the basic size. A **unilateral tolerance**, such as the tolerance for the 1.250 diameter of the shaft in Figure B-1, varies in only one direction from the basic size. In both cases, the number of digits to the right of the decimal point for the dimension should be the same as the number of digits to the right of the decimal point for the tolerance.

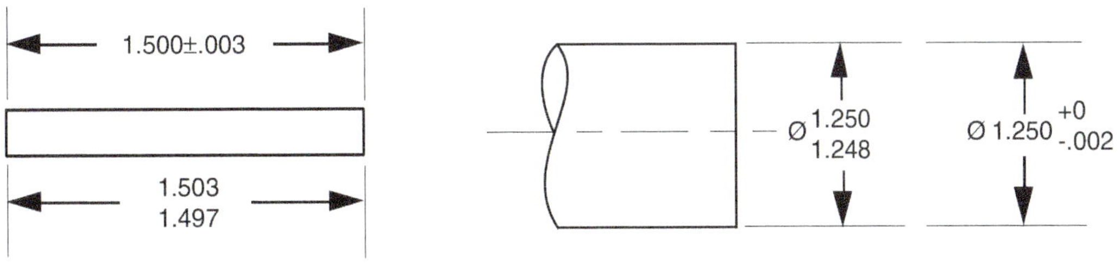

Figure B-1 Plus and minus tolerances and limit tolerances.

single-limit dimension

Sometimes it is not necessary to specify both limits of a tolerance, and only one limit dimension is needed. MIN or MAX is placed after a numerical dimension to indicate that it is a **single-limit dimension**. The depth of holes, length of threads, and radii of corner rounds are often specified using single-limit dimensions.

Many times several methods of representing tolerances are used on a single drawing. For instance, general tolerances may be listed in the title block, and plus and minus dimensions or limit dimensions may be used for a few dimensions to which the general tolerances do not apply.

tolerance accumulation

tolerance stackup

chain dimensioning

datum dimensioning

An important consideration in applying tolerances is the effect of one tolerance on another, especially because tolerances are cumulative. This is known as **tolerance accumulation** or **tolerance stackup** for a chain of dimensions. In the example in Figure B-2, two dimensioning schemes are shown. Above the part, **chain dimensioning** shows the distance from one feature to the next. This would be an appropriate dimensioning scheme if the distance from one hole to the next was critical, because the distance between holes must be between .995 and 1.005. But if the distances of the holes from the left edge of the part were critical, a large tolerance would accumulate. The third hole would be 2.985 from the left edge if all of the dimensions happened to be at the minimum limit of the dimension, or 3.015 from the left edge if all dimensions were at the maximum limit. Thus, the actual dimension of the third hole from the left edge using chain dimensioning is 3.000 ± .015. This tolerance accumulation can be avoided using **datum dimensioning**, where the dimensions are given with respect to a datum, in this case the left edge of the part as shown below the part in Figure B-2. Using this scheme, the third hole is specified to be between 2.995 and 3.005 from the left edge. Of course, the distance between the second and third holes could be as small as .990 if the second hole is at 2.005 and the third hole is at 2.995. Alternatively, the distance between the second and third holes could be as large as 1.010 if the holes are at 1.995 and 3.005. Consequently, the decision about which way to dimension the holes depends on whether the distance between holes is critical or whether the distance from each hole to the left edge is critical.

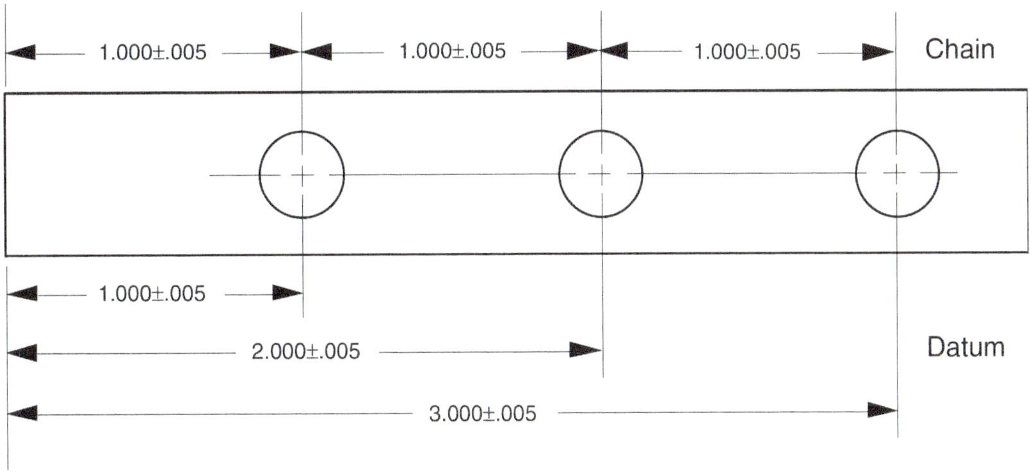

Figure B-2 Chain dimensioning and datum dimensioning.

B.3 How To Determine Tolerances

One of the most difficult aspects of dimensioning and tolerancing for inexperienced engineers and designers is to determine what tolerance to specify on a drawing. In some cases, a specific tolerance is quite clear based on the function of the design. For instance, if a 3/4 inch shaft must rotate at moderate speed in a hole, the tolerances on the shaft and the hole can be determined based on standard classes of fits that are available in many engineering and machinist handbooks. In this particular case, the hole could be between .7500 and .7512, while the shaft could have a diameter of .7484 to .7492. A shaft any bigger or a hole any smaller would result in too tight of a clearance for the shaft to rotate freely, or could result in the shaft being too big for the hole. A smaller shaft or larger hole than the specified limits would result in a sloppy fit between the shaft and the hole. The point here is not the details of how the tolerances are determined. That is too specific to individual cases for this discussion and is presented in many handbooks and texts. Instead, the point is that the function of the system that is being designed drives the tolerances. In this case, *the tolerances drive the manufacturing process* used to machine the hole and the shaft.

But a more common situation is the opposite: *The manufacturing process drives the tolerance*. If the only machine tool that is available to form a hole is a drill press, the 3/4 inch hole that is cut will be between .748 and .754 in diameter based on expected tolerances for the drilling operation. But this means that the hole in the example just presented would have too much variability to ensure that the shaft rotates in it properly. Of course, if only one shaft-hole assembly was being made, the hole could be drilled first and measured. Then the shaft could be turned on a lathe to a diameter just slightly smaller to achieve the fit described. However, this approach will clearly not work for mass-produced, interchangeable parts. Either a more accurate machining method is needed to create the hole or the design must be altered to avoid drilling a hole altogether. (In this example, a standard-size, off-the shelf bearing could be inserted into a larger drilled hole to mate with the shaft.)

When the manufacturing processes that are available drive the tolerances, the tolerances on the drawing must reflect the manufacturing process that is used to make the part, and the design must be altered to ensure that the part can be made with the available machining processes. Guidelines are available for the tolerance range of various common machining processes, including the following:

lapping

Lapping: A process that produces a very smooth and accurate surface by rubbing the surface of the workpiece against a mating form, often using a fine abrasive between the surfaces.

honing

Honing: A process in which a honing stone made of fine abrasive is used to form a surface on a workpiece.

grinding

Grinding: A process that removes a small amount of material from a workpiece with a rapidly rotating grinding wheel.

broaching

Broaching: A process similar to filing in which a cutting tool is reciprocated along its axis to remove a small amount of material from a workpiece.

reaming

Reaming: Removal of a small amount of material from a drilled hole using a rotating cutting tool (reamer).

turning

Turning: Axisymmetric removal of material from a rotating workpiece by feeding a cutting tool into the workpiece using a lathe.

boring

Boring: Axisymmetric removal of material from a workpiece using a tool with a single cutting surface. Either the workpiece or the tool may be rotating.

milling

Milling: Removal of material by feeding a workpiece into a rotating cutting tool.

drilling

Drilling: Creating a hole by feeding the end of a sharpened cylindrical cutting tool (drill bit) into the material.

stamping and punching

Stamping and Punching: Cutting material using a punch and a die to shear the material, much like a paper hole-punch.

The tolerance typically depends on the size of the feature, as well as the quality of the machine tool, sharpness of the cutting tool, and skill of the operator. Figure B-3 indicates the range of tolerances to be expected for various machining operations. To use the table shown in the figure, first consider the size of the feature in the left column. Then go below the sizes in the left column to the type of machining operation used. The bar associated with that process indicates the range of tolerances in the row associated with the feature size. For instance, consider a feature that has a dimension of 2 inches. If broaching were used to create that feature, the chart indicates that the total tolerance would be .0004 to .0015, depending on the machining conditions. Thus, this dimension could be specified as 2.0000 ± .0002 up to 2.0000 ± .0008 (rounding up). If the same piece were created on a milling machine, the tolerances would be .0025 to .010, indicating a dimension of 2.0000 ± .0013 up to 2.000 ± .005. If broaching were not available, we would have to be satisfied with the tolerances of a milling machine and adjust the design accordingly, if the milling tolerances are too large. If both broaching and milling were available, the criticality of the tolerance and the cost of manufacturing would drive the tolerance that would be specified on the drawing. (Broaching, a mass-production process, is usually more expensive than milling, so milling would be preferred if the milling tolerances are acceptable.) Tolerances for metric dimensioning can be determined by converting the values in Figure B-3 to millimeters and rounding off the result to one less place to the right of the decimal point.

Size (in.)	Total Tolerance (in.)								
0.000-0.599	0.00015	0.0002	0.0003	0.0005	0.0008	0.0012	0.002	0.003	0.005
0.600-0.999	0.00015	0.00025	0.0004	0.0006	0.001	0.0015	0.0025	0.004	0.006
1.000-1.499	0.0002	0.0003	0.0005	0.0008	0.0012	0.002	0.003	0.005	0.008
1.500-2.799	0.00025	0.0004	0.0006	0.001	0.0015	0.0025	0.004	0.006	0.010
2.800-4.499	0.0003	0.0005	0.0008	0.0012	0.002	0.003	0.005	0.008	0.012
4.500-7.799	0.0004	0.0006	0.001	0.0015	0.0025	0.004	0.006	0.010	0.015
7.800-13.599	0.0005	0.0008	0.0012	0.002	0.003	0.005	0.008	0.012	0.025

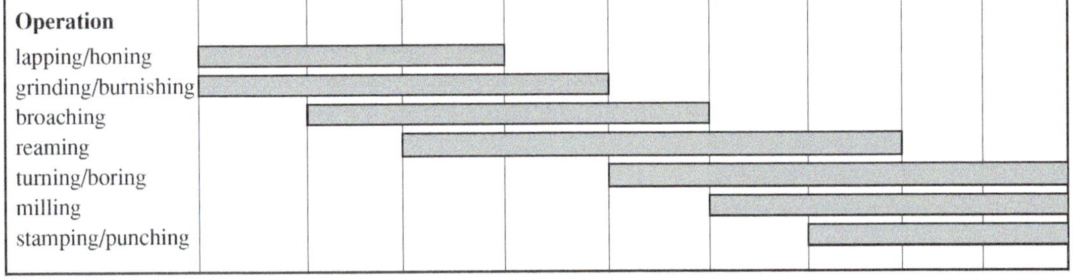

Figure B-3 Range of tolerances for various machining operations.

Typical tolerances on drills depend on the drill size. Standard drill sizes are fractional in increments of 1/64 inch and range from 1/16 inch to 4 inches. In addition, drill sizes are specified according to an alphanumeric code. The coded drill sizes range from 80 (.0135 inch) to 1 (.2280 inch) and from A (.234 inch) to Z (.413 inch). The dimensions for the coded drill sizes are widely available in engineering and machinist handbooks. In addition, the proper drill sizes for drilling holes that will be tapped are listed in these handbooks. The standard tolerances for drilled holes are given in Table B-1.

Table B-1 Standard Tolerances for Drilled Holes

Drill Size (In.)	Tolerance (In.)	
	Plus	Minus
0.0135 (80)–0.185 (13)	0.003	0.002
0.1875–0.246 (D)	0.004	0.002
0.250–0.750	0.005	0.002
0.7656–1.000	0.007	0.003
1.0156–2.000	0.010	0.004
2.0312–3.500	0.015	0.005

B.4 Fits

The degree of tightness or looseness between mating parts is known as the fit. This concept is most easily explained in terms of a shaft and a hole, although the concept can be applied to other mating parts in which an internal member is designed to fit within an external member. For example, a shaft that is .750 inch in diameter will fit into a hole that is .751 inch in diameter with .001 inch of clearance, but it will be too large for a hole that is .749 inch in diameter. But the situation is more complicated than this because the tolerances on the diameters of both the shaft and the hole play roles in the fit. Consider the example shown in Figure B-4 of a hole and a shaft, each with a basic diameter of .750 inch.

clearance fit

clearance

allowance

In a **clearance fit,** the internal component fits in the mating part so that there is always a space or gap between the parts. In Figure B-4, the largest shaft size for a clearance fit is .748 inch and the smallest hole size is .750 inch, so there is always a space or **clearance** of at least +.002 inch. This space is called the **allowance.** It is always positive for a clearance fit. The gap may be even larger. The smallest shaft in the largest hole results in an allowance of +.006 inch. A clearance fit is used when a shaft is meant to rotate or slide within the hole.

interference fit

least material condition

maximum material condition

press fit

shrink fits

An **interference fit** occurs when the internal component is larger than the mating external member, so that the two parts will interfere with one another if they are assembled. For the interference fit in Figure B-4, the smallest shaft, which is .753 inch, is too big to fit into the largest hole, which is only .752 inch. In this case, the allowance is -.001 inch, a negative number to indicate interference. This is known as the **least material condition**, or LMC, for which the least material for the shaft and the block with the hole are present. In other words, both the shaft and the portion of the block left over after the hole is cut are as small as possible. For the block, this corresponds to the largest hole. At the opposite extreme is the **maximum material condition**, or MMC, for which the maximum amount of material is present for the shaft and the block with the hole. For this condition, which corresponds to the largest shaft (.755 inch) and the smallest hole (.750 inch), the allowance is -.005 inch. To assemble these parts requires either using force to press the shaft into the hole, which is called a **press fit**, or using heating or cooling to stretch the hole or shrink the shaft. The latter are called **shrink fits**. They are often accomplished by heating the hole to expand it or cooling the shaft to shrink it, assembling the shaft in the hole, and then allowing the assembly to return to room temperature. Interference fits can be quite useful in design applications to mechanically fasten parts together without the use of fasteners or adhesive.

transition fit

A **transition fit** results when the tolerances are such that the fit can be either a clearance fit or an interference fit when the parts are assembled. The transition fit in Figure B-4 is a clearance fit if the shaft has a dimension of .749 and the hole is .752 (the least material condition), but it is an interference fit if the shaft is .751 and the hole is .750 (the maximum material condition). Transition fits are sometimes used to position or locate one part with respect to another.

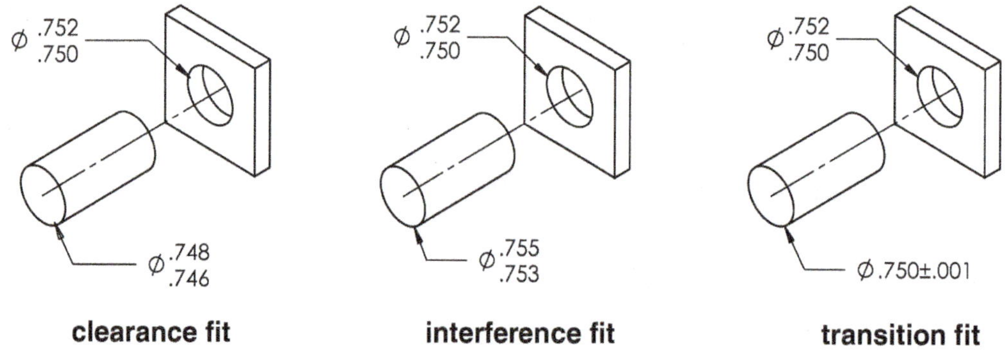

Figure B-4 Clearance, interference, and transition fits.

ANSI, ISO, and other organizations have issued engineering standards for a variety of types of shaft-in-hole fits to aid engineers in designing for particular applications. For instance, "running and sliding fits" prescribe the clearances for situations in which a shaft should rotate (running) or slide in a hole. There are a variety of running and sliding fits ranging from "close sliding fits" in which the clearance is quite small so that there is no perceptible play between the parts to "loose running fits" in which the clearance is quite large when accuracy is not essential. The engineering standard indicates the optimal clearance to be specified for the particular type of running or sliding condition that is required. There are also engineering standards for "locational fits," which are used to position one part with respect to another, and for "force and shrink fits" to fasten parts together.

B.5 Surface Finish

roughness

waviness

lay

All surfaces are rough and irregular on a microscopic scale. The roughness can be described in terms of asperities (peaks or ridges) and valleys. The irregularities of a surface are classified into roughness and waviness. **Roughness** describes the smallscale, somewhat random irregularity of a surface, while **waviness** describes a more regular variation in the surface that typically has a wavelength of 0.8 mm or more. The surface can be thought of as a roughness superimposed on waviness. The roughness and waviness can be measured in several ways, but the most common is with a profilometer, an instrument that drags a fine-pointed stylus over the surface and records the height of the stylus. The roughness is measured in terms of the roughness average (R_a). The roughness average is the average height of peaks above or valleys below the local surface height. More precisely, it is defined as the average of the absolute values of the deviation of the local surface height from the average height of the surface. This is equivalent to half the average of the peak-to-valley height. In addition, the **lay**, or predominant direction of the marks left by the machine tools, may be important in describing the surface.

The surface finish resulting from several machining and forming processes are shown in Figure B-5. This figure shows the range of possible surface roughness averages (R_a) in both micrometers (1 μm = 10^{-6} m) and microinches (1 μin = 10^{-6} in). For instance, surfaces formed by drilling are likely to have a roughness from 12.5 μm down to 0.8 μm (500 μin down to 32 μin). Typically, the roughness is in the middle of the specified range, about 6.3 μm down to 1.6 μm for the case of drilling. In addition to the machining processes described in Section B.3, the following processes are also included in Figure B-5.

sawing

polishing

sand casting

forging

permanent mold casting

Sawing: Cutting material by moving a toothed blade through the material.

Polishing: Removal of scratches and tool marks on a surface using a belt or rotating wheel of soft material such as cloth or leather, often with a fine abrasive.

Sand Casting: Forming a part by pouring molten metal into a mold made of sand.

Forging: Shaping of metal to a desired form by pressing or hammering, usually with the metal hot.

Permanent Mold Casting: Forming a part by pouring molten metal into a permanent mold.

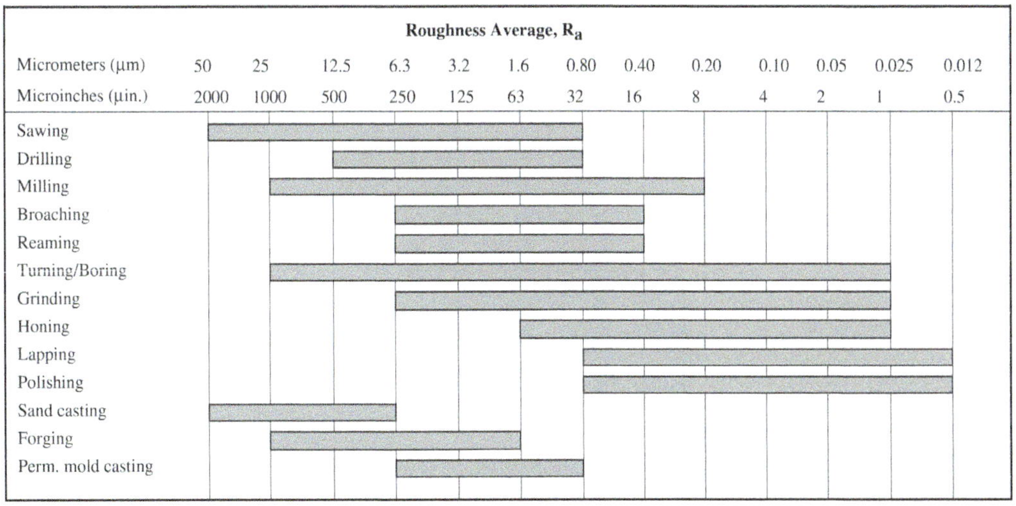

Figure B-5 Surface roughness for various machining operations.

To put the roughness averages in perspective, consider that a 0.10-μm roughness is a mirrorlike surface, free from visible marks of any kind. A 1.6-μm roughness is a high-quality, smooth machined finish. A 6.3-μm roughness is the ordinary finish from typical machining operations.

The surface finish may be specified on a drawing using the special symbols touching the surface or with a leader line to the surface as shown in Figure B-6. For the top surface, the roughness is specified as 1.2-μm corresponding to a high-quality, smooth machine finish. The surface finish symbol on the right side of the object includes a horizontal bar that indicates that material removal by machining is required for this surface. The surface finish symbol on the left side of the object indicates that material removal from this side is prohibited. This surface must be the one resulting from the original process used to produce the surface, such as casting, forging, or injection molding.

Other notations can be added to the surface roughness symbol. A number to the left of the surface finish mark describes the material removal allowance indicating the amount of stock material to be removed by machining. A symbol to the right of the surface roughness mark indicates the lay, or predominant direction of machining marks: perpendicular, parallel, circular, or radial. Other notations can be used to specify the waviness. Details of surface finish notation are available in engineering and machinist handbooks.

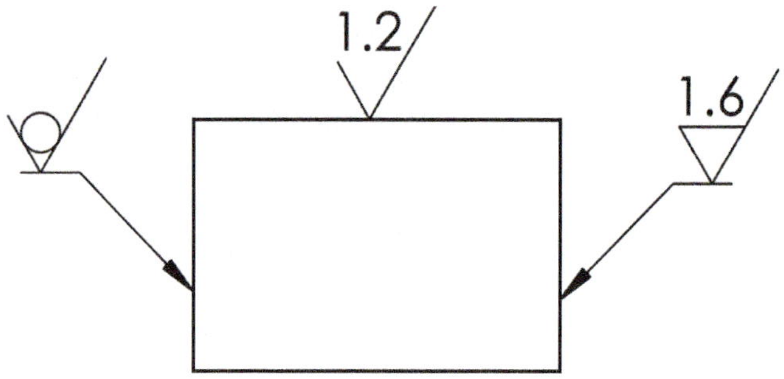

Figure B-6 Surface roughness symbols.

B.6 Geometric DImensioning and Tolerancing

Ideally, engineering drawings should be unambiguous in their interpretation. In many cases, stating dimensions and tolerances provides adequate information so that the part can be manufactured to ensure interchangeability of parts and optimal performance. But sometimes the traditional dimensioning using the plus and minus scheme does not adequately describe the geometry of the part. For instance, consider the part shown in Figure B-7. The dimension shown in the drawing could have several possible interpretations. For instance, the dimension at the top edge could be 14.9 mm while the dimension at the bottom edge could be 15.1 mm, resulting in a trapezoidal part instead of a rectangular part. Or the right edge of the part could be bowed so that the part is 15.1 mm wide at the top and bottom, but 14.9 mm wide at the middle. As a result, the part would not fit flush against another flat surface even though the part is 15 $\pm$.1 mm wide at any particular location. In both of these cases, the geometry of the part is not described fully by the plus and minus tolerances on the dimension.

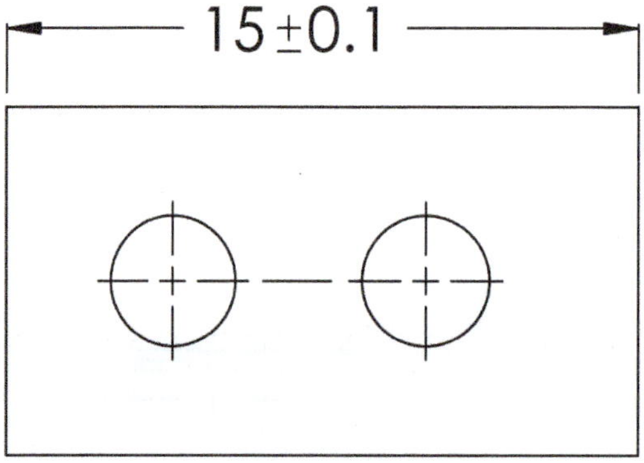

Figure B-7 Ambiguity can result from dimensioning as shown.

geometric dimensioning and tolerancing

A system known as **geometric dimensioning and tolerancing** is used to prevent such ambiguities in the design by presenting the design intent more clearly. Geometric dimensioning and tolerancing supplements the traditional dimensioning system with specifications that describe the geometry of the part to enhance the interpretation of the drawing. This system is used to indicate dimensions critical to a part's function and ensure part interchangeability. Items related to geometric dimensioning and tolerancing are indicated on a drawing by a feature symbol consisting of a frame or box around the information describing the geometry. A complete discussion of geometric dimensioning and tolerancing is beyond the scope of this book—there are entire books on the subject. Nevertheless, several of the key aspects and applications of geometric dimensioning and tolerancing are presented here to familiarize the reader with the basic terminology and symbols that are used.

form tolerances
straightness
flatness
circularity
roundness
cylindricity
profile tolerances

The various types of tolerances and the characteristics that they describe are indicated in Figure B-8. Each has its own symbol that is used on a drawing. Perhaps the easiest type of geometric tolerances to understand are the **form tolerances,** which describe the shape of a single feature. For instance, straightness indicates the limits on how much a surface or axis can bow with respect to a straight line. Figure B-9 shows the **straightness** specification for a surface. A feature control box with the straightness symbol (a horizontal bar) and a numerical dimension are connected by a leader line to the surface. The interpretation is that the upper surface must be straight enough so that all points on the surface are within the tolerance zone of 0.010 mm. An example of a **flatness** specification is shown in Figure B-10. The parallelogram symbol with a numerical dimension in a feature control box indicates that the planar surface must lie between two parallel planes 0.20 mm apart. Other single feature specifications include **circularity** or **roundness** (for circular features to be within a tolerance zone defined by two concentric circles), and **cylindricity** (for cylindrical features to be within a tolerance zone defined by two concentric cylinders). Closely related to form tolerances are **profile tolerances,** which define the acceptable deviation of the outline of an object (profile) from that specified in the drawing. These tolerances define a zone on either side of the true profile shown on the drawing within which the line or surface should remain. Profile tolerances can be applied to either a line or a surface.

orientation tolerances
parallelism

Orientation tolerances prescribe the relation between features. **Parallelism** controls the degree to which a surface is parallel to a reference plane, or datum plane.

	TYPE OF TOLERANCE	CHARACTERISTIC	SYMBOL
FOR INDIVIDUAL FEATURES	FORM	STRAIGHTNESS	—
		FLATNESS	▱
		CIRCULARITY (ROUNDNESS)	○
		CYLINDRICITY	⌭
FOR INDIVIDUAL OR RELATED FEATURES	PROFILE	PROFILE OF A LINE	⌒
		PROFILE OF A SURFACE	⌓
FOR RELATED FEATURES	ORIENTATION	ANGULARITY	∠
		PERPENDICULARITY	⊥
		PARALLELISM	//
	LOCATION	POSITION	⌖
		CONCENTRICITY	◎
		SYMMETRY	⌯
	RUNOUT	CIRCULAR RUNOUT	↗
		TOTAL RUNOUT	⌰

Figure B-8 Types of tolerances and their symbols.

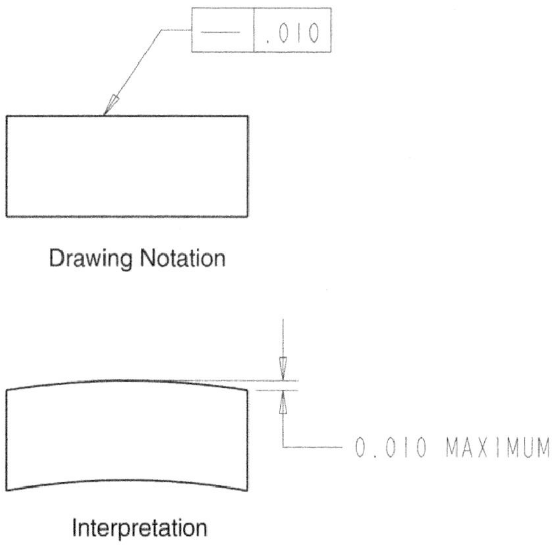

Figure B-9 Straightness specification.

perpendicularity
angularity
concentricity
symmetry
runout

In Figure B-11, the horizontal datum plane that coincides with the bottom edge of the part is identified with a box around B attached to the bottom edge. The parallelism feature symbol (parallel lines), the numerical tolerance, and the reference datum plane are included in a feature control box with a leader line to the surface to which the tolerance applies. In this case, the upper surface must be within 0.10 mm of parallel to datum plane B. **Perpendicularity** with respect to a datum plane is indicated in a similar way, as shown in Figure B-12. Here the vertical plane must be perpendicular to datum plane A to within 0.20 mm. In this example, an alternative notation for specifying datum plane A is used. The datum plane symbol used in Figure B-12 is common, but the datum plane symbol used in Figure B-11 and Figure B-13 is preferred. The datum plane symbol can be applied in several different ways as shown in Figure B-11 and Figure B-13. Several other relations between features can be specified, including the tolerance zone for **angularity** from a datum plane or axis, the tolerance zone for **concentricity** of a surface of revolution with respect to an axis, the tolerance zone for **symmetry** with respect to a datum plane, and the tolerance zone for **runout**, which describes the circularity with respect to an axis of revolution.

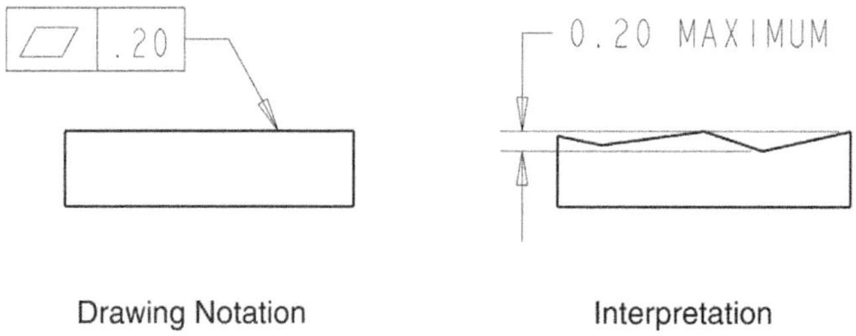

Figure B-10 Flatness specification.

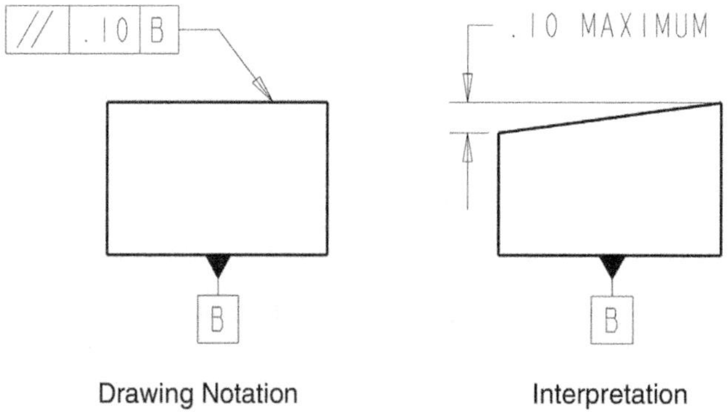

Figure B-11 Parallelism specification.

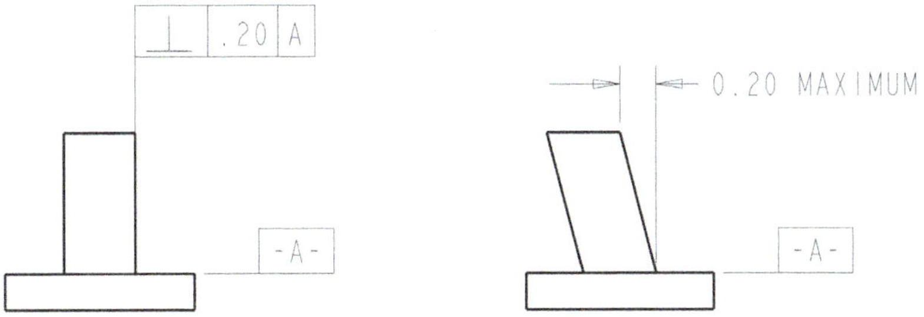

Figure B-12 Perpendicularity specification.

positional tolerances

The most difficult concept in geometric dimensioning and tolerancing is the concept of **positional tolerances**. It is demonstrated most easily in indicating the position of a hole, as shown in Figure B-13a. In this case, the ultimate goal is to have a circular open area (hole) that is at least 25.0 mm in diameter with its center exactly at the position indicated in the drawing. Note that even though we want a 25.0-mm hole, the hole is dimensioned as 25.3 mm in the drawing. The reason for this will shortly become evident.

true-position dimensions

Rather than describing the position of the hole with a plus-or-minus tolerance, the ideal position of the hole is prescribed. The boxes around the 40 mm and 60 mm dimensions indicate that they are basic **true-position dimensions**, which describe the perfect position of the center of the hole. Then the feature control box below the dimension of the hole shows how close the center of the hole should be to the true position. In this case, the hole should be within a circular tolerance zone with a diameter of 0.1 mm of the true position (indicated by the plus in the circle within the feature control box followed by Ø.1). The reference datum planes are also included in the feature control box. Datum plane A is the surface of the part in the plane of the page. Datum planes B and C are references for the true-position dimensions of 40 mm and 60 mm. These three mutually perpendicular datum planes provide a reference point for the positional tolerance of the hole.

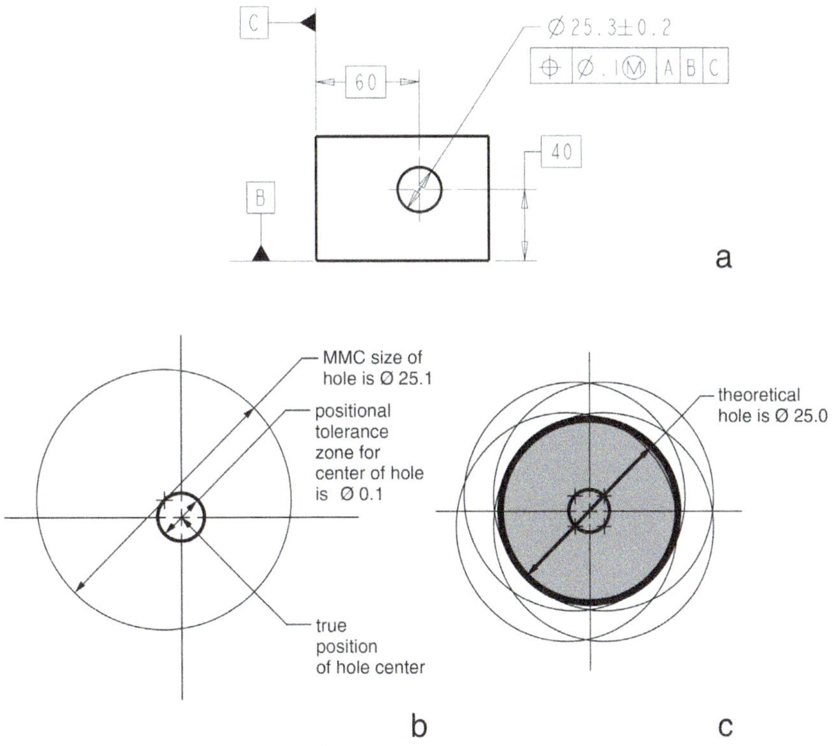

Figure B-13 Positional tolerances..

The positional tolerance zone for the center of the hole is shown in Figure B-13b. The "true position of the hole center" is based on the 40 and 60 mm dimensions. Around this center is a circle with a diameter of 0.1 mm (not drawn to scale in Figure B-13b) that defines the "positional tolerance zone." The center of the actual hole must be within this circle. For instance, an acceptable hole would be a 25.1-mm diameter hole with its center on the edge of this circular tolerance zone, as shown in Figure B-13b.

maximum material condition

The circled M in the feature control box in Figure B-13a indicates that the 0.1-mm circular positional tolerance zone is applied at the **maximum material condition** (MMC). For a hole, the maximum material condition is the smallest acceptable dimension of the hole, because this is the situation in which the most material would be present. For the part in Figure B-13a, the diameter of the hole for the maximum material condition is 25.1 mm, given a dimension of 25.3 ± 0.2 for the diameter of the hole. Thus, the circle representing a hole with its center on the edge of the positional tolerance zone has a diameter of 25.1 mm, as shown in Figure B-13b. Of course, the 25.1-mm-diameter hole could be anywhere within the positional tolerance zone or on its edge. Three more 25.1-mm-diameter circles are drawn in Figure B-13c in addition to the original 25.1-mm circle. The centers of all four circles are on the edge of the positional tolerance zone. Note that there is a circle that lies inside all of the 25.1-mm-diameter circles, which is called the **theoretical hole**. It is relatively straightforward to find the diameter of the theoretical hole, based on the positional tolerance zone for the center of the hole and the MMC size of the hole. The radius of the theoretical hole is the distance from the true position of the hole center to the nearest portion of any one of the MMC circles. Here the theoretical hole has a diameter of 25.0 mm. In fact, the theoretical hole always has a diameter equal to the diameter for the maximum material condition minus the diameter of the positional tolerance (in this case, 25.1 mm – 0.1 mm = 25.0 mm).

theoretical hole

The key result is that if the hole diameter is within the specified tolerance (25.3 ± .2 mm) and the center of the hole is within the positional tolerance zone (.1 mm), there will always be a circular opening in the part with a diameter of at least 25.0 mm. Now assume that the hole must mate with a circular pin that is positioned at the true center of the hole and is exactly 25.0 mm in diameter. Given the dimensioning scheme in Figure B-13a, the pin will always align with the theoretical hole and fit through the theoretical hole, even if the actual hole is not positioned so that its center is exactly at the true position of the hole center. Of course, the theoretical hole is based on the maximum material condition (smallest hole allowed). If the hole is at the large end of the acceptable tolerance, 25.5 mm in diameter, circles drawn in Figure B-13c will be larger. But the theoretical hole will still be open for the pin. Likewise, if the circles drawn in Figure B-13c are not on the edge of the positional tolerance zone, but inside of the zone, the entire theoretical hole will still be open for the pin. Thus, defining the position and diameter of the hole as in Figure B-13a forces the existence of an open area (theoretical hole) that is at least 25.0 mm in diameter and centered at the true position of the hole center.

The maximum material condition reflects the practical aspect of assembly. Consider two parts that are bolted together with several bolts. If the bolt holes are not accurately positioned, not all of the bolts can be inserted. Either drilling out the holes to make them larger or using smaller diameter bolts fixes the problem. The idea of the maximum material condition is to capture these trade-offs. The maximum material condition indicates the inner boundary (the theoretical hole) for the actual hole. The boundary of the actual hole must always be outside of this inner boundary. In this way, a bolt or pin that is positioned at the true position of the hole will always fit through the actual hole. Likewise, the least material condition, or largest hole size, can be used to define a theoretical circle within which the actual hole will be. The boundary for the actual hole must lie inside of the theoretical circle as defined by the LMC. Further information regarding geometric dimensioning and tolerancing is provided in engineering handbooks and graphics textbooks.

Appendix C Technology Timeline

5400 B.C.

3500 B.C.

221 B.C.

Fire and the development of early stone tools	**Stone Age** 1,000,000 B.C.
Early oil lamps that burned fish and animal oil	12,000 B.C.
Wheel	4,000 B.C.
Writing (cuneiform and hieroglyphics)	**Agricultural Age** 3500 B.C.
Tools made from metal	**Bronze Age** 3000 B.C.
Smelting of iron; development of the Greek alphabet	**Iron Age** 1200 B.C.
Irrigation and sewers in Rome	700 B.C.
Construction of the Great Wall of China	221 B.C.
First high-level aqueduct in Rome; use of cement	144 B.C.
Water wheels	100 B.C.
Sailing ships that could travel with and against the wind	A.D. 500
Windmills	A.D. 700
Chinese use first paper currency	A.D. 1024
Chinese use gunpowder in war against Mongols	A.D. 1281
Age of Exploration	A.D. 1400–1600

1450	Gutenberg invents movable type
1450	Printing press (Gutenberg); Anemometer (measuring wind speed)
1600	Magnetism discovered by William Gilbert
1608	Telescope
1610	Galileo's telescope
1628	William Harvey discovers circulation of the blood
1687	Discovery of gravity (Newton)
1698	Steam engine
1712	**Newton's Laws**
1714	Thermometer
1724	Newcomen's steam engine
1752	Fahrenheit thermometer invented
Industrial Age 1759	**Benjamin Franklin proves lightning is electricity**
1760–1840	Industrial Revolution in Britain
1769	**James Watt's steam engine**

1712

1759

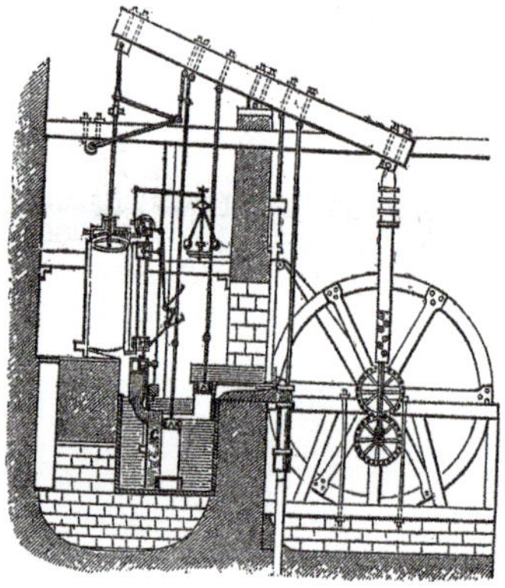

1769

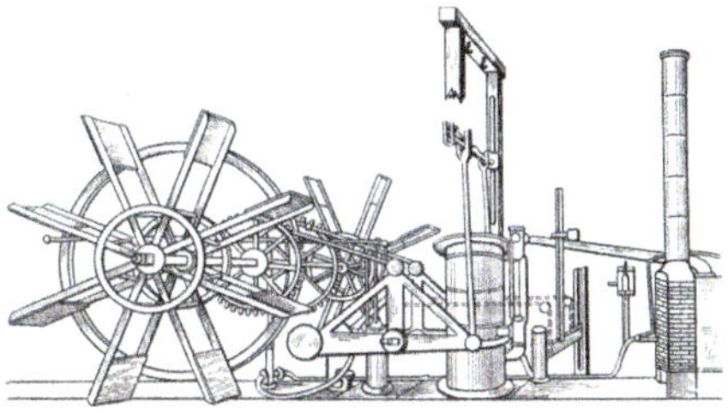

1807

1843

Hargreaves's spinning jenny patented — 1770

Wrought-iron process by Henry Cort — 1784

Cartwright's loom — 1785

LaVoisier's theory of chemical combustion — 1789

Eli Whitney's cotton — 1793

Battery — 1799

Jacquard's loom — 1804

Robert Fulton's steamboat — 1807

First textile mill in United States, in Rhode Island — 1822

The steam locomotive — 1829

Telegraph
Electromagnetic induction — 1830

Bessemer process for making steel — 1835

Photography invented — 1839

Morse begins telegraph line between Baltimore, MD and Washington, DC — 1843

Dynamite — 1867

1869	Periodic table
1876	Alexander Graham Bell invents the telephone
1878	Typewriter
1879	Light bulb (Edison)
1885	First successful gasoline-driven car, by Carl Benz in Germany
1887	Daimler's internal combustion engine automobile
1888	**First Kodak handheld camera**
1893	Diesel engine invented by Dr. Rudolf Diesel
1895	Radio (Marconi) X-rays
1901	Marconi sends first transatlantic radio signals
1903	**Wright brothers' airplane**
1905	Albert Einstein proposes his theory of relativity
1908	Model T Ford built using mass production
1909	Beginning of age of plastics; Synthetic rubber
1915	SONAR invented

1903

Kodak camera (1888: Model shown is from 1934)

Automatic traffic light patented by Garrett A. Morgan — 1923

First round-the-world flight; Kleenex tissues — 1924

First successful experiments with hydroponics — 1925

First television demonstration; pop-up toaster — 1926

Television; Refrigerator — 1927

Penicillin invented by Alexander Fleming — 1928

Electroencephalograph (EEG); foam rubber — 1929

Empire State Building in New York City — 1931

1931

The electron microscope — 1934

Radar developed; nylon patented; B-17 bomber produced — 1935

The ballpoint pen is patented — 1938

First jet aircraft flies in Germany — 1939

First color snapshots; Nuclear reactor — 1942

First automatic digital computer, the Mark I — 1944

ENIAC computer built using 18,000 vacuum tubes; Atomic bomb — 1945

1947 — Microwave oven; first Honda motorcycle; development of the transistor at Bell Laboratories

1950 — **Power steering; start of regular color TV broadcasts; first credit card (Diner's Club)**

1952 — **Salk polio vaccine; 3-D movies; Mr. Potato Head**

1954 — Nuclear-powered submarine "Nautilus"; photocell developed

1955 — Nuclear-powered electricity generation; fiber optics

1956 — First desktop computer (Burroughs E-101)

1957 — Sputnik; Fortran (computer language)

Information Age 1958 — First U.S. satellite

1959 — Russia sends unmanned spacecraft to the moon (Luna 2); Xerox copier

1960 — Lasers; light-emitting diodes (LEDs); weather satellites

1964 — China tests nuclear bomb; IBM word processors

1965 — First space walk

1966 — Electronic fuel injection; Russians make successful soft landing on the moon (Luna 9)

1969 — **American manned moon landing; first Concorde SST flight; Laser**

1971 — Microprocessor developed by Intel; Mariner 9 orbits Mars; Email

1950

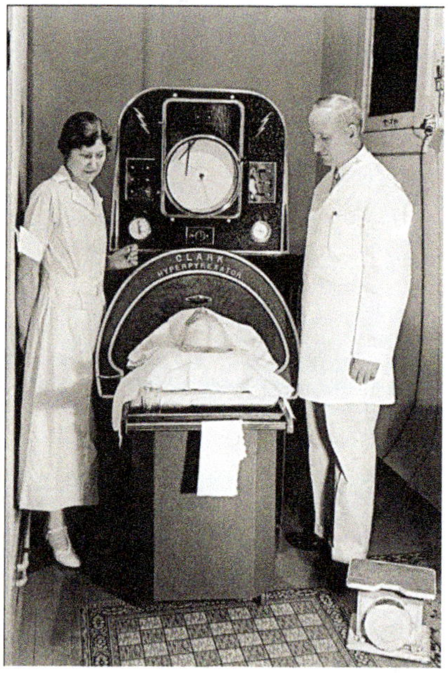

1952

1969

1977

1984

CAT scan; photography from satellites (Landsat); videodiscs; C programming language — **1972**

Genetic engineering; Skylab orbiting space station launched; Mobile phone — **1973**

Toronto Communications Tower (tallest building in the world) — **1974**

Birth of the Apple computer; electronic cameras — **1976**

Trans-Alaska pipeline system completed — **1977**

Skylab, the orbiting space station, falls back to Earth — **1979**

Solar-powered aircraft, First wind farm — **1980**

First launch of a reusable space vehicle (space shuttle *Columbia*) — **1981**

Artificial heart — **1982**

C++ programming language — **1983**

Compact disc player — **1984**

DNA fingerprinting — **1986**

World Wide Web — **1989**

NASA's Hubble Space Telescope deployed by space shuttle *Discovery;* Search engine — **1990**

Playstation — **1994**

1997	eBay online auction and shopping website
1998	Google
1999	First human chromosome completely sequenced
2001	Xbox video game console; iPod and iTunes
2001	Worldwide introduction of the Toyota Prius hybrid car
2002	**First production of Segway electric vehicle**
2003	MySpace website launched
2004	World of Warcraft massively multiplayer online game released; Nintendo DS
2004	**Facebook social networking website launched**
2005	Sony PlayStation Portable released in North America
2005	YouTube.com launches
2006	SpaceX Falcon 1 launched
2007	iPhone
2007	Google search engine becomes the most visited website
2008	Mapping and sequencing of eight human genomes

2002

2004

Tesla Roadster and Chevrolet Volt electric cars developed — **2008**

3D Television; iPad; Instagram — **2010**

Snapchat — **2011**

Raspberry Pi — **2012**

PlayStation 4 (PS4) — **2013**

Driverless cars; Amazon Alexa — **2014**

Drone use for package delivery — **2015**

Pokemon Go — **2016**

Fortnite — **2017**

Oculus 2 virtual reality headset — **2020**

PS5 — **2020**

2008

2015

APPENDIX D Technology Student Association (TSA)

The Technology Student Association (TSA) is a national organization of students engaged in science, technology, engineering, and mathematics (STEM). Open to students enrolled in or who have completed technology education courses, TSA's membership includes over 250,000 middle and high school students in approximately 2,300 schools spanning 48 states. TSA is supported by educators, parents, and business leaders who believe in the need for a technologically literate society. Members learn through exciting competitive events, leadership opportunities, and much more. The diversity of activities makes TSA a positive experience for every student. From engineers to business managers, our alumni credit TSA with a positive influence on their lives.

Chapters

TSA chapters take the study of STEM (science, technology, engineering, and mathematics) beyond the classroom and give students the chance to pursue academic challenges among friends with similar goals and interests. Together, chapter members work on competitive events, attend conferences at the state and national levels, and have a good time raising funds to get there. Chapter organization develops leadership, as members may become officers within their state and then run nationally. Our chapters are committed to a national service project and are among the most service-oriented groups in the community.

To start a TSA chapter, all that is needed is a motivated teacher and an eager group of students. An expensive array of high-tech equipment is not necessary.

Competitions

Imagine an activity so captivating that your students spend hours working on it after school for weeks at a time. That's what happens when TSA members engage in TSA's competitions. Only TSA members could compete at exciting state conferences and then at the annual National Conference. Expert judging by technology educators and industry representatives inspires the best from participants. Members are rewarded not only with medals or trophies, but also with memories of the camaraderie and the challenge of a state or national TSA conference.

Middle School and High School Level Competitions

TSA offers competitions at both the middle and high school level. Students can work individually or as a team in such areas as agriculture and biotechnology, website design, dragster design, flight, geospatial technology, CAD with 3-D modeling, digital video production, extemporaneous speaking, structural design and engineering, and scientific visualization.

Each year TSA brings together leading technology educators to revise, review, and develop competitions at the middle or high school level. Every competitive event is correlated to STEM concepts, and the criteria for evaluating each competition is detailed. TSA competitions (approximately 30 on the middle school level and 30 on the high school level) can be implemented at the local, state, and national level.

Once affiliated, your chapter has access to Total TSA, which includes full guidelines for competitions, as well as correlations to STEM standards; leadership activities, with suggestions for using the activities with specific events; and, membership materials; this section provides the information you need to know for running and maintaining your chapter.

Many state TSA delegations host an annual competitive event conference, as well as a leadership conference. These conferences give student members the opportunity to test their technology skills, and many students then choose to compete at the national level conference. Membership in TSA is required in order to participate at any TSA-sanctioned event at the local, state, or national level.

Whether working on an activity at the local level, or competing at the state or national level, all students who use TSA materials benefit from being a TSA member. Opportunities exist for chapter, state, and national leadership roles through elected offices and working committees. Networking among chapters is encouraged.

Curriculum Integration—Meeting STEM Criteria Through TSA Participation

Education directives at the federal and state levels focus on providing avenues for initiatives such as STEM (Science, Technology, Engineering, and Mathematics) education. TSA helps teachers meet the criteria for STEM education goals through exciting programs that include technology activities, competitions, leadership, and teamwork for students. In addition, all TSA activities are correlated to the Standards for Technological and Engineering Literacy, as set forth by the International Technology and Engineering Educators Association (ITEEA). Teachers can use TSA's activities during the school day in their classroom and extend the learning beyond the classroom through a TSA chapter.

Advantages of TSA Membership

Perhaps the most important benefit of TSA membership is the inspiration and enthusiasm that students gain from receiving recognition for applying their knowledge. Demonstrating skills in a forum beyond school walls motivates students, especially when there is a large, like-minded audience of peers who support them. Other benefits include:

- Enhancing technological literacy and leadership skills
- Working on complex activities from start to finish
- Competing on local, state and national levels
- Developing leadership skills
- Meeting and working with business and industry leaders
- Attending organized meetings
- Networking with students and teachers with similar interests from all over the United States
- Contributing time and effort to a national service project
- Receiving national membership services
- Receiving recognition through TSA's honor society, scholarships, and achievement programs
- Participating in a national organization that is helping to shape the future of technology education

TSA Chapter Advisors

Serving as a TSA chapter advisor at your school means more than having a "technology club." It means your school is a part of a national organization that promotes and recognizes excellence in technology students through applied science, technology, engineering, and mathematics (STEM) competitions and leadership activities.

For most TSA advisors, TSA is not the only organization they are involved with. Many also coach sports, serve as class advisors, or work on other school activities. So why do they also advise a TSA chapter? Most would tell you that they do so because they feel TSA offers:

- Leadership opportunities
- Teamwork experiences
- Challenging competitions
- A sense of belonging to a group of students interested in STEM areas

How Much Time Does It Take?

There is no question that being an advisor for a TSA chapter (or any other student activity) takes time and commitment. How much time is a function of several variables:

- **Co-curricular or extra-curricular.** TSA is designed so that it can be a part of a technology classroom or can be an after-school activity.
- **Program of work.** This is the basis of chapter activities for the year. The program of work is designed to accommodate the amount of time the advisor and members are willing to devote to the endeavor. Considerations for the program of work include competitions, fundraising (if needed), service projects, and social events.
- **Organization.** Organizing your technology program around TSA can save time. TSA is designed to be student run through elected officers—this is often a great classroom management tool.

More information about the Technology Student Association and how to start a chapter can be found at *www.tsaweb.org* or by contacting the national TSA office via email at *general@tsaweb.org* or toll free at 888-860-9010.

Tests of Engineering Aptitude, Mathematics, and Science (TEAMS)

TEAMS is a one-day competition for students in middle and high school that allows them to apply their knowledge of skills in science, technology, engineering, and mathematics (STEM) to issues facing our global society. Focused on an annual theme, original academic and innovative concepts are developed for the competition based on the National Academy of Engineering's Grand Challenges. Engaging in the challenges requires critical job-readiness skills such as teamwork, analytical thinking, and multi-dimensional problem solving.

How TEAMS works:

- Teams of 4 to 8 students work together to answer multiple choice and short essay questions based on a National Academy of Engineering Grand Challenge theme.
- The competitions occur at a host university or at your own school during the competition window in February and March each year.
- Recognition includes participation certificates and state and national awards.
- Top ranking teams qualify to compete in the national TEAMS competitions, held each year during the national TSA conference at the end of June.

Why teachers value TEAMS:

- The TEAMS competition is creative, exciting and theme-based, with problems designed to be solved in teams—the way engineers work.
- TEAMS resources and competition materials are designed and tested by experienced, credentialed engineering professors and professionals.
- There is no need for extra materials, equipment, computers, or supplies.
- Competitions can occur at your own school—there is no need to travel.

Through TEAMS, your students gain:

- Problem solving abilities
- Collaboration and team-building experience
- An understanding of the role of engineering in solving societal, environmental, community, and global problems
- An opportunity to explore engineering as a profession
- Exposure to STEM disciplines outside the classroom

The TEAMS competition is managed by the Technology Student Association (TSA). More information about TEAMS can be found at *teams.tsaweb.org* or via email *teams@tsaweb.org* or toll free at 888-860-9010.

Glossary

GLOSSARY G

A

ABET Accreditation Board for Engineering and Technology

absolute pressure the pressure when one uses a perfect vacuum as a reference point

accreditation process that involves experts in a particular area of engineering visiting an institution that offers the discipline as a major and reviewing the quality of the program

accuracy a measure of how close a calculation or measurement (or an average of a group of measurements) is to the actual value

actions any executable steps in an algorithm that do not require a decision to be made

additive manufacturing involves building a product by adding layer upon layer of material, such as a thermoplastic, representing cross-sections of the product until the product is fully assembled

algorithm step-by-step description of accomplishing a task

aligned parts or sections of drawings that directly line up with each other

alloy a mixture of two or more metals

ampere the unit of measurement for electrical current

angel investors individuals who invest at the very beginning of a company, before stock is even available to the public and other traditional investors or lenders are unwilling to commit resources

aqueduct a channel built to carry water

arguments the list of input variables to a function

Arithmetic Logic Unit (ALU) the part of the Central Processing Unit (CPU) which takes care of math and operations involving integers

artificial intelligence any use of machines to imitate human intelligence

assembly drawings drawings that show how the components of a design fit together

assembly line a system in which each worker performs a specific operation in assembling an item as it is passed along, often on a slowly moving belt, from one workstation to the next

assignment operator tells the computer to put a value into a variable

atmospheric pressure the surface pressure when the gas above an object is the air in the Earth's atmosphere

augmented reality retains the real world in an immersive environment, but augments it with an overlay of additional, valuable information through a digital medium

automation a system in which machines are controlled automatically

auxiliary view a pictorial drawing that shows a straight-on view of one surface of an object

average an estimate of the center value (or typical value) of a data set. The most common average is found by adding all the numbers in the data set and dividing the result by the number of items in the data set.

B

balancing feedback loop a closed-loop system that aims to stabilize the output

behavioral interview question attempts to understand how you addressed different situations in the past, including difficulties at work

Best Practice guidelines a comprehensive list of safety topics provided by the Occupational Safety and Health Administration (OSHA)

bill of materials list of material required for a component or assembly, number required in the assembly, and other information

bin a single column in a histogram

binary behavior any sort of behavior that results in two and only two possible outcomes

biomimetics leveraging ideas from nature to solve engineering design problems and create systems; also referred to as biomimicry

BIOS low-level software that runs with the hardware of a computer system to facilitate data flow between the computer's operating system and attached devices

block comment a block of non-executed text between the special block characters %{ and %} that allow the user to type full length paragraphs as a comment without needing to type the % symbol before each sentence

Boolean logic a form of algebra that allows us to communicate instructions to computer systems; each value is either True or False and can be combined with logical operations to express more complex instructions

brainstorming a method of coming up with alternative designs

breakpoints allow you to check your agreement with the program at various stages when using the Editor/Debugger in MATLAB

Brinell hardness test a mechanical property test measuring hardness; used for materials that have surfaces that are rough or coarse

British thermal unit the amount of heat required to raise the temperature of one pound-mass of water by one degree Fahrenheit

Bronze Age the period during which people began to craft tools and weapons from bronze rather than from stone

business communication platforms tools that include text, voice, and video communication in addition to file sharing

byte 8-bits of data

C

cache a small amount of memory in a Central Processing Unit (CPU) which can be accessed very fast, designed for the registers to pull data from

capacitance units of coulombs per volt

capital anything that increases an individual's, group's, or company's ability to generate value

Celsius a temperature scale developed by Anders Celsius (1701–1744), a Swedish astronomer, that uses a freezing point of 0°C and a boiling point of 100°C

centerlines thin, long-dash/short-dash lines that mark axes of rotationally symmetric parts or features

Central Processing Unit (CPU) the computer's brain, or processor; most likely the largest item on the motherboard

central tendency the center value of a data set (generally, for what's called a probability distribution), which is often the average

charge known as (Q), it is generally quantified using the derived unit coulomb [C] , a property of matter that determines how electric and magnetic fields affect particles

Charpy V-Notch test a mechanical property test measuring a material's toughness, determining if it is brittle or ductile; involves measuring how much energy it takes to fracture a material

chemical property describes how a material changes from one substance to another

closed-loop system a system that can respond and change its inputs or outputs

Code of Ethics a guide to ethical decision-making that is established by a group, such as a professional community, company, or family. These often describe recommended courses of action for common ethical issues.

Command Prompt a place in the MATLAB interface indicated by two greater-than signs (>>) where users can enter commands in real time

comment item placed in a sketch or code for a program, to explain certain pieces

commercial districts business districts, in which a minimum pressure of 75 psi is acceptable at a water hydrant; even higher pressures must be available in districts with tall buildings

comparator an instrument that compares the actual output of a system with the desired output or result

competency-based interview question attempts to understand if you have the appropriate background knowledge to perform the job successfully

compression that state of a member when it is experiencing a compressive force

compressive force a type of force in a truss in which members are compressed and not pulled

computer-aided design (CAD) computer-aided design technology that allows the user to create a design on the computer screen and then send it directly to a machine tool

computer-aided manufacturing (CAM) a software application that uses computers to control factory machines

computer-integrated manufacturing (CIM) uses computers not only for design and manufacturing, but also for business needs

concise brief and comprehensive

conductance the reciprocal of resistance, measured in siemens [S]

conservation of energy the statement that the total energy in an isolated system remains unchanged

conservation of mass the statement that mass cannot be created nor destroyed by different reactions or processes

console-based therapy an area of research about leveraging technology to enhance physical therapy interventions

constraint-based solid modeling overcomes the weakness of constructive solid geometry modeling by making the modeling process more intuitive

constraints limitations, boundaries on a solution; the limits imposed on a design solution

construction lines lines used to guide other lines that are not actually part of the component being drawn

constructive solid geometry based on the combination of geometric primitives such as right rectangular prisms (blocks), right triangular prisms (wedges), spheres, cones, and cylinders

contour dimensioning placing dimensions of a feature in the view where the feature occurs in its most characteristic shape and where it is visible (as opposed to hidden)

control unit the part of the Central Processing Unit (CPU) which controls what is happening inside the CPU itself

control variable part of an experiment that can vary but is held constant to let the experimenter observe the influence of the independent variable on the dependent variable

controller a device that turns a process on or off, or that changes it in some way

corrosive a type of material that can create a chemical reaction with another material and cause damage to that material

cosine written as cos(angle), the trigonometric function defined in the context of a right triangle that is often used to model physical phenomena like sound waves and electric current

coulomb a charge of one coulomb represents the total charge on approximately 6.24 * 1018 protons

counter variable keeps track of the number of times we have repeated a calculation or decision

craft manufacturing the method of making products one at a time, from start to finish

criteria requirements a design needs to meet in order to perform its desired functions

cross section a two-dimensional profile of a part

cumulative distribution function (CDF) a plot in which the values for each bin on the normalized plot are summed from the first bin up to each individual bin

D

data science a field that is concerned with extracting usable information from structured or unstructured data using mathematical, algorithmic, and statistical methods

debug (or debugging) the process of locating and correcting syntax and/or logical errors in a program

decimal places the number of digits to the right of the decimal point

decision matrices a mathematical process for decision-making; can be used by engineers to summarize how each of the concepts they generate and test compare with one another

decisions any executable steps in an algorithm that require the answer to a question with "Yes" or "No"

density the amount of mass per unit of volume

dependability a dependable person is one who can be relied upon

dependent variable in an experiment, a variable that reacts to a change in the independent variable

derived dimension a dimension that is some combination of the seven base dimensions

derived unit a defined unit in the SI system made up of a combination of base SI units that make it easier to convey complicated derived dimensions; often named after a famous scientist or engineer

descriptive geometry the mathematical principles of projection. These principles form the basis of engineering graphics today.

design for manufacturability a type of engineering design process that prioritizes how the item will be manufactured to minimize difficulties in production

design process an iterative series of steps engineers use to arrive at the solution to a problem

design specifications numerical values assigned to criteria and constraints that are the basis for testing if the design meets the established criteria and constraints

detail drawings document the detailed design of individual components using orthographic views

development team members of a team who focus on project deliverables—sharing their progress publicly as often as possible in completed, if minor, advancements

differential pressure the difference in pressure between two points

digital workflow the automation of a company's (or team's) processes in addition to internal steps to accomplish the company's (or team's) goals

dimension a measurable physical idea; it generally consists solely of a word description with no numbers

dimension lines thin lines with arrowheads at each end that indicate sizes in the drawing

dimensional constraints used to specify distances between items in a solid model

distribution data spread

double-main system a design approach to the local neighborhood water distribution strategy, where there is a main on each side of the street

driving force the difference in pressure which impacts fluid flow

ductility a mechanical property of a material describing how well a material can sustain plastic deformation without failing

durable goods products that last a long time

E

efficiency a measure of how much of a quantity, typically energy or power, is lost in a process

elastic region a specific point that unless reached, a material will return to its original length when the material is subject to a load as part of the tensile test or any other externally applied force

electric charge positive or negative charges

electric current a movement of electric charge in a wire or other solid material

electric power the rate at which electrical energy is transferred in an electrical system, like a circuit

electromechanical system a system that combines elements of electrical processes and mechanical processes, such as a pinball machine

energy the ability to do work

energy of translational motion a form of energy possessed by an object in motion

engineering a profession that draws upon knowledge of mathematics and science to create ways to economically use available materials and the forces of nature to benefit the human race through the process of design

engineering graphics a highly stylized way of presenting images of parts or assemblies

engineering notation expressed in the form ###.### × 10^M, where M is an integer multiple of 3, and the number of digits to the left of the decimal point is one, two, or three as needed to yield a power of 10 that is indeed a multiple of 3

ergonomics also called human factors engineering; deals with designing products so they can be used easily and comfortably

ethical dilemma a difficult decision that does not have a clear ethical solution

ethics ethical decisions are those based on a "right" vs "wrong" set of options as defined for a population by an external source

evaluation apprehension the reluctance to share ideas because the individual fears a negative reaction

exploded view reveals details of individual parts, along with showing the order in which they are assembled

explosive a type of material that has the potential of creating a chemical reaction that leads to an excessive release of gas or energy

exponential growth characterizes rapid growth that can be modeled using an exponential function

extension lines thin lines that extend from the object to the dimension line to indicate which feature is associated with the dimension

extreme value minimum and maximum values of data

extruded stretched

F

factory system a system in which people made goods using machines rather than by hand

Fahrenheit a temperature scale developed by Gabriel Fahrenheit (1686–1736), a German physicist and engineer, that uses a freezing point of 32°F and a boiling point of 212°F

farads measure of capacitance; one farad equals one coulomb per volt

feature-based modeling attempts to make the modeling process more efficient by creating and modifying geometric features of a solid model in a way that represents how geometries are created using common manufacturing processes

features characteristics of a device, drawing, part, or assembly

feedback information about the output of a system that can be used to adjust it

Fermi problems problems in which so much information is missing that a solution seems impossible

Floating Point Unit (FPU) the part of the Central Processing Unit (CPU) which tackles floating point math (using real numbers with decimals)

flow rate the volume of fluid that passes through an area per unit of time; the product of cross-sectional area and the fluid speed

flowchart a graphical representation of a written algorithm that describes the sequence of actions, decisions, and path of a process

fluid general term that applies to a gas, such as helium or air, or a liquid, such as water or honey, which continuously flows when external forces are applied

foreshortening lines of an object are changed to produce the illusion of the part extending into the third dimension

form factor the physical size of the hard drive

free body diagram a representation of a single object or subsystem composed of multiple objects that is used to examine the different forces acting on it

free riding when a team member doesn't contribute to the final product

functional block diagram a picture that relates the inputs, functions required to perform the system's processes, and outputs

functional decomposition breaking down a system into its different functionalities, inputs, and outputs

functions smaller programs referenced within a larger program

G

Gantt chart a project schedule, listing each task and showing how long it takes, and assigning tasks to show who has responsibility for them

gas pressure the pressure created by the movement of gas molecules inside a closed container

gauge pressure when measuring pressure, uses the local atmospheric pressure as a reference point

geometric constraints used to define positional relationships between entities in the model in terms of the geometry

global a reserved word used to create global variables in functions; followed by a comma-separated list of the variables that should be considered global variables

global variables the variables in a function that are visible within any scope

Grand Challenges a list of 14 challenges for the 21st century, each of which has the potential to transform the way we live, work, and play

graphical algorithms a pictorial representation of instructions required to solve a problem

gravity on the Earth, gravity is approximately 9.8 meters per second squared [m/s^2]; on the moon, gravity is approximately 1.6 meters per second squared [m/s^2]

H

hard drive device in a computer that stores data

hazardous materials materials that can be harmful to health or the environment but are still in their original, usable form

hazardous waste waste that is potentially harmful to health or the environment

henrys unit of measure of inductance

hidden lines thin, dashed lines that represent features that are hidden behind surfaces in the current view

histogram plot using bars to represent a distribution of numerical data

history-based modeling states that the order in which parts are created is critical

horsepower a unit of power originally used to quantify how the steam engine could replace the work done by a horse

hybrid brainstorming a brainstorming model that splits ideation into two parts, an individual phase followed by a team phase

hybrid electric vehicles combine the engine of a conventional vehicle with the battery and electric motor of an electric vehicle

hydrostatic pressure the pressure exerted on a submerged object by the weight of a liquid above it

I

ideal problems engineering problems that calculate a single, exact answer

ideation the generation of concepts or ideas to solve a design problem

independent variable in an experiment, the variable that is controlled

index variable keeps track of the number of times we have repeated a calculation or decision

inductance a wire in a changing magnetic field such as a collapsing one will have a current induced: this is inductance

inductor component that resists sudden changes to current

Industrial Revolution a period in which human and animal muscle power was replaced by machines

infinite feedback loop a nonterminating feedback loop

Information Age a period that continues today in which the changes caused by machinery and factory production gradually gave way to a new type of technology: information

informed design process a process providing a way to make decisions without complete knowledge and then reassess those decisions later on

infrastructure systems such as roads and waterways that are primarily designed by civil engineers

input the command given to a system; the known quantities used in a design or program

interchangeable part a part that is manufactured to be exactly like another part, so that any one can be substituted for any other

integrated graphics section of a Central Processing Unit (CPU) that is dedicated to graphics, figuring out what needs to be displayed

intellectual property (IP) property such as a written work or art that is owned by someone or some entity; it can be owned by you, your company, or your school

interference when physical parts interfere with the operation of other physical parts

Internet of Things the incorporation of processing abilities, software, or sensors into physical, everyday objects to form a network of connected devices

Iron Age the era when iron came into common use

isometric projection simplifies the pictorial perspective. In an isometric projection, parallel lines remain parallel.

J

joule the amount of energy required to move an object one meter using one Newton of force

just-in-time manufacturing uses careful scheduling to keep materials and products in storage for as little time as possible

K

kelvin a temperature scale named for First Baron William Thomson Kelvin (1824–1907), an English mathematician and physicist, who first proposed the idea of "infinite cold," or absolute zero, in 1848, using the Celsius scale for comparison

kinetic energy energy that results from motion

L

lateral thinking expands in different directions to generate multiple possible ideas for solving problems

leader lines thin, solid lines terminated with arrowheads that are used to indicate a feature with which a dimension or note is associated

line of sight angle at which an object is viewed

linear rate of change a rate that looks like a straight line on a graph

local variables the variables created within the "private workspace" of a function

loops programming structures that allow a group of lines in a program to be executed multiple times instead of just once

lower-yield point the lowest stress that occurs after the yield strength in the stress-strain curve

M

machine learning a field concerned with developing models that use data to improve their performance

manufacturing the process of making goods by hand or machine

mass a fundamental property of materials, so it is imperative to have a standard definition of the scientific unit, kilogram

mass production the production of goods in large quantities by groups of workers in factories

Material Safety Data Sheet (MSDS) paperwork explaining how hazardous materials are to be used and stored

mathematical model a model that uses equations to describe how a product will function or perform

MATLAB a programming language: short for matrix laboratory

mean one measure of central tendency; it is found by adding all the numbers in the data set and dividing the result by the number of items in the data set

mechanical property describes how a material behaves when subjected to an external force (or load)

median one measure of central tendency; it is the value between the lower half and upper half of the population

members long slender objects that are used to form a truss, which could be made of several different types of materials like woods and metals

method of joints one method of determining the forces within the members composing a truss that focuses on the pins holding the members together

metric a measurement system based on powers of 10; a measurable quantity used to assess performance

M-Files MATLAB program files

model a computer representation of an object or part

Modulus of Elasticity a measure of how resistant a material is to elastic deformation (also called Young's Modulus)

moments measures of the tendency for an object to rotate around a specific point

Moore's Law states that the number of transistors within a computer processor doubles every two years

morals personal beliefs about "right" and "wrong," may be in conflict with Code of Ethics

morphological chart a visual method for generating design concept ideas by splitting a potential design into necessary functions and methods of achieving those functions

motherboard the large circuit board inside a computer; everything in the computer connects to the motherboard, passing all information from one component to another

multimeter device capable of measuring current, voltage, or resistance

multiview projection a collection of flat 2-D drawings of the different sides of an object

must criteria qualities that any successful solution must have

N

nanotechnology the creation of materials, devices, and systems at a scale of 1 to 100 nanometers

necking the point before failing, where stress on a material is localized disproportionately in one smaller region

negatively skewed a distribution in which the mean has been pulled lower than the median

Nelson rules a variety of conditions that indicate that a process is out of control

newton the SI unit of force, which is defined as the force required to accelerate a mass of one kilogram at a rate of one meter per second squared

nondurable goods products that are designed to be consumed in a short period of time

normal distribution distribution which follows the 68-95-99.7 rule

normal force the force the surface an object is resting on exerts back to avoid two solids passing through one another; this force is perpendicular to the surface

normalized plot a plot in which the values are divided by the total number of data points

O

oblique projection places the principal face of the object parallel to the plane of the paper with the axes in the plane of the paper perpendicular to one another

Occupational Safety and Health Administration (OSHA) an organization of the federal government that sets many safety regulations

Ohm's Law an equation that describes the relationship between current, voltage, and resistance: $V = R*I$

open source the licensing to the software and internal workings of the platform are free

open-loop system a system that has no way of monitoring and adjusting itself

operating system system software (usually Windows, Linux, or a Mac operating system) that manages all the software and hardware on a computer

orders of magnitude often used when comparing things of very different scales, such as a small rock and a planet

orthographic projection typically shows three views of an object; each view shows a different side

outlier an extreme value

output what is produced by a system; the actual result

overdimensioned having one redundant dimension

P

parallel projection type of drawing in which lines along edges are parallel and do not converge

parametric parameters of the model may be modified to change its geometry

parent-child relation an aspect of solid modeling that states that the child cannot exist without the parent existing first

participatory design involves inviting all possible stakeholders to work with engineers and other members of the team to collaboratively make design decisions and develop ideas throughout the entire design process

parts list includes the part name, identification number, material, number required in the assembly, and other information

pascal in SI units, the unit of pressure, defined as one newton of force acting on an area of one square meter

Pascal's law states that the pressure applied to an enclosed liquid at any point will be transmitted throughout the fluid equally in all directions

perspective projection drawn so that parallel lines converge in the distance

phase the state of a material (liquid, gas, solid, or plasma); a physical property of a material

physical property for a material, something that can be observed or measured without altering the identity of the material

pictorial-perspective drawing this type of drawing presents an object much as it would look to the human eye or in a photograph

plastic region a region of the stress-strain curve in which any further deformation is permanent

Poiseuille's Law helps describe how fluid flows through a long cylindrical pipe in terms of the pressure drop ΔP between the two ends of the pipe, the viscosity μ, the length of the pipe L, flow rate Q, and cross-sectional area A

positively skewed a distribution in which the mean has been pulled higher than the median

potential energy energy that is stored

power the amount of work done during a given period of time

power grid interconnected network of electrical transmission delivering power from suppliers to consumers

power supply the large unit inside the computer where the power cord is attached and converts the supplied power to what is needed by the computer

precedence of lines when two lines in a particular view coincide, this dictates which line is shown: visible lines have precedence over hidden lines, which have precedence over centerlines

precision a combination of accuracy and repeatability

pressure the force acting over an area, where the force is perpendicular to the area

primitive modeling based on the combination of geometric primitives such as right rectangular prisms (blocks), right triangular prisms (wedges), spheres, cones, and cylinders

prior art existing solutions that can be researched to study the way a problem has been solved before

probability associated with assessing the likelihood that an event will or will not occur

process the action part of a system

product life cycle the timeframe that encompasses when an item first is introduced to customers to the point where it becomes obsolete and removed from the market

product owner the primary liaison to customers and clients who represents their voice in the design process

production blocking when a team member drives the conversation, limiting opportunities for others to speak

productivity refers to how quickly and cost-effectively a product is made

programming the process of expressing an algorithm in a language that a computer can interpret; the process in which a problem is formulated in a way that a computer can understand and solve

programs a set of instructions given to a computer to perform a specific task

proportional limit a point in the elastic region of the stress-strain curve for a metal alloy

prototype a full-scale, fully operational version of a solution

punctuality being on time and ready when expected

Q

quality control chart used to show how close to the mean the results of a process are when measured over time

R

radian the angle at the center of a circle formed by an arc (S) equal in length to the radius (r) of that circle

radioactive waste waste material containing unstable atoms that give off radiation

Random Access Memory computer memory found on the motherboard that is mounted on small circuit boards, sometimes called 'sticks' of RAM; the more RAM a computer has, the faster it operates

range restriction rule which prohibits some values, usually based on characteristics of the data

Rankine a temperature scale named for William J. M. Rankine (1820–1872), a Scottish engineer and physicist, who proposed an analogy to the kelvin scale, using the Fahrenheit scale for comparison

rapid prototyping involves the use of computer-aided design data to quickly create a scale model of parts or assemblies before committing resources to manufacturing the product at scale

reactions the forces and moments that an object exerts on its supports

reference dimension provided only for information to help in interpreting the drawing

registers temporary holding positions for data being manipulated in a Central Processing Unit (CPU)

reliability a reliable person can be relied upon for accuracy and honesty each and every time

rendering a virtual light source is assumed to be near the object to provide a three-dimensional lighting effect

renewable energy sources any source of energy that can be replaced

repeatability a measure of how close together multiple measurements of the same parameter are, whether or not they are close to the actual value

residential districts housing districts, in which a minimum pressure of 60 psi at a water hydrant is recommended

resistance opposition to the flow of current through a material

responsibility taking ownership of one's role and deliverables within a project

resultant force the force that is the combination of its purely horizontal and vertical forces; can also be used to refer to the combination of multiple forces that are not necessarily its components

revolving extruding around an axis

risk/benefit trade-off a common trade-off made in solving large problems; to obtain desired benefits, designers must allow for risk while trying to keep the risk as low as possible

robotics the area of artificial intelligence that designs and builds robots

Rockwell hardness test the most common method of measuring a material's hardness, which involves applying a 4 mm (or smaller) steel ball or diamond cone to a material's surface and measuring the depth of the indentation

S

safety factor a multiple of some criteria used in a design

scale indicates the ratio of the size of the drawing to the size of the actual object

scale model a model that is either larger or smaller than the final product

scientific notation typically expressed in the form $\#.\#\#\# \times 10^N$, where the single digit to the left of the decimal point is the most significant nonzero digit of the value being represented; N is an integer and multiplying by 10^N serves to locate the true position of the decimal point

scope the entire description of a project

scrum master a manager who is primarily responsible for ensuring that nothing hinders the team's progress—in a facilitative role

semiconductors a type of material that has useful properties, such as its conductivity falling between a conductor and insulator, for use in electronics like computers

sensor a monitor of a system's output

shading a virtual light source is assumed to be near the object to provide a three-dimensional lighting effect

should criteria specifications that are generally agreed upon as desirable and that help distinguish one solution as better than another

SI prefix in using one of these when reporting numbers, we avoid scientific notation or long strings of zeros

significant figures the digits considered reliable as a result of measurement or calculation

sine written as sin(angle), the trigonometric function defined in the context of a right triangle that is often used to model physical phenomena like sound waves and electric current

single-main system a design approach to the local neighborhood water distribution strategy, where a single large pipe serves both sides of a street

situational interview question asks you to demonstrate how you would handle a specific hypothetical scenario in the future

68-95-99.7 rule a rule that determines if a distribution is normal: 68% of values are within one standard deviation of the mean, 95% are within 2 standard deviations, and 99.7% are within 3 standard deviations

six thinking hats a brainstorming model created by Edward de Bono in which a team engages in viewing a problem through six different perspectives

sketch Arduino's name for files containing code—with a file extension of .ino

skewed a distribution that is not symmetric

slack time a time period when project activities can be completed before the time the activity needs to be completed

smelting a process in which iron was made from iron ore by melting the ore to take out the impurities

software low-level and high-level programs used to perform specific tasks on a computer

solid modeling represents an object in the virtual environment just as it exists in reality, having volume as well as surfaces and edges

SOLVEM stands for Sketch, Observations or Objectives, List, Variables, and Equations Manipulation

specific gravity a dimensionless ratio of the density of an object to the density of water

specific weight the weight of an object (w) divided by the volume the object occupies (V)

speech recognition digitizing the sound waves of speech and then converting them to the basic sounds that are used in language

speech synthesis converting text to speech by breaking down the written words into basic sounds of the language and then converting the sounds into digital audio

spring constant measures the stiffness of a spring

stakeholders who might be affected by the various alternative courses of action that might be implemented

standard deviation found by taking the square root of the variance; measures variation or dispersion in a data set

standpipe an elevated water distribution tank used to gain the necessary hydraulic head if the topography does not permit sufficient head from a surface reservoir

statistical process control (SPC) a method of monitoring, controlling, and improving a manufacturing process

statistics provide quantitative measures to "things" that behave in a random manner

Stone Age the period during which people used stone to make tools

strain the ratio of the change in length to the original length when an object is subjected to an applied force

stress the measure of the force per unit area experienced by a material subjected to an applied force

stress-strain curve plots the strain (the ratio of the change in length to the original length) against the stress (the applied load pulling the material apart)

subsystems smaller systems that often make up other, larger systems

surface model defines the object in terms of surfaces such as plates (flat) and shells (curved) in addition to edges

surface pressure the pressure created by the weight of a gas above an object

symmetry lines thin, long-dash/short-dash lines that mark axes of rotationally symmetric parts or features

system a group of interrelated components designed collectively to achieve a desired goal

T

technological resources physical and intangible resources that enable the creation of a physical product; includes people, capital, time, information, energy, materials, and tools and machines

temperature originally conceived as a description of energy: heat (thermal energy) flows spontaneously from "hot" to "cold"; numerous temperature scales have been developed to give temperature a quantitative meaning

tensile force a type of force in a truss that pulls members in such a way that they would elongate

tensile test a common test used to examine the properties of a material by subjecting it to an incrementally larger applied force

tension that state of a member when it is experiencing a tensile force

thermal energy the energy of the atoms that make up a substance

Thiele tube a special type of glassware that is designed to hold and heat paraffin oil with a uniform temperature; used to determine the melting point of a material

3-D model the most general model used in CAD software; the CAD software electronically stores the entire three-dimensional shape of the object

3-D printing the construction of three-dimensional objects from a digital 3-D model

3-D projection a drawing that looks like a visual representation of an object: drawn "as it looks" with lines that converge in the distance

time management managing the available time for a project or job effectively

title block records important information about a drawing; normally appears in the lower right corner of the drawing

total pressure the summation of surface and hydrostatic pressure

toxic a type of material that poses a serious health issue to the point of possible death, if not helped

truss a common structure you might see across contexts like roofing and on bridges; consists of members connected such that they form a rigid structure

Turing test one of the simplest tests to determine if a machine or artificial intelligence can demonstrate human intelligence

2 1/2-D model has a third dimension that is recognized by the CAD software, but the third dimension is simply an extrusion of a two-dimensional shape

2-D drawing a compact way to depict all the detail of an object, as in a blueprint, using two coordinate directions

2-D model the computer graphics equivalent of an orthographic projection created using a pencil and paper

U

ultimate strength the maximum stress that a material can withstand in the stress-strain curve

Ultimate Tensile Strength (UTS) the maximum stress we can apply before material failure

unidirectional having the same orientation

unit allows us to quantify a dimension—to state a number describing how much of that dimension exists in a specific situation. Units are defined by convention and related to an accepted standard.

V

vacuum pressure when a gauge pressure being measured is less than the local atmospheric pressure

variance To determine a variance (σ2), the difference between each point and the mean is determined, and each difference is squared to keep all terms positive. This sum is then divided by one less than the number of data points.

vertical thinking involves using a linear, sequential process for solving problems

virtual interviews job interviews conducted using an online platform, such as video conferencing software

virtual reality using artificial intelligence to create a computer-generated environment

visible lines thick, solid lines that represent the outline of the object that can be seen in the current view

voltage a measure of the force with which electric current is pushed through wires

W

watt the SI unit of power

weight the weight (w) of an object is a force equal to the mass of the object (m) times the acceleration of gravity (g)

weighted decision matrix a decision matrix that adds a weighting system to the attributes used in the comparison, allowing engineers to mathematically value one criteria over another (e.g., a should criterion could be weighted less than a must criterion)

wireframe model the simplest 3-D model; only edges of the object are represented

witness lines thin lines that extend from the object to the dimension line to indicate which feature is associated with the dimension

work something produced or accomplished through effort or activity

workspace the smaller pane on the right side of the MATLAB interface window; shows all the variables in use at the current time

written algorithms a narrative set of instructions required to solve a problem

Y

yield strength the point of the stress-strain curve where a material will no longer return to its original length and will keep a 0.2% elongation even after the applied force is removed

yield stress point on the stress-strain curve where a material enters a plastic region

Young's Modulus a measure of how resistant a material is to elastic deformation

Index

Index

Index

Index

Index

Index

Index

Index

Index

Photo Credits

Page	Figure Number	Credit line
1	0.1	Blend Images/ Shutterstock
2	0.3	Warren Goldswain/ Shutterstock
2	0.2	goodluz / 123RF
2	0.4	Tribalium/ Shutterstock
3	0.5	Evlakhov Valeriy/ Shutterstock
3	0.6	Kletr/ Shutterstock
4	0.7	Albert Lozano/ Shutterstock
4	0.11	Roxana Bashyrova/ Shutterstock
4	0.12	Sarah Cates/ Shutterstock
4	0.13	Sean Bolt/ Shutterstock
4	0.15	STILLFX/ Shutterstock
4	0.9	Coprid/ Shutterstock
4	0.8	Pradeep Rawat/ Pearson India Education Services Pvt/ Ltd
4	0.1	nulinukas/ Shutterstock
4	0.14	Coleman Yuen/ Pearson Education Asia Ltd
4	0.17	Lost Mountain Studio/ Shutterstock
4	0.16	Дмитрий Гооль / 123RF
11	1.1	Michael Ebardt/ Shutterstock
11	1.2	Kompaniets Taras/ Shutterstock
12	1.3	The World Traveller/ Shutterstock
13	1.4	Filip Fuxa/ Shutterstock
14	1.5	Everett Collection/ Shutterstock
14	1.6	Morphart Creation/ Shutterstock
15	1.7b	amasterphotographer/ Shutterstock
15	1.8	Kenneth Reid
15	1.7a	gallofoto/ Shutterstock
17	1.9	Kenneth Reid
18	1.1	Ericsmandes/ Shutterstock
18		Dariush M/ Shutterstock
19	1.11	Better Stock/ Shutterstock
22	1.12	Pearson Education Ltd
26	1.13	Elizabeth A Stephan
26	1.14	Elizabeth A Stephan
27	1.15	Jenson/ Shutterstock
29	1.16	Elizabeth A Stephan
32	1.17	Mark Agnor / 123RF
33		areebarbar/ Shutterstock
34	1.18	pictur / 123RF
34	1.19	pickingpok/ Shutterstock

Photo Credits

Page	Figure Number	Credit line
35	1.2	Tomas Mikula/ Shutterstock
36		Scharfsinn / Shutterstock
42	2.1	Jack Frog. Shutterstock
43	2.2	E. Petersen. Shutterstock
43	2.3	Fenton one. Shutterstock
44	2.4	Lightwave Stock Media / 123RF
50	2.5	RGtimeline. Shutterstock
52	2.6	Ken Reid
52	2.8	mikeledray/ Shutterstock
52	2.7	Istvan Csak. Shutterstock
53	2.9	Travis Klein/ Shutterstock
64	2.1	Pressmaster/ Shutterstock
65		sirtravelalot/ Shutterstock
65		Faraways/ Shutterstock
81	3.2	Everett Historical/ Shutterstock
82	3.3	ELIZABETH A STEPHAN
110	4.3	Janaka Dharmasena/ Shutterstock
111	4.5	studioVin/ Shutterstock
112	4.4	Dja65/ Shutterstock
112	4.6	Ken Reid
114	4.7	Andriy Popov / 123RF
116	4.8	Vaniato/ Shutterstock
122		NASA
134		ultimathule/ Shutterstock
127	5.1	David Carillet/ Shutterstock
127	5.1	dreamnikon / 123RF
131		Ken Reid
131	5.2	Ken Reid
134	5.3	Ken Reid
148	6.1	Ken Reid
148	6.2	Tashsat/ Shutterstock
149	6.3	Everett - Art/ Shutterstock
149	6.4	cybrain/ Shutterstock
149	6.5	Ken Reid
175		Ken Reid
207	7.12	Elizabeth A Stephan
209	7.1	MeeRok/ Shutterstock
236		W Park
236		W Park

Photo Credits

Page	Figure Number	Credit line
236		W Park
236		W Park
238	8.1	GTibbetts/ Shutterstock
239	8.2	destinacigdem / 123RF
244	8.3	Mrs_ya/ Shutterstock
245	8.5	Megan Marshall
250		Elizabeth A Stephan
252	8.6	Ermolaev Alexander/ Shutterstock
253	8.7	John Dorado/ Shutterstock
255	8.8	fotosav/ Shutterstock
306	9.1	BonD80/ Shutterstock
307	9.3	Jim Guy/ Shutterstock
307	9.4	Andrii A/ Shutterstock
308	9.5	Oleksiy Mark/ Shutterstock
308	9.5	George Dolgikh/ Shutterstock
309	9.7	Michael C/ Gray/ Shutterstock
327	10.1	Alexander Tolstykh/ Shutterstock
327	10.2	Shutterstock
328	10.4	Rainer Plendl / 123RF
328	10.3	Kirill Makarov / 123RF
330	10.5	esp2k/ Shutterstock
331	10.6	Ken Reid
333	10.8	Ken Reid
338	10.12	Ken Reid
338	10.13	Ken Reid
342	10.15	Ken Reid
342	10.16	Ken Reid
348		Elizabeth A Stephan
348		Martyn F. Chillmaid. Pearson Education Ltd
348		Viktor Chursin. Shutterstock
348		Martyn F. Chillmaid. Pearson Education Ltd
358		Ken Reid
380	12.2	jupeart/ Shutterstock
380	12.1	antonioscarpi / 123RF
380		STILLFX. Shutterstock
380		Coprid. Shutterstock
380		Roxana Bashyrova. Shutterstock
380		idea for life. Shutterstock

Photo Credits

Page	Figure Number	Credit line
Appendix C		Jens Stolt / 123RF
Appendix C		Fedor Selivanov. Shutterstock
Appendix C		Hung Chung Chih. Shutterstock
Appendix C		Georgios Kollidas. Shutterstock
Appendix C		Fer Gregory. Shutterstock
Appendix C		Morphart Creation. Shutterstock
Appendix C		IgorGolovniov. Shutterstock
Appendix C		Anneka. Shutterstock
Appendix C		jennyt. Shutterstock
Appendix C		Jason Page. Shutterstock
Appendix C		Ronald Sumners. Shutterstock
Appendix C		Dario Sabljak. Shutterstock
Appendix C		chippix. Shutterstock
Appendix C		Castleski. Shutterstock
Appendix C		Albert Lozano. Shutterstock
Appendix C		Tatiana Popova. Shutterstock
Appendix C		Dima Sidelnikov. Shutterstock
Appendix C		Sarawut Aiemsinsuk. Shutterstock
Appendix C		Matej Kastelic. Shutterstock
Appendix C		texelart / 123RF
Appendix D	D1	Ken Reid
Appendix D	D2	Ken Reid
Appendix D	D3	Ken Reid
Cover		Hero Images Inc. / Alamay Stock Photo
Cover		Hero Images Inc. / Alamay Stock Photo
Cover		lucadp/123rf.com
Cover		stockbroker/123rf.com
Cover		stockbroker/123rf.com
Cover		Gorodenkoff/Shutterstock
Cover		Triff/Shutterstock